Lecture Notes in Physics

The Lecture Notes in Physics

The series Lecture Notes in Physics (LNP), founded in 1969, reports new developments in physics research and teaching – quickly and informally, but with a high quality and the explicit aim to summarize and communicate current knowledge in an accessible way. Books published in this series are conceived as bridging material between advanced graduate textbooks and the forefront of research and to serve three purposes:

- to be a compact and modern up-to-date source of reference on a well-defined topic

- to serve as an accessible introduction to the field to postgraduate students and nonspecialist researchers from related areas

- to be a source of advanced teaching material for specialized seminars, courses and schools

Both monographs and multi-author volumes will be considered for publication. Edited volumes should, however, consist of a very limited number of contributions only. Proceedings will not be considered for LNP.

Volumes published in LNP are disseminated both in print and in electronic formats, the electronic archive being available at springerlink.com. The series content is indexed, abstracted and referenced by many abstracting and information services, bibliographic networks, subscription agencies, library networks, and consortia.

Proposals should be send to a member of the Editorial Board, or directly to the managing editor at Springer:

Dr. Christian Caron
Springer Heidelberg
Physics Editorial Department I
Tiergartenstrasse 17
69121 Heidelberg / Germany
christian.caron@springer.com

Volker Eyert

The Augmented Spherical Wave Method

A Comprehensive Treatment

 Springer

Author

Volker Eyert
Center for Electronic Correlations and Magnetism
Institute of Physics, University of Augsburg
86135 Augsburg, Germany
eyert@physik.uni-augsburg.de

V. Eyert, *The Augmented Spherical Wave Method*, Lect. Notes Phys. 719 (Springer, Berlin Heidelberg 2007), DOI 10.1007/978-3-540-71007-3

Library of Congress Control Number: 2007923153

ISSN 0075-8450
ISBN 978-3-540-71006-6 Springer Berlin Heidelberg New York

Springer is a part of Springer Science+Business Media
springer.com
© Springer-Verlag Berlin Heidelberg 2007

Typesetting: by the authors and Integra using a Springer LaTeX macro package
Cover design: WMXDesign GmbH, Heidelberg

Printed on acid-free paper   SPIN: 12023240   5 4 3 2 1 0

To my children, Florian and Carolin

Preface

The origin of this book dates back to the beginning of the year 1987, when I started to work on my PhD in the group of Prof. J. Kübler at the Technical University of Darmstadt. The discussions in those days were much influenced by the discovery of the high-T_c superconductors and, hence, it became clear quite early that a full-potential augmented spherical wave (ASW) code capable of calculating elastic properties and phonon frequencies via the frozen-phonon approach was desirable. The development and implementation of such a code became the subject of my thesis.

Yet, learning the basic notions of the ASW method was hampered by the fact that review articles were not available and the original work by Williams, Kübler, and Gelatt, while being very concise, did not answer the simple questions a beginner would ask. Benefiting from the diploma theses of D. Hackenbracht and M. Methfessel, I started to write down a first detailed description of both the standard and the full-potential ASW method, which eventually formed the backbone of my PhD thesis. It laid ground for the present notes, which by now cover many aspects of the ASW method.

This book addresses all those readers who want to learn the basic functionality of methods for electronic structure calculations in general and of the ASW method in particular. In addition, being quite detailed, it tries to capture many of the above-mentioned beginners' and non-specialists' questions. Moreover, it provides a guiding hand to the many practiners who started using the ASW method and want to learn more about the details. Of course, the large amount of background material should also content the experts in the field. Finally, since the ASW method shares much of the basic formalism with other spherical-wave-based schemes as the Korringa–Kohn–Rostoker (KKR) and the linear muffin-tin orbital (LMTO) method, the book may also be valuable for researchers familiar with these.

In writing this book and setting up a completely new implementation of the ASW program package, I have much benefited from various support and numerous discussions. My memory is with my friend and colleague Dr. Jürgen Sticht, who deceased much too early one month ago. Jürgen introduced me into the mysteries of the ASW method and with him I share a very fruitful time of code optimization and vectorization. My thanks include many other people, who in one way or the other had a strong impact on my work. Without being complete, I am particularly grateful to Prof. O. K. Andersen, Prof. R. Claessen, Prof. U. Eckern, Prof. R. Frésard, PD Dr. K.-H. Höck, Prof. S. Horn, Prof. T. Kopp, Prof. J. Kübler, Prof. J. Mannhart, Prof. S. F. Matar, Dr. T. Maurer, Dr. A. Mavromaras, Dr. M. S. Methfessel, Prof. W. Nolting, Prof. W. Scherer, Prof. P. C. Schmidt, Prof. K. Schwarz, Dr. M. Stephan, Prof. D. Vollhardt, and Dr. E. Wimmer. Last but not least, it is a great pleasure to thank Dr. C. Caron, Mrs. G. Hakuba, and Mrs. J. Lenz of the Springer-Verlag for their professional help during the final phase of this book. This project has been partially supported by the Deutsche Forschungsgemeinschaft through Sonderforschungsbereich 252 and 484.

Potsdam, January 2007   *Volker Eyert*

Contents

1

Introduction

1.1 Overview

Since its introduction in the 1960's by Hohenberg, Kohn, and Sham, density-functional theory (DFT) [18, 19] has become one of the major tools for understanding materials properties. DFT not only had a decisive impact on the field of electronic structure calculations but also has strongly influenced condensed matter physics as a whole. Of course, this is due to the incredible success in accurately predicting ground state properties of numerous materials. Nevertheless, the widespread use of DFT is also a consequence of its conceptual simplicity, which allows for a very efficient use of computational resources [11, 13, 16, 22]. Nowadays systems comprising several hundreds or even thousands of atoms can be accessed by DFT calculations. Although most applications have been for solids the theory has been likewise successfully used for atoms and molecules, where it competes with standard quantum chemical methods [20, 25].

In practice, the application of density-functional theory is usually based on formulating the variational principle for the determination of the single-particle wave function in terms of a set of basis functions. While being finite, such a basis set cannot be complete and, hence, the shape of the basis functions is crucial to both the accuracy and the efficiency of a first-principles method. Choices to be made are essentially between plane-wave and spherical-wave based schemes as well as between pseudopotential (PP) and partial-wave approaches. The latter grew out of Slater's muffin-tin approximation and build on the continuous and differentiable matching of solutions to Schrödinger's equation evaluated in different portions of space. This class of schemes comprises, e.g. the augmented plane wave (APW) method, the Korringa-Kohn-Rostoker (KKR) method, the linear augmented plane wave (LAPW) method, the linear muffin-tin orbitals (LMTO) method, and the augmented spherical wave (ASW) method [8–10, 14, 20].

In particular, the invention of the linear methods by Andersen in the first half of the 1970s has initiated a still lasting period of very fruitful materials research [1–3]. Due to the tremendous reduction of computer requirements brought by these methods, elemental solids and compounds with increasing complexity could be

V. Eyert: *Introduction*, Lect. Notes Phys. **719**, 1–4 (2007)
DOI 10.1007/978-3-540-71007-3_1   © Springer-Verlag Berlin Heidelberg 2007

investigated with so far unknown accuracy [5–7]. The work of Andersen had a great impact on the development of both plane-wave and spherical-wave methods [3, 4]. Preferring the latter type of methods, Andersen himself pushed forward the development of the linear muffin-tin orbitals (LMTO) method. In contrast, the plane-wave approach led to the linear augmented plane wave (LAPW) method as well as its full-potential generalization, the full-potential linear plane wave (FLAPW) method.

Inspired by the work of Andersen, Williams and coworkers, while starting from the concept of renormalized atoms as proposed by Watson, Ehrenreich, Hodges, and Gelatt [15, 17, 26], arrived at the augmented spherical wave (ASW) method, which actually differed only slightly from the LMTO method [27].

While many other methods have found detailed representations in the literature [12, 21, 23, 24], the widespread and even commercial use of the ASW method is contrasted by an apparent lack of a comprehensive description. This is where the present book tries to close a gap.

The organization of the book is as follows: Chapter 2 is devoted to the original standard ASW method, which is based on the atomic sphere approximation (ASA) and, hence, does not include any full-potential terms. Chapter 3 deals with the envelope functions as well as the structure constants, which build the analytical anchor of the method. Although this chapter covers exclusively mathematical aspects of the ASW method it does, nevertheless, form the basis for the standard method as well as all further developments. We will benefit from these results especially while introducing in Chap. 4 the full-potential ASW method, which is much more involved than the standard method. Nevertheless, we will not discuss issues regarding numerics or the implementation.

1.2 A Note on Units

Following common practice we will use in the present work atomic (Rydberg) units and thus measure lengths in units of the Bohr radius

$$a_B = \frac{\alpha}{4\pi R_\infty} = \frac{4\pi\epsilon_0\hbar^2}{m_e e^2} = 0.5291772108 \times 10^{-10}\,\mathrm{m}\,, \tag{1.2.1}$$

where α is the dimensionless fine structure constant,

$$\alpha^{-1} = \frac{4\pi\epsilon_0\hbar c}{e^2} = 137.03599911 \tag{1.2.2}$$

and R_∞ the Rydberg constant,

$$R_\infty = \frac{\alpha^2 m_e c}{2h} = 10973731.568527\,\mathrm{m}^{-1}\,. \tag{1.2.3}$$

Energies are given in units of Rydberg

$$\mathrm{Ry} = \frac{e^2}{2 \cdot 4\pi\epsilon_0 a_B} = \frac{m_e e^4}{2 \cdot (4\pi\epsilon_0)^2\hbar^2} = R_\infty hc = \frac{1}{2}\alpha^2 m_e c^2$$
$$= 2.179872085 \times 10^{-18}\,\mathrm{J}\,. \tag{1.2.4}$$

With
$$e = 1.60217653 \times 10^{-19} \, \text{C} \qquad (1.2.5)$$

for the fundamental unit of charge we note the familiar result

$$\text{Ry} = 13.60569225 \, \text{eV} \,. \qquad (1.2.6)$$

All values for the fundamental constants have been taken from the web site of the National Institute of Standards and Technology, http://physics.nist.gov/.

In order to arrive at atomic units one may simply set

$$\hbar = 1 \,, \qquad e^2 = 2 \,, \qquad m_e = \frac{1}{2} \,, \qquad \text{and} \qquad 4\pi\epsilon_0 = 1 \qquad (1.2.7)$$

in equations, which are formulated in SI or cgs units [20].

Note that our choice of atomic (Rydberg) units leads to the following value for the velocity of light, which is needed for relativistic calculations

$$c = \frac{e^2}{\hbar\alpha} = 274.07199822 \,. \qquad (1.2.8)$$

In passing we mention atomic (Hartree) units, which are arrived at by setting

$$\hbar = 1 \,, \qquad e^2 = 1 \,, \qquad m_e = 1 \,, \qquad \text{and} \qquad 4\pi\epsilon_0 = 1 \qquad (1.2.9)$$

in equations, which are formulated in SI or cgs units [20] and which lead to the Hartree as the unit of energy with

$$\text{H} = 2 \, \text{Ry} = 27.2113845 \, \text{eV} \,. \qquad (1.2.10)$$

References

1. O. K. Andersen, Comments on the KKR-Wavefunction; Extension of the Spherical Wave Expansion beyond the Muffin-Tins. In: *Computational Methods in Band Theory*, ed by P. M. Marcus, J. F. Janak, and A. R. Williams (Plenum Press, New York 1971) pp 178–182
2. O. K. Andersen, Solid State Commun. **13**, 133 (1973)
3. O. K. Andersen, Phys. Rev. B **12**, 3060 (1975)
4. O. K. Andersen, Linear Methods in Band Theory. In: *The Electronic Structure of Complex Systems*, ed by P. Phariseau and W. Temmerman (Plenum Press, New York 1984) pp 11–66
5. O. K. Andersen, O. Jepsen, and D. Glötzel, Canonical Description of the Band Structures of Metals. In: *Highlights of Condensed-Matter Theory*, Proceedings of the International School of Physics "Enrico Fermi", Course LXXXIX, ed by F. Bassani, F. Fumi, and M. P. Tosi (North-Holland, Amsterdam 1985) pp 59–176

6. O. K. Andersen, O. Jepsen, and M. Sob, Linearized Band Structure Methods. In: *Electronic Band Structure and its Applications*, ed by M. Yussouff (Springer, Berlin 1986) pp 1–57

7. O. K. Andersen, A. P. Postnikov, and S. Yu. Savrasov, The Muffin-Tin Orbital Point of View. In: *Applications of Multiple Scattering Theory to Materials Science*, ed by W. H. Butler, P. H. Dederichs, A. Gonis, and R. L. Weaver, Mat. Res. Soc. Symp. Proc. **253**, 37 (1992)

8. N. W. Ashcroft and N. D. Mermin, *Solid State Physics* (Holt-Saunders, Philadelphia 1976)

9. J. Callaway, *Energy Band Theory* (Academic Press, New York 1964)

10. J. Callaway, *Quantum Theory of the Solid State* (Academic Press, Boston 1991)

11. R. M. Dreizler and E. K. U. Gross, *Density-Functional Theory* (Springer, Berlin 1990)

12. H. Eschrig, *Optimized LCAO Method and the Electronic Structure of Extended Systems* (Springer, Berlin 1989)

13. H. Eschrig, *The Fundamentals of Density-Functional Theory* (Edition am Gutenbergplatz, Leipzig 2003)

14. V. Eyert, *Electronic Structure of Crystalline Materials*, 2nd edn (University of Augsburg, Augsburg 2005)

15. C. D. Gelatt, Jr., H. Ehrenreich, and R. E. Watson, Phys. Rev. B **15**, 1613 (1977)

16. E. K. U. Gross and R. M. Dreizler, *Density-Functional Theory* (Plenum Press, New York 1995)

17. L. Hodges, R. E. Watson, H. Ehrenreich, Phys. Rev. B **5**, 3953 (1972)

18. P. Hohenberg and W. Kohn, Phys. Rev. **136**, B864 (1964)

19. W. Kohn and L. J. Sham, Phys. Rev. **140**, A1133 (1965)

20. J. Kübler and V. Eyert, Electronic structure calculations. In: *Electronic and Magnetic Properties of Metals and Ceramics*, ed by K. H. J. Buschow (VCH Verlagsgesellschaft, Weinheim 1992) pp 1–145; vol 3A of *Materials Science and Technology*, ed by R. W. Cahn, P. Haasen, and E. J. Kramer (VCH Verlagsgesellschaft, Weinheim 1991-1996)

21. T. Loucks, *Augmented Plane Wave Method* (Benjamin, New York 1967)

22. R. G. Parr and W. Yang, *Density-Functional Theory of Atoms and Molecules* (Oxford University Press, Oxford 1989)

23. D. J. Singh, *Planewaves, Pseudopotentials and the LAPW Method* (Kluwer Academic Publishers, Boston 1994)

24. H. L. Skriver, *The LMTO Method* (Springer, Berlin 1984)

25. M. Springborg, *Density-Functional Methods: Applications in Chemistry and Materials Science*, (Wiley, Chichester 1997)

26. R. E. Watson, H. Ehrenreich, and L. Hodges, Phys. Rev. Lett. **15**, 829 (1970)

27. A. R. Williams, J. Kübler, and C. D. Gelatt, Jr., Phys. Rev. B **19**, 6094 (1979)

2

The Standard ASW Method

Our survey of the ASW method starts with an outline of the standard scheme, which, within the framework of density-functional theory and the local density approximation, allows for both fast and conceptionally simple calculations of the ground state properties of solids [23, 25, 43, 73]. After defining the ASW basis set we will construct the elements of the secular matrix. Via the variational principle the latter allows to numerically determine the expansion coefficients of the ground state wave function as expanded in terms of the basis functions. Knowledge of the wave function enables for a calculation of the electron density and the effective single-particle potential, the latter of which closes the self-consistency cycle. Finally, we sketch the calculation of the total energy. Since the standard ASW method lays the basis for further developments our presentation will be a compromise between being simple and concise.

2.1 Setup of Basis Functions

The goal of any first-principles method is to determine the single-particle wave function from Schrödinger's equation

$$[-\Delta + v_\sigma(\mathbf{r}) - \varepsilon]\,\psi_\sigma(\varepsilon, \mathbf{r}) = 0\,, \qquad (2.1.1)$$

where, as usual, $-\Delta$, the negative of the Laplacian, designates the operator of the kinetic energy. $\psi_\sigma(\varepsilon, \mathbf{r})$ is a spinor component of the wave function and σ labels the spin index. In addition, $v_\sigma(\mathbf{r})$ denotes the spin-dependent, effective single-particle potential. It may be provided by density functional theory, in which case (2.1.1) is just the Kohn-Sham equation. As a matter of fact, solving Schrödinger's equation in the general form (2.1.1) is usually far from straightforward. However, we may employ the variational principle and expand the wave function in a set of basis functions. The stationary states are then found by varying the expansion coefficients and possibly the shape of the basis functions themselves. Since in practice we are forced to use a finite and, hence, incomplete set of functions

V. Eyert: *The Standard ASW Method*, Lect. Notes Phys. **719**, 5–45 (2007)
DOI 10.1007/978-3-540-71007-3_2

the choice of basis functions is very important as concerns both the quality of the final solution and the numerical effort to find it. As already mentioned in the introduction, it is this selection of a basis set, which distinguishes the different methods nowadays used for electronic structure calculations. The ASW method belongs to the partial-wave methods and is based on Slater's muffin-tin approximation. In addition, it uses spherical waves as well as Andersen's linearization.

Obviously, the basis functions themselves should at best also fulfill Schrödinger's equation (2.1.1). Yet, this would again lead to the above mentioned difficulties and concessions are thus necessary. A very popular one is based on Slater's observation that the crystal potential, while being dominated by the purely atomic and, hence, spherical symmetric contributions near the nuclei, is rather flat in the regions far away from the atomic centers [66]. Idealizing this situation, Slater proposed the muffin-tin approximation (MTA), which adopts the geometry of a muffin-tin as visualized in Fig. 2.1. To be specific, the MTA replaces the full crystal potential by its spherical symmetric average within the non-overlapping muffin-tin spheres and a constant value, the so-called muffin-tin zero v_{MTZ}, in the remaining interstitial region. The effective single-particle potential is thus modelled as

$$v_\sigma^{MT}(\mathbf{r}) = v_{MTZ}\Theta_I + \sum_{\mu i} v_{i\sigma}^{MT}(\mathbf{r}_{\mu i})\Theta_{\mu i}$$

$$= v_{MTZ} + \sum_{\mu i} \hat{v}_{i\sigma}^{MT}(\mathbf{r}_{\mu i}) \,. \tag{2.1.2}$$

Fig. 2.1. Muffin-tin

Here we have defined

$$\mathbf{r}_{\mu i} := \mathbf{r} - \mathbf{R}_{\mu i} \quad \text{and} \quad \mathbf{R}_{\mu i} := \mathbf{R}_\mu + \boldsymbol{\tau}_i \,, \qquad (2.1.3)$$

where $\mathbf{R}_\mu$ is a lattice vector and $\boldsymbol{\tau}_i$ the position of the nucleus labeled i in the unit cell; σ again denotes the spin index. The step functions Θ_I and $\Theta_{\mu i}$ limit the range of the functions, which are attached to them, to the interstitial region and the muffin-tin sphere of radius S_i centered at site $\mathbf{R}_{\mu i}$, respectively. In the second line of (2.1.2) we have added an alternative notation using the potential

$$\hat{v}_{i\sigma}^{MT}(\mathbf{r}_{\mu i}) := \left[v_{i\sigma}^{MT}(\mathbf{r}_{\mu i}) - v_{MTZ} \right] \Theta_{\mu i} \,. \qquad (2.1.4)$$

It just represents a constant shift of all potentials inside the atomic spheres but may be likewise interpreted as a splitting of the total potential into two parts, one of which extends through all space and is rather flat (or constant, as in the present case). Below, we will denote this contribution as the pseudo part, since inside the muffin-tin spheres it has no meaning in itself. In contrast, the remaining local parts contain all the intraatomic details of the potential but are confined to the respective muffin-tin spheres. The total potential is then arrived at by adding, inside the spheres, the local contributions to the pseudo part.

Within the MTA the choice of the sphere radii is restricted by the plausible requirement that the spheres must not overlap. A somewhat different point of view is taken by the atomic-sphere approximation (ASA) invented by Andersen [5, 6], which requires the sum of the atomic sphere volumes to equal the cell volume,

$$\sum_i \tilde{\Omega}_i := \sum_i \frac{4\pi}{3} \tilde{S}_i^3 \overset{!}{=} \Omega_c \,. \qquad (2.1.5)$$

Here $\tilde{\Omega}_i$ and Ω_c denote the volumes of the atomic sphere i and the unit cell, respectively. In general, we will use a tilde for quantities connected to the space-filling atomic spheres. As it stands, the ASA leads to a formal elimination of the interstitial region. This serves the purpose of minimizing the errors coming with the muffin-tin approximation as well as with the linearization to be described below, which both are largest for the interstitial region. Furthermore, it has been demonstrated that for moderate overlaps of the spheres the ASA provides a better prescription of the potential in the bonding region between the atoms and thus mimics the full potential in a better way than the MTA does [8]. In the following, we will opt for the ASA but still want to keep things on general grounds as long as possible. Thus, we formally stay with the muffin-tin form (2.1.2) of the potential without explicitly making use of the atomic-sphere approximation.

Having modelled the effective single-particle potential in the way just described we will next construct the basis functions from the muffin-tin potential. This is done by first solving Schrödinger's equation separately in the atomic spheres as well as in the interstitial region and then matching the resulting partial waves continuously and differentially at the sphere boundaries. This will lead to a single basis function, an *augmented spherical wave*. To be concrete, the basis functions have to obey Schrödinger's equation (2.1.1) with the potential replaced by the muffin-tin potential (2.1.2),

$$\left[-\Delta + v_\sigma^{MT}(\mathbf{r}) - \varepsilon\right] \varphi_\sigma(\varepsilon, \mathbf{r}) =$$
$$\left[-\Delta + v_\sigma^{MT}(\mathbf{r}) - v_{MTZ} - \kappa^2\right] \varphi_\sigma(\varepsilon, \mathbf{r}) = 0 , \qquad (2.1.6)$$

separately in the interstitial region and the atomic spheres. Here, we have in the second step referred the potential to the muffin-tin zero and defined

$$\kappa^2 = \varepsilon - v_{MTZ} . \qquad (2.1.7)$$

By construction, the basis functions are only approximate solutions of the correct Schrödinger equation (2.1.1), where the potential was still unrestricted. This is a consistent approximation within the standard ASW method, which uses the atomic-sphere approximation throughout. But even for a full-potential method use of the ASA for the construction of the basis functions is justified by the fact that the basis functions enter only as part of the wave function and inaccuracies of the former are most likely cured by the variational determination of the latter. Use of the MTA (or ASA) for the construction of the basis functions may thus be regarded as an optimal balance between effort and accuracy.

Starting the setup of the partial waves in the interstitial region we combine (2.1.2) and (2.1.6) and arrive at Helmholtz's equation

$$\left[-\Delta - \kappa^2\right] \varphi_\sigma(\kappa^2, \mathbf{r}) = 0 . \qquad (2.1.8)$$

As well known, this differential equation is trivially solved by plane waves. However, we may likewise choose spherical waves, which are centered at the atomic sites and consist of products of spherical harmonics and radial functions. The latter are solutions of the free-particle radial Schrödinger equation for energy κ^2 and, hence, well known as spherical Bessel, Neumann or Hankel functions. Note that the singularities of the latter two functions are located at the respective origins, i.e. at the centers of the atomic spheres and thus do not spoil the analytic behaviour of these functions in the interstitial region.

The choice of spherical waves is common to the Korringa-Kohn-Rostoker (KKR) [38, 41], linear muffin-tin orbitals (LMTO) [6], and ASW methods. In particular, the KKR method opts for spherical Neumann functions for the radial part. The interstitial part of the basis function reads as

$$N_{L\kappa\sigma}^\infty(\mathbf{r}_{\mu i})\Theta_I := N_{L\kappa}^I(\mathbf{r}_{\mu i}) := N_{L\kappa}(\mathbf{r}_{\mu i})\Theta_I , \qquad (2.1.9)$$

where

$$N_{L\kappa}(\mathbf{r}_{\mu i}) := -\kappa^{l+1} n_l(\kappa r_{\mu i}) Y_L(\hat{\mathbf{r}}_{\mu i}) . \qquad (2.1.10)$$

Here, $n_l(\kappa r_{\mu i})$ denotes the spherical Neumann function in the notation of Abramowitz and Stegun [1, Chap. 10]. The index $L = (l, m)$ is a compose of the angular momentum and the magnetic quantum numbers. $Y_L(\hat{\mathbf{r}}_{\mu i})$ denotes a spherical harmonic or rather a cubic harmonic in the form proposed in App. B.2 and

$\hat{\mathbf{r}}_{\mu i}$ stands for a unit vector. The prefactor κ^{l+1} serves the purpose of cancelling the leading κ-dimension of the spherical Neumann function and making the radial part of the function $N_{L\kappa}(\mathbf{r}_{\mu i})$ a real function, hence, easy to deal with in computer implementations. In passing, we mention that, in the context of the ASW method, the latter function is commonly also designated as the Neumann function.

In principle, we could have alternatively opted for spherical Hankel functions for the radial part of the basis functions. In this case the interstitial part of the basis function would have assumed the form

$$H_{L\kappa\sigma}^{\infty}(\mathbf{r}_{\mu i})\Theta_I := H_{L\kappa}^{I}(\mathbf{r}_{\mu i}) := H_{L\kappa}(\mathbf{r}_{\mu i})\Theta_I , \qquad (2.1.11)$$

with

$$H_{L\kappa}(\mathbf{r}_{\mu i}) := i\kappa^{l+1}h_l^{(1)}(\kappa r_{\mu i})Y_L(\hat{\mathbf{r}}_{\mu i}) , \qquad (2.1.12)$$

where $h_l^{(1)}(\kappa r_{\mu i})$ is the spherical Hankel function in the notation of Abramowitz and Stegun and, as before, the function $H_{L\kappa}(\mathbf{r}_{\mu i})$ is also named a Hankel function. Again, the prefactor serves the purpose of cancelling the leading κ-dimension and making the radial part a real function. Yet, the latter is true for negative energies κ^2 only, where, however, the spherical Hankel functions offer the additional great advantage of decaying exponentially in space. This is an appealing behaviour especially if non-crystalline or finite systems are dealt with. Nevertheless, the fact that the radial part in general is a complex function makes the Hankel functions rather unsuitable for the KKR method.

As was first pointed out by Andersen at the beginning of the 1970s the energy (κ-) dependence of the final basis function turns out to be rather weak. This observation led Andersen to argue that it is sufficient to retain only the first two terms of the corresponding Taylor series [4, 65]. Formally suppressing the interstitial region by introducing the atomic-sphere approximation, Andersen was able to remove even the linear term for the interstitial functions and to arrive at energy-independent functions with a fixed value of κ^2 for the radial part. Of course, this value was expected to be in the region of interest, i.e. between $-1\,\mathrm{Ryd}$ and $+1\,\mathrm{Ryd}$. With energy-independent basis functions at hand it was possible to replace the time consuming root tracing needed in the KKR method for calculating the single-particle energies by a simple eigenvalue problem. This meant a tremendous cut down of computation times since the eigenvalue problem is by at least an order of magnitude faster to solve. Andersen's work thus marked the starting point for a number of new, the so-called linear methods among them the linear muffin-tin orbitals (LMTO) method, the linear augmented plane wave (LAPW) method, and the ASW method, which in the sense just mentioned are superior to the classical schemes such as the augmented plane wave (APW) method and the Korringa-Kohn-Rostoker (KKR) method [6, 7, 9, 43, 65]. The albeit small limitations due to the energy linearization could be cured by employing so-called multiple-κ sets, which use basis functions with two or three different values of the interstitial energy κ^2. However, for the standard LMTO and ASW methods, which use the atomic-sphere approximation, the effect is almost negligible and multiple-κ sets are thus not used. Yet, they will be needed in those full-potential methods, which use the muffin-tin approximation, hence, a finite interstitial region. In order to prepare for later developments we include the energy parameter κ already in the present formulation.

With the energy parameter κ fixed, a greater flexibility for the choice of basis functions is obtained since there is no longer the need to use a single type of function for the whole energy range. In contrast, we are by now able to opt for that type of function, which offers the best properties and then fix the energy parameter κ^2 to a suitable value within the above proposed range of $-1\,\mathrm{Ryd}$ and $+1\,\mathrm{Ryd}$. To be specific, in the ASW method we set $\kappa^2 = -0.015\,\mathrm{Ryd}$ and opt for the Hankel function $H_{L\kappa\sigma}^{\infty}(\mathbf{r}_{\mu i})$, which for this energy offers the above mentioned advantages of being real and decaying exponentially. Nevertheless, we could have likewise chosen a positive energy parameter, in which case the ASW method uses the Neumann function $N_{L\kappa\sigma}^{\infty}(\mathbf{r}_{\mu i})$ as the interstitial part of the basis function. In passing, we mention that the LMTO method uses $\kappa^2 = 0\,\mathrm{Ryd}$, in which case both types of functions are identical and assume a particularly simple form as will be discussed in Chap. 3 [6].

While staying with the negative energy case and, hence, using the Hankel function (2.1.11) as the interstitial part of the basis function we will not utilize the simplifications offered by the asymptotic decay of the spherical Hankel functions for negative energies but rather stay with a more general derivation. This will save us an extra discussion for the Neumann functions. In practice, this means that all the derivations formulated below with Hankel functions will also hold for Neumann functions. Note that, in contrast to previous definitions of the ASW basis functions [73], the functions (2.1.10) and (2.1.12) do not contain an extra i^l-factor this making them (as well as the real-space structure constants to be defined lateron) real quantities in the respective energy range. In addition, our new choice leads to considerable numerical simplifications for the full-potential ASW method to be derived in Chap. 4.

Having discussed the interstitial basis functions we next turn to the regions inside the atomic spheres. Here, the potential shows all the intraatomic details and, hence, solving Schrödinger's equation is more complicated. However, within the muffin-tin approximation the full crystal potential is reduced to its spherical symmetric average inside the atomic spheres this allowing for a separation of Schrödinger's equation into angular and radial parts. While the former again leads to spherical harmonics the latter can be determined numerically subject to the conditions of continuous and differentiable matching to the interstitial basis function. As a consequence, in addition to being used in the interstitial region only, the interstitial basis functions serve the purpose of providing boundary conditions for the solution of Schrödinger's equation inside the spheres. In the atomic-sphere approximation, due to the formal elimination of the interstitial region, it is the second aspect alone, which survives. For this reason, the interstitial basis functions are also called envelope functions. We will discuss especially the mathematical aspects of the envelope functions in more detail in Chap. 3.

Turning to the region inside the spheres, we first concentrate on that single atomic sphere where the spherical wave is centered. In this so-called on-center sphere spherical symmetry is obvious and we readily arrive at the intraatomic basis function, which is called the augmented Hankel function,

$$H_{L\kappa\sigma}^{\infty}(\mathbf{r}_{\mu i})\Theta_{\mu i} := \tilde{H}_{L\kappa\sigma}(\mathbf{r}_{\mu i}) := \tilde{h}_{l\kappa\sigma}(r_{\mu i})Y_L(\hat{\mathbf{r}}_{\mu i})\Theta_{\mu i} \ . \qquad (2.1.13)$$

Its radial part obeys the radial Schrödinger equation

$$\left[-\frac{1}{r_{\mu i}}\frac{\partial^2}{\partial r_{\mu i}^2}r_{\mu i} + \frac{l(l+1)}{r_{\mu i}^2} + v_{i\sigma}^{MT}(r_{\mu i}) - v_{MTZ} - E_{l\kappa i\sigma}^{(H)}\right]\tilde{h}_{l\kappa\sigma}(r_{\mu i}) =$$
$$\left[-\frac{1}{r_{\mu i}}\frac{\partial^2}{\partial r_{\mu i}^2}r_{\mu i} + \frac{l(l+1)}{r_{\mu i}^2} + \hat{v}_{i\sigma}^{MT}(r_{\mu i}) - E_{l\kappa i\sigma}^{(H)}\right]\tilde{h}_{l\kappa\sigma}(r_{\mu i}) = 0 \ .$$
$$(2.1.14)$$

The numerical solution of this equation as well as the determination of the Hankel energy $E_{l\kappa i\sigma}^{(H)}$ are subject to three boundary conditions, namely, the regularity of the radial function $\tilde{h}_{l\kappa\sigma}(r_{\mu i})$ at its origin and the continuous and differentiable matching at the sphere boundary,

$$\left[(\frac{\partial}{\partial r_{\mu i}})^n\left(\tilde{h}_{l\kappa\sigma}(r_{\mu i}) - i\kappa^{l+1}h_l^{(1)}(\kappa r_{\mu i})\right)\right]_{r_{\mu i}=\tilde{S}_i} = 0 \ , \qquad n = 0, 1 \ . \qquad (2.1.15)$$

Still, in order to solve (2.1.14) numerically, we have to specify the principal quantum numbers and the angular momenta l of all states to be taken into account. For the angular momenta we include all l values ranging from 0 to some maximum value l_{low}. Usually, the latter is set to 2 or 3 depending on the respective atom. While for main group elements we include s and p states, we account also for d and possibly f orbitals in case of transition metal and rare earth atoms. All these states form the ASW basis set, which, following Andersen, we call the set of *lower waves*.

The principal quantum number of each partial wave, which corresponds to the number of nodes once the angular momentum of a lower wave has been fixed, is selected in accordance with the outer electrons of each atom. To be specific, we first divide all states of an atom into core and valence states. The former are characterized by vanishing value and slope at the sphere boundary and, hence, confined to the inner region of the atom. As a consequence, while contributing to the electron density the core states are not included in the ASW basis set, which comprises only the valence states. We will come back to the core states at the end of this section. In general, only the outermost state for each atom and $l \leq l_{low}$ is chosen as a valence state and included in the basis set. This is done implicitly and the respective principal quantum number thus suppressed from the notation. Of course, there might be situations, where, in addition to the outermost valence states, high lying core states should also be included in the ASW basis set. However, in order to take the notation simple, we will not deal with such cases in the present context.

By now we have augmented the spherical wave centered at $\mathbf{R}_{\mu i}$ inside its on-center sphere. In all other atomic spheres of the crystal, which are centered at sites $\mathbf{R}_{\nu j} \neq \mathbf{R}_{\mu i}$, and referred to as the off-center spheres, augmentation is not as straightforward. This is due to the fact that the envelope function centered at $\mathbf{R}_{\mu i}$ lacks spherical symmetry relative to the centers $\mathbf{R}_{\nu j}$ of the off-center spheres. Yet, the problem can be resolved by applying the expansion theorem for the envelope functions, which allows to expand the Hankel function centered at $\mathbf{R}_{\mu i}$ as defined by (2.1.12) in Bessel functions centered at $\mathbf{R}_{\nu j}$,

$$H_{L\kappa}(\mathbf{r}_{\mu i}) = \sum_{L'} J_{L'\kappa}(\mathbf{r}_{\nu j}) B_{L'L\kappa}(\mathbf{R}_{\nu j} - \mathbf{R}_{\mu i}) . \qquad (2.1.16)$$

The expansion theorem is valid for all vectors $\mathbf{r}$, which fulfill the condition $|\mathbf{r}_{\nu j}| < |\mathbf{R}_{\nu j} - \mathbf{R}_{\mu i}|$, thus for all $\mathbf{r}$ lying within a sphere of radius equal to the distance between the two centers and centered at $\mathbf{R}_{\nu j}$. The Bessel function

$$J_{L'\kappa}(\mathbf{r}_{\nu j}) := \kappa^{-l'} j_{l'}(\kappa r_{\nu j}) Y_{L'}(\hat{\mathbf{r}}_{\nu j}) , \qquad (2.1.17)$$

like the Hankel function given by (2.1.12), arises from solving Helmholtz's equation (2.1.8). $j_{l'}(\kappa r_{\nu j})$ denotes a spherical Bessel function. The prefactor again was introduced to cancel the leading κ-dimension and to make the radial part a real quantity for any value of κ^2.

The expansion coefficients entering (2.1.16) are the structure constants known from the KKR method [30, 62]. Their calculation as well as the proof of the expansion theorem will be given in full detail in Sects. 3.6 and 3.7. Here we present the result

$$B_{L'L\kappa}(\mathbf{R}_{\nu j} - \mathbf{R}_{\mu i}) = 4\pi \sum_{L''} i^{l-l'-l''} \kappa^{l+l'-l''} c_{LL'L''} H_{L''\kappa}(\mathbf{R}_{\nu j} - \mathbf{R}_{\mu i}) , \qquad (2.1.18)$$

which is valid for $\mathbf{R}_{\nu j} \neq \mathbf{R}_{\mu i}$ and where

$$c_{LL'L''} = \int d^2\hat{\mathbf{r}} \, Y_L^*(\hat{\mathbf{r}}) Y_{L'}(\hat{\mathbf{r}}) Y_{L''}(\hat{\mathbf{r}}) , \qquad (2.1.19)$$

denotes a (real) Gaunt coefficient.

With the expansion theorem (2.1.16) at hand we are ready to augment the spherical wave in any of the off-center spheres. This is due to the fact that the Bessel functions given by (2.1.17) are centered at the center of this sphere and thus reflect the spherical symmetry relative to the respective atomic site. As a consequence, the basis function in the off-center sphere centered at $\mathbf{R}_{\nu j}$ can be expressend in terms of augmented Bessel functions

$$\tilde{J}_{L'\kappa\sigma}(\mathbf{r}_{\nu j}) := \tilde{j}_{l'\kappa\sigma}(r_{\nu j}) Y_{L'}(\hat{\mathbf{r}}_{\nu j}) \Theta_{\nu j} . \qquad (2.1.20)$$

Their radial parts obey the radial Schrödinger equation

$$\left[-\frac{1}{r_{\nu j}} \frac{\partial^2}{\partial r_{\nu j}^2} r_{\nu j} + \frac{l'(l'+1)}{r_{\nu j}^2} + v_{j\sigma}^{MT}(r_{\nu j}) - v_{MTZ} - E_{l'\kappa j\sigma}^{(J)} \right] \tilde{j}_{l'\kappa\sigma}(r_{\nu j}) =$$

$$\left[-\frac{1}{r_{\nu j}} \frac{\partial^2}{\partial r_{\nu j}^2} r_{\nu j} + \frac{l'(l'+1)}{r_{\nu j}^2} + \hat{v}_{j\sigma}^{MT}(r_{\nu j}) - E_{l'\kappa j\sigma}^{(J)} \right] \tilde{j}_{l'\kappa\sigma}(r_{\nu j}) = 0 .$$

$$(2.1.21)$$

As for the augmented Hankel function, the numerical solution of this differential equation as well as the determination of the Bessel energy $E^{(J)}_{l'\kappa j\sigma}$ are subject to the conditions of continuous and differentiable matching at the sphere boundary,

$$\left[\left(\frac{\partial}{\partial r_{\nu j}}\right)^n \left(\tilde{j}_{l'\kappa\sigma}(r_{\nu j}) - \kappa^{-l'} j_{l'}(\kappa r_{\nu j})\right)\right]_{r_{\nu j}=S_j} = 0, \qquad n = 0, 1, \qquad (2.1.22)$$

as well as the regularity of the radial function $\tilde{j}_{l'\kappa\sigma}(r_{\nu j})$ at the origin. Note that the radial equations (2.1.14) and (2.1.21) are identical and, hence, the difference between the augmented functions $\tilde{h}_{l\kappa\sigma}$ and $\tilde{j}_{l\kappa\sigma}$ as well as the energies $E^{(H)}_{l\kappa i\sigma}$ and $E^{(J)}_{l\kappa i\sigma}$ results alone from the different boundary conditions (2.1.15) and (2.1.22).

With the radial part at hand we arrive at the following expression for the basis function centered at $\mathbf{R}_{\mu i}$ inside the off-center sphere centered at $\mathbf{R}_{\nu j}$

$$H^{\infty}_{L\kappa\sigma}(\mathbf{r}_{\mu i})\Theta_{\nu j} = \sum_{L'} \tilde{J}_{L'\kappa\sigma}(\mathbf{r}_{\nu j})B_{L'L\kappa}(\mathbf{R}_{\nu j} - \mathbf{R}_{\mu i}). \qquad (2.1.23)$$

Note that, due to the identity of $H_{L\kappa}$ and $\tilde{H}_{L\kappa\sigma}$ as well as $J_{L\kappa}$ and $\tilde{J}_{L\kappa\sigma}$ at the sphere boundary, the expansion coefficients in (2.1.16) and (2.1.23) are the same. In passing, we mention that (2.1.16) and (2.1.23) are also called the one-center expansion of the Hankel envelope function and the basis function in terms of Bessel envelope functions and augmented Bessel functions, respectively. To be more detailed, they constitute the one-center expansions in the off-center spheres. In contrast, the one-center expansion of a Hankel envelope function and an augmented Hankel function, respectively, in their on-center sphere are just these functions themselves. This has important implications for the L' summation entering (2.1.16) and (2.1.23).

As for the augmented Hankel functions we still need to specify the maximum angular momentum as well as the principal quantum numbers to be used for solving (2.1.21). As a matter of fact, the one-center expansion in the off-center spheres, i.e. expression of the Hankel function in terms of an angular momentum series over Bessel functions via the expansions (2.1.16) and (2.1.16) does not converge. For that reason, we would in principle have to include infinitely many Bessel functions. However, as we will see in the subsequent section, we will be able, by combining two such series expansions, to end up with only few terms in the series. We denote the corresponding angular momentum cutoff as l_{int}, which is usually chosen as $l_{int} = l_{low} + 1$. Again following Andersen, we call the partial waves with $l_{low} < l \leq l_{int}$ intermediate waves. In contrast, partial waves with $l > l_{int}$ are named higher waves. The principal quantum number required for solving (2.1.21) is chosen in accordance with the valence states for the corresponding atom just in the same manner as for the augmented Hankel function above.

Finally, combining (2.1.11), (2.1.13), and (2.1.23), we construct the full basis function, the *augmented spherical wave*

$$H^{\infty}_{L\kappa\sigma}(\mathbf{r}_{\mu i}) = H^{I}_{L\kappa}(\mathbf{r}_{\mu i}) + \tilde{H}_{L\kappa\sigma}(\mathbf{r}_{\mu i})$$
$$+ \sum_{L'\nu j}(1 - \delta_{\mu\nu}\delta_{ij})\tilde{J}_{L'\kappa\sigma}(\mathbf{r}_{\nu j})B_{L'L\kappa}(\mathbf{R}_{\nu j} - \mathbf{R}_{\mu i}), \qquad (2.1.24)$$

which is completely specified by its center, $\mathbf{R}_{\mu i}$, the composite angular momentum index $L = (l, m)$, and the spin index σ; note that the energy κ^2 is kept only as a parameter. In addition, the augmented spherical wave is continuous and differentiable in all space. Since the ASW basis functions arise from solutions of Schrödinger's equation in the respective portions of space, each of these functions is quite well adapted to the actual problem and only a relatively small number of basis functions is needed for the expansion of the final wave function; as already mentioned above, typically 9 states (s, p, d) per atom are sufficient and for atoms with f electrons we end up with 16 states. Such a compact basis set is usually called a minimal basis set. As an additional bonus, the ASW basis functions, being closely related to atomic-like functions, allow for a natural interpretation of the calculated results.

In a final step, we take into account crystal translational symmetry and use Bloch sums of the basis functions defined as

$$D^{\infty}_{L\kappa\sigma}(\mathbf{r}_i, \mathbf{k}) := \sum_{\mu} e^{i\mathbf{k}\mathbf{R}_{\mu}} H^{\infty}_{L\kappa\sigma}(\mathbf{r}_{\mu i}) , \qquad (2.1.25)$$

where

$$\mathbf{r}_i = \mathbf{r} - \boldsymbol{\tau}_i , \qquad (2.1.26)$$

and the symbol D is convention. Note that we will refer to both the functions (2.1.24) and (2.1.25) as basis functions but the actual meaning will be always clear from the context.

In order to prepare for forthcoming sections we define, in addition, the Bloch sum of the envelope function (2.1.11) by

$$D_{L\kappa}(\mathbf{r}_i, \mathbf{k}) := \sum_{\mu} (1 - \delta_{\mu 0}\delta(\mathbf{r}_i)) e^{i\mathbf{k}\mathbf{R}_{\mu}} H_{L\kappa}(\mathbf{r}_{\mu i}) . \qquad (2.1.27)$$

In complete analogy to (2.1.11) its interstitial part reads as

$$D^{I}_{L\kappa}(\mathbf{r}_i, \mathbf{k}) := \sum_{\mu} e^{i\mathbf{k}\mathbf{R}_{\mu}} H^{I}_{L\kappa}(\mathbf{r}_{\mu i}) = D_{L\kappa}(\mathbf{r}_i, \mathbf{k})\Theta_I . \qquad (2.1.28)$$

Moreover, we define the Bloch-summed structure constants as

$$B_{L'L\kappa}(\boldsymbol{\tau}_j - \boldsymbol{\tau}_i, \mathbf{k}) := \sum_{\mu} (1 - \delta_{\mu\nu}\delta_{ij}) e^{i\mathbf{k}\mathbf{R}_{\mu}} B_{L'L\kappa}(\boldsymbol{\tau}_j - \boldsymbol{\tau}_i - \mathbf{R}_{\mu})$$

$$= 4\pi \sum_{L''} i^{l-l'-l''} \kappa^{l+l'-l''} c_{LL'L''} D_{L''\kappa}(\boldsymbol{\tau}_j - \boldsymbol{\tau}_i, \mathbf{k}) , \qquad (2.1.29)$$

where we have used the identities (2.1.18) and (2.1.27) in the second step. Hence, we get for the Bloch-summed basis function the result

$$D^{\infty}_{L\kappa\sigma}(\mathbf{r}_i, \mathbf{k}) = D^{I}_{L\kappa}(\mathbf{r}_i, \mathbf{k}) + \tilde{H}_{L\kappa\sigma}(\mathbf{r}_i)$$
$$+ \sum_{L'j} \tilde{J}_{L'\kappa\sigma}(\mathbf{r}_j) B_{L'L\kappa}(\boldsymbol{\tau}_j - \boldsymbol{\tau}_i, \mathbf{k}) . \qquad (2.1.30)$$

Finally, combining the previous identities we write the wave function entering the variational procedure as a linear combination of the just defined Bloch sums

$$
\begin{aligned}
\psi_{\mathbf{k}\sigma}(\mathbf{r}) &= \sum_{L\kappa i} c_{L\kappa i\sigma}(\mathbf{k}) D_{L\kappa\sigma}^{\infty}(\mathbf{r}_i, \mathbf{k}) \\
&= \sum_{L\kappa i} c_{L\kappa i\sigma}(\mathbf{k}) D_{L\kappa}^{I}(\mathbf{r}_i, \mathbf{k}) \\
&\quad + \sum_{L\kappa i} c_{L\kappa i\sigma}(\mathbf{k}) \tilde{H}_{L\kappa\sigma}(\mathbf{r}_i) \\
&\quad + \sum_{L'\kappa j} a_{L'\kappa j\sigma}(\mathbf{k}) \tilde{J}_{L'\kappa\sigma}(\mathbf{r}_j) \,,
\end{aligned}
\tag{2.1.31}
$$

where, in the second step, we have used the representation (2.1.30) as well as the abbreviation

$$
a_{L'\kappa j\sigma}(\mathbf{k}) = \sum_{Li} c_{L\kappa i\sigma}(\mathbf{k}) B_{L'L\kappa}(\boldsymbol{\tau}_j - \boldsymbol{\tau}_i, \mathbf{k}) \,.
\tag{2.1.32}
$$

Note that the variational procedure is of the Raleigh-Ritz type with the basis functions being fixed and the stationary states arrived at by varying the coefficients $c_{L\kappa i\sigma}(\mathbf{k})$ [26]. A more detailed description will be given in the following section.

In closing this section, we turn to the determination of the core states. One way to do so is the so-called frozen-core approximation, which uses core states as arising from an atomic calculation without allowing them to respond to the changes of the valence states coming with the formation of chemical bonds between the atoms during the iteration towards self-consistency. This offers the advantage that the large core state energies are fixed, which makes the evaluation of small total energy differences as they arise, e.g. in frozen-phonon calculations, much easier. Nevertheless, in the ASW method we opt for the correct approach and evaluate the core states from the same radial Schrödinger equation as used for the valence states, i.e. from (2.1.14) or (2.1.21). Since these equations contain the actual intraatomic potential the core states feel any changes of the potential due to the valence states and rearrange their shape during the iterations towards self-consistency. The ASW methods thus belongs to the so-called all-electron methods.

Denoting the radial part of the core states as $\varphi_{nli\sigma}(E_{nli\sigma}, r_{\mu i})$, where n is the principal quantum number, we write down the radial Schrödinger equation

$$
\begin{aligned}
&\left[-\frac{1}{r_{\mu i}} \frac{\partial^2}{\partial r_{\mu i}^2} r_{\mu i} + \frac{l(l+1)}{r_{\mu i}^2} + v_{i\sigma}^{MT}(r_{\mu i}) - v_{MTZ} - E_{nli\sigma} \right] \varphi_{nli\sigma}(E_{nli\sigma}, r_{\mu i}) = \\
&\left[-\frac{1}{r_{\mu i}} \frac{\partial^2}{\partial r_{\mu i}^2} r_{\mu i} + \frac{l(l+1)}{r_{\mu i}^2} + \hat{v}_{i\sigma}^{MT}(r_{\mu i}) - E_{nli\sigma} \right] \varphi_{nli\sigma}(E_{nli\sigma}, r_{\mu i}) = 0 \,.
\end{aligned}
\tag{2.1.33}
$$

As already mentioned above, the major difference to the valence states is due to the boundary conditions for the core states, which are also subject to the regularity of the radial function at the origin. Yet, in contrast to the valence states, both value and slope at the sphere boundary must vanish since the core states do not take part in the chemical bonding and thus are confined to their respective atomic sphere.

Methods using augmentation for the construction of basis functions offer the special advantage that by construction the valence states are orthogonal to the core states and thus an explicit orthogonalization is not necessary. This traces back to the fact that the radial parts of the basis functions, i.e. of the augmented Hankel and Bessel functions, obey the same radial Schrödinger equation as the radial parts of the core states. As a consequence, the radial overlap integral is just the Wronskian of the two functions involved [49,52] (see also (3.3.14)). We thus obtain, e.g. for the augmented Hankel function

$$\left(E_{l\kappa i\sigma}^{(H)} - E_{nli\sigma}\right) \int_0^{S_i} dr_{\mu i}\, r_{\mu i}^2 \tilde{h}_{l\kappa\sigma}(r_{\mu i})\varphi_{nli\sigma}(E_{nli\sigma}, r_{\mu i})$$

$$= W\{r_{\mu i}\tilde{h}_{l\kappa\sigma}(r_{\mu i}), r_{\mu i}\varphi_{nli\sigma}(E_{nli\sigma}, r_{\mu i})\, r_{\mu i}\}\, \Big|_0^{S_i}. \tag{2.1.34}$$

The Wronskian in turn is defined by

$$W\{r_{\mu i}\tilde{h}_{l\kappa\sigma}(r_{\mu i}), r_{\mu i}\varphi_{nli\sigma}(E_{nli\sigma}, r_{\mu i})\, r_{\mu i}\}$$

$$= r_{\mu i}^2 \tilde{h}_{l\kappa\sigma}(r_{\mu i})\frac{\partial}{\partial r_{\mu i}}\varphi_{nli\sigma}(E_{nli\sigma}, r_{\mu i}) - r_{\mu i}^2 \varphi_{nli\sigma}(E_{nli\sigma}, r_{\mu i})\frac{\partial}{\partial r_{\mu i}}\tilde{h}_{l\kappa\sigma}(r_{\mu i}). \tag{2.1.35}$$

Combining (2.1.34) and (2.1.35) we find that the contribution to the integral from the origin vanishes due to the regularity of all the functions involved. In contrast, at the sphere boundary $r_{\mu i} = S_i$ the core states do not contribute because of their values and slopes being zero. Since the Hankel energy $E_{l\kappa i\sigma}^{(H)}$ is always higher than any of the core state energies we arrive at the conclusion that the integral in (2.1.34) vanishes. The same holds for the augmented Bessel function and thus the core states are orthogonal to the valence states. Note that this has been achieved just by augmentation without any explicit orthogonalization procedure.

2.2 The Secular Matrix

With the basis functions at hand we proceed to determining the coefficients entering the expansion (2.1.31) of the ground state wave function in terms of the Bloch-summed basis functions (2.1.30). Eventually, these coefficients grow out of the Rayleigh-Ritz variational procedure built on the minimization of the total energy with respect to the many-body wave function. Within the Hartree- and Hartree-Fock schemes the many-body wave function is expressed in terms of the single-particle wave functions and, hence, the variational procedure leads directly to single-particle equations. This is different in density-functional theory, where the variational principle is initially formulated in terms of the density. However, by reintroducing single-particle states and by formally identifying the density of the interacting many-body system with that of a fictitious non-interacting system built on these single-particle states, Kohn and Sham were able to derive a set of single-particle Schrödinger-like equations,

$$[H_\sigma - \varepsilon_{\mathbf{k}\sigma}] |\psi_{\mathbf{k}\sigma}\rangle = 0 , \tag{2.2.1}$$

with the effective single-particle Hamiltonian

$$H_\sigma = -\Delta + v_\sigma(\mathbf{r}) - v_{MTZ} , \tag{2.2.2}$$

already known from (2.1.1). Here, we have explicitly referred the potential to the muffin-tin zero and implied Bloch symmetry by attaching the $\mathbf{k}$-label to the single-particle energies and states. Within the variational principle the single-particle energies $\varepsilon_{\mathbf{k}\sigma}$ are just Lagrange multipliers, which guarantee normalization of the single-particle wave function $\psi_{\mathbf{k}\sigma}$. We thus arrive at the Euler-Lagrange variational equation

$$\delta\left[\langle\psi_{\mathbf{k}\sigma}|H_\sigma|\psi_{\mathbf{k}\sigma}\rangle - \varepsilon_{\mathbf{k}\sigma}\langle\psi_{\mathbf{k}\sigma}|\psi_{\mathbf{k}\sigma}\rangle\right] = 0 . \tag{2.2.3}$$

In writing this expression we have implicitly assumed that the Hamiltonian is diagonal in spin. In particular, this implies the neglect of spin-orbit coupling. In addition, the decoupling of the spin states is based on the assumption that the system possesses a global spin quantization axis. As a consequence, we exclude those cases, which show a more general spin arrangement and which are covered by the so-called "non-collinear" ASW method [44, 45, 68].

Next, defining a bra-ket notation via

$$|L\kappa i\rangle^\infty := D^\infty_{L\kappa\sigma}(\mathbf{r}_i, \mathbf{k}) , \tag{2.2.4}$$

for the Bloch-summed basis functions and using the expansion (2.1.31) of the wave function in terms of basis functions we rewrite (2.2.3) as

$$\delta\left[\sum_{L\kappa_1 i L'\kappa_2 j} c^*_{L\kappa_1 i\sigma}(\mathbf{k})c_{L'\kappa_2 j\sigma}(\mathbf{k})\left[{}^\infty\langle L\kappa_1 i|H_\sigma|L'\kappa_2 j\rangle^\infty_c\right.\right.$$
$$\left.\left. -\varepsilon_{\mathbf{k}\sigma}{}^\infty\langle L\kappa_1 i|L'\kappa_2 j\rangle^\infty_c\right]\right] = 0 . \tag{2.2.5}$$

Here, the matrix elements are real-space integrals and the index c indicates integration over the unit cell. Since within a linear method the energy corresponding to the state (2.1.31) can be minimized only by variation of the expansion coefficients we differentiate (2.2.5) with respect to $c_{L'\kappa_2 j\sigma}(\mathbf{k})$ or else with respect to $c^*_{L\kappa_1 i\sigma}(\mathbf{k})$ and arrive at the linear equation system

$$\sum_{L'\kappa_2 m} c_{L'\kappa_2 j\sigma}(\mathbf{k})\left[{}^\infty\langle L\kappa_1 i|H_\sigma|L'\kappa_2 j\rangle^\infty_c - \varepsilon_{\mathbf{k}\sigma}{}^\infty\langle L\kappa_1 i|L'\kappa_2 j\rangle^\infty_c\right] = 0 . \tag{2.2.6}$$

It allows to determine the band energies $\varepsilon_{\mathbf{k}n\sigma}$ from the condition that the coefficient determinant has to vanish for (2.2.6) to have a nontrivial solution. Here n labels the N different solutions, where N is the number of basis functions included in the expansion (2.1.31) and, hence, the rank of the coefficient matrix. In addition, the solution of (2.2.6) yields the wave function $\psi_{\mathbf{k}n\sigma}$ in terms of the expansion coefficients $c_{L\kappa_1 i\sigma}(\mathbf{k})$ of the respective band. The generalized eigenvalue problem posed by (2.2.6) can be easily transformed into a standard eigenvalue problem with the help of the Cholesky decomposition, which is equivalent to a Gram-Schmidt orthogonalization. Obviously, it can be written as

$$(\mathcal{H} - \varepsilon \mathcal{S})\mathbf{C} = \mathbf{0} , \tag{2.2.7}$$

where $\mathcal{H}$ and $\mathcal{S}$ denote the Hamiltonian and overlap matrix, respectively. The latter is not necessarily the unit matrix since the basis functions usually are not orthonormalized. However, since the overlap matrix is positive definite, it can be decomposed into the product of an upper and a lower triangular matrices [60],

$$\mathcal{S} = \mathcal{U}^+ \mathcal{U} , \tag{2.2.8}$$

where $\mathcal{U}$ is an upper triagonal matrix. Inserting this into (2.2.7) we arrive at the standard eigenvalue problem

$$\left[(\mathcal{U}^+)^{-1} \mathcal{H} (\mathcal{U})^{-1} - \varepsilon \mathcal{E}\right] \mathcal{U} \mathbf{C} = \mathbf{0} , \tag{2.2.9}$$

where $\mathcal{E}$ is the unit matrix. This standard eigenvalue problem can be routinely solved by mathematical-library routines but needs a backtransformation of the eigenvectors to those of the original generalized eigenvalue problem. Nowadays, there exist even library routines for the latter problem and we do not have to care about the details of the Cholesky decomposition.

Having sketched the general procedure for setting up and diagonalizing the secular matrix entering (2.2.6) we will, on the remaining pages of this section, derive the explicit form of the elements of both the Hamiltonian and overlap matrix. To this end, we complement (2.2.4) by the definition of a bra-ket notation for the Bloch-summed envelope function

$$|L\kappa i\rangle := D_{L\kappa}(\mathbf{r}_i, \mathbf{k}) . \tag{2.2.10}$$

We may then write the general element of the Hamiltonian matrix as

$$
\begin{aligned}
&{}^\infty\langle L\kappa_1 i | H_\sigma | L'\kappa_2 j\rangle^\infty_c \\
&= \langle L\kappa_1 i | - \Delta | L'\kappa_2 j\rangle_c \\
&\quad + \sum_m \left[{}^\infty\langle L\kappa_1 i | H_\sigma | L'\kappa_2 j\rangle^\infty_{A(m)} - \langle L\kappa_1 i | - \Delta | L'\kappa_2 j\rangle_{A(m)} \right] ,
\end{aligned} \tag{2.2.11}
$$

where the integrals containing $-\Delta$ are to be taken with the non-augmented envelope functions and the index $A(m)$ denotes integration over the atomic sphere centered at $\boldsymbol{\tau}_m$. However, (2.2.11) is still valid for any other choice of sphere radii.

As it stands, (2.2.11) seems to be an *ad hoc* approach to the Hamiltonian matrix elements. Obviously, it is a first step beyond a formulation solely based on the ASA, which would only include the first term in square brackets without taking the integrals with the Laplacian operator into account. Insofar, (2.2.11) is well grounded. The terms adding to the pure ASA matrix elements considerably increase the accuracy of the method. In the LMTO method these terms are explicitly distinguished from the ASA terms and are referred to as the "combined correction". This is different in the ASW method, where they are built in from the very beginning. However, while a formal justification of (2.2.11) is still missing, we will make up for this in Sect. 4.2. For the time being, we note the similarity to the alternative representation of the effective single-particle potential as given in the second line of (2.1.2). There we separated the potential into a so-called pseudo part, which displayed, if at all, only small spatial variations and extended over all space, and additional local parts, which were confined to the atomic spheres and contained the intraatomic strong variations of the potential. Equation (2.2.11) is in the same spirit with the first term being the pseudo contribution and the square brackets covering the local add-ons.

In order to evaluate the integral (2.2.11) we use the fact that in the respective regions of integration the functions entering are eigenfunctions of either the full Hamiltonian H_σ or else the free-electron Hamiltonian $-\Delta$. We may therefore reduce the integrals with these operators to overlap integrals times the respective eigenenergies and note, e.g. for the Laplacian operator

$$\langle L\kappa_1 i| - \Delta |L'\kappa_2 j\rangle_c = \kappa_2^2 \langle L\kappa_1 i|L'\kappa_2 j\rangle_c \ . \tag{2.2.12}$$

We are now able to evaluate the integrals extending over the atomic spheres to be built with the full Hamiltonian. Using (2.1.30) for the Bloch-summed basis function we write

$$
\begin{aligned}
&{}^\infty\langle L\kappa_1 i|H_\sigma|L'\kappa_2 j\rangle^\infty_{A(m)} \\
&= \delta_{im} E^{(H)}_{l\kappa_2 j\sigma} \langle \tilde{H}_{L\kappa_1\sigma}|\tilde{H}_{L'\kappa_2\sigma}\rangle_{A(m)} \delta_{LL'}\delta_{mj} \\
&\quad + \delta_{im} E^{(J)}_{l''\kappa_2 m\sigma} \langle \tilde{H}_{L\kappa_1\sigma}|\tilde{J}_{L''\kappa_2\sigma}\rangle_{A(m)} \delta_{LL''} B_{L''L'\kappa_2}(\boldsymbol{\tau}_m - \boldsymbol{\tau}_j, \mathbf{k}) \\
&\quad + B^*_{L''L\kappa_1}(\boldsymbol{\tau}_m - \boldsymbol{\tau}_i, \mathbf{k}) E^{(H)}_{l'\kappa_2 m\sigma} \delta_{L''L'} \langle \tilde{J}_{L''\kappa_1\sigma}|\tilde{H}_{L'\kappa_2\sigma}\rangle_{A(m)} \delta_{mj} \\
&\quad + \sum_{L''}\sum_{L'''} B^*_{L''L\kappa_1}(\boldsymbol{\tau}_m - \boldsymbol{\tau}_i, \mathbf{k}) E^{(J)}_{l'''\kappa_2 m\sigma} \langle \tilde{J}_{L''\kappa_1\sigma}|\tilde{J}_{L'''\kappa_2\sigma}\rangle_{A(m)} \\
&\qquad\qquad\qquad\qquad \delta_{L''L'''} B_{L'''L'\kappa_2}(\boldsymbol{\tau}_m - \boldsymbol{\tau}_j, \mathbf{k}) \\
&= \delta_{im} E^{(H)}_{l\kappa_2 j\sigma} \langle \tilde{H}_{L\kappa_1\sigma}|\tilde{H}_{L\kappa_2\sigma}\rangle_{A(m)} \delta_{LL'}\delta_{mj} \\
&\quad + \delta_{im} E^{(J)}_{l\kappa_2 m\sigma} \langle \tilde{H}_{L\kappa_1\sigma}|\tilde{J}_{L\kappa_2\sigma}\rangle_{A(m)} B_{LL'\kappa_2}(\boldsymbol{\tau}_m - \boldsymbol{\tau}_j, \mathbf{k}) \\
&\quad + B^*_{L'L\kappa_1}(\boldsymbol{\tau}_m - \boldsymbol{\tau}_i, \mathbf{k}) E^{(H)}_{l'\kappa_2 m\sigma} \langle \tilde{J}_{L'\kappa_1\sigma}|\tilde{H}_{L'\kappa_2\sigma}\rangle_{A(m)} \delta_{mj} \\
&\quad + \sum_{L''} B^*_{L''L\kappa_1}(\boldsymbol{\tau}_m - \boldsymbol{\tau}_i, \mathbf{k}) E^{(J)}_{l''\kappa_2 m\sigma} \langle \tilde{J}_{L''\kappa_1\sigma}|\tilde{J}_{L''\kappa_2\sigma}\rangle_{A(m)} \\
&\qquad\qquad\qquad\qquad B_{L''L'\kappa_2}(\boldsymbol{\tau}_m - \boldsymbol{\tau}_j, \mathbf{k}) \ . \tag{2.2.13}
\end{aligned}
$$

Here we have taken into account the Schrödinger equations (2.1.14) and (2.1.21) as well as the orthogonality of the spherical harmonics, which leaves us with the Kronecker-δ's with respect to angular momenta and reduces the intraatomic integrals to only their radial contributions. Of the radial integrals those, which contain only one type of function for $\kappa_1^2 = \kappa_2^2$, are calculated numerically. These integrals are defined in (A.1.1) and (A.1.2) as $S_{l\kappa i\sigma}^{(H)} = \langle \tilde{H}_{L\kappa\sigma} | \tilde{H}_{L\kappa\sigma} \rangle_i$ and $S_{l\kappa i\sigma}^{(J)} = \langle \tilde{J}_{L\kappa\sigma} | \tilde{J}_{L\kappa\sigma} \rangle_i$ and designated as the Hankel and Bessel integrals, respectively. The mixed integrals $\langle \tilde{H}_{L\kappa_1\sigma} | \tilde{H}_{L\kappa_2\sigma} \rangle_{A(m)}$, $\langle \tilde{H}_{L\kappa_1\sigma} | \tilde{J}_{L\kappa_2\sigma} \rangle_{A(m)}$ and $\langle \tilde{J}_{L\kappa_1\sigma} | \tilde{J}_{L\kappa_2\sigma} \rangle_{A(m)}$ are then expressed in terms of the Hankel and Bessel energies and integrals with the help of the identities derived in App. A.1.

In (2.2.13) we distinguish three different types of integrals. They are called one-, two-, and three-center integrals depending on the number of sites involved, where counting includes the site of the atomic sphere, over which the integration extends. One-center integrals contain only augmented Hankel functions, two-center integrals both a Hankel and a Bessel function and, finally, three-center integrals contain only augmented Bessel functions. Note that due to the orthogonality of the spherical harmonics the limitation to the low-l Bessel functions is exact for the two-center integrals. In contrast, the three-center integrals still contain an in principle infinite summation over L''. However, in this respect the formulation (2.2.11) of the general matrix element offers a distinct advantage. Due to this formulation, three-center integrals enter only in the square bracket term on the right-hand side of (2.2.11) and they always enter in combination with the corresponding integrals to be built with the envelope functions. Since for high angular momenta the muffin-tin potential in the Hamiltonian H_σ is dominated by the centrifugal term the augmented functions become identical to the envelope functions. As a consequence, the difference of the two terms in square brackets vanishes and it is well justified to omit the difference of three-center terms already for quite low angular momenta. This is why the angular momentum cutoff for the intermediate waves, l_{int}, can be fixed to a rather low value, which, in practice, is $l_{low} + 1$.

The second term in the square bracket on the right-hand side of (2.2.11), i.e. the integral over the atomic spheres, which contains envelope rather than augmented functions, simply results from (2.2.13) by replacing the Hankel and Bessel energies by κ_2^2 and the augmented functions by the envelope functions. We thus note

$$
\begin{aligned}
\langle L\kappa_1 i | &- \Delta | L'\kappa_2 j \rangle_{A(m)} \\
= &\, \delta_{im} \kappa_2^2 \langle H_{L\kappa_1\sigma} | H_{L\kappa_2\sigma} \rangle_{A(m)} \delta_{LL'} \delta_{mj} \\
&+ \delta_{im} \kappa_2^2 \langle H_{L\kappa_1\sigma} | J_{L\kappa_2\sigma} \rangle_{A(m)} B_{LL'\kappa_2}(\boldsymbol{\tau}_m - \boldsymbol{\tau}_j, \mathbf{k}) \\
&+ B_{L'L\kappa_1}^*(\boldsymbol{\tau}_m - \boldsymbol{\tau}_i, \mathbf{k}) \kappa_2^2 \langle J_{L'\kappa_1\sigma} | H_{L'\kappa_2\sigma} \rangle_{A(m)} \delta_{mj} \\
&+ \sum_{L''} B_{L''L\kappa_1}^*(\boldsymbol{\tau}_m - \boldsymbol{\tau}_i, \mathbf{k}) \kappa_2^2 \langle J_{L''\kappa_1\sigma} | J_{L''\kappa_2\sigma} \rangle_{A(m)} B_{L''L'\kappa_2}(\boldsymbol{\tau}_m - \boldsymbol{\tau}_j, \mathbf{k}) \, ,
\end{aligned}
$$

(2.2.14)

where we have again used the orthogonality of the spherical harmonics. The radial integrals are calculated using the identities given in Sect. 3.8.

Obviously, difficulties may arise from the first, second, and third term on the right-hand side of (2.2.14) due to the singularity of the Hankel envelope functions at their respective origins. While the product of a Hankel envelope function and a Bessel envelope function can still be integrated the difficulty remains in the first term. However, as we will see in the next paragraph, this integral is only formally included here in order to compensate part of the first term of (2.2.11), which holds the integral over the product of two Hankel envelope functions extending over all space. In particular, if both basis functions are centered at the same site, the integration over the product of the two Hankel envelope functions centered at this site extends only over the region outside their on-center sphere and the singularities do not enter.

Next, we turn to the first term in (2.2.11), which, according to (2.2.12), can again be reduced to an overlap integral. Calculation of the latter is likewise the subject of Sect. 3.8. In particular, (3.8.39) gives the overlap integral of two Bloch-summed envelope functions extending over the interstitial region, which is identical to the difference of the first term and the second term in square brackets on the right-hand side of (2.2.11). Using these results and combining (2.2.11) to (2.2.13) as well as (3.8.39) we are eventually able to write down the general matrix element of the Hamiltonian as

$$
\begin{aligned}
{}^{\infty}\langle L\kappa_1 i|H_\sigma|L'\kappa_2 j\rangle^{\infty}_c &\\
= \Big[& E^{(H)}_{l\kappa_2 j\sigma}\langle \tilde{H}_{L\kappa_1\sigma}|\tilde{H}_{L\kappa_2\sigma}\rangle_{A(j)} + \kappa_2^2 \langle H_{L\kappa_1}|H_{L\kappa_2}\rangle'_{A(j)} \Big] \delta_{LL'}\delta_{ij} \\
&+ \frac{\kappa_2^2}{\kappa_2^2 - \kappa_1^2} \Big[B_{LL'\kappa_2}(\boldsymbol{\tau}_i - \boldsymbol{\tau}_j, \mathbf{k}) - B^*_{L'L\kappa_1}(\boldsymbol{\tau}_j - \boldsymbol{\tau}_i, \mathbf{k}) \Big] \big(1 - \delta(\kappa_1^2 - \kappa_2^2) \big) \\
&+ \kappa_2^2 \dot{B}_{LL'\kappa_1}(\boldsymbol{\tau}_i - \boldsymbol{\tau}_j, \mathbf{k})\delta(\kappa_1^2 - \kappa_2^2) \\
&+ \Big[E^{(J)}_{l\kappa_2 i\sigma}\langle \tilde{H}_{L\kappa_1\sigma}|\tilde{J}_{L\kappa_2\sigma}\rangle_{A(i)} - \kappa_2^2\langle H_{L\kappa_1}|J_{L\kappa_2}\rangle_{A(i)} \Big] B_{LL'\kappa_2}(\boldsymbol{\tau}_i - \boldsymbol{\tau}_j, \mathbf{k}) \\
&+ B^*_{L'L\kappa_1}(\boldsymbol{\tau}_j - \boldsymbol{\tau}_i, \mathbf{k}) \Big[E^{(H)}_{l'\kappa_2 j\sigma}\langle \tilde{J}_{L'\kappa_1\sigma}|\tilde{H}_{L'\kappa_2\sigma}\rangle_{A(j)} - \kappa_2^2\langle J_{L'\kappa_1}|H_{L'\kappa_2}\rangle_{A(j)} \Big] \\
&+ \sum_m \sum_{L''} B^*_{L''L\kappa_1}(\boldsymbol{\tau}_m - \boldsymbol{\tau}_i, \mathbf{k}) \\
&\qquad \Big[E^{(J)}_{l''\kappa_2 m\sigma}\langle \tilde{J}_{L''\kappa_1\sigma}|\tilde{J}_{L''\kappa_2\sigma}\rangle_{A(m)} - \kappa_2^2\langle J_{L''\kappa_1}|J_{L''\kappa_2}\rangle_{A(m)} \Big] \\
&\qquad B_{L''L'\kappa_2}(\boldsymbol{\tau}_m - \boldsymbol{\tau}_j, \mathbf{k}) \, . \qquad (2.2.15)
\end{aligned}
$$

In particular, we point to the two terms in the first square bracket, which comprise the integral over the product of two augmented Hankel functions extending over their on-center sphere and the integral over the product of two Hankel envelope functions extending over all space outside the on-center sphere, as indicated by the prime. Thus, as mentioned in connection with (2.2.14), the singularities of the Hankel envelope functions do not enter.

It is useful to define the following abbreviations for the one-, two-, and three-center contributions

$$
X^{(H,1)}_{L\kappa_1\kappa_2 i\sigma} = E^{(H)}_{l\kappa_2 i\sigma}\langle \tilde{H}_{L\kappa_1\sigma}|\tilde{H}_{L\kappa_2\sigma}\rangle_{A(i)} + \kappa_2^2\langle H_{L\kappa_1}|H_{L\kappa_2}\rangle'_{A(i)} \, , \qquad (2.2.16)
$$

$$X^{(H,2)}_{L\kappa_1\kappa_2 i\sigma} = E^{(J)}_{l\kappa_2 i\sigma}\langle\tilde{H}_{L\kappa_1\sigma}|\tilde{J}_{L\kappa_2\sigma}\rangle_{A(i)} - \kappa_2^2\langle H_{L\kappa_1}|J_{L\kappa_2}\rangle_{A(i)}$$

$$+\frac{\kappa_2^2}{\kappa_2^2 - \kappa_1^2}\left(1 - \delta(\kappa_1^2 - \kappa_2^2)\right)\,, \tag{2.2.17}$$

$$X^{(H,3)}_{L\kappa_1\kappa_2 i\sigma} = E^{(J)}_{l\kappa_2 i\sigma}\langle\tilde{J}_{L\kappa_1\sigma}|\tilde{J}_{L\kappa_2\sigma}\rangle_{A(i)} - \kappa_2^2\langle J_{L\kappa_1}|J_{L\kappa_2}\rangle_{A(i)}\,. \tag{2.2.18}$$

Furthermore, combining (2.2.17), (3.3.37), and (A.1.9) we derive the identity

$$E^{(H)}_{l\kappa_2 i\sigma}\langle\tilde{J}_{L\kappa_1\sigma}|\tilde{H}_{L\kappa_2\sigma}\rangle_{A(i)} - \kappa_2^2\langle J_{L\kappa_1}|H_{L\kappa_2}\rangle_{A(i)} - \frac{\kappa_2^2}{\kappa_2^2 - \kappa_1^2}\left(1 - \delta(\kappa_1^2 - \kappa_2^2)\right)$$

$$= X^{(H,2)}_{L\kappa_2\kappa_1 i\sigma} + \delta(\kappa_1^2 - \kappa_2^2)\,, \tag{2.2.19}$$

which complements (2.2.17). The Hamiltonian matrix element then assumes the form

$$^\infty\langle L\kappa_1 i|H_\sigma|L'\kappa_2 j\rangle^\infty_c$$

$$= X^{(H,1)}_{L\kappa_1\kappa_2 i\sigma}\delta_{LL'}\delta_{ij} + \kappa_2^2\dot{B}_{LL'\kappa_1}(\boldsymbol{\tau}_i - \boldsymbol{\tau}_j, \mathbf{k})\delta(\kappa_1^2 - \kappa_2^2)$$

$$+X^{(H,2)}_{L\kappa_1\kappa_2 i\sigma}B_{LL'\kappa_2}(\boldsymbol{\tau}_i - \boldsymbol{\tau}_j, \mathbf{k}) + B^*_{L'L\kappa_2}(\boldsymbol{\tau}_j - \boldsymbol{\tau}_i, \mathbf{k})\left(X^{(H,2)}_{L'\kappa_2\kappa_1 j\sigma} + \delta(\kappa_1^2 - \kappa_2^2)\right)$$

$$+\sum_m\sum_{L''}B^*_{L''L\kappa_1}(\boldsymbol{\tau}_m - \boldsymbol{\tau}_i, \mathbf{k})X^{(H,3)}_{L''\kappa_1\kappa_2 m\sigma}B_{L''L'\kappa_2}(\boldsymbol{\tau}_m - \boldsymbol{\tau}_j, \mathbf{k})\,. \tag{2.2.20}$$

Here we have, in addition, used the identity (A.1.8). Finally, working with cubic rather than spherical harmonics and, using the identity (3.7.18), we get for the Hamiltonian matrix element the final result

$$^\infty\langle L\kappa_1 i|H_\sigma|L'\kappa_2 j\rangle^\infty_c$$

$$= X^{(H,1)}_{L\kappa_1\kappa_2 i\sigma}\delta_{LL'}\delta_{ij} + \kappa_2^2\dot{B}_{LL'\kappa_1}(\boldsymbol{\tau}_i - \boldsymbol{\tau}_j, \mathbf{k})\delta(\kappa_1^2 - \kappa_2^2)$$

$$+X^{(H,2)}_{L\kappa_1\kappa_2 i\sigma}B_{LL'\kappa_2}(\boldsymbol{\tau}_i - \boldsymbol{\tau}_j, \mathbf{k}) + B_{LL'\kappa_1}(\boldsymbol{\tau}_i - \boldsymbol{\tau}_j, \mathbf{k})\left(X^{(H,2)}_{L'\kappa_2\kappa_1 j\sigma} + \delta(\kappa_1^2 - \kappa_2^2)\right)$$

$$+\sum_m\sum_{L''}B^*_{L''L\kappa_1}(\boldsymbol{\tau}_m - \boldsymbol{\tau}_i, \mathbf{k})X^{(H,3)}_{L''\kappa_1\kappa_2 m\sigma}B_{L''L'\kappa_2}(\boldsymbol{\tau}_m - \boldsymbol{\tau}_j, \mathbf{k})\,. \tag{2.2.21}$$

With the Hamiltonian matrix element at hand evaluation of the elements of the overlap matrix turns out to be quite easy. In accordance with (2.2.11) these matrix elements are defined as

$$^\infty\langle L\kappa_1 i|L'\kappa_2 j\rangle^\infty_c$$

$$= \langle L\kappa_1 i|L'\kappa_2 j\rangle_c$$

$$+ \sum_m\left[^\infty\langle L\kappa_1 i|L'\kappa_2 j\rangle^\infty_{A(m)} - \langle L\kappa_1 i|L'\kappa_2 j\rangle_{A(m)}\right]\,. \tag{2.2.22}$$

Since all the functions entering this expression are eigenfunctions of either the full Hamiltonian H_σ or else the free-electron Hamiltonian $-\Delta$ we may fall back on the just derived results for the Hamiltonian matrix. To be specific, we start from (2.2.15) and replace all energies, which appear in a numerator, by unity. We thus obtain as an intermediate result for the general matrix element of the overlap matrix

$$
\begin{aligned}
{}^{\infty}&\langle L\kappa_1 i | L'\kappa_2 j\rangle_c^{\infty}\\
&= \left[\langle \tilde{H}_{L\kappa_1\sigma}|\tilde{H}_{L\kappa_2\sigma}\rangle_{A(j)} + \langle H_{L\kappa_1}|H_{L\kappa_2}\rangle'_{A(j)}\right]\delta_{LL'}\delta_{ij}\\
&\quad + \frac{1}{\kappa_2^2 - \kappa_1^2}\left[B_{LL'\kappa_2}(\boldsymbol{\tau}_i - \boldsymbol{\tau}_j,\mathbf{k}) - B^*_{L'L\kappa_1}(\boldsymbol{\tau}_j - \boldsymbol{\tau}_i,\mathbf{k})\right]\left(1 - \delta(\kappa_1^2 - \kappa_2^2)\right)\\
&\quad + \dot{B}_{LL'\kappa_1}(\boldsymbol{\tau}_i - \boldsymbol{\tau}_j,\mathbf{k})\delta(\kappa_1^2 - \kappa_2^2)\\
&\quad + \left[\langle \tilde{H}_{L\kappa_1\sigma}|\tilde{J}_{L\kappa_2\sigma}\rangle_{A(i)} - \langle H_{L\kappa_1}|J_{L\kappa_2}\rangle_{A(i)}\right]B_{LL'\kappa_2}(\boldsymbol{\tau}_i - \boldsymbol{\tau}_j,\mathbf{k})\\
&\quad + B^*_{L'L\kappa_1}(\boldsymbol{\tau}_j - \boldsymbol{\tau}_i,\mathbf{k})\left[\langle \tilde{J}_{L'\kappa_1\sigma}|\tilde{H}_{L'\kappa_2\sigma}\rangle_{A(j)} - \langle J_{L'\kappa_1}|H_{L'\kappa_2}\rangle_{A(j)}\right]\\
&\quad + \sum_m\sum_{L''} B^*_{L''L\kappa_1}(\boldsymbol{\tau}_m - \boldsymbol{\tau}_i,\mathbf{k})\left[\langle \tilde{J}_{L''\kappa_1\sigma}|\tilde{J}_{L''\kappa_2\sigma}\rangle_{A(m)} - \langle J_{L''\kappa_1}|J_{L''\kappa_2}\rangle_{A(m)}\right]\\
&\qquad\qquad B_{L''L'\kappa_2}(\boldsymbol{\tau}_m - \boldsymbol{\tau}_j,\mathbf{k})\,.
\end{aligned}
\tag{2.2.23}
$$

As before we abbreviate the one-, two-, and three-center contributions to the elements of the overlap matrix by

$$
X^{(S,1)}_{L\kappa_1\kappa_2 i\sigma} = \langle \tilde{H}_{L\kappa_1\sigma}|\tilde{H}_{L\kappa_2\sigma}\rangle_{A(i)} + \langle H_{L\kappa_1}|H_{L\kappa_2}\rangle'_{A(i)}\,,
\tag{2.2.24}
$$

$$
X^{(S,2)}_{L\kappa_1\kappa_2 i\sigma} = \langle \tilde{H}_{L\kappa_1\sigma}|\tilde{J}_{L\kappa_2\sigma}\rangle_{A(i)} - \langle H_{L\kappa_1}|J_{L\kappa_2}\rangle_{A(i)} + \frac{1}{\kappa_2^2 - \kappa_1^2}\left(1 - \delta(\kappa_1^2 - \kappa_2^2)\right)\,,
\tag{2.2.25}
$$

$$
X^{(S,3)}_{L\kappa_1\kappa_2 i\sigma} = \langle \tilde{J}_{L\kappa_1\sigma}|\tilde{J}_{L\kappa_2\sigma}\rangle_{A(i)} - \langle J_{L\kappa_1}|J_{L\kappa_2}\rangle_{A(i)}\,.
\tag{2.2.26}
$$

Combining (2.2.23) to (2.2.26) we note the result

$$
\begin{aligned}
{}^{\infty}&\langle L\kappa_1 i|L'\kappa_2 j\rangle_c^{\infty}\\
&= X^{(S,1)}_{L\kappa_1\kappa_2 i\sigma}\delta_{LL'}\delta_{ij} + \dot{B}_{LL'\kappa_1}(\boldsymbol{\tau}_i - \boldsymbol{\tau}_j,\mathbf{k})\delta(\kappa_1^2 - \kappa_2^2)\\
&\quad + X^{(S,2)}_{L\kappa_1\kappa_2 i\sigma}B_{LL'\kappa_2}(\boldsymbol{\tau}_i - \boldsymbol{\tau}_j,\mathbf{k}) + B^*_{L'L\kappa_1}(\boldsymbol{\tau}_j - \boldsymbol{\tau}_i,\mathbf{k})X^{(S,2)}_{L'\kappa_2\kappa_1 j\sigma}\\
&\quad + \sum_m\sum_{L''} B^*_{L''L\kappa_1}(\boldsymbol{\tau}_m - \boldsymbol{\tau}_i,\mathbf{k})X^{(S,3)}_{L''\kappa_1\kappa_2 m\sigma}B_{L''L'\kappa_2}(\boldsymbol{\tau}_m - \boldsymbol{\tau}_j,\mathbf{k})\,,
\end{aligned}
\tag{2.2.27}
$$

which for cubic harmonics reads as

$$
\begin{aligned}
{}^{\infty}&\langle L\kappa_1 i|L'\kappa_2 j\rangle_c^{\infty}\\
&= X^{(S,1)}_{L\kappa_1\kappa_2 i\sigma}\delta_{LL'}\delta_{ij} + \dot{B}_{LL'\kappa_1}(\boldsymbol{\tau}_i - \boldsymbol{\tau}_j,\mathbf{k})\delta(\kappa_1^2 - \kappa_2^2)\\
&\quad + X^{(S,2)}_{L\kappa_1\kappa_2 i\sigma}B_{LL'\kappa_2}(\boldsymbol{\tau}_i - \boldsymbol{\tau}_j,\mathbf{k}) + B_{LL'\kappa_1}(\boldsymbol{\tau}_i - \boldsymbol{\tau}_j,\mathbf{k})X^{(S,2)}_{L'\kappa_2\kappa_1 j\sigma}\\
&\quad + \sum_m\sum_{L''} B^*_{L''L\kappa_1}(\boldsymbol{\tau}_m - \boldsymbol{\tau}_i,\mathbf{k})X^{(S,3)}_{L''\kappa_1\kappa_2 m\sigma}B_{L''L'\kappa_2}(\boldsymbol{\tau}_m - \boldsymbol{\tau}_j,\mathbf{k})\,.
\end{aligned}
\tag{2.2.28}
$$

By now, we have arrived at a formulation for the elements of the secular matrix, which offers several advantages. First off all, we have separated structural information from intraatomic information by writing each term as a structure constant times intraatomic radial integrals. The latter need to be calculated only once in an iteration before the time consuming loop over **k**-points starts. Moreover, as can be read off from the definitions (2.2.16) to (2.2.18), (2.2.24) to (2.2.26) and (A.1.8), the intraatomic contributions depend exclusively on integrals over envelope functions, which can be performed analytically, as well as on the Hankel and Bessel energies $E_{l\kappa i\sigma}^{(H)}$ and $E_{l\kappa i\sigma}^{(J)}$ and on the Hankel and Bessel integrals $S_{l\kappa i\sigma}^{(H)}$ and $S_{l\kappa i\sigma}^{(J)}$. Hence, except for the crystal structure information, the secular matrix is completely specified by four numbers per basis state, which contain all information about the shape of the crystal potential. Second, the present formulation of the secular matrix allows for a very efficient computation. In practice, (2.2.21) and (2.2.28) enable for both a high degree of vectorization and low memory costs and thus contribute a lot to the high computational efficiency of the ASW method.

2.3 Electron Density

Having described the setup of ASW basis functions as well as of the secular matrix we next turn to the calculation of the spin-dependent electron density, which comprises contributions from both the valence and core electrons.

Within the framework of density-functional theory, the spin-dependent valence electron density is given by

$$\rho_{val,\sigma}(\mathbf{r}) = \sum_{\mathbf{k}n} |\psi_{\mathbf{k}n\sigma}(\mathbf{r})|^2 \Theta(E_F - \varepsilon_{\mathbf{k}n\sigma}) . \qquad (2.3.1)$$

Here we have implied zero-temperature Fermi statistics by summing over the occupied states up to the Fermi energy E_F. Note that, according to the definition (2.2.2), the band energies $\varepsilon_{\mathbf{k}n\sigma}$ as well as the Fermi energy are referred to the muffin-tin zero, v_{MTZ}. The wave function is that defined by (2.1.31) with the coefficients determined by the solution of the generalized eigenvalue problem (2.2.7); n labels the different eigenstates. The spin-dependent electron density is then arrived at by combining (2.3.1) and (2.1.31) with the expression (2.1.30) for the ASW basis function. However, within the standard ASW method it is possible to follow a different route, which is based on the density of states and which will be used in the present context.

To this end, we first recall the fact that we started out from a muffin-tin potential in Sect. 2.1. For this reason, due to the self-consistency condition growing out of density-functional theory the potential to be eventually extracted from the electron density also must be a muffin-tin potential. Hence, we may already reduce the electron density to a muffin-tin form. Within this so-called shape approximation it is sufficient to calculate only the spherical symmetric contributions inside the atomic spheres. However, the spherical symmetric valence electron density may be equally well extracted from the electronic density of states. It has been argued that this offers the additional advantage of an increased accuracy because the Rayleigh-Ritz variational procedure gives higher accuracy to the eigenenergies than to the eigenfunctions [6, 7, 73].

Following the previous arguments, we concentrate from now on on the spherical symmetric spin-dependent electron density within the atomic spheres. In addition, we strictly enforce the ASA, i.e. we ignore the interstitial region completely and treat the atomic spheres as non-overlapping. Note that at this point we lay ground for the systematic error of the total energy expression of the standard ASW method. This will be explicited in Sect. 2.4.

To be concrete, we rewrite the wave function as given by (2.1.30) and (2.1.31) for the situation that the vector $\mathbf{r}$ lies within an atomic sphere as

$$\psi_{\mathbf{k}\sigma}(\mathbf{r}) = \sum_{L\kappa i} \left[c_{L\kappa i\sigma}(\mathbf{k})\tilde{H}_{L\kappa\sigma}(\mathbf{r}_i) + a_{L\kappa i\sigma}(\mathbf{k})\tilde{J}_{L\kappa\sigma}(\mathbf{r}_i) \right] , \qquad (2.3.2)$$

where we have used the abbreviation (2.1.32) in the form

$$a_{L\kappa i\sigma}(\mathbf{k}) = \sum_{L'j} c_{L'\kappa j\sigma}(\mathbf{k})B_{LL'\kappa}(\boldsymbol{\tau}_i - \boldsymbol{\tau}_j, \mathbf{k}) . \qquad (2.3.3)$$

Inserting this into (2.3.1) we get for the spin-dependent valence electron density

$$\rho_{val,\sigma}(\mathbf{r}) = \sum_i \rho_{val,i\sigma}(\mathbf{r}_i) , \qquad (2.3.4)$$

with the spherical symmetric part given by

$$\begin{aligned}
&\rho_{val,i\sigma}(r_i) \\
&= \frac{1}{4\pi} \int d\Omega \, \rho_{val,i\sigma}(\mathbf{r}_i) \\
&= \frac{1}{4\pi} \int d\Omega \sum_{\mathbf{k}} \Theta(E_F - \varepsilon_{\mathbf{k}\sigma}) \\
&\qquad \sum_L | \sum_\kappa \left(c_{L\kappa i\sigma}(\mathbf{k})\tilde{H}_{L\kappa\sigma}(\mathbf{r}_i) + a_{L\kappa i\sigma}(\mathbf{k})\tilde{J}_{L\kappa\sigma}(\mathbf{r}_i) \right) |^2 \\
&= \frac{1}{4\pi} \int d\Omega \int_{-\infty}^{E_F} dE \sum_{\mathbf{k}} \delta(E - \varepsilon_{\mathbf{k}\sigma}) \\
&\qquad \sum_L | \sum_\kappa \left(c_{L\kappa i\sigma}(\mathbf{k})\tilde{H}_{L\kappa\sigma}(\mathbf{r}_i) + a_{L\kappa i\sigma}(\mathbf{k})\tilde{J}_{L\kappa\sigma}(\mathbf{r}_i) \right) |^2. \qquad (2.3.5)
\end{aligned}$$

Here we have formally absorbed the band index n into the $\mathbf{k}$-point label. Note that the sum over δ functions is just the electronic density of states

$$\rho_\sigma(E) = \sum_{\mathbf{k}} \delta(E - \varepsilon_{\mathbf{k}\sigma}) . \qquad (2.3.6)$$

The valence electron density as given by (2.3.5) thus arises as a sum over weighted or partial densities of states. This representation is based on the orthonormality of the spherical harmonics as well as the fact that within the ASA the overlap of the atomic spheres is ignored. As a consequence, the valence electron density is diagonal in both the atomic sphere and the angular momentum index and can be written as a single sum over these indices.

Nevertheless, due to the energy linearization it is not straightforward to perform the energy integration on the right-hand side of (2.3.5). Instead of the augmented Hankel and Bessel functions, which are inside their respective spheres solutions of Schrödinger's equation for two particular energies (the Hankel and Bessel energies), we would seemingly need the full energy dependence of the wave function. Actually, as we will see below, the energy linearization leads a way out of this problem. Nevertheless, it pays to check what the valence electron density would look like if the full energy dependence were taken into account. For this reason, we take a sideglance to the KKR method [38, 41, 43, 48], where the wave function reads as

$$\psi_{\mathbf{k}\sigma E}(\mathbf{r}) = \sum_{Li} \beta_{Li\sigma}(\mathbf{k}) R_{l\sigma}(E, r_i) Y_L(\hat{\mathbf{r}}_i) . \qquad (2.3.7)$$

The functions $R_{l\sigma}(E, r_i)$ are normalized regular solutions of the radial Schrödinger equation (2.1.14) for energy E. Hence, they embrace the set of augmented Hankel and Bessel functions, which are solutions of the same radial equation albeit for only the Hankel and Bessel energies. The coefficients $\beta_{Li\sigma}(\mathbf{k})$ are equivalent to the expansion coefficients $c_{L\kappa i\sigma}(\mathbf{k})$ entering (2.1.31). Using the KKR wave function (2.3.7) we readily arrive at the following expression for the radial part of the valence electron density

$$
\begin{aligned}
\rho_{val,i\sigma}(r_i) &= \frac{1}{4\pi} \int d\Omega \, \rho_{val,i\sigma}(\mathbf{r}_i) \\
&= \frac{1}{4\pi} \sum_{\mathbf{k}} \sum_{L} |\beta_{Li\sigma}(\mathbf{k})|^2 R_{l\sigma}^2(E_\sigma(\mathbf{k}), r_i) \Theta(E_F - \varepsilon_{\mathbf{k}\sigma}) \\
&= \frac{1}{4\pi} \int_{-\infty}^{E_F} dE \sum_{l} \sum_{\mathbf{k}} \delta(E - \varepsilon_{\mathbf{k}\sigma}) \sum_{m} |\beta_{lmi\sigma}(\mathbf{k})|^2 R_{l\sigma}^2(E, r_i) .
\end{aligned}
$$

$$(2.3.8)$$

In contrast to (2.3.5) this expression preserves the full energy dependence of the radial function and thus the energy integration can be easily performed. Using the decomposition of the norm of the wave functions into contributions from atomic spheres and angular momenta,

$$\int_{\Omega_c} d^3\mathbf{r} \, |\psi_{\mathbf{k}\sigma E}(\mathbf{r})|^2 = \sum_{lmi} |\beta_{lmi\sigma}(\mathbf{k})|^2 \overset{!}{=} 1 , \qquad (2.3.9)$$

we can define partial densities of states by

$$\rho_{li\sigma}(E) := \sum_{\mathbf{k}} \delta(E - \varepsilon_{\mathbf{k}\sigma}) \sum_{m} |\beta_{lmi\sigma}(\mathbf{k})|^2 , \qquad (2.3.10)$$

and obtain for the KKR valence electron density

$$\rho_{val,i\sigma}(r_i) = \frac{1}{4\pi} \int_{-\infty}^{E_F} dE \sum_l \rho_{li\sigma}(E) R_{l\sigma}^2(E, r_i) . \qquad (2.3.11)$$

This result will be useful in setting up the valence electron density within a linear basis set.

With the just derived representation of the KKR valence electron density in mind we turn back to the ASW method and aim at the definition of partial densities of states via the decomposition of the norm. The latter is obtained from the expansion (2.1.31) of the wave functions in terms of the basis functions as

$$\int_{\Omega_c} d^3\mathbf{r} |\psi_{\mathbf{k}\sigma}(\mathbf{r})|^2 = \sum_{L\kappa_1 i} \sum_{L'\kappa_2 j} c_{L\kappa_1 i\sigma}^*(\mathbf{k}) c_{L'\kappa_2 j\sigma}(\mathbf{k}) {}^\infty\langle L\kappa_1 i | L'\kappa_2 j \rangle_c^\infty . \qquad (2.3.12)$$

Inserting into this expression (2.2.23) for the overlap matrix and using the definition (2.3.3) we get

$$\int_{\Omega_c} d^3\mathbf{r} |\psi_{\mathbf{k}\sigma}(\mathbf{r})|^2$$

$$= \sum_{L\kappa_1\kappa_2 i} \Bigg[c_{L\kappa_1 i\sigma}^*(\mathbf{k}) c_{L\kappa_2 i\sigma}(\mathbf{k}) \left[\langle \tilde{H}_{L\kappa_1\sigma} | \tilde{H}_{L\kappa_2\sigma} \rangle_i + \langle H_{L\kappa_1} | H_{L\kappa_2} \rangle_i' \right]$$

$$+ c_{L\kappa_1 i\sigma}^*(\mathbf{k}) \sum_{L'j} c_{L'\kappa_2 j\sigma}(\mathbf{k}) \frac{1}{\kappa_2^2 - \kappa_1^2}$$

$$\left[B_{LL'\kappa_2}(\boldsymbol{\tau}_i - \boldsymbol{\tau}_j, \mathbf{k}) - B_{L'L\kappa_1}^*(\boldsymbol{\tau}_j - \boldsymbol{\tau}_i, \mathbf{k}) \right] \left(1 - \delta(\kappa_1^2 - \kappa_2^2) \right)$$

$$+ c_{L\kappa_1 i\sigma}^*(\mathbf{k}) \sum_{L'j} c_{L'\kappa_2 j\sigma}(\mathbf{k}) \dot{B}_{LL'\kappa_1}(\boldsymbol{\tau}_i - \boldsymbol{\tau}_j, \mathbf{k}) \delta(\kappa_1^2 - \kappa_2^2)$$

$$+ c_{L\kappa_1 i\sigma}^*(\mathbf{k}) a_{L\kappa_2 i\sigma}(\mathbf{k}) \left[\langle \tilde{H}_{L\kappa_1\sigma} | \tilde{J}_{L\kappa_2\sigma} \rangle_i - \langle H_{L\kappa_1} | J_{L\kappa_2} \rangle_i \right]$$

$$+ a_{L\kappa_1 i\sigma}^*(\mathbf{k}) c_{L\kappa_2 i\sigma}(\mathbf{k}) \left[\langle \tilde{J}_{L\kappa_1\sigma} | \tilde{H}_{L\kappa_2\sigma} \rangle_i - \langle J_{L\kappa_1} | H_{L\kappa_2} \rangle_i \right]$$

$$+ a_{L\kappa_1 i\sigma}^*(\mathbf{k}) a_{L\kappa_2 i\sigma}(\mathbf{k}) \left[\langle \tilde{J}_{L\kappa_1\sigma} | \tilde{J}_{L\kappa_2\sigma} \rangle_i - \langle J_{L\kappa_1} | J_{L\kappa_2} \rangle_i \right] \Bigg] . \qquad (2.3.13)$$

As already mentioned in the previous sections, the Hankel and Bessel functions are included up to l_{low} and l_{int}, respectively. Note that (2.3.13) is the exact representation of the norm. However, it does not allow for a straightforward decomposition into partial densities of states due to the double sum in the second and third term on the right-hand side. These terms originate from the so-called "combined correction" terms, i.e. from the difference of the first term and the second term in square brackets on the right-hand side of (2.2.11). Since we already opted for the ASA in the present context we replace the exact expression by the following ASA expression for the norm

$$\int_{\Omega_c} d^3\mathbf{r} |\psi_{\mathbf{k}\sigma}(\mathbf{r})|^2 = \sum_{L\kappa_1\kappa_2 i} \left[c^*_{L\kappa_1 i\sigma}(\mathbf{k})c_{L\kappa_2 i\sigma}(\mathbf{k})\langle \tilde{H}_{L\kappa_1\sigma}|\tilde{H}_{L\kappa_2\sigma}\rangle_i \right.$$

$$+ c^*_{L\kappa_1 i\sigma}(\mathbf{k})a_{L\kappa_2 i\sigma}(\mathbf{k})\langle \tilde{H}_{L\kappa_1\sigma}|\tilde{J}_{L\kappa_2\sigma}\rangle_i$$

$$+ a^*_{L\kappa_1 i\sigma}(\mathbf{k})c_{L\kappa_2 i\sigma}(\mathbf{k})\langle \tilde{J}_{L\kappa_1\sigma}|\tilde{H}_{L\kappa_2\sigma}\rangle_i$$

$$\left. + a^*_{L\kappa_1 i\sigma}(\mathbf{k})a_{L\kappa_2 i\sigma}(\mathbf{k})\langle \tilde{J}_{L\kappa_1\sigma}|\tilde{J}_{L\kappa_2\sigma}\rangle_i \right] + \Delta_\sigma(\mathbf{k})$$

$$=: \sum_{li} q_{li\sigma}(\mathbf{k})$$

$$\stackrel{!}{=} 1 . \qquad (2.3.14)$$

Here we have eventually added a small quantity $\Delta_\sigma(\mathbf{k})$, which accounts for the difference between the expressions (2.3.13) and (2.3.14). Thus, it can in principle be calculated exactly. Yet, since this quantity makes only a small contribution to the norm it is omitted in practice and the resulting deviation of the norm from unity is cured by a renormalization of the remaining terms in (2.3.14).

As indicated in the second but last line of (2.3.14) we then arrive at the desired unique decomposition of the norm, which still allows to define **k**-dependent partial occupation numbers. Moreover, we now have a simple recipe for the calculation of partial densities of states at hand, which read as

$$\rho_{li\sigma}(E) := \sum_{\mathbf{k}} \delta(E - \varepsilon_{\mathbf{k}\sigma})q_{li\sigma}(\mathbf{k}) . \qquad (2.3.15)$$

Still, we are not able to perform the energy integration needed for the calculation of the spin-dependent valence electron density. In order to do so we would need the full energy dependence of the radial function, which in case of the KKR method is contained in the solution $R_{l\sigma}(E, r_j)$ of the radial Schrödinger equation. However, we point out that the full energy dependence is not really omitted in a linear method but is just replaced by a linear function [6, 65, 73]. Under the assumption of a perfectly linear dependence the partial density of states would be completely specified by its first three moments, which are defined by

$$M_{li\sigma}^{(k)} = \int_{-\infty}^{E_F} dE \, E^k \rho_{li\sigma}(E) , \qquad k = 0, 1, 2 . \qquad (2.3.16)$$

Even if small deviations from the linear behaviour are included it will be sufficient to take additionally the fourth moment into account. This allows to define a new expression for the density of states

$$\tilde{\rho}_{li\sigma}(E) := \sum_{\alpha=1}^{2} \delta(E - E_{li\sigma}^{(\alpha)})Q_{li\sigma}^{(\alpha)} , \qquad (2.3.17)$$

where the two energies $E_{li\sigma}^{(\alpha)}$ and the particle numbers $Q_{li\sigma}^{(\alpha)}$ are subject to the condition that they yield the same first four moments as the true density of states does, i.e.

$$\sum_{\alpha=1}^{2} \left(E_{li\sigma}^{(\alpha)}\right)^k Q_{li\sigma}^{(\alpha)} \stackrel{!}{=} \int_{-\infty}^{E_F} dE \, E^k \rho_{li\sigma}(E) \,, \qquad k = 0, 1, 2, 3 \,. \tag{2.3.18}$$

The calculation of the moments of the partial densities of states as well as the details of the moment analysis (2.3.18) are the subject of Apps. A.2 and A.4, respectively.

Finally, with the true density of states replaced by (2.3.17) we arrive at the representation of the spin-dependent valence electron density as

$$\rho_{val,i\sigma}(r_i) = \frac{1}{4\pi} \sum_{l} \sum_{\alpha=1}^{2} Q_{li\sigma}^{(\alpha)} R_{l\sigma}^2(E_{li\sigma}^{(\alpha)}, r_i) = \sum_{l} \rho_{val,li\sigma}(r_i) \,. \tag{2.3.19}$$

It does not require the full energy dependence of the solution of the radial Schrödinger equation. Instead, we have to solve this equation for two energies only. Nevertheless, the first four moments of the true density of states are fully reproduced.

The previous results reveal an additional "symmetry" of the standard ASW method. Obviously, the intraatomic valence electron density by now is completely specified by only two energies $E_{li\sigma}^{(\alpha)}$ and particle numbers $Q_{li\sigma}^{(\alpha)}$. In return, as outlined at the end of Sect. 2.2, specification of the secular matrix likewise required four quantities per orbital, namely, the Hankel and Bessel energies and integrals.

As a consequence, the iteration process can be cut into two parts. While the band calculation starts from the Hankel and Bessel energies and integrals and returns the above energies $E_{li\sigma}^{(\alpha)}$ and particle numbers $Q_{li\sigma}^{(\alpha)}$ things are reversed for the intraatomic calculations. Eventually, this bouncing back and forth of different quantities allows for an additional speedup of the method, which will be outlined in more detail in the following section.

Having performed the calculation of the valence electron density we turn to the core electrons. Their density arises from a sum over all core states, which have wave functions $\varphi_{nli\sigma}(E_{nli\sigma}, r_i)$ and energies $E_{nli\sigma}$ resulting from the radial Schrödinger equation (2.1.33) subject to the conditions of vanishing value and slope at the sphere boundary. Hence, we get for the spin-dependent core electron density

$$\rho_{core,\sigma}(\mathbf{r}) = \sum_{i} \rho_{core,\sigma}(\mathbf{r}_i) = \sum_{nlmi} \varphi_{nli\sigma}^2(E_{nli\sigma}, r_i) |Y_L(\hat{\mathbf{r}}_i)|^2$$

$$= \sum_{nli} \frac{2l+1}{4\pi} \varphi_{nli\sigma}^2(E_{nli\sigma}, r_i)$$

$$= \sum_{i} \rho_{core,\sigma}(r_i) \,. \tag{2.3.20}$$

Finally, combining (2.3.19) and (2.3.20) we arrive at the following result for the spin-dependent electron density

$$\rho_{el,\sigma}(\mathbf{r}) = \sum_{i} \left(\rho_{val,\sigma}(\mathbf{r}_i) + \rho_{core,\sigma}(\mathbf{r}_i)\right) =: \sum_{i} \rho_{el,\sigma}(\mathbf{r}_i) \,. \tag{2.3.21}$$

2.4 The Effective Potential

In order to close the self-consistency cycle we still have to calculate the effective single-particle potential. Again, we rely on density-functional theory and the local-density approximation, which in an approximate manner cast the full many-body problem into a single-particle self-consistent field problem. The potential is then represented as the sum of the external, Hartree and exchange-correlation potential [10, 11, 33, 39, 43]. The external potential is the Coulomb potential originating from the nuclei and possibly from external electromagnetic fields. In contrast, the Hartree potential comprises the classical contribution to the electron-electron Coulomb interaction. Finally, the non-classical contributions are covered by the exchange-correlation potential.

For practical calculations it is useful to first combine the electron density of the electrons as calculated in the previous section with the density of the nuclei, which is given by

$$\rho_{nucl}(\mathbf{r}) = -\sum_i \delta(\mathbf{r}_i) Z_i \ . \tag{2.4.1}$$

The classical potential due to the resulting total density then contains both the external and the Hartree potential. It can be written as

$$\begin{aligned} v_{es}(\mathbf{r}) &= 2 \int d^3 r' \frac{\rho_{el}(\mathbf{r}')}{|\mathbf{r} - \mathbf{r}'|} - 2 \sum_{\mu i} \frac{Z_i}{|\mathbf{r} - \mathbf{R}_{\mu i}|} \\ &= 2 \sum_{\mu i} \int_{\Omega_i} d^3 r'_{\mu i} \frac{\rho_{el}(\mathbf{r}'_{\mu i})}{|\mathbf{r}_{\mu i} - \mathbf{r}'_{\mu i}|} - 2 \sum_{\mu i} \frac{Z_i}{|\mathbf{r}_{\mu i}|} \ . \end{aligned} \tag{2.4.2}$$

The prefactor 2 entering here is due to our choice of atomic units, where it is identical to e^2. It reflects the fact that so far we have calculated electron and nuclear densities rather than charge densities. Since the Coulomb potential requires a charge density an extra factor e has to be added.

Next we assume the position $\mathbf{r}$ to lie in the atomic sphere centered at the site $\mathbf{R}_{\nu j}$. Using the identity [34, (3.70)]

$$\frac{1}{|\mathbf{r} - \mathbf{r}'|} = \sum_L \frac{4\pi}{2l+1} \frac{r_<^l}{r_>^{l+1}} Y_L^*(\hat{\mathbf{r}}) Y_L(\hat{\mathbf{r}}') \ , \tag{2.4.3}$$

which follows from (3.6.8) for $\kappa \to 0$, we transform (2.4.2) into the following expression

$$\begin{aligned} v_{es}(\mathbf{r}) \big|_{|\mathbf{r}_{\nu j}| \le S_j} &= 8\pi \frac{1}{r_{\nu j}} \int_0^{r_{\nu j}} dr'_{\nu j} \, r'^2_{\nu j} \rho_{el}(r'_{\nu j}) + 8\pi \int_{r_{\nu j}}^{S_j} dr'_{\nu j} \, r'_{\nu j} \rho_{el}(r'_{\nu j}) \\ &\quad - 2 \frac{Z_i}{|\mathbf{r}_{\nu j}|} + 2 \sum_{\mu i} (1 - \delta_{\mu\nu}\delta_{ij}) \frac{Q_i - Z_i}{|\mathbf{r}_{\mu i}|} \ . \end{aligned} \tag{2.4.4}$$

It is based on the shape approximation, i.e. on the assumption of spherical symmetric electron densities confined to atomic spheres, which we used already in Sect. 2.3

for the electron density. In (2.4.4), we have furthermore separated all charges contained in the sphere centered at the site $\mathbf{R}_{\nu j}$ from those falling outside this sphere. In doing so, we have used the definition of point charges

$$Q_i := 4\pi \int_0^{S_i} dr'_{\mu i} \, r'^2_{\mu i} \rho_{el}(r'_{\mu i}) \,, \tag{2.4.5}$$

as resulting from integrating all the electronic charge within an atomic sphere. This is motivated by the fact that, if viewed from outside a sphere, a spherical symmetric charge density within a sphere acts like a point charge located at its center. If combined with the nuclear charges these "electronic point charges" generate the Madelung potential, which is represented by the last term in (2.4.4). However, it is important to note that this treatment neglects the overlap of the atomic spheres, which is substantial especially in the atomic-sphere approximation. This leads to an error in the potential and, in particular, in the total energy to be calculated in the Sect. 2.5. The error is rather small in closed packed solids, where the overlap can be minimized [28, 50, 63]. Yet, it is large enough to spoil the calculation of phonon frequencies and forces. It it the aim of a full-potential ASW method to overcome the shape approximation and to provide higher accuracy for the total energy and the potential.

Still, we can substantially simplify the Madelung term in (2.4.4) by consequently using the muffin-tin shape of the potential. Due to the spherical symmetry of the potential within the atomic spheres, which suppresses the angular degrees of freedom, the Madelung potential, e.g. in the sphere at site $\mathbf{R}_{\nu j}$, can in no way depend on the position $\mathbf{r}_{\nu j}$ within that sphere. For that reason, the Madelung potential reduces to

$$v_{Mad,j} = 2 \sum_{\mu i} (1 - \delta_{0\mu}\delta_{ij}) \frac{Q_i - Z_i}{|\boldsymbol{\tau}_j - \boldsymbol{\tau}_i - \mathbf{R}_\mu|} \,, \tag{2.4.6}$$

which is just a constant shift of the potential within each atomic sphere.

The lattice sum entering (2.4.6) is closely related to the $\mathbf{k} = \mathbf{0}$-Bloch-summed Hankel function discussed in Sect. 3.5. Using the asymptotic form of the barred Hankel function (3.1.45) and the relation (3.1.17) we write

$$\bar{h}_0^{(1)}(\kappa r) \overset{\kappa r \to 0}{\sim} \frac{1}{r} \,, \tag{2.4.7}$$

and

$$H_{L\kappa}(\boldsymbol{\tau}_j - \boldsymbol{\tau}_i - \mathbf{R}_\mu)|_{l=0, \kappa \to 0} = \frac{1}{\sqrt{4\pi}} \frac{1}{|\boldsymbol{\tau}_j - \boldsymbol{\tau}_i - \mathbf{R}_\mu|} \,. \tag{2.4.8}$$

Comparing this to the definition (3.5.2) of the Bloch-summed Hankel envelope function we obtain for the Madelung potential (2.4.6) the result

$$v_{Mad,j} = 2 \sum_{\mu i} (1 - \delta_{0\mu}\delta_{ij}) \sqrt{4\pi} H_{L\kappa}(\boldsymbol{\tau}_j - \boldsymbol{\tau}_i - \mathbf{R}_\mu)(Q_i - Z_i)|_{l=0, \kappa \to 0}$$

$$= 2 \sum_i \sqrt{4\pi} D_{L\kappa}(\boldsymbol{\tau}_j - \boldsymbol{\tau}_i, \mathbf{0})(Q_i - Z_i)|_{l=0, \kappa \to 0} \,. \tag{2.4.9}$$

Note that those parts of the Bloch-summed Hankel function $D_{L\kappa}$, which do not depend on the vector $\boldsymbol{\tau}_j - \boldsymbol{\tau}_i$, cancel out due to charge neutrality of the unit cell,

$$\sum_i (Q_i - Z_i) \overset{!}{=} 0 \,. \tag{2.4.10}$$

This holds especially for the term with $\mathbf{K}_n = \mathbf{0}$ in the reciprocal lattice sum of the function $D_{L\kappa}^{(1)}$, which is defined by (3.5.19). Since this term does not depend on the index j for $\mathbf{k} = \mathbf{0}$ the sum over all charges vanishes and the singular behavior for $\mathbf{K}_n = \mathbf{0}$ [22] is avoided.

Having calculated the classical parts of the single-particle potential we turn to the exchange-correlation potential. As already mentioned, we use the local-density approximation, which returns the exchange-correlation potential as a spin-dependent local function of the spin-dependent electronic densities [10, 11, 39, 40, 43, 72], and write

$$\begin{aligned} v_{xc,\sigma}(\mathbf{r}) &= v_{xc,\sigma}\left[\rho_{el,\sigma}(\mathbf{r}), \rho_{el,-\sigma}(\mathbf{r})\right] \\ &= \sum_i v_{xc,i\sigma}\left[\rho_{el,\sigma}(\mathbf{r}_i), \rho_{el,-\sigma}(\mathbf{r}_i)\right] \\ &= \sum_i v_{xc,i\sigma}(\mathbf{r}_i) \,. \end{aligned} \tag{2.4.11}$$

In the past, several parametrizations for the density dependence of the exchange-correlation potential have been derived from different treatments of the homogeneous electron gas among them those proposed by Hedin and Lundqvist, von Barth and Hedin, Moruzzi, Williams and Janak, Vosko, Wilk and Nusair, and Perdew and Zunger [10, 12, 26, 31, 43, 51, 59, 70]. A complete overview of all these parametrizations was given by MacLaren et al. [47]. Alternatively, we may employ the generalized gradient approximation, for which parametrizations were given by Perdew and coworkers, Engel and Vosko, and Zhang and Yang [20, 46, 54–58, 74].

Combining (2.4.4), (2.4.9) and (2.4.11) we arrive at the following result for the effective single particle potential

$$\begin{aligned} v_\sigma(\mathbf{r}) &= v_{es}(\mathbf{r}) + v_{xc,\sigma}(\mathbf{r}) \\ &= \sum_i \left[v_{es,i}(\mathbf{r}_i) + v_{xc,i\sigma}(\mathbf{r}_i) + v_{Mad,i}\right] \\ &= \sum_i v_\sigma(\mathbf{r}_i) \,. \end{aligned} \tag{2.4.12}$$

Here

$$v_{es,i}(\mathbf{r}_i) = 8\pi \frac{1}{r_i} \int_0^{r_i} dr_i' \, r_i'^2 \rho_{el}(r_i') + 8\pi \int_{r_i}^{S_i} dr_i' \, r_i' \rho_{el}(r_i') - 2\frac{Z_i}{|\mathbf{r}_i|} \,. \tag{2.4.13}$$

is the electrostatic potential generated by all charges within the same sphere, which deviates from the expression (2.4.13) by the Madelung term representing the potential generated by all charges outside the respective sphere. Due to this separation of intraatomic contributions from the outside charges the calculations are significantly simplified. In a last step, we have to specify the muffin-tin zero, which is usually set to the average of the potentials at the surfaces of the atomic spheres.

By now, we have closed the self-consistency cycle since the just derived potential can be inserted into Schrödinger's equations (2.1.14) and (2.1.21) this allowing for the calculation of new augmented Hankel and Bessel functions. Nevertheless, there is an additional bonus due to the previous separation of the Madelung potential, which induces only a constant potential shift in each atomic sphere. As a consequence, all the intraatomic problems are completely decoupled, and we are able to calculate the intraatomic potential according to (2.4.12) without taking the Madelung term into account. Having finished this step we do not directly pass the potential to the following band calculation but instead insert it into Schrödinger's equation to calculate new radial functions and a new electron density. This establishes an intraatomic self-consistency cycle, which allows for a self-consistently calculated potential within each atom before all these atomic potentials are combined for the following step of the band iteration. This might save a lot of computation time.

To be specific, we describe the sequence of intraatomic calculations in more detail. Following the flow diagram given in Fig. 2.2 we start out from the energies $E_{li\sigma}^{(\alpha)}$ and electron numbers $Q_{li\sigma}^{(\alpha)}$ as resulting from the momentum analysis of the partial densities of states. The energies are then transformed to the local energy scale by

$$\bar{E}_{li\sigma}^{(\alpha)} = E_{li\sigma}^{(\alpha)} + v_{MTZ} - v_{Mad,i} \, . \tag{2.4.14}$$

After this we solve the radial Schrödinger equation

$$\left[-\frac{1}{r_i} \frac{\partial^2}{\partial r_i^2} r_i + \frac{l(l+1)}{r_i^2} + v_{es,i}(\mathbf{r}_i) + v_{xc,i\sigma}(\mathbf{r}_i) - v_{MTZ} - E_{li\sigma}^{(\alpha)} \right] R_{l\sigma}(\bar{E}_{li\sigma}^{(\alpha)}, r_i) =$$
$$\left[-\frac{1}{r_i} \frac{\partial^2}{\partial r_i^2} r_i + \frac{l(l+1)}{r_i^2} + v_{es,i}(\mathbf{r}_i) + v_{xc,i\sigma}(\mathbf{r}_i) - v_{Mad,i} - \bar{E}_{li\sigma}^{(\alpha)} \right] R_{l\sigma}(\bar{E}_{li\sigma}^{(\alpha)}, r_i) = 0 \, ,$$
$$\tag{2.4.15}$$

where $R_{l\sigma}(\bar{E}_{li\sigma}^{(\alpha)}, r_i)$ is a real and regular function normalized to

$$\int_0^{S_i} dr_i \, r_i^2 |R_{l\sigma}(\bar{E}_{li\sigma}^{(\alpha)}, r_i)|^2 = Q_{li\sigma}^{(\alpha)} \, . \tag{2.4.16}$$

From the solution of the radial (2.4.15) we then get the logarithmic derivatives at the sphere boundary

$$D_{li\sigma}^{(\alpha)} = r_i \left[R_{l\sigma}(\bar{E}_{li\sigma}^{(\alpha)}, r_i) \right]^{-1} \frac{\partial}{\partial r_i} R_{l\sigma}(\bar{E}_{li\sigma}^{(\alpha)}, r_i) \, |_{r_i=S_i} \, . \tag{2.4.17}$$

This way we arrive at a set of four new quantities per basis state, namely, the logarithmic derivatives $D_{li\sigma}^{(\alpha)}$ and the electron numbers $Q_{li\sigma}^{(\alpha)}$. As experience has

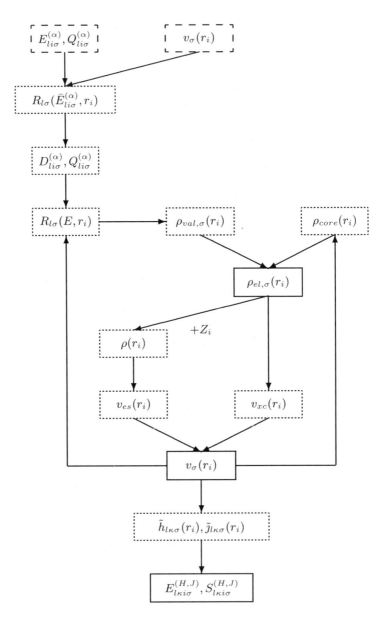

Fig. 2.2. Flow diagram of the intraatomic calculations of the standard ASW method. Input and output variables are highlighted by *dashed* and *solid boxes*, respectively

shown, the logarithmic derivatives are superiour to the energies $E_{li\sigma}^{(\alpha)}$ as they allow for a more stable acceleration of the iteration process. In the course of the latter the logarithmic derivatives and the electron numbers of the actual iteration as well as previous ones are used to make a good guess for the following iteration. There exist many different methods for the acceleration of iteration processes in electronic structure calculations and a large number of original papers dealing with the issue [2, 3, 13, 15–18, 27, 36, 37, 42, 61, 64, 67, 69]. For a recent overview see [24].

In addition to this rather numerical advantage of the logarithmic derivatives they allow for a better illustration of the interplay between intraatomic and band calculations. This has been used already in the concept of "renormalized atoms". It corresponds to atomic calculations subject to boundary conditions, which reflect the influence of the surrounding atoms [29, 32, 71, 73]. It is this concept, which, according to Williams, Kübler, and Gelatt, initiated the development of the ASW method [73].

Finally, with the logarithmic derivatives at hand we enter the intraatomic iteration cycle as sketched in Fig. 2.2. After self-consistency has been achieved, the muffin-tin potential is used in an additional step with the radial Schrödinger equations (2.1.14) and (2.1.21) as well as the boundary conditions (2.1.15) and (2.1.22) to evaluate the Hankel and Bessel energies $E_{l\kappa i\sigma}^{(H)}$ and $E_{l\kappa i\sigma}^{(J)}$ as well as the Hankel and Bessel integrals $S_{l\kappa i\sigma}^{(H)}$ and $S_{l\kappa i\sigma}^{(J)}$, which enter the subsequent band calculation.

2.5 Total Energy

In the previous sections we have learned about the major steps of the ASW method. This included the calculation of partial densities of states and partial occupation numbers, which already give a lot of information about a materials properties. However, in many cases the total energy is also of great interest. We just mention binding energies, bulk moduli, elastic constants or the stability of different magnetic structures. While some of these quantities are beyond the scope of the ASA and need a more accurate treatment as provided by a full-potential scheme, we will, nevertheless, in this last section of the present chapter sketch the evaluation of the total energy.

Within density-functional theory, the total energy is given as the sum of the kinetic energy, the electrostatic energy arising from the classical part of the electron-electron interaction, and the exchange-correlation energy [10, 11, 19, 21, 26, 43, 53].

An additional contribution stems from the external potential, which, in the simplest case, is just the electrostatic potential generated by the nuclei. Finally, we have to add the energy due to the electrostatic interaction between the nuclei. We thus note

$$E_T = E_T \left[\rho_{el,\uparrow}(\mathbf{r}), \rho_{el,\downarrow}(\mathbf{r}) \right]$$
$$= T \left[\rho_{el,\uparrow}(\mathbf{r}), \rho_{el,\downarrow}(\mathbf{r}) \right] + \sum_\sigma \int d^3\mathbf{r} \, v_{ext}(\mathbf{r}) \rho_{el,\sigma}(\mathbf{r})$$

$$+\frac{1}{N}\int\int d^3r d^3\mathbf{r}'\,\frac{\rho_{el}(\mathbf{r})\rho_{el}(\mathbf{r}')}{|\mathbf{r}-\mathbf{r}'|}+E_{xc}\left[\rho_{el,\uparrow}(\mathbf{r}),\rho_{el,\downarrow}(\mathbf{r})\right]$$

$$+\frac{1}{N}\sum_{\mu i}\sum_{\nu j}(1-\delta_{\mu\nu}\delta_{ij})\frac{Z_i Z_j}{|\mathbf{R}_{\nu j}-\mathbf{R}_{\mu i}|}$$

$$=T\left[\rho_{el,\uparrow}(\mathbf{r}),\rho_{el,\downarrow}(\mathbf{r})\right]+E_{es}\left[\rho_{el}(\mathbf{r})\right]+E_{xc}\left[\rho_{el,\uparrow}(\mathbf{r}),\rho_{el,\downarrow}(\mathbf{r})\right],\quad(2.5.1)$$

where all integrations extend over the whole crystal unless otherwise indicated and N denotes the number of unit cells. In addition, we have used atomic units as defined in Sect. 1.2, i.e. $e^2 = 2$. Furthermore, we have in the last step combined all the contributions arising from the classical part of the Coulomb interaction into the total electrostatic energy. The three terms on the right-hand side are now subject to further inspection.

An expression for the kinetic energy grows out of the Kohn-Sham equations by multiplying them with the complex conjugate of the respective wave function, summing over all eigenstates and integrating over all space [19,21,26,53]. From this we obtain

$$T\left[\rho_{el,\uparrow}(\mathbf{r}),\rho_{el,\downarrow}(\mathbf{r})\right]=\sum_{\mathbf{k}\sigma}E_\sigma(\mathbf{k})\Theta(E_F-E_\sigma(\mathbf{k}))+\sum_{nlmi\sigma}E_{nli\sigma}+v_{MTZ}\sum_i Q_i$$

$$-\sum_\sigma\int d^3\mathbf{r}\,v_\sigma(\mathbf{r})\rho_{el,\sigma}(\mathbf{r})\,.\qquad(2.5.2)$$

Here, the third term takes care of the fact that the single-particle energies were referred to the muffin-tin zero whereas the potential is not. The latter is given by

$$v_\sigma(\mathbf{r})=v_{es}^{(el)}(\mathbf{r})+v_{xc,\sigma}(\mathbf{r})+v_{ext}(\mathbf{r})\,,\qquad(2.5.3)$$

where

$$v_{es}^{(el)}(\mathbf{r})=2\int d^3\mathbf{r}'\,\frac{\rho_{el}(\mathbf{r}')}{|\mathbf{r}-\mathbf{r}'|}\,,\qquad(2.5.4)$$

is the electrostatic potential generated by the electronic density. As well known, (2.5.2) can be used to write the total energy as the sum of all single-particle energies (valence and core) minus the so-called "double-counting" terms. Eventually, we will also arrive at an expression of this kind. However, we point out that the effective potential entering the kinetic energy is the potential *making* the single-particle energies and, hence, the electron density, whereas the potential entering the electrostatic energy is the potential *made* by the density. These two potentials should be distinguished as long as full self-consistency has not yet been reached. Nevertheless, according to (2.5.3) the potential entering the kinetic energy also contains the external potential, which does not change during the iterations and thus cancels the corresponding term in (2.5.1) exactly. Combining (2.5.1) to (2.5.3) we thus note

$$E_T=\sum_{\mathbf{k}\sigma}E_\sigma(\mathbf{k})\Theta(E_F-E_\sigma(\mathbf{k}))+\sum_{nlmi\sigma}E_{nli\sigma}+v_{MTZ}\sum_i Q_i$$

$$-\sum_\sigma \int d^3\mathbf{r} \left[v_{es}^{(el)}(\mathbf{r}) + v_{xc,\sigma}(\mathbf{r}) + v_{ext}(\mathbf{r}) \right] \rho_{el,\sigma}(\mathbf{r})$$

$$+\sum_\sigma \int d^3\mathbf{r}\, v_{ext}(\mathbf{r})\rho_{el,\sigma}(\mathbf{r})$$

$$+\frac{1}{N} \int\int d^3\mathbf{r}\, d^3\mathbf{r}'\, \frac{\rho_{el}(\mathbf{r})\rho_{el}(\mathbf{r}')}{|\mathbf{r}-\mathbf{r}'|} + E_{xc}\left[\rho_{el,\uparrow}(\mathbf{r}),\rho_{el,\downarrow}(\mathbf{r})\right]$$

$$+\frac{1}{N} \sum_{\mu i}\sum_{\nu j} (1-\delta_{\mu\nu}\delta_{ij}) \frac{Z_i Z_j}{|\mathbf{R}_{\nu j} - \mathbf{R}_{\mu i}|}$$

$$= \sum_{\mathbf{k}\sigma} E_\sigma(\mathbf{k})\Theta(E_F - E_\sigma(\mathbf{k})) + \sum_{nlmi\sigma} E_{nli\sigma} + v_{MTZ}\sum_i Q_i$$

$$-\sum_\sigma \int d^3\mathbf{r} \left[v_{es}^{(el,in)}(\mathbf{r}) + v_{xc,\sigma}(\mathbf{r}) \right] \rho_{el,\sigma}(\mathbf{r})$$

$$+\frac{1}{2}\sum_\sigma \int d^3\mathbf{r}\, v_{es}^{(el,out)}(\mathbf{r})\rho_{el,\sigma}(\mathbf{r}) + E_{xc}\left[\rho_{el,\uparrow}(\mathbf{r}),\rho_{el,\downarrow}(\mathbf{r})\right]$$

$$+\frac{1}{N} \sum_{\mu i}\sum_{\nu j} (1-\delta_{\mu\nu}\delta_{ij}) \frac{Z_i Z_j}{|\mathbf{R}_{\nu j} - \mathbf{R}_{\mu i}|}$$

$$= \sum_{\mathbf{k}\sigma} E_\sigma(\mathbf{k})\Theta(E_F - E_\sigma(\mathbf{k})) + \sum_{nlmi\sigma} E_{nli\sigma} + v_{MTZ}\sum_i Q_i$$

$$-\frac{1}{2}\sum_\sigma \int d^3\mathbf{r}\, v_{es}^{(el)}(\mathbf{r})\rho_{el,\sigma}(\mathbf{r}) + \frac{1}{N}\sum_{\mu i}\sum_{\nu j}(1-\delta_{\mu\nu}\delta_{ij})\frac{Z_i Z_j}{|\mathbf{R}_{\nu j}-\mathbf{R}_{\mu i}|}$$

$$+E_{xc}\left[\rho_{el,\uparrow}(\mathbf{r}),\rho_{el,\downarrow}(\mathbf{r})\right] - \sum_\sigma \int d^3\mathbf{r}\, v_{xc,\sigma}(\mathbf{r})\rho_{el,\sigma}(\mathbf{r})\,, \tag{2.5.5}$$

where we have omitted the integrals over the external potential in the second step. While at that stage we still have distinguished the potentials $v_{es}^{(el,in)}(\mathbf{r})$ *making* the electronic density and $v_{es}^{(el,out)}(\mathbf{r})$ *made from* the electronic density, these potentials were combined in the third step leading to the representation of the total energy as the sum of single-particle energies minus the double-counting terms.

In further evaluating the total energy we aim at a formulation, which allows for both a simple calculation and interpretation of the results. Such a formulation is indeed possible and consists of a sum of atomic contributions plus the Madelung energy, which includes a sum over pairs of atomic sites,

$$E_T = \sum_i E_{T,i} + E_{Mad}\,. \tag{2.5.6}$$

In order to derive this representation we start from the band energy contribution to the total energy, which is the first term on the right-hand side of (2.5.5). It may be simplified a lot with the help of the definition (2.3.15) of the partial densities of states as well as the alternative representation (2.3.17) arising from the moment analysis. We thus note

$$\sum_{\mathbf{k}\sigma} E_\sigma(\mathbf{k})\Theta(E_F - E_\sigma(\mathbf{k})) = \sum_\sigma \int_{-\infty}^{E_F} dE \sum_{\mathbf{k}} E_\sigma(\mathbf{k})\delta(E - E_\sigma(\mathbf{k}))$$

$$= \sum_\sigma \int_{-\infty}^{E_F} dE \; E\rho_\sigma(E)$$

$$= \sum_{li\sigma} \int_{-\infty}^{E_F} dE \; E\rho_{li\sigma}(E)$$

$$= \sum_{li\sigma} \sum_{\alpha=1}^{2} E_{li\sigma}^{(\alpha)} Q_{li\sigma}^{(\alpha)} \; . \tag{2.5.7}$$

Using (2.4.14) we can still split off the local Madelung potential. Defining an analogous energy shift for the core states by

$$\bar{E}_{nli\sigma} := E_{nli\sigma} + v_{MTZ} - v_{Mad,i} \tag{2.5.8}$$

we obtain

$$\sum_{\mathbf{k}\sigma} E_\sigma(\mathbf{k})\Theta(E_F - E_\sigma(\mathbf{k})) + \sum_{nlmi\sigma} E_{nli\sigma} + v_{MTZ} \sum_i Q_i$$

$$= \sum_{li\sigma} \left[\sum_{\alpha=1}^{2} E_{li\sigma}^{(\alpha)} Q_{li\sigma}^{(\alpha)} + (2l+1) \sum_n E_{nli\sigma} \right]$$

$$= \sum_{li\sigma} \left[\sum_{\alpha=1}^{2} \bar{E}_{li\sigma}^{(\alpha)} Q_{li\sigma}^{(\alpha)} + (2l+1) \sum_n \bar{E}_{nli\sigma} \right]$$

$$+ \sum_i v_{Mad,i} \left[\sum_{l\sigma} \sum_{\alpha-1}^{2} Q_{li\sigma}^{(\alpha)} + 2\sum_{nl}(2l+1) \right] \; . \tag{2.5.9}$$

The last square-bracket term on the right-hand side is the total electron number inside the atomic sphere at site τ_i, which we defined as Q_i in (2.4.5). In addition, we have from (2.3.19) and (2.3.20)

$$\sum_{l\sigma} \sum_{\alpha=1}^{2} Q_{li\sigma}^{(\alpha)} = 4\pi \sum_\sigma \int_0^{S_i} \rho_{val,i\sigma}(r_i) \; , \tag{2.5.10}$$

and

$$2\sum_{n_i l}(2l+1) = 4\pi \sum_\sigma \int_0^{S_i} \rho_{core,\sigma}(r_i) \; . \tag{2.5.11}$$

Combining (2.5.7) to (2.5.9) we arrive at the following expression for the sum of the single-particle energies

$$\sum_{\mathbf{k}\sigma} E_\sigma(\mathbf{k})\Theta(E_F - E_\sigma(\mathbf{k})) + \sum_{nlmi\sigma} E_{nli\sigma} + v_{MTZ} \sum_i Q_i$$

$$= \sum_{li\sigma} \left[\sum_{\alpha=1}^{2} \bar{E}_{li\sigma}^{(\alpha)} Q_{li\sigma}^{(\alpha)} + (2l+1) \sum_n \bar{E}_{nli\sigma} \right] + \sum_i v_{Mad,i} Q_i \; . \tag{2.5.12}$$

Next we turn to the electronic contribution to the electrostatic energy. Due to our previous choice of the ASA and the resulting neglect of the interstitial region the integrals over the whole crystal turn into sums of integrals over atomic spheres. However, as for the construction of the Hartree potential, the latter integrals can be replaced by point charges located at the sphere centers whenever the twofold integral extends over different spheres. We thus note

$$
-\frac{1}{2} \sum_\sigma \int d^3\mathbf{r} \, v_{es}^{(el)}(\mathbf{r}) \rho_{el,\sigma}(\mathbf{r})
$$

$$
= -\frac{1}{N} \int\int d^3\mathbf{r} d^3\mathbf{r}' \, \frac{\rho_{el}(\mathbf{r})\rho_{el}(\mathbf{r}')}{|\mathbf{r}-\mathbf{r}'|}
$$

$$
= -\frac{1}{N} \sum_{\mu\nu}\sum_{ij} \int_{\Omega_i}\int_{\Omega_j} d^3\mathbf{r}_{\mu i} d^3\mathbf{r}'_{\nu j} \, \frac{\rho_{el,i}(\mathbf{r}_i)\rho_{el,j}(\mathbf{r}'_j)}{|\mathbf{r}_{\mu i}-\mathbf{r}'_{\nu j}|}
$$

$$
= -\sum_i \int_{\Omega_i}\int_{\Omega_i} d^3\mathbf{r}_i d^3\mathbf{r}'_i \, \frac{\rho_{el,i}(\mathbf{r}_i)\rho_{el,i}(\mathbf{r}'_i)}{|\mathbf{r}_i-\mathbf{r}'_i|}
$$

$$
\quad -\frac{1}{N} \sum_{\mu\nu}\sum_{ij} (1-\delta_{\mu\nu}\delta_{ij}) \frac{Q_i Q_j}{|\mathbf{R}_{\mu i}-\mathbf{R}_{\nu j}|}
$$

$$
=: -\frac{1}{2} \sum_i \int_{\Omega_i} d^3\mathbf{r}_i \, \bar{v}_{es,i}^{(el)}(\mathbf{r}_i)\rho_{el,i}(\mathbf{r}_i)
$$

$$
\quad -\frac{1}{N} \sum_{\mu\nu}\sum_{ij} (1-\delta_{\mu\nu}\delta_{ij}) \frac{Q_i Q_j}{|\mathbf{R}_{\mu i}-\mathbf{R}_{\nu j}|} \,, \tag{2.5.13}
$$

where $\bar{v}_{es,i}(\mathbf{r}_i)$ denotes the electrostatic potential in the atomic sphere centered at τ_i and arising only from the electron density inside this sphere. In contrast, all intersphere interactions have been replaced by the electrostatic interaction of point charges located at the atomic sites. Of course, as already discussed in Sect. 2.4, this treatment means a severe approximation as soon as the atomic spheres overlap, which they do in the ASA. As a consequence of this and the complete neglect of the interstitial region, highly accurate total energies are beyond the ASA and can only be evaluated within a full-potential method.

Note that the electrostatic potential introduced in the last step of (2.5.13) is referred to the local potential zero. Defining the potential

$$
v_{es,i}^{(el)}(\mathbf{r}) = \bar{v}_{es,i}^{(el)}(\mathbf{r}) + v_{Mad,i} \tag{2.5.14}
$$

we may thus rewrite (2.5.13) as

$$
-\frac{1}{2} \sum_\sigma \int d^3\mathbf{r} \, v_{es}^{(el)}(\mathbf{r}) \rho_{el,\sigma}(\mathbf{r})
$$

$$
= -\frac{1}{N} \int\int d^3\mathbf{r} d^3\mathbf{r}' \, \frac{\rho_{el}(\mathbf{r})\rho_{el}(\mathbf{r}')}{|\mathbf{r}-\mathbf{r}'|}
$$

$$
= \frac{1}{2} \sum_i \int_{\Omega_i} d^3\mathbf{r}_i \, \bar{v}_{es,i}^{(el)}(\mathbf{r}_i)\rho_{el,i}(\mathbf{r}_i)
$$

$$-\sum_i \int_{\Omega_i} d^3\mathbf{r}_i \, v_{es,i}^{(el)}(\mathbf{r}_i)\rho_{el,i}(\mathbf{r}_i) - \sum_i v_{Mad,i} \int_{\Omega_i} d^3\mathbf{r}_i \, \rho_{el,i}(\mathbf{r}_i)$$

$$-\frac{1}{N}\sum_{\mu\nu}\sum_{ij}(1-\delta_{\mu\nu}\delta_{ij})\frac{Q_iQ_j}{|\mathbf{R}_{\mu i}-\mathbf{R}_{\nu j}|}$$

$$=\frac{1}{2}\sum_i \int_{\Omega_i} d^3\mathbf{r}_i \, \bar{v}_{es,i}^{(el)}(\mathbf{r}_i)\rho_{el,i}(\mathbf{r}_i)$$

$$-\sum_i \int_{\Omega_i} d^3\mathbf{r}_i \, v_{es,i}^{(el)}(\mathbf{r}_i)\rho_{el,i}(\mathbf{r}_i) + \sum_i v_{Mad,i}Q_i$$

$$-\frac{1}{N}\sum_{\mu\nu}\sum_{ij}(1-\delta_{\mu\nu}\delta_{ij})\frac{Q_iQ_j}{|\mathbf{R}_{\mu i}-\mathbf{R}_{\nu j}|} \; . \tag{2.5.15}$$

Here we have divided the intraatomic contributions into two integrals, where the potential is referred either to the local Madelung potential or else to the global potential zero. Combining (2.5.15) with the term comprising the interaction between the nuclei we arrive at the result

$$-\frac{1}{2}\sum_\sigma \int d^3\mathbf{r} \, v_{es}^{(el)}(\mathbf{r})\rho_{el,\sigma}(\mathbf{r}) + \frac{1}{N}\sum_{\mu i}\sum_{\nu j}(1-\delta_{\mu\nu}\delta_{ij})\frac{Z_iZ_j}{|\mathbf{R}_{\nu j}-\mathbf{R}_{\mu i}|}$$

$$=\frac{1}{2}\sum_i \int_{\Omega_i} d^3\mathbf{r}_i \, \bar{v}_{es,i}^{(el)}(\mathbf{r}_i)\rho_{el,i}(\mathbf{r}_i) - \sum_i \int_{\Omega_i} d^3\mathbf{r}_i \, v_{es,i}^{(el)}(\mathbf{r}_i)\rho_{el,i}(\mathbf{r}_i)$$

$$-\frac{1}{N}\sum_{\mu\nu}\sum_{ij}(1-\delta_{\mu\nu}\delta_{ij})\frac{Q_iQ_j}{|\mathbf{R}_{\mu i}-\mathbf{R}_{\nu j}|} + \sum_i v_{Mad,i}Q_i$$

$$+\frac{1}{N}\sum_{\mu i}\sum_{\nu j}(1-\delta_{\mu\nu}\delta_{ij})\frac{Z_iZ_j}{|\mathbf{R}_{\nu j}-\mathbf{R}_{\mu i}|} \; . \tag{2.5.16}$$

The last three terms can be combined into the Madelung energy, which, using (2.4.6), we rewrite as

$$E_{Mad} = -\frac{1}{N}\sum_{\mu\nu}\sum_{ij}(1-\delta_{\mu\nu}\delta_{ij})\frac{Q_iQ_j}{|\mathbf{R}_{\mu i}-\mathbf{R}_{\nu j}|} + \sum_i v_{Mad,i}Q_i$$

$$+\frac{1}{N}\sum_{\mu i}\sum_{\nu j}(1-\delta_{\mu\nu}\delta_{ij})\frac{Z_iZ_j}{|\mathbf{R}_{\nu j}-\mathbf{R}_{\mu i}|}$$

$$=\sum_\mu\sum_{ij}(1-\delta_{\mu 0}\delta_{ij})\frac{[Z_iZ_j - Q_iQ_j + Q_i(Q_j-Z_j) + Q_j(Q_i-Z_i)]}{|\boldsymbol{\tau}_j - \boldsymbol{\tau}_i - \mathbf{R}_\mu|}$$

$$=\sum_\mu\sum_{ij}(1-\delta_{\mu 0}\delta_{ij})\frac{(Q_i-Z_i)(Q_j-Z_j)}{|\boldsymbol{\tau}_j - \boldsymbol{\tau}_i - \mathbf{R}_\mu|}$$

$$=\frac{1}{2}\sum_i v_{Mad,i}(Q_i - Z_i) \; . \tag{2.5.17}$$

As a consequence, we obtain for the sum of all electrostatic contributions the result

$$-\frac{1}{2} \sum_{\sigma} \int d^3\mathbf{r} \, v_{es}^{(el)}(\mathbf{r}) \rho_{el,\sigma}(\mathbf{r}) + \frac{1}{N} \sum_{\mu i} \sum_{\nu j} (1 - \delta_{\mu\nu}\delta_{ij}) \frac{Z_i Z_j}{|\mathbf{R}_{\nu j} - \mathbf{R}_{\mu i}|}$$

$$= \frac{1}{2} \sum_i \int_{\Omega_i} d^3\mathbf{r}_i \, \bar{v}_{es,i}^{(el)}(\mathbf{r}_i) \rho_{el,i}(\mathbf{r}_i) - \sum_i \int_{\Omega_i} d^3\mathbf{r}_i \, v_{es,i}^{(el)}(\mathbf{r}_i) \rho_{el,i}(\mathbf{r}_i) + E_{Mad}$$

$$= -\frac{1}{2} \sum_i \int_{\Omega_i} d^3\mathbf{r}_i \, \bar{v}_{es,i}^{(el)}(\mathbf{r}_i) \rho_{el,i}(\mathbf{r}_i) - \sum_i v_{Mad,i} Q_i + E_{Mad} \,. \tag{2.5.18}$$

Finally, the exchange-correlation energy is evaluated using the local density approximation [19, 21, 26, 53]

$$E_{xc,\sigma}\left[\rho_{el,\sigma}(\mathbf{r}), \rho_{el,-\sigma}(\mathbf{r})\right] = \frac{1}{N} \int d^3\mathbf{r} \, \varepsilon_{xc,\sigma}\left[\rho_{el,\sigma}(\mathbf{r}), \rho_{el,-\sigma}(\mathbf{r})\right] \rho_{el,\sigma}(\mathbf{r}) \,, \tag{2.5.19}$$

according to which the function $\varepsilon_{xc,\sigma}$ is a local function of the spin-dependent densities. In complete analogy to (2.4.11) we write

$$\varepsilon_{xc,\sigma}(\mathbf{r}) = \varepsilon_{xc,\sigma}\left[\rho_{el,\sigma}(\mathbf{r}) \,, \rho_{el,-\sigma}(\mathbf{r})\right]$$

$$= \sum_i \varepsilon_{xc,i\sigma}\left[\rho_{el,\sigma}(\mathbf{r}_i) \,, \rho_{el,-\sigma}(\mathbf{r}_i)\right]$$

$$= \sum_i \varepsilon_{xc,i\sigma}(\mathbf{r}_i) \,, \tag{2.5.20}$$

again using one of the parametrizations mentioned in Sect. 2.4. With the local density approximation at hand the terms in the respective last lines of (2.5.5) can be written as

$$E_{xc,\sigma}\left[\rho_{el,\sigma}(\mathbf{r}), \rho_{el,-\sigma}(\mathbf{r})\right] - \frac{1}{N} \int d^3\mathbf{r} \, v_{xc,\sigma}(\mathbf{r})\rho_{el,\sigma}(\mathbf{r})$$

$$= \sum_i \int_{\Omega_i} d^3\mathbf{r}_i \, \left[\varepsilon_{xc,i\sigma}(\mathbf{r}_i) - v_{xc,i\sigma}(\mathbf{r}_i)\right] \rho_{el,i\sigma}(\mathbf{r}_i) \,. \tag{2.5.21}$$

Inserting the intermediate results (2.5.12), (2.5.18), and (2.5.21) into the initial formula (2.5.5) we arrive at two alternative final expressions for the total energy

$$E_T = \sum_{li\sigma} \left[\sum_{\alpha=1}^{2} \bar{E}_{li\sigma}^{(\alpha)} Q_{li\sigma}^{(\alpha)} + (2l+1) \sum_n \bar{E}_{nli\sigma}\right]$$

$$- \frac{1}{2} \sum_i \int_{\Omega_i} d^3\mathbf{r}_i \, \bar{v}_{es,i}^{(el)}(\mathbf{r}_i) \rho_{el,i}(\mathbf{r}_i) + E_{Mad}$$

$$+ \sum_{i\sigma} \int_{\Omega_i} d^3\mathbf{r}_i \, \left[\varepsilon_{xc,i\sigma}(\mathbf{r}_i) - v_{xc,i\sigma}(\mathbf{r}_i)\right] \rho_{el,i\sigma}(\mathbf{r}_i)$$

$$= \sum_{li\sigma} \left[\sum_{\alpha=1}^{2} E_{li\sigma}^{(\alpha)} Q_{li\sigma}^{(\alpha)} + (2l+1) \sum_n E_{nli\sigma}\right] + v_{MTZ} \sum_i Q_i$$

$$
+ \frac{1}{2} \sum_i \int_{\Omega_i} d^3\mathbf{r}_i \, \bar{v}_{es,i}^{(el)}(\mathbf{r}_i) \rho_{el,i}(\mathbf{r}_i) - \sum_i \int_{\Omega_i} d^3\mathbf{r}_i \, v_{es,i}^{(el)}(\mathbf{r}_i) \rho_{el,i}(\mathbf{r}_i) + E_{Mad}
$$

$$
+ \sum_{i\sigma} \int_{\Omega_i} d^3\mathbf{r}_i \, \left[\varepsilon_{xc,i\sigma}(\mathbf{r}_i) - v_{xc,i\sigma}(\mathbf{r}_i) \right] \rho_{el,i\sigma}(\mathbf{r}_i)
$$

$$
= \sum_{li\sigma} \left[\sum_{\alpha=1}^{2} E_{li\sigma}^{(\alpha)} Q_{li\sigma}^{(\alpha)} + (2l+1) \sum_n E_{nli\sigma} \right] + v_{MTZ} \sum_i Q_i
$$

$$
- \sum_{i\sigma} \int_{\Omega_i} d^3\mathbf{r}_i \, \left[v_{es,i}^{(el)}(\mathbf{r}_i) + v_{xc,i\sigma}(\mathbf{r}_i) \right] \rho_{el,i\sigma}(\mathbf{r}_i)
$$

$$
+ \frac{1}{2} \sum_{i\sigma} \int_{\Omega_i} d^3\mathbf{r}_i \, \bar{v}_{es,i}^{(el)}(\mathbf{r}_i) \rho_{el,i\sigma}(\mathbf{r}_i) + \sum_{i\sigma} \int_{\Omega_i} d^3\mathbf{r}_i \, \varepsilon_{xc,i\sigma}(\mathbf{r}_i) \rho_{el,i\sigma}(\mathbf{r}_i) + E_{Mad} \ ,
$$

$$
(2.5.22)
$$

where, in the last step, we have explicitly noted the double-counting terms. From (2.5.22) the separation into a single and a double sum over atomic sites as already outlined in (2.5.6) is obvious. The atomic contributions are explicitly given by

$$
E_{T,i} = \sum_{l\sigma} \left[\sum_{\alpha=1}^{2} \bar{E}_{li\sigma}^{(\alpha)} Q_{li\sigma}^{(\alpha)} + (2l+1) \sum_n \bar{E}_{nli\sigma} \right]
$$

$$
- \frac{1}{2} \int_{\Omega_i} d^3\mathbf{r}_i \, \bar{v}_{es,i}^{(el)}(\mathbf{r}_i) \rho_{el,i}(\mathbf{r}_i)
$$

$$
+ \sum_\sigma \int_{\Omega_i} d^3\mathbf{r}_i \, \left[\varepsilon_{xc,i\sigma}(\mathbf{r}_i) - v_{xc,i\sigma}(\mathbf{r}_i) \right] \rho_{el,i\sigma}(\mathbf{r}_i)
$$

$$
= \sum_{l\sigma} \left[\sum_{\alpha=1}^{2} E_{li\sigma}^{(\alpha)} Q_{li\sigma}^{(\alpha)} + (2l+1) \sum_n E_{nli\sigma} \right] + v_{MTZ} Q_i
$$

$$
+ \frac{1}{2} \int_{\Omega_i} d^3\mathbf{r}_i \, \bar{v}_{es,i}^{(el)}(\mathbf{r}_i) \rho_{el,i}(\mathbf{r}_i) - \int_{\Omega_i} d^3\mathbf{r}_i \, v_{es,i}^{(el)}(\mathbf{r}_i) \rho_{el,i}(\mathbf{r}_i)
$$

$$
+ \sum_\sigma \int_{\Omega_i} d^3\mathbf{r}_i \, \left[\varepsilon_{xc,i\sigma}(\mathbf{r}_i) - v_{xc,i\sigma}(\mathbf{r}_i) \right] \rho_{el,i\sigma}(\mathbf{r}_i)
$$

$$
= \sum_{l\sigma} \left[\sum_{\alpha=1}^{2} E_{li\sigma}^{(\alpha)} Q_{li\sigma}^{(\alpha)} + (2l+1) \sum_n E_{nli\sigma} \right] + v_{MTZ} Q_i
$$

$$
- \sum_\sigma \int_{\Omega_i} d^3\mathbf{r}_i \, \left[v_{es,i}^{(el)}(\mathbf{r}_i) + v_{xc,i\sigma}(\mathbf{r}_i) \right] \rho_{el,i\sigma}(\mathbf{r}_i)
$$

$$
+ \frac{1}{2} \sum_\sigma \int_{\Omega_i} d^3\mathbf{r}_i \, \bar{v}_{es,i}^{(el)}(\mathbf{r}_i) \rho_{el,i\sigma}(\mathbf{r}_i) + \sum_\sigma \int_{\Omega_i} d^3\mathbf{r}_i \, \varepsilon_{xc,i\sigma}(\mathbf{r}_i) \rho_{el,i\sigma}(\mathbf{r}_i) \ ,
$$

$$
(2.5.23)
$$

and the Madelung energy follows from (2.5.17).

References

1. M. Abramowitz and I. A. Stegun, *Handbook of Mathematical Functions* (Dover, New York 1972)
2. H. Akai and P. H. Dederichs, J. Phys. C **18**, 2455 (1985)
3. D. G. Anderson, J. Assoc. Comput. Mach. **12**, 547 (1965)
4. O. K. Andersen, Comments on the KKR-Wavefunction; Extension of the Spherical Wave Expansion beyond the Muffin-Tins. In: *Computational Methods in Band Theory*, ed by P. M. Marcus, J. F. Janak, and A. R. Williams (Plenum Press, New York 1971) pp 178–182
5. O. K. Andersen, Solid State Commun. **13**, 133 (1973)
6. O. K. Andersen, Phys. Rev. B **12**, 3060 (1975)
7. O. K. Andersen, Linear Methods in Band Theory. In: *The Electronic Structure of Complex Systems*, ed by P. Phariseau and W. Temmerman (Plenum Press, New York 1984) pp 11–66
8. O. K. Andersen, Muffin-Tin Orbital Theory. Lecture Notes from the ICTP workshop on *Methods on Electronic Structure Calculations*, (International Center for Theoretical Physics, Trieste 1992)
9. O. K. Andersen, O. Jepsen, and D. Glötzel, Canonical Description of the Band Structures of Metals. In: *Highlights of Condensed-Matter Theory*, Proceedings of the International School of Physics "Enrico Fermi", Course LXXXIX, ed by F. Bassani, F. Fumi, and M. P. Tosi (North-Holland, Amsterdam 1985) pp 59–176
10. U. von Barth, Density-Functional Theory for Solids. In: *The Electronic Structure of Complex Systems*, ed by P. Phariseau and W. Temmerman (Plenum Press, New York 1984) pp 67–140
11. U. von Barth, An Overview of Density-Functional Theory. In: *Many-Body Phenomena at Surfaces*, ed by D. Langreth and H. Suhl (Academic Press, Orlando 1984) pp 3–50
12. U. von Barth and L. Hedin, J. Phys. C **5**, 1629 (1972)
13. P. Bendt and A. Zunger, Phys. Rev. B **26**, 3114 (1982)
14. P. E. Blöchl, Gesamtenergien, Kräfte und Metall-Halbleiter Grenzflächen. PhD thesis, Universität Stuttgart (1989)
15. C. G. Broyden, Math. Comput. **19**, 577 (1965)
16. C. G. Broyden, Math. Comput. **21**, 368 (1966)
17. P. H. Dederichs and R. Zeller, Phys. Rev. B **28**, 5462 (1983)
18. J. E. Dennis, Jr., and J. J. Moré, SIAM Review **19**, 46 (1977)
19. R. M. Dreizler and E. K. U. Gross, *Density-Functional Theory* (Springer, Berlin 1990)
20. E. Engel and S. H. Vosko, Phys. Rev. B **47**, 13164 (1993)
21. H. Eschrig, *The Fundamentals of Density-Functional Theory* (Edition am Gutenbergplatz, Leipzig 2003)
22. P. P. Ewald, Ann. Phys. **64**, 253 (1921)
23. V. Eyert, Entwicklung und Implementation eines Full-Potential-ASW-Verfahrens. PhD thesis, Technische Hochschule Darmstadt (1991)
24. V. Eyert, J. Comput. Phys. **124**, 271 (1996)
25. V. Eyert, *Basic notions and applications of the augmented spherical wave method*, Int. J. Quantum Chem. **77**, 1007 (2000)
26. V. Eyert, *Electronic Structure of Crystalline Materials*, 2nd edn (University of Augsburg, Augsburg 2005)
27. L. G. Ferreira, J. Comput. Phys. **36**, 198 (1980)

28. E. R. Fuller and E. R. Naimon, Phys. Rev. B **6**, 3609 (1972)
29. C. D. Gelatt, Jr., H. Ehrenreich, and R. E. Watson, Phys. Rev. B **15**, 1613 (1977)
30. F. S. Ham and B. Segall, Phys. Rev. **124**, 1786 (1961)
31. L. Hedin and B. I. Lundqvist, J. Phys. C **4**, 2064 (1971)
32. L. Hodges, R. E. Watson, H. Ehrenreich, Phys. Rev. B **5**, 3953 (1972)
33. P. Hohenberg and W. Kohn, Phys. Rev. **136**, B864 (1964)
34. J. D. Jackson, *Classical Electrodynamics* (Wiley, New York 1975)
35. J. F. Janak, V. L. Moruzzi, and A. R. Williams, Phys. Rev. B **12**, 1257 (1975)
36. D. D. Johnson, Phys. Rev. B **38**, 12807 (1988)
37. G. P. Kerker, Phys. Rev. B **23**, 3082 (1981)
38. W. Kohn and N. Rostoker, Phys. Rev. **94**, 1111 (1954)
39. W. Kohn and L. J. Sham, Phys. Rev. **140**, A1133 (1965)
40. W. Kohn and P. Vashishta, General Density-Functional Theory. In: *Theory of the Inhomogeneous Electron Gas*, ed by S. Lundqvist and N. H. March (Plenum Press, New York 1983) pp 79–147
41. J. Korringa, Physica **13**, 392 (1947)
42. G. Kresse and J. Furthmüller, Phys. Rev. B **54**, 11169 (1996)
43. J. Kübler and V. Eyert, Electronic structure calculations. In: *Electronic and Magnetic Properties of Metals and Ceramics*, ed by K. H. J. Buschow (VCH Verlagsgesellschaft, Weinheim 1992) pp 1–145; vol 3A of *Materials Science and Technology*, ed by R. W. Cahn, P. Haasen, and E. J. Kramer (VCH Verlagsgesellschaft, Weinheim 1991–1996)
44. J. Kübler, K.-H. Höck, J. Sticht, and A. R. Williams, J. Phys. F **18**, 469 (1988)
45. J. Kübler, K.-H. Höck, J. Sticht, and A. R. Williams, J. Appl. Phys. **63**, 3482 (1988)
46. M. Levy and J. P. Perdew, Phys. Rev. B **48**, 11638 (1993)
47. J. M. MacLaren, D. P. Clougherty, M. E. McHenry, and M. M. Donovan, Comput. Phys. Commun. **66**, 383 (1991)
48. P. M. Marcus, J. F. Janak, and A. R. Williams, *Computational Methods in Band Theory* (Plenum Press, New York 1971)
49. A. Messiah, *Quantum Mechanics*, vol 1 (North Holland, Amsterdam 1976)
50. M. S. Methfessel, Zur Berechnung der Lösungswärme von Metallhydriden. Diploma thesis, Ruhr-Universität Bochum (1980)
51. V. L. Moruzzi, J. F. Janak, and A. R. Williams, *Calculated Electronic Properties of Metals* (Pergamon Press, New York 1978)
52. W. Nolting, *Grundkurs: Theoretische Physik*, vol 5, part 1: *Quantenmechanik – Grundlagen* (Springer, Berlin, 2004)
53. R. G. Parr and W. Yang, *Density-Functional Theory of Atoms and Molecules* (Oxford University Press, Oxford 1989)
54. J. P. Perdew, Unified Theory of Exchange and Correlation Beyond the Local Density Approximation. In: *Electronic Structure of Solids '91*, ed by P. Ziesche and H. Eschrig (Akademie Verlag, Berlin 1991) pp 11–20
55. J. P. Perdew, K. Burke, and M. Ernzerhof, Phys. Rev. Lett. **77**, 3865 (1996)
56. J. P. Perdew, K. Burke, and Y. Wang, Phys. Rev. B **54**, 16533 (1996)
57. J. P. Perdew, J. A. Chevary, S. H. Vosko, K. A. Jackson, M. R. Pederson, D. J. Singh, and C. Fiolhais, Phys. Rev. B **46**, 6671 (1992)
58. J. P. Perdew and Y. Wang, Phys. Rev. B **33**, 8800 (1986)
59. J. P. Perdew and A. Zunger, Phys. Rev. B **23**, 5048 (1981)
60. W. H. Press, B. P. Flannery, S. A. Teukolsky, and W. T. Vetterling, *Numerical Recipes – The Art of Scientific Computing* (Cambridge University Press, Cambridge 1989)

61. P. Pulay, Chem. Phys. Lett. **73**, 393 (1980)
62. B. Segall and F. S. Ham, In: *Methods in Computational Physics*, ed by B. Alder, S. Fernbach, and M. Rotenberg (Academic Press, New York 1968) pp 251–293
63. C. A. Sholl, Proc. Phys. Soc. **92**, 434 (1967)
64. D. Singh, H. Krakauer, and C. S. Wang, Phys. Rev. B **34**, 8391 (1986)
65. H. L. Skriver, *The LMTO Method* (Springer, Berlin 1984)
66. J. C. Slater, Phys. Rev. **51**, 846 (1937)
67. G. P. Srivastava, J. Phys. A **17**, L317 (1984)
68. J. Sticht, K.-H. Höck, and J. Kübler, J. Phys.: Cond. Matt. **1**, 8155 (1989)
69. D. Vanderbilt and S. G. Louie, Phys. Rev. B **30**, 6118 (1984)
70. S. H. Vosko, L. Wilk, and M. Nusair, Can. J. Phys. **58**, 1200 (1980)
71. R. E. Watson, H. Ehrenreich, and L. Hodges, Phys. Rev. Lett. **15**, 829 (1970)
72. A. R. Williams and U. von Barth, Applications of Density-Functional Theory to Atoms, Molecules and Solids. In: *Theory of the Inhomogeneous Electron Gas*, ed by S. Lundqvist and N. H. March (Plenum Press, New York 1983) pp 189–307
73. A. R. Williams, J. Kübler, and C. D. Gelatt, Jr., Phys. Rev. B **19**, 6094 (1979)
74. Y. Zhang and W. Yang, Phys. Rev. Lett. **80**, 890 (1998)

3

Envelope Functions and Structure Constants

Having outlined the basics of the standard ASW method we will in the present chapter discuss in more detail those aspects, which are based on the explicit form of the envelope functions. As already mentioned at the beginning of Chap. 2, the properties of the envelope functions as built on Hankel, Bessel, and Neumann functions as well as spherical and cubic harmonics can be studied by making use of the wealth of exact relations existing for all these functions. This is important since the envelope functions via the boundary conditions at the atomic spheres have a large influence on the whole set of basis functions.

The chapter is organized as follows: We start discussing the envelope functions and their Bloch sums. Having derived the expansion theorems and the explicit form of the structure constants we will turn to the calculation of exact overlap integrals of the envelope functions. Finally, we will, by defining pseudo functions, introduce a new type of functions and discuss some of their properties. This set of functions will turn out to be very useful within the framework of the plane-wave based full-potential ASW method but it will also be very useful for understanding other implementations. Still there exist many relations and computational details, which are more technical than others and which are shifted to the App. B in order to keep things clear.

3.1 Envelope Functions: Basic Properties

In this section we start writing down the basic definitions of the envelope functions and add a lot of useful relations for these functions as well as the pure spherical Bessel functions of all three kinds.

As outlined in Sect. 2.1 the envelope functions may be defined in terms of Neumann functions as

$$N_{L\kappa}(\mathbf{r}) := -\kappa^{l+1} n_l(\kappa r) Y_L(\hat{\mathbf{r}}) , \qquad (3.1.1)$$

where $L = (l, m)$ and $n_l(\kappa r)$ is a spherical Neumann function or spherical Bessel function of the second kind [1, (10.1.1)],

V. Eyert: *Envelope Functions and Structure Constants*, Lect. Notes Phys. **719**, 47–115 (2007)
DOI 10.1007/978-3-540-71007-3_3 © Springer-Verlag Berlin Heidelberg 2007

$$n_l(\rho) = \sqrt{\frac{\pi}{2\rho}} N_{l+\frac{1}{2}}(\rho) .$$ (3.1.2)

Here, $N_{l+\frac{1}{2}}(\rho)$ is an ordinary Neumann function of fractional order and we used $\rho = \kappa r$. The above and all following formulas are based on the notation given by Abramowitz and Stegun [1, Chap. 10] (where the spherical Neumann function is designated by the letter y rather than n) or Jackson [8, Sect. 16.1]. This notation slightly differs from that used by Messiah [9, App. B.6]. Finally, in (3.1.1), $Y_L(\hat{\mathbf{r}})$ denotes a spherical harmonic or rather a cubic harmonic in the form proposed in App. B.2 and $\hat{\mathbf{r}}$ stands for a unit vector.

Again, we point out that in the linear methods the κ-dependence of the envelope function is just a formal one. It reflects the well justified assumption that it is sufficient to replace the full energy dependence of the wave function by only the leading terms of a Taylor expansion about a somehow fixed energy. If additionally the atomic-sphere approximation is used and the interstitial region formally eliminated the energy dependence of the envelope functions is reduced to its zeroth order and κ enters only as a fixed parameter. In contrast, if the muffin-tin approximation with touching spheres and, hence, a finite interstitial region is used, it might be necessary to invent a second set of envelope functions with a different energy parameter in order to better account for the energy dependence of the wave function.

The prefactors in (3.1.1) are merely convention. κ^{l+1} serves the purpose of cancelling the leading κ-dimension in the spherical Neumann function and has the additional effect that $\kappa^{l+1}n_l(\kappa r)$ is a real function for any energy κ^2. Note again that, in contrast to previous definitions [14], the envelope function does not contain an extra i^l-factor this making the function (as well as the real-space structure constants to be defined lateron) a real quantity.

The envelope function $N_{L\kappa}(\mathbf{r})$ as defined in (3.1.1) may be expanded via structure constants in spherical Bessel functions centered at a different site. We will not deal with the expansion theorem in this section but complement (3.1.1) by the definition

$$J_{L\kappa}(\mathbf{r}) := \kappa^{-l} j_l(\kappa r) Y_L(\hat{\mathbf{r}}) ,$$ (3.1.3)

where $j_l(\kappa r)$ is the spherical Bessel function (of the first kind) [1, (10.1.1)],

$$j_l(\rho) = \sqrt{\frac{\pi}{2\rho}} J_{l+\frac{1}{2}}(\rho) ,$$ (3.1.4)

and $J_{l+\frac{1}{2}}(\rho)$ is an ordinary Bessel function of fractional order. The prefactor κ^{-l} again cancels the leading κ-dimension of the spherical Bessel function and makes $\kappa^{-l}j_l(\kappa r)$ a real function for any value of κ.

As already discussed in Sect. 2.1, the ASW method may be formulated with Hankel envelope functions rather than with Neumann functions. Due to the exponential decay of the former we opt for negative values of the energy parameter κ^2 and prefer to work with Hankel envelope functions. They read as

$$H_{L\kappa}(\mathbf{r}) := i\kappa^{l+1} h_l^{(1)}(\kappa r) Y_L(\hat{\mathbf{r}}) ,$$ (3.1.5)

where $h_l^{(1)}(\kappa r)$ is the outgoing spherical Hankel function or one of the spherical Bessel functions of the third kind as defined by [1, (10.1.1)]

$$h_l^{(1,2)}(\rho) := j_l(\rho) \pm i n_l(\rho)$$

$$= \sqrt{\frac{\pi}{2\rho}} H_{l+\frac{1}{2}}^{(1,2)}(\rho) . \tag{3.1.6}$$

The prefactors once more guarantee that $i\kappa^{l+1} h_l^{(1)}(\kappa r)$ is a real function for negative energies κ^2.

With the basic definitions of the ASW envelope functions at hand we turn to listing many useful relations for these functions. The spherical Bessel functions n_l, j_l and $h_l^{(1,2)}$ as defined by (3.1.2), (3.1.4) and (3.1.6) all grow out the same differential equation, namely [1, (10.1.1)]

$$\left[\frac{\partial^2}{\partial\rho^2} + \frac{2}{\rho} \frac{\partial}{\partial\rho} + 1 - \frac{l(l+1)}{\rho^2} \right] f_l(\rho) = 0 \qquad \text{for } l = 0, \pm 1, \pm 2, \dots . \tag{3.1.7}$$

By virtue of the well known operator identity

$$\frac{\partial^2}{\partial\rho^2} + \frac{2}{\rho} \frac{\partial}{\partial\rho} = \frac{1}{\rho^2} \frac{\partial}{\partial\rho} \left(\rho^2 \frac{\partial}{\partial\rho} \right) = \frac{1}{\rho} \frac{\partial^2}{\partial\rho^2} \rho , \tag{3.1.8}$$

it may be cast into the form

$$\left[\frac{\partial^2}{\partial\rho^2} + 1 - \frac{l(l+1)}{\rho^2} \right] \rho f_l(\rho) = 0 \qquad \text{for } l = 0, \pm 1, \pm 2, \dots , \tag{3.1.9}$$

which, in addition, defines the Riccati-Bessel functions $\rho f_l(\rho)$.

Finally, for constant κ we arrive at the free-particle radial Schrödinger equation

$$\left[\frac{\partial^2}{\partial r^2} + \frac{2}{r} \frac{\partial}{\partial r} + \kappa^2 - \frac{l(l+1)}{r^2} \right] f_l(\kappa r) = 0 \qquad \text{for } l = 0, \pm 1, \pm 2, \dots , \tag{3.1.10}$$

or

$$\left[\frac{\partial^2}{\partial r^2} + \kappa^2 - \frac{l(l+1)}{r^2} \right] r f_l(\kappa r) = 0 \qquad \text{for } l = 0, \pm 1, \pm 2, \dots . \tag{3.1.11}$$

The latter form is especially useful for the calculation of radial integrals of products of spherical Bessel functions as will be evaluated in Sect. 3.3.

In addition to the spherical Bessel functions of all three kinds we define a set of — as we call them — barred functions by

$$\bar{n}_l(\kappa r) := -\kappa^{l+1} n_l(\kappa r) , \tag{3.1.12}$$

$$\bar{j}_l(\kappa r) := \kappa^{-l} j_l(\kappa r) , \tag{3.1.13}$$

$$\bar{h}_l^{(1)}(\kappa r) := i\kappa^{l+1} h_l^{(1)}(\kappa r)$$

$$= \bar{n}_l(\kappa r) + i\kappa^{2l+1} \bar{j}_l(\kappa r) . \tag{3.1.14}$$

These functions are just the radial parts of the envelope functions defined in (3.1.1), (3.1.3) and (3.1.5), which now read as

$$N_{L\kappa}(\mathbf{r}) = \bar{n}_l(\kappa r) Y_L(\hat{\mathbf{r}}) \,, \tag{3.1.15}$$

$$J_{L\kappa}(\mathbf{r}) = \bar{j}_l(\kappa r) Y_L(\hat{\mathbf{r}}) \,, \tag{3.1.16}$$

$$H_{L\kappa}(\mathbf{r}) = \bar{h}_l^{(1)}(\kappa r) Y_L(\hat{\mathbf{r}})$$
$$= N_{L\kappa}(\mathbf{r}) + i\kappa^{2l+1} J_{L\kappa}(\mathbf{r}) \,. \tag{3.1.17}$$

For both the spherical Bessel functions of all three kinds and the barred functions we will next derive a number of identities. However, calculational details will not be dealt with in the present section but rather deferred to the App. B.1. As already mentioned above all equations are based on the conventions of Abramowitz and Stegun [1]. In addition, for the calculation of cubic harmonics we refer the reader to the App. B.2.

In order to get used to the spherical Bessel functions as well as their barred counterparts we start noting their explicit expressions for $l = 0$ and 1 [1, (10.1.11/12)]

$$n_0(\rho) = -\frac{\cos\rho}{\rho} \,, \tag{3.1.18}$$

$$n_1(\rho) = -\frac{\cos\rho}{\rho^2} - \frac{\sin\rho}{\rho} \,, \tag{3.1.19}$$

$$j_0(\rho) = \frac{\sin\rho}{\rho} \,, \tag{3.1.20}$$

$$j_1(\rho) = \frac{\sin\rho}{\rho^2} - \frac{\cos\rho}{\rho} \,, \tag{3.1.21}$$

$$h_0^{(1)}(\rho) = \frac{e^{i\rho}}{i\rho} \,, \tag{3.1.22}$$

$$h_1^{(1)}(\rho) = -\frac{e^{i\rho}}{\rho}\left(1 + \frac{i}{\rho}\right) \,. \tag{3.1.23}$$

From this we get for the barred functions

$$\bar{n}_0(\kappa r) = \frac{\cos\kappa r}{r} \,, \tag{3.1.24}$$

$$\bar{n}_1(\kappa r) = \frac{\cos\kappa r}{r^2} + \kappa\frac{\sin\kappa r}{r} \,, \tag{3.1.25}$$

$$\bar{j}_0(\kappa r) = \frac{\sin\kappa r}{\kappa r} \,, \tag{3.1.26}$$

$$\bar{j}_1(\kappa r) = \frac{1}{\kappa}\frac{\sin\kappa r}{(\kappa r)^2} - \frac{\cos\kappa r}{\kappa^2 r} \,, \tag{3.1.27}$$

$$\bar{h}_0^{(1)}(\kappa r) = \frac{e^{i\kappa r}}{r} \,, \tag{3.1.28}$$

$$\bar{h}_1^{(1)}(\kappa r) = -\frac{e^{i\kappa r}}{r}\left(i\kappa - \frac{1}{r}\right), \tag{3.1.29}$$

which are real functions for any value of κ except for the spherical Hankel functions, which are real for negative and zero κ^2 only.

For general values of l the explicit formulas assume a more complicated form. In particular, we note for the spherical Hankel function $h_l^{(1)}(\rho)$ [1, (10.1.16)]

$$h_l^{(1)}(\rho) = (-i)^l \frac{e^{i\rho}}{i\rho} \sum_{k=0}^{l} a_{lk}(-i\rho)^{-k} , \tag{3.1.30}$$

where

$$a_{lk} = \frac{(l+k)!}{2^k k!(l-k)!} . \tag{3.1.31}$$

Defining

$$R_l(i\rho) = \sum_{k=0}^{l} a_{lk}(-i\rho)^{-k} , \tag{3.1.32}$$

we thus write

$$h_l^{(1)}(\rho) = (-i)^l \frac{e^{i\rho}}{i\rho} R_l(i\rho) . \tag{3.1.33}$$

For the spherical Hankel function $h_l^{(2)}(\rho)$ the corresponding identity reads

$$h_l^{(2)}(\rho) = i^l \frac{e^{-i\rho}}{-i\rho} R_l(-i\rho) . \tag{3.1.34}$$

Next, using the identity (3.1.6) we arrive at the explicit formulas for the spherical Bessel and Neumann functions

$$j_l(\rho) = \frac{1}{2}(-i)^l \left[\frac{e^{i\rho}}{i\rho} R_l(i\rho) + (-)^l \frac{e^{-i\rho}}{-i\rho} R_l(-i\rho) \right] , \tag{3.1.35}$$

and

$$n_l(\rho) = \frac{1}{2i}(-i)^l \left[\frac{e^{i\rho}}{i\rho} R_l(i\rho) - (-)^l \frac{e^{-i\rho}}{-i\rho} R_l(-i\rho) \right] . \tag{3.1.36}$$

Again, we aim at analogous expressions for the barred functions. With their definitions (3.1.12) to (3.1.14) at hand we note

$$\begin{aligned}
\bar{h}_l^{(1)}(\kappa r) &= (-i\kappa)^l \frac{e^{i\kappa r}}{r} \sum_{k=0}^{l} a_{lk}(-i\kappa r)^{-k} \\
&= \sum_{k=0}^{l} a_{lk}(-i\kappa)^{l-k} \frac{e^{i\kappa r}}{r^{k+1}} ,
\end{aligned} \tag{3.1.37}$$

$$\begin{aligned}
\bar{j}_l(\kappa r) &= \frac{1}{2}(-)^{l+1}(-i\kappa)^{-l-1} \frac{e^{i\kappa r}}{r} \sum_{k=0}^{l} a_{lk}(-i\kappa r)^{-k} \\
&+ \frac{1}{2}(-)^{l+1}(i\kappa)^{-l-1} \frac{e^{-i\kappa r}}{r} \sum_{k=0}^{l} a_{lk}(i\kappa r)^{-k}
\end{aligned}$$

$$= \frac{1}{2}(-)^{l+1} \sum_{k=0}^{l} a_{lk}(-i\kappa)^{-l-k-1} \frac{e^{i\kappa r}}{r^{k+1}}$$

$$+ \frac{1}{2}(-)^{l+1} \sum_{k=0}^{l} a_{lk}(i\kappa)^{-l-k-1} \frac{e^{-i\kappa r}}{r^{k+1}} , \qquad (3.1.38)$$

and

$$\bar{n}_l(\kappa r) = \frac{1}{2}(-i\kappa)^l \frac{e^{i\kappa r}}{r} \sum_{k=0}^{l} a_{lk}(-i\kappa r)^{-k} + \frac{1}{2}(i\kappa)^l \frac{e^{-i\kappa r}}{r} \sum_{k=0}^{l} a_{lk}(i\kappa r)^{-k}$$

$$= \frac{1}{2} \sum_{k=0}^{l} a_{lk}(-i\kappa)^{l-k} \frac{e^{i\kappa r}}{r^{k+1}} + \frac{1}{2} \sum_{k=0}^{l} a_{lk}(i\kappa)^{l-k} \frac{e^{-i\kappa r}}{r^{k+1}}$$

$$= \frac{1}{2} \left(\bar{h}_l^{(1)}(\kappa r) + \bar{h}_l^{(1)}(-\kappa r) \right) . \qquad (3.1.39)$$

Next, we investigate the behaviour of these functions for the limits of both small and large arguments. In the former case we note for the spherical Bessel functions [8, (16.12)]

$$n_l(\rho) \overset{\rho \to 0}{\sim} -\frac{(2l-1)!!}{\rho^{l+1}} \left[1 + \frac{\rho^2}{2(2l-1)} + \ldots \right] , \qquad (3.1.40)$$

$$j_l(\rho) \overset{\rho \to 0}{\sim} \frac{\rho^l}{(2l+1)!!} \left[1 - \frac{\rho^2}{2(2l+3)} + \ldots \right] , \qquad (3.1.41)$$

$$h_l^{(1)}(\rho) \overset{\rho \to 0}{\sim} -i\frac{(2l-1)!!}{\rho^{l+1}} \left[1 + \frac{\rho^2}{2(2l-1)} + \ldots \right] , \qquad (3.1.42)$$

and for the barred functions

$$\bar{n}_l(\kappa r) \overset{\kappa r \to 0}{\sim} \frac{(2l-1)!!}{r^{l+1}} \left[1 + \frac{(\kappa r)^2}{2(2l-1)} + \ldots \right] , \qquad (3.1.43)$$

$$\bar{j}_l(\kappa r) \overset{\kappa r \to 0}{\sim} \frac{r^l}{(2l+1)!!} \left[1 - \frac{(\kappa r)^2}{2(2l+3)} + \ldots \right] , \qquad (3.1.44)$$

$$\bar{h}_l^{(1)}(\kappa r) \overset{\kappa r \to 0}{\sim} \frac{(2l-1)!!}{r^{l+1}} \left[1 + \frac{(\kappa r)^2}{2(2l-1)} + \ldots \right] . \qquad (3.1.45)$$

Thus, by the introduction of the barred functions we have circumvented the irregular behaviour of the spherical Neumann and Hankel functions for small κ^2.

For large arguments the spherical Bessel functions reduce to the form [8, (16.13)]

$$n_l(\rho) \overset{\rho \to \infty}{\sim} \frac{1}{\rho} \cos(\rho - l\frac{\pi}{2}) , \qquad (3.1.46)$$

$$j_l(\rho) \overset{\rho \to \infty}{\sim} \frac{1}{\rho} \sin(\rho - l\frac{\pi}{2}) , \qquad (3.1.47)$$

$$h_l^{(1)}(\rho) \overset{\rho \to \infty}{\sim} \frac{1}{i\rho} e^{i(\rho - l\frac{\pi}{2})} , \qquad (3.1.48)$$

from which the corresponding equations for the barred functions may be easily derived. Note the aforementioned exponential decay of the spherical Hankel function $h_l^{(1)}(\rho)$ for negative ρ^2 in (3.1.48).

Of special importance for the calculation of the spherical Bessel functions and their derivatives are the identities [1, (10.1.15)]

$$n_l(\rho) = (-)^{l+1} j_{-l-1}(\rho) \qquad \text{for } l = 0, \pm 1, \pm 2, \ldots \qquad (3.1.49)$$

and

$$h_l^{(1)}(\rho) = i(-)^{l+1} h_{-l-1}^{(1)}(\rho) \qquad \text{for } l = 0, 1, 2, \ldots . \qquad (3.1.50)$$

From these we derive for the barred functions

$$\bar{n}_l(\kappa r) = (-)^l \bar{j}_{-l-1}(\kappa r) \qquad \text{for } l = 0, \pm 1, \pm 2, \ldots \qquad (3.1.51)$$

and

$$\bar{h}_l^{(1)}(\kappa r) = i(-)^{l+1} \kappa^{2l+1} \bar{h}_{-l-1}^{(1)}(\kappa r) \qquad \text{for } l = 0, 1, 2, \ldots . \qquad (3.1.52)$$

Probably even more important is the recursion relation connecting spherical Bessel functions of all three kinds for different l, which reads as [1, (10.1.19)]

$$(2l + 1) f_l(\rho) = \rho [f_{l-1}(\rho) + f_{l+1}(\rho)] . \qquad (3.1.53)$$

Here, f_l may be any linear combination of the functions j_l and n_l. However, for the barred functions differences arise from the presence of the different prefactors. We note

$$(2l + 1) \bar{n}_l(\kappa r) = r \left[\kappa^2 \bar{n}_{l-1}(\kappa r) + \bar{n}_{l+1}(\kappa r) \right] , \qquad (3.1.54)$$

$$(2l + 1) \bar{j}_l(\kappa r) = r \left[\bar{j}_{l-1}(\kappa r) + \kappa^2 \bar{j}_{l+1}(\kappa r) \right] , \qquad (3.1.55)$$

$$(2l + 1) \bar{h}_l^{(1)}(\kappa r) = r \left[\kappa^2 \bar{h}_{l-1}^{(1)}(\kappa r) + \bar{h}_{l+1}^{(1)}(\kappa r) \right] . \qquad (3.1.56)$$

In passing, we mention some symmetry relations for the envelope functions. Essentially, they are based on the parity of the spherical harmonics,

$$Y_L(\widehat{-\mathbf{r}}) = (-)^l Y_L(\hat{\mathbf{r}}) , \qquad (3.1.57)$$

and lead to

$$F_{L\kappa}(-\mathbf{r}) = (-)^l F_{L\kappa}(\mathbf{r}) , \qquad (3.1.58)$$

where $F_{L\kappa}(\mathbf{r})$ is any of the envelope functions $N_{L\kappa}(\mathbf{r})$, $J_{L\kappa}(\mathbf{r})$ or $H_{L\kappa}(\mathbf{r})$ as defined in (3.1.15) to (3.1.17). Since the radial parts of the envelope functions are defined as real quantities in the respective ranges of the energy parameter κ^2, we note, in addition, the identity

$$F_{L\kappa}^*(\mathbf{r}) = F_{L\kappa}(\mathbf{r}) , \qquad (3.1.59)$$

which, however, is valid only when the spherical harmonics are replaced by the cubic harmonics as they are defined in App. B.2. Of course, the previous identities hold also for the energy derivatives of the envelope functions.

3.2 Envelope Functions: Derivatives and Wronskians

Having established the basic identities for the spherical Bessel functions of all three kinds we turn to the calculation of their derivatives with respect to both r and κ^2. Starting with the radial derivatives we first note the identity [1, (10.1.22)]

$$f_{l+1}(\rho) = \left(-\frac{\partial}{\partial\rho} + \frac{l}{\rho}\right) f_l(\rho) , \qquad (3.2.1)$$

which is valid for any linear combination of j_l and n_l. Since $\rho = \kappa r$, we have for fixed κ

$$\frac{\partial}{\partial\rho} = \frac{1}{\kappa} \frac{\partial}{\partial r} , \qquad (3.2.2)$$

and thus

$$\frac{\partial}{\partial r} f_l(\kappa r) = \frac{l}{r} f_l(\kappa r) - \kappa f_{l+1}(\kappa r) . \qquad (3.2.3)$$

From this we get for the barred functions

$$\frac{\partial}{\partial r} \bar{n}_l(\kappa r) = \frac{l}{r} \bar{n}_l(\kappa r) - \bar{n}_{l+1}(\kappa r) , \qquad (3.2.4)$$

$$\frac{\partial}{\partial r} \bar{j}_l(\kappa r) = \frac{l}{r} \bar{j}_l(\kappa r) - \kappa^2 \bar{j}_{l+1}(\kappa r) , \qquad (3.2.5)$$

$$\frac{\partial}{\partial r} \bar{h}_l^{(1)}(\kappa r) = \frac{l}{r} \bar{h}_l^{(1)}(\kappa r) - \bar{h}_{l+1}^{(1)}(\kappa r) . \qquad (3.2.6)$$

Equation (3.2.1) can be combined with the recursion relation (3.1.53) to yield the alternative identity [1, (10.1.21)]

$$f_{l-1}(\rho) = \left(\frac{\partial}{\partial\rho} + \frac{l+1}{\rho}\right) f_l(\rho) . \qquad (3.2.7)$$

From this we get with the help of (3.2.2)

$$\frac{\partial}{\partial r} f_l(\kappa r) = -\frac{l+1}{r} f_l(\kappa r) + \kappa f_{l-1}(\kappa r), \qquad (3.2.8)$$

and, finally,

$$\frac{\partial}{\partial r} \bar{n}_l(\kappa r) = -\frac{l+1}{r} \bar{n}_l(\kappa r) + \kappa^2 \bar{n}_{l-1}(\kappa r) , \qquad (3.2.9)$$

$$\frac{\partial}{\partial r} \bar{j}_l(\kappa r) = -\frac{l+1}{r} \bar{j}_l(\kappa r) + \bar{j}_{l-1}(\kappa r) , \qquad (3.2.10)$$

$$\frac{\partial}{\partial r} \bar{h}_l^{(1)}(\kappa r) = -\frac{l+1}{r} \bar{h}_l^{(1)}(\kappa r) + \kappa^2 \bar{h}_{l-1}^{(1)}(\kappa r) . \qquad (3.2.11)$$

In passing, we mention that the corresponding identities for the radial functions used by Williams, Kübler, and Gelatt, which carry the above mentioned extra i^l factor [14], grow out of (3.2.4) to (3.2.6) and (3.2.9) to (3.2.11) by replacing the minus and plus sign, respectively, in front of the second term of these equations by $+i$.

Equation (3.2.1) gives rise to the following useful relation

$$\frac{\partial}{\partial\rho}\rho^{-l}f_l(\rho) = -l\rho^{-l-1}f_l(\rho) + \rho^{-l}\frac{\partial}{\partial\rho}f_l(\rho)$$

$$= -l\rho^{-l-1}f_l(\rho) + \rho^{-l}\left[\frac{l}{\rho}f_l(\rho) - f_{l+1}(\rho)\right]$$

$$= -\rho^{-l}f_{l+1}(\rho)\ . \tag{3.2.12}$$

Using (3.2.2) for constant κ we obtain

$$\frac{\partial}{\partial r}r^{-l}f_l(\kappa r) = -\kappa r^{-l}f_{l+1}(\kappa r)\ , \tag{3.2.13}$$

from which we get for the barred functions

$$\frac{\partial}{\partial r}r^{-l}\bar{n}_l(\kappa r) = -r^{-l}\bar{n}_{l+1}(\kappa r)\ , \tag{3.2.14}$$

$$\frac{\partial}{\partial r}r^{-l}\bar{j}_l(\kappa r) = -\kappa^2 r^{-l}\bar{j}_{l+1}(\kappa r)\ , \tag{3.2.15}$$

$$\frac{\partial}{\partial r}r^{-l}\bar{h}_l^{(1)}(\kappa r) = -r^{-l}\bar{h}_{l+1}^{(1)}(\kappa r)\ . \tag{3.2.16}$$

In contrast, (3.2.7) leads to the alternative expression

$$\frac{\partial}{\partial\rho}\rho^{l+1}f_l(\rho) = (l+1)\rho^l f_l(\rho) + \rho^{l+1}\frac{\partial}{\partial\rho}f_l(\rho)$$

$$= (l+1)\rho^l f_l(\rho) + \rho^{l+1}\left[f_{l-1}(\rho) - \frac{l+1}{\rho}f_l(\rho)\right]$$

$$= \rho^{l+1}f_{l-1}(\rho)\ . \tag{3.2.17}$$

Again using (3.2.2) for constant κ we obtain

$$\frac{\partial}{\partial r}r^{l+1}f_l(\kappa r) = \kappa r^{l+1}f_{l-1}(\kappa r)\ , \tag{3.2.18}$$

from which we get for the barred functions

$$\frac{\partial}{\partial r}r^{l+1}\bar{n}_l(\kappa r) = \kappa^2 r^{l+1}\bar{n}_{l-1}(\kappa r)\ , \tag{3.2.19}$$

$$\frac{\partial}{\partial r}r^{l+1}\bar{j}_l(\kappa r) = r^{l+1}\bar{j}_{l-1}(\kappa r)\ , \tag{3.2.20}$$

$$\frac{\partial}{\partial r}r^{l+1}\bar{h}_l^{(1)}(\kappa r) = \kappa^2 r^{l+1}\bar{h}_{l-1}^{(1)}(\kappa r)\ . \tag{3.2.21}$$

Having discussed the dependence on r we turn to derivatives of the spherical Bessel functions with respect to the energy argument κ^2 and note

$$\frac{\partial}{\partial \kappa^2} = \frac{1}{2\kappa} \frac{\partial}{\partial \kappa} = \frac{r}{2\kappa} \frac{\partial}{\partial \rho} , \tag{3.2.22}$$

where $\rho = \kappa r$. Combining (3.2.22) with the identities (3.2.1) and (3.2.7) we get for any of the spherical Bessel functions

$$\begin{aligned}
\frac{\partial}{\partial \kappa^2} f_l(\kappa r) &= \frac{l}{2\kappa^2} f_l(\kappa r) - \frac{r}{2\kappa} f_{l+1}(\kappa r) \\
&= -\frac{l+1}{2\kappa^2} f_l(\kappa r) + \frac{r}{2\kappa} f_{l-1}(\kappa r) \\
&= \frac{1}{4\kappa^2} f_l(\kappa r) - \frac{r}{4\kappa} f_{l+1}(\kappa r) + \frac{r}{4\kappa} f_{l-1}(\kappa r) ,
\end{aligned} \tag{3.2.23}$$

where, in the last line, we have combined the two previous lines. We thus get for the barred functions

$$\begin{aligned}
\frac{\partial}{\partial \kappa^2} \bar{n}_l(\kappa r) &= -\frac{1}{2\kappa} \left(\frac{\partial}{\partial \kappa} \kappa^{l+1} \right) n_l(\kappa r) \\
&\quad + \frac{l+1}{2} \kappa^{l-1} n_l(\kappa r) - \frac{r}{2} \kappa^l n_{l-1}(\kappa r) \\
&= \frac{r}{2} \bar{n}_{l-1}(\kappa r) ,
\end{aligned} \tag{3.2.24}$$

$$\begin{aligned}
\frac{\partial}{\partial \kappa^2} \bar{j}_l(\kappa r) &= \frac{1}{2\kappa} \left(\frac{\partial}{\partial \kappa} \kappa^{-l} \right) j_l(\kappa r) \\
&\quad + \frac{l}{2} \kappa^{-l-2} j_l(\kappa r) - \frac{r}{2} \kappa^{-l-1} j_{l+1}(\kappa r) \\
&= -\frac{r}{2} \bar{j}_{l+1}(\kappa r) ,
\end{aligned} \tag{3.2.25}$$

$$\begin{aligned}
\frac{\partial}{\partial \kappa^2} \bar{h}_l^{(1)}(\kappa r) &= \frac{i}{2\kappa} \left(\frac{\partial}{\partial \kappa} \kappa^{l+1} \right) h_l^{(1)}(\kappa r) \\
&\quad - i \frac{l+1}{2} \kappa^{l-1} h_l^{(1)}(\kappa r) + i \frac{r}{2} \kappa^l h_{l-1}^{(1)}(\kappa r) \\
&= \frac{r}{2} \bar{h}_{l-1}^{(1)}(\kappa r) .
\end{aligned} \tag{3.2.26}$$

Special care needs the case $l = 0$ for the spherical Neumann and Hankel functions, for which we use the identities (3.1.51) and (3.1.52) to obtain

$$\frac{\partial}{\partial \kappa^2} \bar{n}_0(\kappa r) = -\frac{r}{2} \bar{j}_0(\kappa r) \tag{3.2.27}$$

and

$$\frac{\partial}{\partial \kappa^2} \bar{h}_0^{(1)}(\kappa r) = \frac{r}{2} \bar{h}_{-1}^{(1)}(\kappa r) = \frac{r}{2} i \kappa^{-1} \bar{h}_0^{(1)}(\kappa r) . \tag{3.2.28}$$

Note that the latter expression diverges for $\kappa^2 \to 0$.

In addition to the derivatives with respect to r and κ^2 we need also the combined derivatives $\frac{\partial}{\partial r} \frac{\partial}{\partial \kappa^2}$. They enter the radial integrals over products of envelope functions to be evaluated in Sect. 3.3. For simplicity we write down these derivatives for the barred functions only. Starting with the barred Neumann functions we fall back on (3.2.24) and (3.2.4) to note

$$
\begin{aligned}
\frac{\partial}{\partial \kappa^2} \frac{\partial}{\partial r} \bar{n}_l(\kappa r) &= \frac{\partial}{\partial r} \frac{\partial}{\partial \kappa^2} \bar{n}_l(\kappa r) \\
&= \frac{\partial}{\partial r} \left(\frac{r}{2} \bar{n}_{l-1}(\kappa r) \right) \\
&= \frac{1}{2} \bar{n}_{l-1}(\kappa r) + \frac{r}{2} \left[\frac{l-1}{r} \bar{n}_{l-1}(\kappa r) - \bar{n}_l(\kappa r) \right] \\
&= \frac{l}{2} \bar{n}_{l-1}(\kappa r) - \frac{r}{2} \bar{n}_l(\kappa r) .
\end{aligned}
\tag{3.2.29}
$$

In order to derive an analogous expression for the barred Bessel functions we use (3.2.25) and (3.2.10) this leading to

$$
\begin{aligned}
\frac{\partial}{\partial \kappa^2} \frac{\partial}{\partial r} \bar{j}_l(\kappa r) &= \frac{\partial}{\partial r} \frac{\partial}{\partial \kappa^2} \bar{j}_l(\kappa r) \\
&= \frac{\partial}{\partial r} \left(-\frac{r}{2} \bar{j}_{l+1}(\kappa r) \right) \\
&= -\frac{1}{2} \bar{j}_{l+1}(\kappa r) - \frac{r}{2} \left[-\frac{l+2}{r} \bar{j}_{l+1}(\kappa r) + \bar{j}_l(\kappa r) \right] \\
&= \frac{l+1}{2} \bar{j}_{l+1}(\kappa r) - \frac{r}{2} \bar{j}_l(\kappa r) .
\end{aligned}
\tag{3.2.30}
$$

Finally, for the barred Hankel function we have to combine (3.2.26) and (3.2.6), which are formally identical to (3.2.24) and (3.2.4) for the barred Neumann function. Hence, we are able to arrive at the expression for the barred Hankel function by just rewriting (3.2.29) and get

$$
\begin{aligned}
\frac{\partial}{\partial \kappa^2} \frac{\partial}{\partial r} \bar{h}_l^{(1)}(\kappa r) &= \frac{\partial}{\partial r} \frac{\partial}{\partial \kappa^2} \bar{h}_l^{(1)}(\kappa r) \\
&= \frac{\partial}{\partial r} \left(\frac{r}{2} \bar{h}_{l-1}^{(1)}(\kappa r) \right) \\
&= \frac{1}{2} \bar{h}_{l-1}^{(1)}(\kappa r) + \frac{r}{2} \left[\frac{l-1}{r} \bar{h}_{l-1}^{(1)}(\kappa r) - \bar{h}_l^{(1)}(\kappa r) \right] \\
&= \frac{l}{2} \bar{h}_{l-1}^{(1)}(\kappa r) - \frac{r}{2} \bar{h}_l^{(1)}(\kappa r) .
\end{aligned}
\tag{3.2.31}
$$

Note that the extra treatment for the case $l = 0$, which was necessary in the calculation of the energy derivative of the barred Neumann and Hankel functions, is not needed here. This is due to the fact that the functions with index $l - 1$ in the last lines of (3.2.29) and (3.2.31) are preceeded with the factor l which lets the corresponding terms vanish for $l = 0$.

While dealing with a set of functions, which all solve the same differential equation, we should also discuss their Wronskians. These are defined by

$$W\{f_l(\rho), g_l(\rho); \rho\} = f_l(\rho)\frac{\partial}{\partial\rho}g_l(\rho) - g_l(\rho)\frac{\partial}{\partial\rho}f_l(\rho), \tag{3.2.32}$$

where f_l and g_l again denote any linear combination of the spherical Bessel functions. Since the Wronskian $W\{f(\rho), f(\rho); \rho\}$ vanishes trivially we are left with only a few nonzero, fundamental Wronskians, for which we note (see [1, (10.1.6/7)] or [8, (16.15)])

$$W\{j_l(\rho), n_l(\rho); \rho\} = \frac{1}{i}W\{j_l(\rho), h_l^{(1)}(\rho); \rho\} = \frac{i}{2}W\{h_l^{(1)}(\rho), h_l^{(2)}(\rho); \rho\} = \frac{1}{\rho^2}. \tag{3.2.33}$$

The $\frac{1}{\rho^2}$-factor on the right-hand side of (3.2.33) can still be removed by passing over to the Riccati-Bessel functions. For these functions the (constant) fundamental Wronskians read as

$$W\{\rho j_l(\rho), \rho n_l(\rho); \rho\} = \frac{1}{i}W\{\rho j_l(\rho), \rho h_l^{(1)}(\rho); \rho\}$$
$$= \frac{i}{2}W\{\rho h_l^{(1)}(\rho), \rho h_l^{(2)}(\rho); \rho\} = 1. \tag{3.2.34}$$

Combination of the Wronskians (3.2.32) with the differentiation formulas (3.2.1) and (3.2.7) leads to the so-called cross-product relation

$$W\{f_l(\rho), g_l(\rho); \rho\} = f_l(\rho)g_{l-1}(\rho) - f_{l-1}(\rho)g_l(\rho)$$
$$= f_{l+1}(\rho)g_l(\rho) - f_l(\rho)g_{l+1}(\rho), \tag{3.2.35}$$

which again are valid for any linear combination of j_l and n_l.

Even more important for our purpose are the Wronskians of the spherical Bessel functions based on derivatives with respect to r

$$W\{f_l(\kappa_1 r), g_l(\kappa_2 r); r\} = f_l(\kappa_1 r)\frac{\partial}{\partial r}g_l(\kappa_2 r) - g_l(\kappa_2 r)\frac{\partial}{\partial r}f_l(\kappa_1 r). \tag{3.2.36}$$

Again we combine this definition with the differentiation formulas (3.2.3) and (3.2.8) to derive the cross-product relation

$$W\{f_l(\kappa_1 r), g_l(\kappa_2 r); r\} = \kappa_2 f_l(\kappa_1 r)g_{l-1}(\kappa_2 r) - \kappa_1 f_{l-1}(\kappa_1 r)g_l(\kappa_2 r)$$
$$= \kappa_1 f_{l+1}(\kappa_1 r)g_l(\kappa_2 r) - \kappa_2 f_l(\kappa_1 r)g_{l+1}(\kappa_2 r). \tag{3.2.37}$$

Expressions for the Wronskians of the Riccati-Bessel functions, i.e. for the functions $r f_l(\kappa_1 r)$ and $r g_l(\kappa_2 r)$ arise from (3.2.36) and (3.2.37) simply by multiplying all functions by r since the cross terms on the right-hand side of (3.2.36) cancel out. Just for completeness we note

$$W\{rf_l(\kappa_1 r), rg_l(\kappa_2 r); r\} = r^2 W\{f_l(\kappa_1 r), g_l(\kappa_2 r); r\}$$
$$= rf_l(\kappa_1 r)\frac{\partial}{\partial r}\left(rg_l(\kappa_2 r)\right) - rg_l(\kappa_2 r)\frac{\partial}{\partial r}\left(rf_l(\kappa_1 r)\right)$$
$$= \kappa_2 rf_l(\kappa_1 r)rg_{l-1}(\kappa_2 r) - \kappa_1 rf_{l-1}(\kappa_1 r)rg_l(\kappa_2 r)$$
$$= \kappa_1 rf_{l+1}(\kappa_1 r)rg_l(\kappa_2 r) - \kappa_2 rf_l(\kappa_1 r)rg_{l+1}(\kappa_2 r) .$$

$$(3.2.38)$$

From this and (3.2.34) we get for the special case $\kappa = \kappa_1 = \kappa_2$ the following fundamental Wronskians

$$W\{rj_l(\kappa r), rn_l(\kappa r); r\} = \frac{1}{i}W\{rj_l(\kappa r), rh_l^{(1)}(\kappa r); r\} = \frac{1}{\kappa} , \qquad (3.2.39)$$

which for the barred functions leads to

$$W\{r\bar{n}_l(\kappa r), r\bar{j}_l(\kappa r); r\} = W\{r\bar{h}_l^{(1)}(\kappa r), r\bar{j}_l(\kappa r); r\} = 1 . \qquad (3.2.40)$$

Again we point out that due to the r-factors appended to the spherical Bessel functions we arrive at constant Wronskians.

In the last part of this section we aim at calculating gradients of the envelope functions. This is most easily done by using the representation (B.3.30) of the gradient in spherical coordinates,

$$\nabla = \nabla_r + \nabla_{\vartheta,\varphi} = \hat{\mathbf{u}}_r \frac{\partial}{\partial r} + \hat{\mathbf{u}}_\vartheta \frac{1}{r}\frac{\partial}{\partial \vartheta} + \hat{\mathbf{u}}_\varphi \frac{1}{r\sin\vartheta}\frac{\partial}{\partial \varphi} , \qquad (3.2.41)$$

as well as the definition (B.3.31) of the three standard unit vectors, also in spherical coordinates. In addition, we fall back on the expansions (B.3.35) and (B.3.38), i.e.

$$\hat{\mathbf{u}}_r \mathcal{Y}_{L'}(\vartheta, \varphi) = \sum_{L''} \mathbf{G}_{L''L'} \mathcal{Y}_{L''}(\vartheta, \varphi) , \qquad (3.2.42)$$

and

$$r\nabla_{\vartheta,\varphi} \mathcal{Y}_{L'}(\vartheta, \varphi) = \sum_{L''} \mathbf{D}_{L''L'} \mathcal{Y}_{L''}(\vartheta, \varphi) , \qquad (3.2.43)$$

with the coefficient matrices given by (B.3.36) and (B.3.37), i.e.

$$\mathbf{G}_{LL'} = \int d^2\hat{\mathbf{r}}\, \mathcal{Y}_L(\hat{\mathbf{r}})\hat{\mathbf{u}}_r \mathcal{Y}_{L'}(\hat{\mathbf{r}}) , \qquad (3.2.44)$$

and

$$\mathbf{D}_{LL'} = \int d^2\hat{\mathbf{r}}\, \mathcal{Y}_L(\hat{\mathbf{r}})\left(r\nabla_{\vartheta,\varphi}\right)\mathcal{Y}_{L'}(\hat{\mathbf{r}}) . \qquad (3.2.45)$$

As outlined in App. B.3, according to (B.3.39) the vector $\hat{\mathbf{u}}_r$ can be expressed in terms of the cubic harmonics for $l = 1$ and the matrices $\mathbf{G}_{LL'}$ reduce to Gaunt coefficients with non-vanishing contributions only for $l' = l \pm 1$ as well as $m' = m$

or $m' = m \pm 1$; the terms with $l' = l$ vanish since the sum of all three angular momenta must be even for parity reasons. In contrast, as was also discussed in App. B.3, the matrices $\mathbf{D}_{LL'}$ can then be easily evaluated with the help of the Wigner-Eckart theorem and reduce to

$$\mathbf{D}_{LL'} = [\delta_{l'l-1}(-l') + \delta_{l'l+1}(l'+1)]\,\mathbf{G}_{LL'}\,. \qquad (3.2.46)$$

With these identities at hand we are able to evaluate the gradients of the envelope functions. Starting with the Hankel envelope functions (3.1.17) we write

$$\nabla H_{L\kappa}(\mathbf{r}) = (\nabla_r + \nabla_{\vartheta,\varphi})\,\bar{h}_l^{(1)}(\kappa r)\mathcal{Y}_L(\hat{\mathbf{r}})$$

$$= \frac{\partial}{\partial r}\bar{h}_l^{(1)}(\kappa r)\hat{\mathbf{u}}_r\mathcal{Y}_L(\hat{\mathbf{r}}) + \frac{1}{r}\bar{h}_l^{(1)}(\kappa r)r\nabla_{\vartheta,\varphi}\mathcal{Y}_L(\hat{\mathbf{r}})$$

$$= \frac{\partial}{\partial r}\bar{h}_l^{(1)}(\kappa r)\sum_{L''}\mathbf{G}_{L''L}\mathcal{Y}_{L''}(\hat{\mathbf{r}}) + \frac{1}{r}\bar{h}_l^{(1)}(\kappa r)\sum_{L''}\mathbf{D}_{L''L}\mathcal{Y}_{L''}(\hat{\mathbf{r}})$$

$$= \sum_{L''}\mathbf{G}_{L''L}\left[\delta_{ll''-1}\left(\frac{\partial}{\partial r}\bar{h}_l^{(1)}(\kappa r) - \frac{l}{r}\bar{h}_l^{(1)}(\kappa r)\right)\right.$$

$$\left. + \delta_{ll''+1}\left(\frac{\partial}{\partial r}\bar{h}_l^{(1)}(\kappa r) + \frac{l+1}{r}\bar{h}_l^{(1)}(\kappa r)\right)\right]\mathcal{Y}_{L''}(\hat{\mathbf{r}})$$

$$= \sum_{L''}\mathbf{G}_{L''L}\left[\delta_{ll''-1}\left(-\bar{h}_{l+1}^{(1)}(\kappa r)\right) + \delta_{ll''+1}\left(\kappa^2\bar{h}_{l-1}^{(1)}(\kappa r)\right)\right]\mathcal{Y}_{L''}(\hat{\mathbf{r}})$$

$$= \sum_{L''}\mathbf{G}_{L''L}\left[-\delta_{ll''-1} + \delta_{ll''+1}\kappa^2\right]\bar{h}_{l''}^{(1)}(\kappa r)\mathcal{Y}_{L''}(\hat{\mathbf{r}})$$

$$= \sum_{L''}\mathbf{G}_{L''L}\left[-\delta_{ll''-1} + \delta_{ll''+1}\kappa^2\right]H_{L''\kappa}(\mathbf{r})$$

$$= -\sum_{L''}\mathbf{G}_{L''L}(i\kappa)^{l-l''+1}H_{L''\kappa}(\mathbf{r})\,. \qquad (3.2.47)$$

Here we have used the identities (B.2.6) and (B.2.11) as well as a more compact formulation in the last line.

For the Neumann functions (3.1.15) we evaluate the radial derivatives with the help of the identities (B.2.4) and (B.2.9), which are formally identical to those for the barred Hankel functions. We thus note as a final result

$$\nabla N_{L\kappa}(\mathbf{r}) = \sum_{L''}\mathbf{G}_{L''L}\left[-\delta_{ll''-1} + \delta_{ll''+1}\kappa^2\right]N_{L''\kappa}(\mathbf{r})$$

$$= -\sum_{L''}\mathbf{G}_{L''L}(i\kappa)^{l-l''+1}N_{L''\kappa}(\mathbf{r})\,. \qquad (3.2.48)$$

In contrast, for the Bessel envelope functions (3.1.16) the radial derivatives of the barred Bessel functions are evaluated with the help of (B.2.5) and (B.2.10) and we obtain

$$\nabla J_{L\kappa}(\mathbf{r}) = (\nabla_r + \nabla_{\vartheta,\varphi}) \, \bar{j}_l(\kappa r) \mathcal{Y}_L(\hat{\mathbf{r}})$$

$$= \frac{\partial}{\partial r} \bar{j}_l(\kappa r) \hat{\mathbf{u}}_r \mathcal{Y}_L(\hat{\mathbf{r}}) + \frac{1}{r} \bar{j}_l(\kappa r) r \nabla_{\vartheta,\varphi} \mathcal{Y}_L(\hat{\mathbf{r}})$$

$$= \frac{\partial}{\partial r} \bar{j}_l(\kappa r) \sum_{L''} \mathbf{G}_{L''L} \mathcal{Y}_{L''}(\hat{\mathbf{r}}) + \frac{1}{r} \bar{j}_l(\kappa r) \sum_{L''} \mathbf{D}_{L''L} \mathcal{Y}_{L''}(\hat{\mathbf{r}})$$

$$= \sum_{L''} \mathbf{G}_{L''L} \left[\delta_{ll''-1} \left(\frac{\partial}{\partial r} \bar{j}_l(\kappa r) - \frac{l}{r} \bar{j}_l(\kappa r) \right) \right.$$

$$\left. + \delta_{ll''+1} \left(\frac{\partial}{\partial r} \bar{j}_l(\kappa r) + \frac{l+1}{r} \bar{j}_l(\kappa r) \right) \right] \mathcal{Y}_{L''}(\hat{\mathbf{r}})$$

$$= \sum_{L''} \mathbf{G}_{L''L} \left[\delta_{ll''-1} \left(-\kappa^2 \bar{j}_{l+1}(\kappa r) \right) + \delta_{ll''+1} \left(\bar{j}_{l-1}(\kappa r) \right) \right] \mathcal{Y}_{L''}(\hat{\mathbf{r}})$$

$$= \sum_{L''} \mathbf{G}_{L''L} \left[-\delta_{ll''-1}\kappa^2 + \delta_{ll''+1} \right] \bar{j}_{l''}(\kappa r) \mathcal{Y}_{L''}(\hat{\mathbf{r}})$$

$$= \sum_{L''} \mathbf{G}_{L''L} \left[-\delta_{ll''-1}\kappa^2 + \delta_{ll''+1} \right] J_{L''\kappa}(\mathbf{r})$$

$$= \sum_{L''} \mathbf{G}_{L''L} (i\kappa)^{l''-l+1} J_{L''\kappa}(\mathbf{r}) \, . \tag{3.2.49}$$

In the present context, it is interesting to note how the results (3.2.47) to (3.2.49) would look like if we had opted for the original definition of the radial part of the envelope functions as proposed by Williams, Kübler, and Gelatt, which contains an additional i^l factor. According to the discussion following (3.2.11), in this case the minus and plus signs before the first and second Kronecker-δ, respectively, of these equations would both turn into $+i$. In addition, the i in the $i\kappa$ factors as well as the minus signs in the last lines of (3.2.47) to (3.2.49) would cancel.

3.3 Integrals Involving Envelope Functions

Using the spherical Bessel functions of all three kinds as parts of the ASW envelope functions we will be faced with the calculation of integrals over these functions or their products. In the present context we discuss especially two cases.

Two types of integrals involving a single envelope function follow directly from (3.2.12) and (3.2.17). Integrating the former we obtain

$$\int^z d\rho \, \rho^{-l+1} f_l(\rho) = -z^{-l+1} f_{l-1}(z) \, , \tag{3.3.1}$$

where f_l is any linear combination of spherical Bessel functions. For the barred functions (3.3.1) reduces to

$$\int^s dr\, r^{-l+1}\bar{n}_l(\kappa r) = -s^{-l+1}\bar{n}_{l-1}(\kappa s) \ , \tag{3.3.2}$$

$$\int^s dr\, r^{-l+1}\bar{j}_l(\kappa r) = -\frac{1}{\kappa^2}s^{-l+1}\bar{j}_{l-1}(\kappa s) \ , \tag{3.3.3}$$

$$\int^s dr\, r^{-l+1}\bar{h}_l^{(1)}(\kappa r) = -s^{-l+1}\bar{h}_{l-1}^{(1)}(\kappa s) \ . \tag{3.3.4}$$

In contrast, integrating (3.2.17) we arrive at

$$\int^z d\rho\, \rho^{l+2} f_l(\rho) = z^{l+2} f_{l+1}(z) \ , \tag{3.3.5}$$

which leads to the following identities for the barred functions

$$\int^s dr\, r^{l+2}\bar{n}_l(\kappa r) = \frac{1}{\kappa^2}s^{l+2}\bar{n}_{l+1}(\kappa s) \ , \tag{3.3.6}$$

$$\int^s dr\, r^{l+2}\bar{j}_l(\kappa r) = s^{l+2}\bar{j}_{l+1}(\kappa s) \ , \tag{3.3.7}$$

$$\int^s dr\, r^{l+2}\bar{h}_l^{(1)}(\kappa r) = \frac{1}{\kappa^2}s^{l+2}\bar{h}_{l+1}^{(1)}(\kappa s) \ . \tag{3.3.8}$$

As an application of the previous identities we calculate the integral

$$\int_{\Omega_S} d^3\mathbf{r}\, e^{i\mathbf{kr}} \ ,$$

where Ω_S designates a sphere of radius S centered at the origin. For the exponential we use the expansion [9, App. B] or [8, Sect. 16.8]

$$e^{i\mathbf{kr}} = 4\pi \sum_L i^l j_l(kr) Y_L^*(\hat{\mathbf{k}}) Y_L(\hat{\mathbf{r}}) \ , \tag{3.3.9}$$

which will be discussed in more detail in Sect. 3.6 and where $L = (l, m)$ as usual. Inserting (3.3.9) into the above integral we obtain

$$\int_{\Omega_S} d^3\mathbf{r}\, e^{i\mathbf{kr}} = 4\pi \sum_L i^l Y_L^*(\hat{\mathbf{k}}) \int_{\Omega_S} d^3\mathbf{r}\, j_l(kr) Y_L(\hat{\mathbf{r}})$$

$$= 4\pi \int_0^S dr\, r^2 j_0(kr) \ . \tag{3.3.10}$$

According to (3.1.13) the spherical Bessel function $j_0(kr)$ is identical to its barred counterpart $\bar{j}_0(kr)$ and so we may use (3.3.7) to note

$$\int_{\Omega_S} d^3\mathbf{r}\, e^{i\mathbf{kr}} = 4\pi S^2 \bar{j}_1(kS) = 4\pi S^3 \frac{j_1(kS)}{kS} \ , \tag{3.3.11}$$

where we have used (3.1.44) and once again (3.1.13).

Even more important than the just mentioned integral are the integrals over products of spherical Bessel functions of the general type

$$\int^s dr\, r^2 f_l(\kappa_1 r) g_l(\kappa_2 r) \,, \tag{3.3.12}$$

where again f_l and g_l may be any linear combination of the spherical Bessel functions. As already mentioned at the beginning of this section these integrals may be calculated quite conveniently by use of the generating differential equation. Applying, in particular, the radial Schrödinger equation (3.1.11) to both $f_l(\kappa_1 r)$ and $g_l(\kappa_2 r)$, multiplying with the respective other function and subtracting both equations we get

$$r f_l(\kappa_1 r)\frac{\partial^2}{\partial r^2} r g_l(\kappa_2 r) - r g_l(\kappa_2 r)\frac{\partial^2}{\partial r^2} r f_l(\kappa_1 r) = (\kappa_1^2 - \kappa_2^2)\, r^2 f_l(\kappa_1 r) g_l(\kappa_2 r) \,. \tag{3.3.13}$$

Integrating the left-hand side by parts we immediately arrive at the result

$$
\begin{aligned}
(\kappa_1^2 - \kappa_2^2) &\int^s dr\, r^2 f_l(\kappa_1 r) g_l(\kappa_2 r) \\
&= \left[r f_l(\kappa_1 r)\frac{\partial}{\partial r} r g_l(\kappa_2 r) - r g_l(\kappa_2 r)\frac{\partial}{\partial r} r f_l(\kappa_1 r) \right]_{r=s} \\
&\quad - \int^s dr \left[\left(\frac{\partial}{\partial r} r f_l(\kappa_1 r)\right)\left(\frac{\partial}{\partial r} r g_l(\kappa_2 r)\right) - \left(\frac{\partial}{\partial r} r g_l(\kappa_2 r)\right)\left(\frac{\partial}{\partial r} r f_l(\kappa_1 r)\right) \right] \\
&= \left[r f_l(\kappa_1 r)\frac{\partial}{\partial r} r g_l(\kappa_2 r) - r g_l(\kappa_2 r)\frac{\partial}{\partial r} r f_l(\kappa_1 r) \right]_{r=s} \\
&= W\{r f_l(\kappa_1 r), r g_l(\kappa_2 r); r\}\,|_{r=s} \,, \tag{3.3.14}
\end{aligned}
$$

where in the last step we have used the definition (3.2.38) of the Wronskian. Obviously, (3.3.14) is equivalent to the following important identity

$$
\begin{aligned}
\int^s & dr\, r^2 f_l(\kappa_1 r) g_l(\kappa_2 r) \\
&= \frac{1}{\kappa_1^2 - \kappa_2^2} W\{r f_l(\kappa_1 r), r g_l(\kappa_2 r); r\}|_{r=s} \\
&= \frac{s^2}{\kappa_1^2 - \kappa_2^2}\left[f_l(\kappa_1 r)\frac{\partial}{\partial r} g_l(\kappa_2 r) - g_l(\kappa_2 r)\frac{\partial}{\partial r} f_l(\kappa_1 r) \right]_{r=s} \\
&= \frac{s^2}{\kappa_1^2 - \kappa_2^2}\left[\kappa_2 f_l(\kappa_1 s) g_{l-1}(\kappa_2 s) - \kappa_1 f_{l-1}(\kappa_1 s) g_l(\kappa_2 s) \right] \\
&= \frac{s^2}{\kappa_1^2 - \kappa_2^2}\left[\kappa_1 f_{l+1}(\kappa_1 s) g_l(\kappa_2 s) - \kappa_2 f_l(\kappa_1 s) g_{l+1}(\kappa_2 s) \right] \,. \tag{3.3.15}
\end{aligned}
$$

Here we have used (3.2.38) for the relation of the Wronskians and the cross-product relations. However, note that we have taken the s-factors out of the square brackets.

Next we turn to the "single κ" case $\kappa_1 = \kappa_2$. In this case it is not possible to evaluate the integral using (3.3.15) directly since the energy denominator vanishes. L'Hospitals rule does not work here either since the numerator does *not* vanish for $\kappa_1 = \kappa_2$. Instead, as we have mentioned in connection with (3.2.40) and as can be read off from (3.3.14), in this case the numerator is a constant, which, however, can be neglected in a definite integral.

Nevertheless, being a bit formal for the time being we turn to the definite integral arising from (3.3.15)

$$
\int_{s_1}^{s_2} dr \, r^2 f_l(\kappa_1 r) g_l(\kappa_2 r)
$$
$$
= \frac{1}{\kappa_1^2 - \kappa_2^2} \left[W\{r f_l(\kappa_1 r), r g_l(\kappa_2 r); r\} \mid_{r=s_2} - W\{r f_l(\kappa_1 r), r g_l(\kappa_2 r); r\} \mid_{r=s_1} \right].
$$
(3.3.16)

From (3.3.16) it is now explicit that for $\kappa_1 = \kappa_2$ *both* the numerator and denominator trivially vanish because the Wronskian is a constant with respect to s for equal values of the energy parameter. Hence, we may now apply l'Hospitals rule and differentiate both the numerator and the denominator with respect to κ_1^2 at $\kappa_1 = \kappa_2$ this resulting in

$$
\int_{s_1}^{s_2} dr \, r^2 f_l(\kappa_1 r) g_l(\kappa_2 r) \mid_{\kappa_1 = \kappa_2}
$$
$$
= \frac{\partial}{\partial \kappa_1^2} W\{r f_l(\kappa_1 r), r g_l(\kappa_2 r); r\} \mid_{\kappa_1 = \kappa_2, r = s_2}
$$
$$
- \frac{\partial}{\partial \kappa_1^2} W\{r f_l(\kappa_1 r), r g_l(\kappa_2 r); r\} \mid_{\kappa_1 = \kappa_2, r = s_1} .
$$
(3.3.17)

Finally, since (3.3.17) holds for any pair of values s_1 and s_2, we may go back to the indefinite integral and write

$$
\int^s dr \, r^2 f_l(\kappa_1 r) g_l(\kappa_2 r) \mid_{\kappa_1 = \kappa_2}
$$
$$
= \frac{\partial}{\partial \kappa_1^2} \left[W\{r f_l(\kappa_1 r), r g_l(\kappa_2 r); r\} \right]_{\kappa_1 = \kappa_2, r = s}
$$
$$
= s^2 \left[\frac{\partial}{\partial \kappa_1^2} (f_l(\kappa_1 r)) \frac{\partial}{\partial r} (g_l(\kappa_2 r)) - g_l(\kappa_2 r) \frac{\partial}{\partial \kappa_1^2} \frac{\partial}{\partial r} (f_l(\kappa_1 r)) \right]_{r = s, \kappa_1 = \kappa_2}
$$
$$
= s^2 \left[\kappa_2 \frac{\partial}{\partial \kappa_1^2} f_l(\kappa_1 s) g_{l-1}(\kappa_2 s) - \frac{\partial}{\partial \kappa_1^2} (\kappa_1 f_{l-1}(\kappa_1 s)) g_l(\kappa_2 s) \right]_{\kappa_1 = \kappa_2}
$$
$$
= s^2 \left[\frac{\partial}{\partial \kappa_1^2} (\kappa_1 f_{l+1}(\kappa_1 s)) g_l(\kappa_2 s) - \kappa_2 \frac{\partial}{\partial \kappa_1^2} f_l(\kappa_1 s) g_{l+1}(\kappa_2 s) \right]_{\kappa_1 = \kappa_2} ,
$$
(3.3.18)

where we have additionally appended the expressions corresponding to the third to fifth line of (3.3.15).

It is quite instructive to study an alternative derivation of the previous identities, which is based on the observation that the energy derivative of the Wronskian acts onto one of the two functions only. Hence, we may likewise start from the energy derivative of the radial Schrödinger equation (3.1.11)

$$\frac{\partial}{\partial \kappa^2} \left[\frac{\partial^2}{\partial r^2} + \kappa^2 - \frac{l(l+1)}{r^2} \right] (r f_l(\kappa r))$$

$$= \left[\frac{\partial^2}{\partial r^2} + \kappa^2 - \frac{l(l+1)}{r^2} \right] \frac{\partial}{\partial \kappa^2} (r f_l(\kappa r)) + r f_l(\kappa r) = 0 . \qquad (3.3.19)$$

From this and the radial Schrödinger equation for the function $g_l(\kappa r)$ we get

$$r f_l(\kappa r) r g_l(\kappa r) = -r g_l(\kappa r) \left[\frac{\partial^2}{\partial r^2} + \kappa^2 - \frac{l(l+1)}{r^2} \right] \frac{\partial}{\partial \kappa^2} (r f_l(\kappa r)) , \qquad (3.3.20)$$

and

$$0 = \frac{\partial}{\partial \kappa^2} (r f_l(\kappa r)) \left[\frac{\partial^2}{\partial r^2} + \kappa^2 - \frac{l(l+1)}{r^2} \right] (r g_l(\kappa r)) . \qquad (3.3.21)$$

Adding both identities and integrating by parts we arrive at

$$\int^s dr \, r^2 f_l(\kappa r) g_l(\kappa r)$$

$$= \int^s dr \left[\left(\frac{\partial}{\partial \kappa^2} r f_l(\kappa r) \right) \left(\frac{\partial^2}{\partial r^2} r g_l(\kappa r) \right) - r g_l(\kappa r) \frac{\partial^2}{\partial r^2} \left(\frac{\partial}{\partial \kappa^2} r f_l(\kappa r) \right) \right]$$

$$= \left[\frac{\partial}{\partial \kappa^2} (r f_l(\kappa r)) \frac{\partial}{\partial r} (r g_l(\kappa r)) - r g_l(\kappa r) \frac{\partial}{\partial r} \frac{\partial}{\partial \kappa^2} (r f_l(\kappa r)) \right]_{r=s} . \qquad (3.3.22)$$

Obviously, (3.3.22) is identical to the third line in (3.3.18).

Although the just outlined second way of calculating the integral (3.3.12) is equivalent to the first one it is more straightforward. This is due to the fact that it employs the "single κ" case as well as the energy derivative from the very beginning and thus avoids to take the limit $\kappa_1 \to \kappa_2$ half the way. In Sect. 3.8 we will learn about the analogous calculation of the integral when the two functions involved are centered at different sites. Insofar our previous excursion served as a preparation, which will make things easier to understand lateron.

In order to further evaluate the integral (3.3.18) we use (3.2.23) and note

$$s^2 \frac{\partial}{\partial \kappa_1^2} \left(-\kappa_1 f_{l-1}(\kappa_1 s) g_l(\kappa_2 s) \right) \Bigg|_{\kappa_1 = \kappa_2} = \frac{s^3}{2} f_l(\kappa_2 s) g_l(\kappa_2 s) - \frac{s^2}{2\kappa_2} l f_{l-1}(\kappa_2 s) g_l(\kappa_2 s) ,$$

$$(3.3.23)$$

as well as

$$s^2 \frac{\partial}{\partial \kappa_1^2} \left(\kappa_2 f_l(\kappa_1 s) g_{l-1}(\kappa_2 s) \right) \Bigg|_{\kappa_1 = \kappa_2}$$

$$= -\frac{s^3}{4} f_{l+1}(\kappa_2 s) g_{l-1}(\kappa_2 s) - \frac{s^2}{4\kappa_2} f_l(\kappa_2 s) g_{l-1}(\kappa_2 s)$$

$$+ \frac{s^3}{4} f_{l-1}(\kappa_2 s) g_{l-1}(\kappa_2 s) . \qquad (3.3.24)$$

Using the recursion relation (3.1.53) we obtain from this

$$
s^2 \frac{\partial}{\partial \kappa_1^2} \left[\kappa_2 f_l(\kappa_1 s) g_{l-1}(\kappa_2 s) - \kappa_1 f_{l-1}(\kappa_1 s) g_l(\kappa_2 s) \right]_{\kappa_1 = \kappa_2}
$$

$$
= \frac{s^3}{2} f_l(\kappa_2 s) g_l(\kappa_2 s) - \frac{s^3}{4} f_{l+1}(\kappa_2 s) g_{l-1}(\kappa_2 s) - \frac{s^3}{4} f_{l-1}(\kappa_2 s) g_{l+1}(\kappa_2 s)
$$

$$
+ \frac{s^3}{4} \frac{1}{\kappa_2 s} \left(f_{l-1}(\kappa_2 s) g_l(\kappa_2 s) - f_l(\kappa_2 s) g_{l-1}(\kappa_2 s) \right) . \tag{3.3.25}
$$

The last term on the right-hand side is just $-\frac{1}{4\kappa_2}$ times the (constant) Wronskian. Since we aim at definite integrals this term can be omitted. Hence, inserting (3.3.25) into the fourth line of (3.3.18) we are left with the result

$$
\int^s dr\, r^2 f_l(\kappa r) g_l(\kappa r)
$$

$$
= \frac{s^3}{4} \left[2 f_l(\kappa s) g_l(\kappa s) - f_{l+1}(\kappa s) g_{l-1}(\kappa s) - f_{l-1}(\kappa s) g_{l+1}(\kappa s) \right] , \tag{3.3.26}
$$

which complements (3.3.18) and (3.3.22).

Equation (3.3.26) may be identically formulated with the barred functions, which, according to their definitions (3.1.12) to (3.1.14), only contain extra κ-factors as compared to the spherical Bessel functions. However, this step is not as straightforward when we start out from the third line of (3.3.18) or from (3.3.22) due to the presence of the energy derivative. Hence, before turning to the calculation of integrals over products of the barred functions we will reconsider the integral (3.3.13) for equal values of the energy parameter with the function $f_l(\kappa r)$ replaced by $u(\kappa) f_l(\kappa r)$. $u(\kappa)$ may be any continuous and differentiable function. From (3.3.22) we read off

$$
\int^s dr\, r^2 u(\kappa) f_l(\kappa r) g_l(\kappa r)
$$

$$
= u(\kappa) \int^s dr\, r^2 f_l(\kappa r) g_l(\kappa r)
$$

$$
= u(\kappa) \left[\frac{\partial}{\partial \kappa^2} (r f_l(\kappa r)) \frac{\partial}{\partial r} (r g_l(\kappa r)) - r g_l(\kappa r) \frac{\partial}{\partial r} \frac{\partial}{\partial \kappa^2} (r f_l(\kappa r)) \right]_{r=s} . \tag{3.3.27}
$$

The right-hand side can be rewritten with the help of the identity

$$
\left[\frac{\partial}{\partial \kappa^2} (u(\kappa) r f_l(\kappa r)) \frac{\partial}{\partial r} (r g_l(\kappa r)) - r g_l(\kappa r) \frac{\partial}{\partial r} \frac{\partial}{\partial \kappa^2} (u(\kappa) r f_l(\kappa r)) \right]_{r=s}
$$

$$
= \frac{\partial}{\partial \kappa^2} u(\kappa) \left[r f_l(\kappa r) \frac{\partial}{\partial r} (r g_l(\kappa r)) - r g_l(\kappa r) \frac{\partial}{\partial r} (r f_l(\kappa r)) \right]_{r=s}
$$

$$
+ u(\kappa) \left[\frac{\partial}{\partial \kappa^2} (r f_l(\kappa r)) \frac{\partial}{\partial r} (r g_l(\kappa r)) - r g_l(\kappa r) \frac{\partial}{\partial r} \frac{\partial}{\partial \kappa^2} (r f_l(\kappa r)) \right]_{r=s} , \tag{3.3.28}
$$

this resulting in

$$\int^s dr\, r^2 u(\kappa) f_l(\kappa r) g_l(\kappa r)$$

$$= \left[\frac{\partial}{\partial \kappa^2} \left(u(\kappa) r f_l(\kappa r) \right) \frac{\partial}{\partial r} \left(r g_l(\kappa r) \right) - r g_l(\kappa r) \frac{\partial}{\partial r} \frac{\partial}{\partial \kappa^2} \left(u(\kappa) r f_l(\kappa r) \right) \right]_{r=s}$$

$$- \frac{\partial}{\partial \kappa^2} u(\kappa) W \{ s f_l(\kappa s), s g_l(\kappa s); s \} . \tag{3.3.29}$$

Here we have used (3.2.38) for the Wronskian. Again we take advantage of the fact that the Wronskian is constant since then we may omit the last term of (3.3.29) and thus formulate all above results for the integrals over products of spherical Bessel functions identically with the barred functions.

After these preparations we next apply the identities (3.3.15), (3.3.18), and (3.3.26) to the calculation of integrals over products of the barred functions. However, in this context we concentrate on a few definite integrals, which appear when the combined correction to the atomic-sphere approximation is used. Considering the "double κ" case first, we fall back on (3.3.15) and note

$$\int_{\Omega_s} d^3\mathbf{r}\, J^*_{L\kappa_1}(\mathbf{r}) J_{L\kappa_2}(\mathbf{r})$$

$$= \int_0^s dr\, r^2 \bar{\jmath}_l(\kappa_1 r) \bar{\jmath}_l(\kappa_2 r)$$

$$= \frac{1}{\kappa_1^2 - \kappa_2^2} W \{ r \bar{\jmath}_l(\kappa_1 r), r \bar{\jmath}_l(\kappa_2 r); r \} |_{r=s}$$

$$= \frac{s^2}{\kappa_1^2 - \kappa_2^2} \left[\bar{\jmath}_l(\kappa_1 r) \frac{\partial}{\partial r} \bar{\jmath}_l(\kappa_2 r) - \bar{\jmath}_l(\kappa_2 r) \frac{\partial}{\partial r} \bar{\jmath}_l(\kappa_1 r) \right]_{r=s}$$

$$= \frac{s^2}{\kappa_1^2 - \kappa_2^2} \left[\bar{\jmath}_l(\kappa_1 s) \bar{\jmath}_{l-1}(\kappa_2 s) - \bar{\jmath}_{l-1}(\kappa_1 s) \bar{\jmath}_l(\kappa_2 s) \right] , \tag{3.3.30}$$

$$\int_{\Omega_s} d^3\mathbf{r}\, N^*_{L\kappa_1}(\mathbf{r}) J_{L\kappa_2}(\mathbf{r})$$

$$= \int_0^s dr\, r^2 \bar{n}_l(\kappa_1 r) \bar{\jmath}_l(\kappa_2 r)$$

$$= \frac{1}{\kappa_1^2 - \kappa_2^2} W \{ r \bar{n}_l(\kappa_1 r), r \bar{\jmath}_l(\kappa_2 r); r \} |_{r=s} - \frac{1}{\kappa_1^2 - \kappa_2^2}$$

$$= \frac{s^2}{\kappa_1^2 - \kappa_2^2} \left[\bar{n}_l(\kappa_1 r) \frac{\partial}{\partial r} \bar{\jmath}_l(\kappa_2 r) - \bar{\jmath}_l(\kappa_2 r) \frac{\partial}{\partial r} \bar{n}_l(\kappa_1 r) \right]_{r=s} - \frac{1}{\kappa_1^2 - \kappa_2^2}$$

$$= \frac{s^2}{\kappa_1^2 - \kappa_2^2} \left[\bar{n}_l(\kappa_1 s) \bar{\jmath}_{l-1}(\kappa_2 s) - \kappa_1^2 \bar{n}_{l-1}(\kappa_1 s) \bar{\jmath}_l(\kappa_2 s) \right] - \frac{1}{\kappa_1^2 - \kappa_2^2} , \tag{3.3.31}$$

and

$$
\int_{\Omega_s} d^3\mathbf{r}\, H^*_{L\kappa_1}(\mathbf{r}) J_{L\kappa_2}(\mathbf{r})
$$

$$
= \int_0^s dr\, r^2 \bar{h}_l^{(1)}(\kappa_1 r)\bar{\jmath}_l(\kappa_2 r)
$$

$$
= \frac{1}{\kappa_1^2 - \kappa_2^2} W\{r\bar{h}_l(\kappa_1 r), r\bar{\jmath}_l(\kappa_2 r); r\}|_{r=s} - \frac{1}{\kappa_1^2 - \kappa_2^2}
$$

$$
= \frac{s^2}{\kappa_1^2 - \kappa_2^2}\left[\bar{h}_l^{(1)}(\kappa_1 r)\frac{\partial}{\partial r}\bar{\jmath}_l(\kappa_2 r) - \bar{\jmath}_l(\kappa_2 r)\frac{\partial}{\partial r}\bar{h}_l^{(1)}(\kappa_1 r)\right]_{r=s} - \frac{1}{\kappa_1^2 - \kappa_2^2}
$$

$$
= \frac{s^2}{\kappa_1^2 - \kappa_2^2}\left[\bar{h}_l^{(1)}(\kappa_1 s)\bar{\jmath}_{l-1}(\kappa_2 s) - \kappa_1^2\bar{h}_{l-1}^{(1)}(\kappa_1 s)\bar{\jmath}_l(\kappa_2 s)\right] - \frac{1}{\kappa_1^2 - \kappa_2^2}. \qquad (3.3.32)
$$

Here we have used the definitions (3.1.15) to (3.1.17) of the envelope functions, the orthonormality of the spherical harmonics, the definitions (3.1.12) to (3.1.14) of the barred functions, and the formulas (3.1.43) to (3.1.45) for the values of these functions at the lower boundary $r = 0$. Ω_s denotes the volume of a sphere with radius s. We furthermore write down the integrals

$$
\int_{\Omega_c - \Omega_s} d^3\mathbf{r}\, H^*_{L\kappa_1}(\mathbf{r}) H_{L\kappa_2}(\mathbf{r})
$$

$$
= \int_s^\infty dr\, r^2 \bar{h}_l^{(1)}(\kappa_1 r)\bar{h}_l^{(1)}(\kappa_2 r)
$$

$$
= -\frac{1}{\kappa_1^2 - \kappa_2^2} W\{r\bar{h}_l(\kappa_1 r), r\bar{h}_l(\kappa_2 r); r\}|_{r=s}
$$

$$
= -\frac{s^2}{\kappa_1^2 - \kappa_2^2}\left[\bar{h}_l^{(1)}(\kappa_1 r)\frac{\partial}{\partial r}\bar{h}_l^{(1)}(\kappa_2 r) - \bar{h}_l^{(1)}(\kappa_2 r)\frac{\partial}{\partial r}\bar{h}_l^{(1)}(\kappa_1 r)\right]_{r=s}
$$

$$
= -\frac{s^2}{\kappa_1^2 - \kappa_2^2}\left[\kappa_2^2\bar{h}_l^{(1)}(\kappa_1 s)\bar{h}_{l-1}^{(1)}(\kappa_2 s) - \kappa_1^2\bar{h}_{l-1}^{(1)}(\kappa_1 s)\bar{h}_l^{(1)}(\kappa_2 s)\right]
$$

$$
= \frac{s^2}{\kappa_1^2 - \kappa_2^2}\left[\bar{h}_l^{(1)}(\kappa_1 s)\bar{h}_{l+1}^{(1)}(\kappa_2 s) - \bar{h}_{l+1}^{(1)}(\kappa_1 s)\bar{h}_l^{(1)}(\kappa_2 s)\right], \qquad (3.3.33)
$$

and

$$
\int_{\Omega_c - \Omega_s} d^3\mathbf{r}\, H^*_{L\kappa_1}(\mathbf{r}) J_{L\kappa_2}(\mathbf{r})
$$

$$
= \int_s^\infty dr\, r^2 \bar{h}_l^{(1)}(\kappa_1 r)\bar{\jmath}_l(\kappa_2 r)
$$

$$
= -\frac{1}{\kappa_1^2 - \kappa_2^2} W\{r\bar{h}_l(\kappa_1 r), r\bar{\jmath}_l(\kappa_2 r); r\}|_{r=s}
$$

$$
= -\frac{s^2}{\kappa_1^2 - \kappa_2^2}\left[\bar{h}_l^{(1)}(\kappa_1 r)\frac{\partial}{\partial r}\bar{\jmath}_l(\kappa_2 r) - \bar{\jmath}_l(\kappa_2 r)\frac{\partial}{\partial r}\bar{h}_l^{(1)}(\kappa_1 r)\right]_{r=s}
$$

$$
= -\frac{s^2}{\kappa_1^2 - \kappa_2^2}\left[\bar{h}_l^{(1)}(\kappa_1 s)\bar{\jmath}_{l-1}(\kappa_2 s) - \kappa_1^2\bar{h}_{l-1}^{(1)}(\kappa_1 s)\bar{\jmath}_l(\kappa_2 s)\right]
$$

$$= \frac{s^2}{\kappa_1^2 - \kappa_2^2} \left[\kappa_2^2 \bar{h}_l^{(1)}(\kappa_1 s)\bar{j}_{l+1}(\kappa_2 s) - \bar{h}_{l+1}^{(1)}(\kappa_1 s)\bar{j}_l(\kappa_2 s) \right], \qquad (3.3.34)$$

where Ω_c denotes all space or else the volume of the unit cell in case of crystal translational symmetry. In the last step we have used the recursion relation (3.1.56). In addition, we point out that, according to (3.1.48), the contributions from the upper boundaries vanish as long as one of the two κ^2-values is negative. This does not hold if both κ_1^2 and κ_2^2 are positive as well as for Neumann functions. However, in these cases we can circumvent problems by introducing Bloch symmetry before performing the integrals this as well leaving only the contribution from the lower boundary.

In order to prepare for the calculation of Hamiltonian matrix elements in Sect. 2.2 we complement the previous integrals with the following identities

$$\kappa_1^2 \int_{\Omega_s} d^3\mathbf{r}\, J_{L\kappa_1}^*(\mathbf{r}) J_{L\kappa_2}(\mathbf{r})$$
$$= \kappa_1^2 \frac{1}{\kappa_1^2 - \kappa_2^2} W\{r\bar{j}_l(\kappa_1 r), r\bar{j}_l(\kappa_2 r); r\}|_{r=s}$$
$$= \left[\kappa_2^2 \frac{1}{\kappa_1^2 - \kappa_2^2} + 1 \right] W\{r\bar{j}_l(\kappa_1 r), r\bar{j}_l(\kappa_2 r); r\}|_{r=s}$$
$$= \kappa_2^2 \int_{\Omega_s} d^3\mathbf{r}\, J_{L\kappa_1}^*(\mathbf{r}) J_{L\kappa_2}(\mathbf{r}) + W\{r\bar{j}_l(\kappa_1 r), r\bar{j}_l(\kappa_2 r); r\}|_{r=s}, \qquad (3.3.35)$$

$$\kappa_1^2 \int_{\Omega_s} d^3\mathbf{r}\, N_{L\kappa_1}^*(\mathbf{r}) J_{L\kappa_2}(\mathbf{r})$$
$$= \kappa_1^2 \frac{1}{\kappa_1^2 - \kappa_2^2} W\{r\bar{n}_l(\kappa_1 r), r\bar{j}_l(\kappa_2 r); r\}|_{r=s} - \kappa_1^2 \frac{1}{\kappa_1^2 - \kappa_2^2}$$
$$= \left[\kappa_2^2 \frac{1}{\kappa_1^2 - \kappa_2^2} + 1 \right] W\{r\bar{n}_l(\kappa_1 r), r\bar{j}_l(\kappa_2 r); r\}|_{r=s} - \left[\kappa_2^2 \frac{1}{\kappa_1^2 - \kappa_2^2} + 1 \right]$$
$$= \kappa_2^2 \int_{\Omega_s} d^3\mathbf{r}\, N_{L\kappa_1}^*(\mathbf{r}) J_{L\kappa_2}(\mathbf{r}) + W\{r\bar{n}_l(\kappa_1 r), r\bar{j}_l(\kappa_2 r); r\}|_{r=s} - 1,$$
$$(3.3.36)$$

$$\kappa_1^2 \int_{\Omega_s} d^3\mathbf{r}\, H_{L\kappa_1}^*(\mathbf{r}) J_{L\kappa_2}(\mathbf{r})$$
$$= \kappa_1^2 \frac{1}{\kappa_1^2 - \kappa_2^2} W\{r\bar{h}_l(\kappa_1 r), r\bar{j}_l(\kappa_2 r); r\}|_{r=s} - \kappa_1^2 \frac{1}{\kappa_1^2 - \kappa_2^2}$$
$$= \left[\kappa_2^2 \frac{1}{\kappa_1^2 - \kappa_2^2} + 1 \right] W\{r\bar{h}_l(\kappa_1 r), r\bar{j}_l(\kappa_2 r); r\}|_{r=s} - \left[\kappa_2^2 \frac{1}{\kappa_1^2 - \kappa_2^2} + 1 \right]$$
$$= \kappa_2^2 \int_{\Omega_s} d^3\mathbf{r}\, H_{L\kappa_1}^*(\mathbf{r}) J_{L\kappa_2}(\mathbf{r}) + W\{r\bar{h}_l(\kappa_1 r), r\bar{j}_l(\kappa_2 r); r\}|_{r=s} - 1,$$
$$(3.3.37)$$

and

$$\kappa_1^2 \int_{\Omega_c - \Omega_s} d^3\mathbf{r} \, H_{L\kappa_1}^*(\mathbf{r}) H_{L\kappa_2}(\mathbf{r})$$

$$= -\kappa_1^2 \frac{1}{\kappa_1^2 - \kappa_2^2} W\{r\bar{h}_l(\kappa_1 r), r\bar{h}_l(\kappa_2 r); r\}|_{r=s}$$

$$= -\left[\kappa_2^2 \frac{1}{\kappa_1^2 - \kappa_2^2} + 1\right] W\{r\bar{h}_l(\kappa_1 r), r\bar{h}_l(\kappa_2 r); r\}|_{r=s}$$

$$= \kappa_2^2 \int_{\Omega_c - \Omega_s} d^3\mathbf{r} \, H_{L\kappa_1}^*(\mathbf{r}) H_{L\kappa_2}(\mathbf{r}) - W\{r\bar{h}_l(\kappa_1 r), r\bar{h}_l(\kappa_2 r); r\}|_{r=s} \,.$$

$$(3.3.38)$$

Again we explicitly note the integrals over products of the barred functions for the case $\kappa_1 = \kappa_2$. Here we have from (3.3.18) and (3.3.26)

$$\int_{\Omega_s} d^3\mathbf{r} \, J_{L\kappa}^*(\mathbf{r}) J_{L\kappa}(\mathbf{r})$$

$$= \int_0^s dr \, r^2 \left[\bar{j}_l(\kappa r)\right]^2$$

$$= s^2 \left[\frac{\partial}{\partial \kappa^2}\left(\bar{j}_l(\kappa r)\right) \frac{\partial}{\partial r}\left(\bar{j}_l(\kappa r)\right) - \bar{j}_l(\kappa r)\frac{\partial}{\partial \kappa^2}\frac{\partial}{\partial r}\left(\bar{j}_l(\kappa r)\right)\right]_{r=s}$$

$$= \frac{s^3}{2}\left[\left[\bar{j}_l(\kappa s)\right]^2 - \bar{j}_{l-1}(\kappa s)\bar{j}_{l+1}(\kappa s)\right], \qquad (3.3.39)$$

$$\int_{\Omega_s} d^3\mathbf{r} \, N_{L\kappa}^*(\mathbf{r}) J_{L\kappa}(\mathbf{r})$$

$$= \int_0^s dr \, r^2 \bar{n}_l(\kappa r)\bar{j}_l(\kappa r)$$

$$= s^2 \left[\frac{\partial}{\partial \kappa^2}\left(\bar{n}_l(\kappa r)\right) \frac{\partial}{\partial r}\left(\bar{j}_l(\kappa r)\right) - \bar{j}_l(\kappa r)\frac{\partial}{\partial \kappa^2}\frac{\partial}{\partial r}\left(\bar{n}_l(\kappa r)\right)\right]_{r=s}$$

$$= \frac{s^3}{4}\left[2\bar{n}_l(\kappa s)\bar{j}_l(\kappa s) - \kappa^2 \bar{n}_{l-1}(\kappa s)\bar{j}_{l+1}(\kappa s) - \frac{1}{\kappa^2}\bar{n}_{l+1}(\kappa s)\bar{j}_{l-1}(\kappa s)\right] + \frac{2l+1}{4\kappa^2} \,,$$

$$(3.3.40)$$

and

$$\int_{\Omega_s} d^3\mathbf{r} \, H_{L\kappa}^*(\mathbf{r}) J_{L\kappa}(\mathbf{r})$$

$$= \int_0^s dr \, r^2 \bar{h}_l^{(1)}(\kappa r)\bar{j}_l(\kappa r)$$

$$= s^2 \left[\frac{\partial}{\partial \kappa^2}\left(\bar{h}_l^{(1)}(\kappa r)\right) \frac{\partial}{\partial r}\left(\bar{j}_l(\kappa r)\right) - \bar{j}_l(\kappa r)\frac{\partial}{\partial \kappa^2}\frac{\partial}{\partial r}\left(\bar{h}_l^{(1)}(\kappa r)\right)\right]_{r=s}$$

$$= \frac{s^3}{4} \left[2\bar{h}_l^{(1)}(\kappa s)\bar{j}_l(\kappa s) - \kappa^2 \bar{h}_{l-1}^{(1)}(\kappa s)\bar{j}_{l+1}(\kappa s) - \frac{1}{\kappa^2}\bar{h}_{l+1}^{(1)}(\kappa s)\bar{j}_{l-1}(\kappa s) \right] + \frac{2l+1}{4\kappa^2} .$$

$$(3.3.41)$$

Again, we have used the formulas (3.1.43) to (3.1.45) for the functions at the lower boundary $r = 0$. In the evaluation of the latter two integrals in the limit $\kappa^2 \to 0$ we apply these expansions for small arguments to the last term in brackets in the respective second lines. This then reduces to $\frac{4}{s^3}\frac{2l+1}{\kappa^2}$ and thus cancels the last term on the right-hand side. Nevertheless, note that in the respective first lines of (3.3.40) and (3.3.41) there is no contribution from the lower boundary. This difference can be traced back to (3.3.25) where in the last step we omitted the (constant) Wronskian, which now appears in the previous two equations. Hence, the respective first lines offer a simpler way to evaluate the integrals and are thus preferred.

Finally we write down the integral

$$\int_{\Omega_c - \Omega_s} d^3\mathbf{r}\, H_{L\kappa}^*(\mathbf{r})H_{L\kappa}(\mathbf{r})$$

$$= \int_s^\infty dr\, r^2 \left[\bar{h}_l^{(1)}(\kappa r) \right]^2$$

$$= -s^2 \left[\frac{\partial}{\partial\kappa^2}\left(\bar{h}_l^{(1)}(\kappa r)\right)\frac{\partial}{\partial r}\left(\bar{h}_l^{(1)}(\kappa r)\right) - \bar{h}_l^{(1)}(\kappa r)\frac{\partial}{\partial\kappa^2}\frac{\partial}{\partial r}\left(\bar{h}_l^{(1)}(\kappa r)\right) \right]_{r=s}$$

$$= -\frac{s^3}{2} \left[\left[\bar{h}_l^{(1)}(\kappa s)\right]^2 - \bar{h}_{l-1}^{(1)}(\kappa s)\bar{h}_{l+1}^{(1)}(\kappa s) \right] ,$$

$$(3.3.42)$$

where again according to (3.1.48) the contribution from the upper boundary vanishes for negative κ^2. Like for the "double κ" case this is not true for positive κ^2 as well as for Neumann functions but still the above mentioned arguments hold.

Another quite important application of the general integral formula (3.3.15) is the evaluation of the following integral

$$\int_0^\infty dr\, r^2 j_l(\kappa_1 r)j_l(\kappa_2 r) = \frac{s^2}{\kappa_1^2 - \kappa_2^2} \left[\kappa_2 j_l(\kappa_1 s)j_{l-1}(\kappa_2 s) - \kappa_1 j_{l-1}(\kappa_1 s)j_l(\kappa_2 s) \right]_0^\infty ,$$

$$(3.3.43)$$

since it leads to the orthonormality relation for the Riccati-Bessel functions and thereby lays ground for the Hankel transform. Using (3.1.41) and (3.1.47) for the lower and upper boundary, respectively, we are left with the contribution from the latter only. Furthermore, applying elementary relations for the trigonometric functions we get

$$\int_0^\infty dr\, r^2 j_l(\kappa_1 r)j_l(\kappa_2 r)$$

$$= \lim_{s\to\infty} \frac{1}{\kappa_1^2 - \kappa_2^2}\frac{1}{\kappa_1\kappa_2} \left[\kappa_2 \sin(\kappa_1 s - l\frac{\pi}{2})\cos(\kappa_2 s - l\frac{\pi}{2}) \right.$$

$$\left. -\kappa_1 \cos(\kappa_1 s - l\frac{\pi}{2})\sin(\kappa_2 s - l\frac{\pi}{2}) \right]$$

$$= \lim_{s\to\infty} \frac{1}{\kappa_1^2 - \kappa_2^2} \frac{1}{2\kappa_1\kappa_2} \left[(\kappa_2 - \kappa_1)\sin(\kappa_1 s + \kappa_2 s + l\pi) + (\kappa_1 + \kappa_2)\sin(\kappa_1 s - \kappa_2 s) \right]$$

$$= \frac{1}{2\kappa_1\kappa_2} \lim_{s\to\infty} \left[\frac{\sin(\kappa_1 s - \kappa_2 s)}{\kappa_1 - \kappa_2} - (-)^l \frac{\sin(\kappa_1 s + \kappa_2 s)}{\kappa_1 + \kappa_2} \right] . \tag{3.3.44}$$

According to the identity

$$\lim_{s\to\infty} \frac{\sin \kappa_1 s}{\kappa_1} = \pi\delta(\kappa_1) , \tag{3.3.45}$$

the terms in brackets are both representations of the δ distribution. Hence, we arrive at the final result

$$\int_0^\infty dr\, r^2 j_l(\kappa_1 r) j_l(\kappa_2 r) = \frac{\pi}{2\kappa_1\kappa_2} \left[\delta(\kappa_1 - \kappa_2) - (-)^l \delta(\kappa_1 + \kappa_2) \right] , \tag{3.3.46}$$

which, for κ_1 and κ_2 greater than zero, indeed makes the orthonormality of the Riccati-Bessel functions explicit.

3.4 Integral Representation of Hankel Functions

On the following pages we derive an integral representation of the Hankel envelope function. It provides considerable insight into the connection of this function to other basic mathematical functions and is especially useful for calculating Bloch sums of envelope functions with the help of the Ewald method as will be done in Sect. 3.5.

We start from the identity $[1, (7.4.1)]$

$$\int_0^\infty e^{-t^2} dt = \frac{\sqrt{\pi}}{2} , \tag{3.4.1}$$

and define a new integration variable by

$$t = r\xi + i\frac{\kappa}{2\xi}, \text{ i.e. } \frac{dt}{d\xi} = r - i\frac{\kappa}{2\xi^2} , \tag{3.4.2}$$

where κ may be any complex quantity and r is assumed to be real. For the integration limits coming with the integral (3.4.1) we note

$$t = 0 \quad\Longleftrightarrow\quad \xi_0^2 = -i\frac{\kappa}{2r} , \tag{3.4.3}$$

$$t \to \infty \quad\Longleftrightarrow\quad \xi \to \infty . \tag{3.4.4}$$

With this substitution we obtain for the integral (3.4.1) the expression

$$\int_{\xi_0}^\infty d\xi \left(r - i\frac{\kappa}{2\xi^2} \right) e^{-r^2\xi^2 + \frac{\kappa^2}{4\xi^2} - i\kappa r} = \frac{\sqrt{\pi}}{2} , \tag{3.4.5}$$

and, using (3.4.3), from this the intermediate result

$$\frac{e^{i\kappa r}}{r} = \frac{2}{\sqrt{\pi}} \int_{\xi_0}^{\infty} d\xi \left(1 + \frac{\xi_0^2}{\xi^2}\right) e^{-r^2\xi^2 + \frac{\kappa^2}{4\xi^2}}$$

$$= \frac{2}{\sqrt{\pi}} \int_{\xi_0}^{\infty} d\xi\, e^{-r^2\xi^2 + \frac{\kappa^2}{4\xi^2}} + \frac{2}{\sqrt{\pi}} \int_{\xi_0}^{\infty} d\xi\, \frac{\xi_0^2}{\xi^2} e^{-r^2\xi^2 + \frac{\kappa^2}{4\xi^2}}. \qquad (3.4.6)$$

For the second integral on the right-hand side we introduce another change of variables by

$$\bar{\xi} = \frac{\xi_0^2}{\xi}, \quad \text{i.e.} \qquad \frac{d\bar{\xi}}{d\xi} = -\frac{\xi_0^2}{\xi^2}, \qquad (3.4.7)$$

and note

$$\int_{\xi_0}^{\infty} d\xi\, \frac{\xi_0^2}{\xi^2} e^{-r^2\xi^2 + \frac{\kappa^2}{4\xi^2}} = \int_0^{\xi_0} d\bar{\xi}\, e^{-r^2 \frac{\xi_0^4}{\bar{\xi}^2} + \frac{\kappa^2 \bar{\xi}^2}{4\xi_0^4}}$$

$$= \int_0^{\xi_0} d\bar{\xi}\, e^{+\frac{\kappa^2}{4\bar{\xi}^2} - r^2\bar{\xi}^2}. \qquad (3.4.8)$$

Finally, combining (3.4.6) and (3.4.8) we obtain the result

$$\frac{e^{i\kappa r}}{r} = \frac{2}{\sqrt{\pi}} \int_0^{\infty} d\xi\, e^{-r^2\xi^2 + \frac{\kappa^2}{4\xi^2}}. \qquad (3.4.9)$$

Still, we have to specify the contour, along which the integration has to be performed. Special care is needed at the end points $\xi \to 0$ and $\xi \to \infty$. In order to prevent the integrand from diverging we impose the following conditions on the complex integration variable ξ (note that $\Im r = 0$)

- For $|\xi|$ large we require

$$\Re\xi^2 > 0 \qquad \Longleftrightarrow \qquad |2\arg\xi| < \frac{\pi}{2}, \qquad (3.4.10)$$

and the integration variable ξ must be within the shaded region of Fig. 3.1. In particular, for large enough $|\xi|$ the integration can always be performed along the real axis.

- For $|\xi|$ small we require

$$\Re\frac{\kappa^2}{\xi^2} < 0 \qquad \Longleftrightarrow \qquad |2\arg\kappa - 2\arg\xi| > \frac{\pi}{2}$$

$$\Longleftrightarrow \qquad |2\arg\kappa - 2\arg\xi - \pi| < \frac{\pi}{2} \qquad (3.4.11)$$

and the integration variable ξ must be within the shaded region of Fig. 3.2. In particular, if $\Im\kappa > \Re\kappa$ the integration can be throughout performed along the real axis.

Following convention [13, p. 179] we next introduce a parameter ω such that

$$2\arg\xi = \begin{cases} \arg\kappa - \omega & \text{for } \xi \text{ large} \\ \arg\kappa + \omega - \pi & \text{for } \xi \text{ small} \end{cases}. \qquad (3.4.12)$$

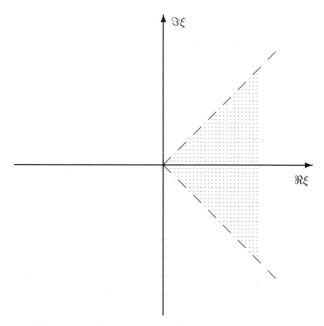

Fig. 3.1. Allowed ξ values for large $|\xi|$ according to condition (3.4.10)

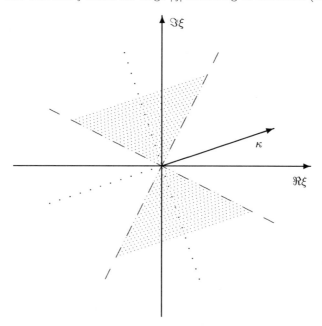

Fig. 3.2. Allowed ξ values for small $|\xi|$ according to condition (3.4.11)

Combining this with the conditions (3.4.10) and (3.4.11) we arrive at the single condition

$$|\arg \kappa - \omega| < \frac{\pi}{2} \quad \Longleftrightarrow \quad \arg \kappa - \frac{\pi}{2} < \omega < \arg \kappa + \frac{\pi}{2} . \qquad (3.4.13)$$

For a particular value of the parameter ω the contour of the integral (3.4.9) is shown in Fig. 3.3. Of course, the most simple choice consists of setting $\omega = \arg \kappa$, in which case we have

$$\arg \xi = \begin{cases} 0 & \text{for } \xi \text{ large} \\ \arg \kappa - \frac{\pi}{2} & \text{for } \xi \text{ small} \end{cases} . \qquad (3.4.14)$$

Hence, for purely imaginary κ the contour of the integral (3.4.9) coincides with the positive real axis. In contrast, for real κ, i.e. $\arg \kappa = 0$, we have to require $\omega < \frac{\pi}{2}$ and thus the contour can never be a straight line. This can also be read off from the condition (3.4.11), which in this case reduces to $|\arg \xi| > \frac{\pi}{4}$. As a consequence, we must not deform the contour to lie completely on the real axis, since this would lead to a singularity in the integrand for small $|\xi|$, which can be traced back to the fact that the second part of the exponential in the integral (3.4.9) diverges unless its argument has a finite phase. The latter may be supplied by κ or ξ or both.

With the previous considerations on the contour of the integral (3.4.9) in mind we may rewrite it in the more compact form

$$\frac{e^{i\kappa r}}{r} = \frac{2}{\sqrt{\pi}} \int_{0e^{\frac{i}{2}(\arg \kappa - \pi + \omega)}}^{\infty e^{\frac{i}{2}(\arg \kappa - \omega)}} d\xi \, e^{-r^2\xi^2 + \frac{\kappa^2}{4\xi^2}} \qquad \text{for } |\arg \kappa - \omega| < \frac{\pi}{2} , \qquad (3.4.15)$$

which for $\omega = \frac{\pi}{2}$ reduces to the formula given by Ham and Segall [7, (7.2), Fig. 1].

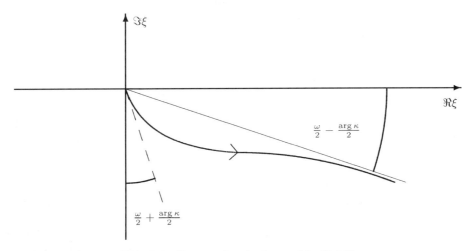

Fig. 3.3. Contour for the integral in (3.4.9)

After these preparations we may now easily formulate the desired integral representation of the Hankel envelope functions. We note the expression

$$\bar{h}_l^{(1)}(\kappa r) = \frac{2}{\sqrt{\pi}} 2^l r^l \int_{0e^{\frac{i}{2}(\arg \kappa - \pi + \omega)}}^{\infty e^{\frac{i}{2}(\arg \kappa - \omega)}} d\xi \, \xi^{2l} e^{(-r^2 \xi^2 + \frac{\kappa^2}{4\xi^2})} \qquad \text{for } |\arg \kappa - \omega| < \frac{\pi}{2} \, ,$$

$$(3.4.16)$$

which is proven by complete induction:
1. $l = 0$

Combining (3.1.28) and (3.4.15) we arrive immediately at (3.4.16).
2. $l \rightarrow l + 1$

Assuming (3.4.16) to be valid and using (3.2.6) we calculate

$$\bar{h}_{l+1}^{(1)}(\kappa r) = \frac{l}{r} \bar{h}_l^{(1)}(\kappa r) - \frac{\partial}{\partial r} \bar{h}_l^{(1)}(\kappa r)$$

$$= \frac{2}{\sqrt{\pi}} 2^l l r^{l-1} \int_{0e^{\frac{i}{2}(\arg \kappa - \pi + \omega)}}^{\infty e^{\frac{i}{2}(\arg \kappa - \omega)}} d\xi \, \xi^{2l} e^{(-r^2 \xi^2 + \frac{\kappa^2}{4\xi^2})}$$

$$- \frac{2}{\sqrt{\pi}} 2^l l r^{l-1} \int_{0e^{\frac{i}{2}(\arg \kappa - \pi + \omega)}}^{\infty e^{\frac{i}{2}(\arg \kappa - \omega)}} d\xi \, \xi^{2l} e^{(-r^2 \xi^2 + \frac{\kappa^2}{4\xi^2})}$$

$$+ \frac{2}{\sqrt{\pi}} 2^{l+1} r^{l+1} \int_{0e^{\frac{i}{2}(\arg \kappa - \pi + \omega)}}^{\infty e^{\frac{i}{2}(\arg \kappa - \omega)}} d\xi \, \xi^{2l+2} e^{(-r^2 \xi^2 + \frac{\kappa^2}{4\xi^2})}$$

$$= \frac{2}{\sqrt{\pi}} 2^{l+1} r^{l+1} \int_{0e^{\frac{i}{2}(\arg \kappa - \pi + \omega)}}^{\infty e^{\frac{i}{2}(\arg \kappa - \omega)}} d\xi \, \xi^{2l+2} e^{(-r^2 \xi^2 + \frac{\kappa^2}{4\xi^2})} \qquad \text{[q.e.d.]} \, .$$

$$(3.4.17)$$

We have thus proven the identity (3.4.16) for arbitrary l and arrived at a general integral representation of the barred Hankel functions.

3.5 Bloch Sums of Envelope Functions

The present section is devoted to the evaluation of Bloch sums of the envelope functions defined in Sect. 3.1. Probably the best known method to perform lattice sums is the method introduced for the case of general lattices by Ewald [4]. Its success is based on both its straightforward formalism, which is independent of the actual geometry of the problem at hand, as well as its wide applicability [11]. In its original version the Ewald method was intended to deal with lattice sums of Coulomb potentials arising from a lattice of monopoles, i.e. Bloch sums of Hankel functions of zero energy and zero angular momentum at the origin of reciprocal space, $\mathbf{k} = \mathbf{0}$. Soon after it was extended to arbitrary power laws by Misra as well as by Born and Bradburn [3, 10] and since then it was adapted to different applications [12]. The only application to the Bloch sum of Hankel functions of arbitrary argument and angular momentum was presented by Ham and Segall [7].

Unfortunately, their derivation was based on the case $l = 0$ and the results were projected in a somewhat elaborate manner to arbitrary l. In view of this fact we will in the present section rederive the Ewald method for Hankel functions without any restriction.

As we will see lateron, calculating the Bloch sum of the envelope functions essentially means to calculate the corresponding sum of the Hankel envelope functions. The latter are given by (3.1.5) or else (3.1.17) and may be written in the form

$$H_{L\kappa}(\tau - \mathbf{R}_\mu) = \bar{h}_l^{(1)}(\kappa|\tau - \mathbf{R}_\mu|)Y_L(\widehat{\tau - \mathbf{R}_\mu}) . \qquad (3.5.1)$$

Here we have changed the notation in order to stress the fact that for crystals any vector may be split uniquely into a lattice vector $\mathbf{R}_\mu$ and a vector τ, which is restricted to the unit cell. With this representation at hand we define the corresponding Bloch sum by

$$D_{L\kappa}(\tau, \mathbf{k}) := \sum_\mu (1 - \delta_{\mu 0}\delta(\tau))e^{i\mathbf{k}\mathbf{R}_\mu}H_{L\kappa}(\tau - \mathbf{R}_\mu) , \qquad (3.5.2)$$

which, for $\tau = \mathbf{r}_i$ reduces to (2.1.27). The index 0 denotes the unit cell with $\mathbf{R}_0 = \mathbf{0}$ and the symbol D has been chosen along with the convention introduced by Ham and Segall [7]. Note that due to the singularity of the Hankel function at the origin we had to restrict the Bloch sum to only those functions, which have a nonvanishing argument. However, it is one of the great advantages of the Ewald method that it allows to evaluate the full lattice sum for $\tau \neq \mathbf{0}$ first and after this to remove the contribution from the function centered at $|\tau - \mathbf{R}_0|$. At the very end the limit $\tau \to \mathbf{0}$ can be performed.

The general philosophy of the Ewald method is essentially based on two ideas: First, it takes advantage of the fact that by virtue of the Poisson transform (see App. B.4) the real-space lattice sum (3.5.2) can likewise be written as a reciprocal-space lattice sum. It thus allows for two alternative representations. Second, if by any means the function to be lattice summed can be split into two parts containing the short and long range contributions, respectively, these may be summed up separately in real and reciprocal space. This way a very rapid and unconditioned convergence of both sums can be achieved. Once these essentials have been accepted the only problem left is to find a proper representation of the function of interest in order to utilize the second idea.

Fortunately, for the envelope function based on spherical Hankel functions as given by (3.5.1) such a description exists in form of the integral representation (3.4.5) noted at the end of the preceding section. Inserting it into (3.5.1) we get for the Bloch sum (3.5.2) the expression

$$D_{L\kappa}(\tau, \mathbf{k}) = \frac{2}{\sqrt{\pi}}2^l \int_{0e^{\frac{i}{2}(\arg \kappa - \pi + \omega)}}^{\infty e^{\frac{i}{2}(\arg \kappa - \omega)}} \xi^{2l} \left[\sum_\mu (1 - \delta_{\mu 0}\delta(\tau))e^{i\mathbf{k}\mathbf{R}_\mu}|\tau_\mu|^l \right.$$
$$\left. e^{-(\tau_\mu)^2\xi^2}Y_L(\hat{\tau}_\mu) \right] e^{\frac{\kappa^2}{4\xi^2}} d\xi$$

$$\text{for } |\omega - \arg \kappa| < \frac{\pi}{2} \, , \tag{3.5.3}$$

where we have used the abbreviation

$$\boldsymbol{\tau}_\mu := \boldsymbol{\tau} - \mathbf{R}_\mu \, . \tag{3.5.4}$$

Splitting the first exponent in (3.5.3) we obtain

$$D_{L\kappa}(\boldsymbol{\tau}, \mathbf{k}) = \frac{2}{\sqrt{\pi}} 2^l e^{i\mathbf{k}\boldsymbol{\tau}} \int_{0e^{\frac{i}{2}(\arg \kappa - \pi + \omega)}}^{\infty e^{\frac{i}{2}(\arg \kappa - \omega)}} \xi^{2l} \left[\sum_\mu (1 - \delta_{\mu 0}\delta(\boldsymbol{\tau}))e^{-i\mathbf{k}\boldsymbol{\tau}_\mu} |\boldsymbol{\tau}_\mu|^l \right.$$

$$\left. e^{-\boldsymbol{\tau}_\mu^2 \xi^2} Y_L(\hat{\boldsymbol{\tau}}_\mu) \right] e^{\frac{\kappa^2}{4\xi^2}} d\xi$$

$$\text{for } |\omega - \arg \kappa| < \frac{\pi}{2} \, . \tag{3.5.5}$$

At this stage we follow the above advice to handle the case $\boldsymbol{\tau} = \mathbf{0}$ at the very end. Hence, we assume for the time being $\boldsymbol{\tau} \neq \mathbf{0}$. As a consequence, we can immediately benefit from the results of App. B.4. In particular, we obtain from (B.4.8) the Poisson transform of the lattice sum in brackets in (3.5.5)

$$\sum_\mu e^{-i\mathbf{k}\boldsymbol{\tau}_\mu} |\boldsymbol{\tau}_\mu|^l e^{-\boldsymbol{\tau}_\mu^2 \xi^2} Y_L(\hat{\boldsymbol{\tau}}_\mu)$$

$$= \frac{\pi^{\frac{3}{2}}}{\Omega_c} (-i)^l 2^{-l} \xi^{-(2l+3)} \sum_n e^{i\mathbf{K}_n \boldsymbol{\tau}} |\mathbf{K}_n + \mathbf{k}|^l Y_L(\widehat{\mathbf{K}_n + \mathbf{k}}) e^{-\frac{(\mathbf{K}_n + \mathbf{k})^2}{4\xi^2}} \, . \tag{3.5.6}$$

Inserting this into (3.5.5) we get immediately

$$D_{L\kappa}(\boldsymbol{\tau}, \mathbf{k}) = \frac{2\pi}{\Omega_c} (-i)^l \sum_n e^{i(\mathbf{K}_n + \mathbf{k})\boldsymbol{\tau}} |\mathbf{K}_n + \mathbf{k}|^l Y_L(\widehat{\mathbf{K}_n + \mathbf{k}})$$

$$\int_{0e^{\frac{i}{2}(\arg \kappa - \pi + \omega)}}^{\infty e^{\frac{i}{2}(\arg \kappa - \omega)}} \xi^{-3} e^{\frac{\kappa^2 - (\mathbf{K}_n + \mathbf{k})^2}{4\xi^2}} d\xi$$

$$\text{for } \boldsymbol{\tau} \neq \mathbf{0} \text{ and } |\omega - \arg \kappa| < \frac{\pi}{2} \, , \tag{3.5.7}$$

which is just the alternative representation of the real-space lattice sum (3.5.2)/(3.5.5) by means of a reciprocal-space lattice sum, which we were looking for.

The only problem left now is to separate the short and long range contributions. However, remembering the fact that the integral representation (3.4.5) of the single Hankel function mainly consists of an integral over Gaussians each with a different width, this separation may be easily achieved by splitting the integral into two, containing the short and long range Gaussians, respectively. Since the integration and the Poisson transform are performed independently we may then use the real-space lattice sum in the integral containing the short range part and the reciprocal-space lattice sum for the integral with the long range Gaussians. Denoting the above mentioned break point as $\frac{1}{2}\eta^{\frac{1}{2}}$, where η is the Ewald parameter, we combine (3.5.5) and (3.5.7) to the following result

$$D_{L\kappa}(\boldsymbol{\tau},\mathbf{k}) = D_{L\kappa}^{(1)}(\boldsymbol{\tau},\mathbf{k}) + D_{L\kappa}^{(2)}(\boldsymbol{\tau},\mathbf{k}) , \qquad (3.5.8)$$

with

$$D_{L\kappa}^{(1)}(\boldsymbol{\tau},\mathbf{k}) = \frac{2\pi}{\Omega_c}(-i)^l \sum_n e^{i(\mathbf{K}_n+\mathbf{k})\boldsymbol{\tau}}|\mathbf{K}_n+\mathbf{k}|^l Y_L(\widehat{\mathbf{K}_n+\mathbf{k}})$$

$$\int_{0e^{\frac{i}{2}(\arg\kappa-\pi+\omega)}}^{\frac{1}{2}\eta^{1/2}} \xi^{-3} e^{\frac{\kappa^2-(\mathbf{K}_n+\mathbf{k})^2}{4\xi^2}} \, d\xi$$

$$\text{for } |\omega - \arg\kappa| < \frac{\pi}{2} , \qquad (3.5.9)$$

and

$$D_{L\kappa}^{(2)}(\boldsymbol{\tau},\mathbf{k}) = \frac{2}{\sqrt{\pi}} 2^l e^{i\mathbf{k}\boldsymbol{\tau}} \sum_\mu e^{-i\mathbf{k}\boldsymbol{\tau}_\mu}|\boldsymbol{\tau}_\mu|^l Y_L(\hat{\boldsymbol{\tau}}_\mu)$$

$$\int_{\frac{1}{2}\eta^{1/2}}^{\infty e^{\frac{i}{2}(\arg\kappa-\omega)}} \xi^{2l} e^{-\boldsymbol{\tau}_\mu^2\xi^2+\frac{\kappa^2}{4\xi^2}} \, d\xi$$

$$\text{for } |\omega - \arg\kappa| < \frac{\pi}{2} , \qquad (3.5.10)$$

for $\boldsymbol{\tau} \neq \mathbf{0}$.

The integral appearing in (3.5.9) can be evaluated analytically as long as the integrand vanishes for $\xi \to 0$. However, as already discussed in Sect. 3.4, this is guaranteed by a proper choice of the phase ω appearing in the integral representations (3.4.16). Hence, we get for the integral

$$\int_{0e^{\frac{i}{2}(\arg\kappa-\pi+\omega)}}^{\frac{1}{2}\eta^{1/2}} \xi^{-3} e^{\frac{\kappa^2-(\mathbf{K}_n+\mathbf{k})^2}{4\xi^2}} \, d\xi = -2\frac{e^{\frac{\kappa^2-(\mathbf{K}_n+\mathbf{k})^2}{\eta}}}{\kappa^2-(\mathbf{K}_n+\mathbf{k})^2} , \qquad (3.5.11)$$

where the contribution from the lower boundary vanishes due to the special choice of the phase ω as outlined in Sect. 3.4. The case $\kappa^2 = (\mathbf{K}_n+\mathbf{k})^2$, which goes under the name free-electron singularity, deserves special attention. We note

$$\int_{0e^{\frac{i}{2}(\arg\kappa-\pi+\omega)}}^{\frac{1}{2}\eta^{1/2}} \xi^{-3} \, d\xi = -\frac{1}{2\xi^2}\Bigg|_{0e^{\frac{i}{2}(\arg\kappa-\pi+\omega)}}^{\frac{1}{2}\eta^{1/2}} = -2\frac{1}{\eta} , \qquad (3.5.12)$$

where again the contribution from the lower boundary vanishes. Alternatively, this result can be derived from a more explicit writing of (3.5.11) as

$$\int_{0e^{\frac{i}{2}(\arg\kappa-\pi+\omega)}}^{\frac{1}{2}\eta^{1/2}} \xi^{-3} e^{\frac{\kappa^2-(\mathbf{K}_n+\mathbf{k})^2}{4\xi^2}} \, d\xi = -2\frac{e^{\frac{\kappa^2-(\mathbf{K}_n+\mathbf{k})^2}{4\xi^2}}}{\kappa^2-(\mathbf{K}_n+\mathbf{k})^2}\Bigg|_{0e^{\frac{i}{2}(\arg\kappa-\pi+\omega)}}^{\frac{1}{2}\eta^{1/2}} , \qquad (3.5.13)$$

and using l'Hospitals rule for $\kappa^2 \to (\mathbf{K}_n + \mathbf{k})^2$.

For later use we calculate, in addition to the integral (3.5.11), its derivative with respect to κ^2 as

$$
\frac{\partial}{\partial \kappa^2} \int_{0e^{\frac{i}{2}(\arg \kappa - \pi + \omega)}}^{\frac{1}{2}\eta^{1/2}} \xi^{-3} e^{\frac{\kappa^2 - (\mathbf{K}_n + \mathbf{k})^2}{4\xi^2}} d\xi
$$

$$
= -2 \frac{\partial}{\partial \kappa^2} \frac{e^{\frac{\kappa^2 - (\mathbf{K}_n + \mathbf{k})^2}{\eta}}}{\kappa^2 - (\mathbf{K}_n + \mathbf{k})^2}
$$

$$
= -2 \frac{e^{\frac{\kappa^2 - (\mathbf{K}_n + \mathbf{k})^2}{\eta}}}{\kappa^2 - (\mathbf{K}_n + \mathbf{k})^2} \left[\frac{1}{\eta} - \frac{1}{\kappa^2 - (\mathbf{K}_n + \mathbf{k})^2} \right] . \tag{3.5.14}
$$

In the limit of $\kappa^2 \to (\mathbf{K}_n + \mathbf{k})^2$ we obtain

$$
\lim_{\kappa^2 \to (\mathbf{K}_n + \mathbf{k})^2} \frac{\partial}{\partial \kappa^2} \int_{0e^{\frac{i}{2}(\arg \kappa - \pi + \omega)}}^{\frac{1}{2}\eta^{1/2}} \xi^{-3} e^{\frac{\kappa^2 - (\mathbf{K}_n + \mathbf{k})^2}{4\xi^2}} d\xi
$$

$$
= \frac{1}{4} \int_{0e^{\frac{i}{2}(\arg \kappa - \pi + \omega)}}^{\frac{1}{2}\eta^{1/2}} \xi^{-5} d\xi = -\frac{1}{16\xi^4} \Big|_{0e^{\frac{i}{2}(\arg \kappa - \pi + \omega)}}^{\frac{1}{2}\eta^{1/2}} = -\frac{1}{\eta^2} , \tag{3.5.15}
$$

where, as before, the contribution from the lower boundary vanishes and which likewise can be derived from a more explicit treatment of the integral (3.5.14) and subsequent application of l'Hospitals rule for $\kappa^2 \to (\mathbf{K}_n + \mathbf{k})^2$.

As concerns evaluation of the integral in (3.5.10) we had already stated in Sect. 3.4 that the phase ω in the integral representation (3.4.16) of the Hankel envelope function, which fixes the contour of the integral, may be set to $\arg \kappa$. This forces the contour to coincide with the real axis for large enough values of ξ. Note that this does *not* conflict with the setting used to calculate the integral (3.5.11). Moreover we may still deform the contour to lie on the real axis at least for all $\xi \geq \frac{1}{2}\eta^{1/2}$ and thus allow for a real Ewald parameter. This simplifies calculation of the Ewald integral

$$
\int_{\frac{1}{2}\eta^{1/2}}^{\infty} \xi^{2l} e^{-\tau_\mu^2 \xi^2 + \frac{\kappa^2}{4\xi^2}} d\xi ,
$$

substantially and makes numerical calculations accessible. Their detailed description is the subject of App. B.5.

A rather special formulation of the Bloch-summed Hankel functions arises from combining (3.5.7) and (3.5.11) and taking the limit $\eta \to \infty$. This results in

$$
D_{L\kappa}(\boldsymbol{\tau}, \mathbf{k}) = -\frac{4\pi}{\Omega_c} (-i)^l \sum_n e^{i(\mathbf{K}_n + \mathbf{k})\boldsymbol{\tau}} |\mathbf{K}_n + \mathbf{k}|^l Y_L(\widehat{\mathbf{K}_n + \mathbf{k}}) \frac{1}{\kappa^2 - (\mathbf{K}_n + \mathbf{k})^2} , \tag{3.5.16}
$$

which is just the Fourier transform of the Hankel envelope function.

Having already done most of the work we are left with the final task to remove from the Bloch sum (3.5.2) that single spherical Hankel function, which causes the divergence in the limit $\boldsymbol{\tau} \to \mathbf{0}$. For that purpose we rewrite (3.5.5) as follows

$$D_{L\kappa}(\boldsymbol{\tau}, \mathbf{k}) = \frac{2}{\sqrt{\pi}} 2^l e^{i\mathbf{k}\boldsymbol{\tau}} \int_{0 e^{\frac{i}{2}(\arg \kappa - \pi + \omega)}}^{\infty e^{\frac{i}{2}(\arg \kappa - \omega)}} \xi^{2l} \left[\sum_\mu e^{-i\mathbf{k}\boldsymbol{\tau}_\mu} |\boldsymbol{\tau}_\mu|^l e^{-\tau_\mu^2 \xi^2} Y_L(\widehat{\boldsymbol{\tau}_\mu}) \right] e^{\frac{\kappa^2}{4\xi^2}} d\xi$$

$$- \frac{1}{\pi} \delta(\boldsymbol{\tau}) \delta_{l0} \int_{0 e^{\frac{i}{2}(\arg \kappa - \pi + \omega)}}^{\infty e^{\frac{i}{2}(\arg \kappa - \omega)}} e^{\frac{\kappa^2}{4\xi^2}} d\xi$$

$$\text{for } |\omega - \arg \kappa| < \frac{\pi}{2} . \tag{3.5.17}$$

where the first term is identical to the formulation (3.5.5) for $\boldsymbol{\tau} \neq \mathbf{0}$ and the second is just the term to be removed. Here we have already used the fact that in the integral representation (3.4.10) for $\boldsymbol{\tau} \to \mathbf{0}$ only the $l = 0$-contribution survives. Next we use again the break point $\frac{1}{2}\eta^{\frac{1}{2}}$ and, in complete analogy to the above treatment of the whole sum, split the integral in the second term of (3.5.17) into two. This corresponds to also splitting the single Hankel function into short and long range contributions. Subtraction of the short range part thus consists merely of removing the respective term from the real-space sum (3.5.10). In contrast, the long range part has to be explicitly subtracted.

Combining (3.5.8) to (3.5.11) and (3.5.17) we arrive at the final result

$$D_{L\kappa}(\boldsymbol{\tau}, \mathbf{k}) = D_{L\kappa}^{(1)}(\boldsymbol{\tau}, \mathbf{k}) + D_{L\kappa}^{(2)}(\boldsymbol{\tau}, \mathbf{k}) + D_{L\kappa}^{(3)}(\boldsymbol{\tau}) , \tag{3.5.18}$$

where

$$D_{L\kappa}^{(1)}(\boldsymbol{\tau}, \mathbf{k}) = -\frac{4\pi}{\Omega_c} (-i)^l \sum_n e^{i(\mathbf{K}_n + \mathbf{k})\boldsymbol{\tau}} |\mathbf{K}_n + \mathbf{k}|^l Y_L(\widehat{\mathbf{K}_n + \mathbf{k}})$$

$$\frac{e^{\frac{\kappa^2 - (\mathbf{K}_n + \mathbf{k})^2}{\eta}}}{\kappa^2 - (\mathbf{K}_n + \mathbf{k})^2} , \tag{3.5.19}$$

$$D_{L\kappa}^{(2)}(\boldsymbol{\tau}, \mathbf{k}) = \frac{2}{\sqrt{\pi}} 2^l e^{i\mathbf{k}\boldsymbol{\tau}} \sum_\mu (1 - \delta_{\mu 0} \delta(\boldsymbol{\tau})) e^{-i\mathbf{k}\boldsymbol{\tau}_\mu} |\boldsymbol{\tau}_\mu|^l Y_L(\widehat{\boldsymbol{\tau}_\mu})$$

$$\int_{\frac{1}{2}\eta^{1/2}}^{\infty} \xi^{2l} e^{-\tau_\mu^2 \xi^2 + \frac{\kappa^2}{4\xi^2}} d\xi , \tag{3.5.20}$$

and

$$D_{L\kappa}^{(3)}(\boldsymbol{\tau}) = -\frac{1}{\pi} \delta(\boldsymbol{\tau}) \delta_{l0} \int_{0 e^{\frac{i}{2}(\arg \kappa - \pi + \omega)}}^{\frac{1}{2}\eta^{1/2}} e^{\frac{\kappa^2}{4\xi^2}} d\xi . \tag{3.5.21}$$

The last term represents the abovementioned long range part of the single Hankel envelope function for $\boldsymbol{\tau} = \mathbf{0}$. Note that, according to (3.5.11) and (3.5.12), the term in the second line of (3.5.19) has to be replaced by $1/\eta$ for $\kappa^2 \to (\mathbf{K}_n + \mathbf{k})^2$.

The detailed calculation of $D_{L\kappa}^{(3)}(\boldsymbol{\tau})$ is left to the App. B.7; here we just note the result (B.7.9)

$$D_{L\kappa}^{(3)}(\boldsymbol{\tau}) = \delta(\boldsymbol{\tau})\delta_{l0}\left[\frac{i\kappa}{4\sqrt{\pi}}\left(\text{erfc}(\frac{i\kappa}{\eta^{1/2}}) - \text{erfc}(-\frac{i\kappa}{\eta^{1/2}})\right) - \frac{\eta^{1/2}}{2\pi}e^{\kappa^2/\eta} - \frac{i\kappa}{2\sqrt{\pi}}\right]$$

$$= -\delta(\boldsymbol{\tau})\delta_{l0}\left[\frac{i\kappa}{2\sqrt{\pi}}\text{erf}(\frac{i\kappa}{\eta^{1/2}}) + \frac{\eta^{1/2}}{2\pi}e^{\kappa^2/\eta} + \frac{i\kappa}{2\sqrt{\pi}}\right]$$

$$= -\delta(\boldsymbol{\tau})\delta_{l0}\left[\frac{\kappa}{2\sqrt{\pi}}e^{\frac{\kappa^2}{\eta}}\Im w(\frac{-\kappa}{\eta^{1/2}}) + \frac{\eta^{1/2}}{2\pi}e^{\kappa^2/\eta} + \frac{i\kappa}{2\sqrt{\pi}}\right] \qquad \text{for} \kappa^2 > 0 \,.$$

$$(3.5.22)$$

The Ewald method as presented above does not make any principal assumptions about the value of the Ewald parameter η. However, the choice of the Ewald parameter crucially affects the convergence of the lattice sums contained in the (3.5.19) and (3.5.20) and may thus make numerical calculations very time consuming if not impossible. For this reason, a recipe to guess a good Ewald parameter is desired. The following has shown to be the best in many cases.

As mentioned above, the integral representation (3.4.5) of the Hankel based envelope function may be viewed as an integral over Gaussians. Thus, to introduce a break point, which splits the integral, means to define a critical width b_c for the Gaussians $e^{-\frac{r^2}{2b^2}}$, beyond which they are summed in reciprocal space. Experience has shown that the best choice consists of setting

$$b_c = \frac{R_{WS}}{\sqrt{3.25}} \,, \qquad (3.5.23)$$

where R_{WS} is the Wigner-Seitz radius of the crystal. From the above definition of the Ewald parameter we read off

$$\eta = \frac{2}{b_c^2} = \frac{6.5}{R_{WS}^2} \,, \qquad (3.5.24)$$

or

$$\eta^{1/2} = \frac{\sqrt{6.5}}{R_{WS}} = \sqrt{6.5}\left(\frac{4\pi}{3}\right)^{\frac{1}{3}}\Omega_c^{-\frac{1}{3}} \approx 4.10979\,\Omega_c^{-\frac{1}{3}} \,. \qquad (3.5.25)$$

Division by 2π finally leads to

$$\frac{\eta^{1/2}}{2\pi} \approx 0.65409\Omega_c^{-\frac{1}{3}} \,. \qquad (3.5.26)$$

Still there might be cases where finetuning of these numbers is necessary.

Having calculated the Bloch sums of the envelope function based on spherical Hankel functions it is a comparatively easy task to evaluate the corresponding Bloch sums of the envelope function arising from spherical Bessel functions. For these we define in complete analogy with (3.5.2)

$$J_{L\kappa}(\boldsymbol{\tau}, \mathbf{k}) := \sum_{\mu} e^{i\mathbf{k}\mathbf{R}_\mu} J_{L\kappa}(\boldsymbol{\tau} - \mathbf{R}_\mu) \,. \qquad (3.5.27)$$

Note that due to the overall regular behaviour of the spherical Bessel function this sum needs not be restricted. Using (3.1.16) we may immediately write down the following result

$$
\begin{aligned}
J_{L\kappa}(\boldsymbol{\tau}, \mathbf{k}) &= \sum_{\mu} e^{i\mathbf{k}\mathbf{R}_{\mu}} \bar{\jmath}_l(\kappa|\boldsymbol{\tau} - \mathbf{R}_{\mu}|) Y_L(\widehat{\boldsymbol{\tau} - \mathbf{R}_{\mu}}) \\
&= \kappa^{-l} e^{i\mathbf{k}\boldsymbol{\tau}} \sum_{\mu} e^{-i\mathbf{k}\boldsymbol{\tau}_{\mu}} j_l(\kappa|\boldsymbol{\tau}_{\mu}|) Y_L(\hat{\boldsymbol{\tau}}_{\mu}) \\
&= \frac{2\pi^2}{\Omega_c} \kappa^{-l-2}(-i)^l \sum_{n} e^{i(\mathbf{K}_n+\mathbf{k})\boldsymbol{\tau}} Y_L(\widehat{\mathbf{K}_n + \mathbf{k}}) \\
&\quad \left[\delta(\kappa - |\mathbf{K}_n + \mathbf{k}|) - (-)^l \delta(\kappa + |\mathbf{K}_n + \mathbf{k}|) \right] .
\end{aligned} \tag{3.5.28}
$$

Here we used the identity (B.4.12) to transform the real-space lattice sum into a reciprocal lattice sum. Note that the δ-distributions entering (3.5.28) correspond to the free-electron singularities.

Combining the results for the Bloch sum of the envelope functions based on the spherical Hankel and Bessel function, i.e. (3.5.18) to (3.5.21) and (3.5.28) we are able to write down the corresponding Bloch sum of the envelope function (3.1.15) based on the spherical Neumann function. Using the definitions (3.5.2) and (3.5.27) as well as (3.1.17) and (3.5.28) we note

$$
\begin{aligned}
K_{L\kappa}(\boldsymbol{\tau}, \mathbf{k}) &:= \sum_{\mu} (1 - \delta_{\mu 0} \delta(\boldsymbol{\tau})) e^{i\mathbf{k}\mathbf{R}_{\mu}} N_{L\kappa}(\boldsymbol{\tau} - \mathbf{R}_{\mu}) \\
&= \sum_{\mu} (1 - \delta_{\mu 0} \delta(\boldsymbol{\tau})) e^{i\mathbf{k}\mathbf{R}_{\mu}} H_{L\kappa}(\boldsymbol{\tau} - \mathbf{R}_{\mu}) \\
&\quad -i\kappa^{2l+1} \sum_{\mu} e^{i\mathbf{k}\mathbf{R}_{\mu}} J_{L\kappa}(\boldsymbol{\tau} - \mathbf{R}_{\mu}) + i\kappa^{2l+1} \delta(\boldsymbol{\tau}) J_{L\kappa}(\boldsymbol{\tau}) \\
&= D_{L\kappa}(\boldsymbol{\tau}, \mathbf{k}) + \delta(\boldsymbol{\tau}) \delta_{l0} \frac{i\kappa}{2\sqrt{\pi}} \\
&\quad - \frac{2\pi^2}{\Omega_c} (-i\kappa)^{l-1} \sum_{n} e^{i(\mathbf{K}_n+\mathbf{k})\boldsymbol{\tau}} Y_L(\widehat{\mathbf{K}_n + \mathbf{k}}) \\
&\quad \left[\delta(\kappa - |\mathbf{K}_n + \mathbf{k}|) - (-)^l \delta(\kappa + |\mathbf{K}_n + \mathbf{k}|) \right] .
\end{aligned} \tag{3.5.29}
$$

Here we applied, in addition, the expansion (3.1.44) for the envelope function $\bar{\jmath}_l(\kappa r)$ for small argument. Finally, we include the constant $i\kappa/2\sqrt{\pi}$ on the right-hand side of (3.5.29) into the $D_{L\kappa}^{(3)}$ term defined in (3.5.21) and obtain

$$
\begin{aligned}
K_{L\kappa}(\boldsymbol{\tau}, \mathbf{k}) &= D_{L\kappa}^{(1)}(\boldsymbol{\tau}, \mathbf{k}) + D_{L\kappa}^{(2)}(\boldsymbol{\tau}, \mathbf{k}) + K_{L\kappa}^{(3)}(\boldsymbol{\tau}) \\
&\quad - \frac{2\pi^2}{\Omega_c} (-i\kappa)^{l-1} \sum_{n} e^{i(\mathbf{K}_n+\mathbf{k})\boldsymbol{\tau}} Y_L(\widehat{\mathbf{K}_n + \mathbf{k}}) \\
&\quad \left[\delta(\kappa - |\mathbf{K}_n + \mathbf{k}|) - (-)^l \delta(\kappa + |\mathbf{K}_n + \mathbf{k}|) \right] ,
\end{aligned} \tag{3.5.30}
$$

where

$$K_{L\kappa}^{(3)}(\boldsymbol{\tau}) := D_{L\kappa}^{(3)}(\boldsymbol{\tau}) + \delta(\boldsymbol{\tau})\delta_{l0}\frac{i\kappa}{2\sqrt{\pi}}$$

$$= -\frac{1}{\pi}\delta(\boldsymbol{\tau})\delta_{l0}\left[\int_{0e^{\frac{i}{2}(\arg\kappa-\pi+\omega)}}^{\frac{1}{2}\eta^{1/2}}e^{\frac{\kappa^2}{4\xi^2}}\,d\xi - \frac{i\kappa\sqrt{\pi}}{2}\right]. \qquad (3.5.31)$$

This last term is readily combined with (3.5.22) to derive an explicit expression for $K_{L\kappa}^{(3)}(\boldsymbol{\tau})$. Note that the last term in (3.5.30) contributes only at the free-electron singularities and may then be combined with the corresponding term in $D_{L\kappa}^{(1)}(\boldsymbol{\tau},\mathbf{k})$.

Finally, we calculate the energy derivatives of the Bloch-summed envelope function, i.e. of $D_{L\kappa}(\boldsymbol{\tau},\mathbf{k})$ and $K_{L\kappa}(\boldsymbol{\tau},\mathbf{k})$. Denoting energy derivatives by a dot we write for the former case

$$\dot{D}_{L\kappa}(\boldsymbol{\tau},\mathbf{k}) = \frac{\partial}{\partial\kappa^2}D_{L\kappa}(\boldsymbol{\tau},\mathbf{k}) = \dot{D}_{L\kappa}^{(1)}(\boldsymbol{\tau},\mathbf{k}) + \dot{D}_{L\kappa}^{(2)}(\boldsymbol{\tau},\mathbf{k}) + \dot{D}_{L\kappa}^{(3)}(\boldsymbol{\tau}), \qquad (3.5.32)$$

where

$$\dot{D}_{L\kappa}^{(1)}(\boldsymbol{\tau},\mathbf{k}) = \frac{\partial}{\partial\kappa^2}D_{L\kappa}^{(1)}(\boldsymbol{\tau},\mathbf{k})$$

$$= -\frac{4\pi}{\Omega_c}(-i)^l\sum_n e^{i(\mathbf{K}_n+\mathbf{k})\boldsymbol{\tau}}|\mathbf{K}_n+\mathbf{k}|^l Y_L(\widehat{\mathbf{K}_n+\mathbf{k}})$$

$$\frac{e^{\frac{\kappa^2-(\mathbf{K}_n+\mathbf{k})^2}{\eta}}}{\kappa^2-(\mathbf{K}_n+\mathbf{k})^2}\left[\frac{1}{\eta} - \frac{1}{\kappa^2-(\mathbf{K}_n+\mathbf{k})^2}\right]$$

$$= \frac{1}{\eta}D_{L\kappa}^{(1)}(\boldsymbol{\tau},\mathbf{k}) + \frac{4\pi}{\Omega_c}(-i)^l\sum_n e^{i(\mathbf{K}_n+\mathbf{k})\boldsymbol{\tau}}|\mathbf{K}_n+\mathbf{k}|^l Y_L(\widehat{\mathbf{K}_n+\mathbf{k}})$$

$$\frac{e^{\frac{\kappa^2-(\mathbf{K}_n+\mathbf{k})^2}{\eta}}}{[\kappa^2-(\mathbf{K}_n+\mathbf{k})^2]^2}, \qquad (3.5.33)$$

$$\dot{D}_{L\kappa}^{(2)}(\boldsymbol{\tau},\mathbf{k}) = \frac{\partial}{\partial\kappa^2}D_{L\kappa}^{(2)}(\boldsymbol{\tau},\mathbf{k})$$

$$= \frac{2}{\sqrt{\pi}}2^l e^{i\mathbf{k}\boldsymbol{\tau}}\sum_\mu(1-\delta_{\mu0}\delta(\boldsymbol{\tau}))e^{-i\mathbf{k}\boldsymbol{\tau}_\mu}|\boldsymbol{\tau}_\mu|^l Y_L(\hat{\boldsymbol{\tau}}_\mu)$$

$$\frac{1}{4}\int_{\frac{1}{2}\eta^{1/2}}^\infty\xi^{2(l-1)}e^{-\tau_\mu^2\xi^2+\frac{\kappa^2}{4\xi^2}}\,d\xi, \qquad (3.5.34)$$

and

$$\dot{D}_{L\kappa}^{(3)}(\boldsymbol{\tau}) = \frac{\partial}{\partial\kappa^2}D_{L\kappa}^{(3)}(\boldsymbol{\tau})$$

$$= \delta(\boldsymbol{\tau})\delta_{l0}\left[\frac{i}{8\sqrt{\pi}\kappa}\left(\text{erfc}(\frac{i\kappa}{\eta^{1/2}}) - \text{erfc}(-\frac{i\kappa}{\eta^{1/2}})\right) - \frac{i}{4\sqrt{\pi}\kappa}\right]$$

$$= -\delta(\boldsymbol{\tau})\delta_{l0}\left[\frac{i}{4\sqrt{\pi}\kappa}\text{erf}(\frac{i\kappa}{\eta^{1/2}}) + \frac{i}{4\sqrt{\pi}\kappa}\right]$$

$$= -\delta(\boldsymbol{\tau})\delta_{l0}\left[\frac{1}{4\sqrt{\pi}\kappa}e^{\frac{\kappa^2}{\eta}}\Im w(\frac{-\kappa}{\eta^{1/2}}) + \frac{i}{4\sqrt{\pi}\kappa}\right] \qquad \text{for} \quad \kappa^2 > 0 .$$

$$(3.5.35)$$

In (3.5.33) the term in the third line has been calculated in (3.5.14). As already discussed in connection with (3.5.14) and (3.5.15) this term has to be replaced by $\frac{1}{2\eta^2}$ for $\kappa^2 \to (\mathbf{K}_n + \mathbf{k})^2$.

The results for $\dot{D}_{L\kappa}^{(3)}(\boldsymbol{\tau})$ were again taken from App. B.7. The corresponding expression for the energy derivative of $K_{L\kappa}^{(3)}(\boldsymbol{\tau})$ trivially follows from (3.5.29) as

$$\dot{K}_{L\kappa}^{(3)}(\boldsymbol{\tau}) = \frac{\partial}{\partial\kappa^2}K_{L\kappa}^{(3)}(\boldsymbol{\tau}) = \frac{\partial}{\partial\kappa^2}D_{L\kappa}^{(3)}(\boldsymbol{\tau}) + \delta(\boldsymbol{\tau})\delta_{l0}\frac{i}{4\sqrt{\pi}\kappa} . \qquad (3.5.36)$$

In the course of numerical calculations it is useful to introduce the following function

$$Z_{l\kappa}(\tau_\mu^2) = \frac{2}{\sqrt{\pi}}2^l\int_{\frac{1}{2}\eta^{1/2}}^\infty \xi^{2l}e^{[-\tau_\mu^2\xi^2 + \frac{\kappa^2}{4\xi^2}]} d\xi . \qquad (3.5.37)$$

We may then write

$$D_{L\kappa}^{(2)}(\boldsymbol{\tau},\mathbf{k}) = e^{i\mathbf{k}\boldsymbol{\tau}}\sum_\mu(1 - \delta_{\mu0}\delta(\boldsymbol{\tau}))e^{-i\mathbf{k}\boldsymbol{\tau}_\mu}|\boldsymbol{\tau}_\mu|^l Y_L(\hat{\boldsymbol{\tau}}_\mu)Z_{l\kappa}(\tau_\mu^2) , \qquad (3.5.38)$$

and

$$\dot{D}_{L\kappa}^{(2)}(\boldsymbol{\tau},\mathbf{k}) = \frac{1}{2}e^{i\mathbf{k}\boldsymbol{\tau}}\sum_\mu(1 - \delta_{\mu0}\delta(\boldsymbol{\tau}))e^{-i\mathbf{k}\boldsymbol{\tau}_\mu}|\boldsymbol{\tau}_\mu|^l Y_L(\hat{\boldsymbol{\tau}}_\mu)Z_{l-1\kappa}(\tau_\mu^2) . \qquad (3.5.39)$$

Note that in the two previous expressions only the spherical harmonics and the exponentials depend on the directions of the vectors $\boldsymbol{\tau}_\mu$.

We close this section with a short discussion on the symmetry properties of the Bloch sums of envelope functions. This is most easily done by going back to the original definition (3.5.2) and using the relations (3.1.57) to (3.1.59). From this we get

$$D_{L\kappa}(-\boldsymbol{\tau},\mathbf{k}) = (-)^l D_{L\kappa}(\boldsymbol{\tau},-\mathbf{k}) , \qquad (3.5.40)$$

where we have used (3.1.58) as well as the fact that in the real-space lattice sum all vectors may be replaced by their negative counterparts. In addition, using cubic harmonics as defined in App. B.2 and staying within the energy ranges where the envelope functions are real we note the identity

$$D_{L\kappa}^*(\boldsymbol{\tau},\mathbf{k}) = D_{L\kappa}(\boldsymbol{\tau},-\mathbf{k}) . \qquad (3.5.41)$$

Finally, combining (3.5.40) and (3.5.41) we obtain

$$D_{L\kappa}(-\boldsymbol{\tau},\mathbf{k}) = (-)^l D^*_{L\kappa}(\boldsymbol{\tau},\mathbf{k}) \ . \tag{3.5.42}$$

Since the previous identities were all derived from very general principles they equally well hold for the Bloch sums of Bessel and Neumann envelope functions. In addition, they are also valid for the respective energy derivatives.

3.6 Expansion Theorems

In this section we will formulate a set of identities for the different types of envelope functions. They are generally known as expansion theorems and allow to expand an envelope function centered at one site in terms of envelope functions centered at a different site.

The derivation of the expansion theorems is essentially based on the equivalence of plane waves and spherical Bessel functions, which both constitute a complete, orthonormal basis set in Hilbert space. For this reason, we may expand the former into the latter using [9, App. B] or [8, Sect. 16.8]

$$e^{i\boldsymbol{\kappa}\mathbf{r}} = 4\pi \sum_L i^l j_l(\kappa r) Y^*_L(\hat{\mathbf{r}}) Y_L(\hat{\boldsymbol{\kappa}}) \tag{3.6.1}$$

with $L = (l, m)$ as usual and the spherical Bessel function $j_l(\kappa r)$ as given by (3.1.4). Note that the spherical harmonics here appear only as parts of the (real) Legendre polynomials. Thus Y_L and Y^*_L may be interchanged in (3.6.1) as well as the following formulas. Application of (3.6.1) to the identity

$$e^{i\boldsymbol{\kappa}(\mathbf{r}-\mathbf{r}')} = e^{i\boldsymbol{\kappa}\mathbf{r}}(e^{i\boldsymbol{\kappa}\mathbf{r}'})^* \ , \tag{3.6.2}$$

yields

$$e^{i\boldsymbol{\kappa}(\mathbf{r}-\mathbf{r}')} = 4\pi \sum_L i^l j_l(\kappa|\mathbf{r}-\mathbf{r}'|) Y^*_L(\widehat{\mathbf{r}-\mathbf{r}'}) Y_L(\hat{\boldsymbol{\kappa}}) \ , \tag{3.6.3}$$

as well as

$$e^{i\boldsymbol{\kappa}(\mathbf{r}-\mathbf{r}')} = (4\pi)^2 \sum_{LL'} i^l j_l(\kappa r) Y^*_L(\hat{\mathbf{r}}) Y_L(\hat{\boldsymbol{\kappa}}) (-i)^{l'} j_{l'}(\kappa r') Y_{L'}(\hat{\mathbf{r}}') Y^*_{L'}(\hat{\boldsymbol{\kappa}})$$

$$= (4\pi)^2 \sum_{LL'} i^{l-l'} j_l(\kappa r) j_{l'}(\kappa r') Y_L(\hat{\boldsymbol{\kappa}}) Y^*_{L'}(\hat{\boldsymbol{\kappa}}) Y^*_L(\hat{\mathbf{r}}) Y_{L'}(\hat{\mathbf{r}}') \ . \tag{3.6.4}$$

Multiplying (3.6.3) by $Y^*_{L''}(\hat{\boldsymbol{\kappa}})$, integrating over $\hat{\boldsymbol{\kappa}}$ and using the orthonormality of the spherical harmonics we get

$$4\pi i^{l''} j_{l''}(\kappa|\mathbf{r}-\mathbf{r}'|) Y^*_{L''}(\widehat{\mathbf{r}-\mathbf{r}'}) = \int d^2\hat{\boldsymbol{\kappa}}\, e^{i\boldsymbol{\kappa}(\mathbf{r}-\mathbf{r}')} Y^*_{L''}(\hat{\boldsymbol{\kappa}}) \ . \tag{3.6.5}$$

From this we have with the help of (3.6.4) the intermediate result

$$j_{l''}(\kappa|\mathbf{r}-\mathbf{r}'|) Y^*_{L''}(\widehat{\mathbf{r}-\mathbf{r}'}) = 4\pi \sum_{LL'} i^{l-l'-l''} C_{LL'L''} j_l(\kappa r) j_{l'}(\kappa r') Y^*_L(\hat{\mathbf{r}}) Y_{L'}(\hat{\mathbf{r}}') \tag{3.6.6}$$

with the (real) Gaunt coefficients

$$c_{LL'L''} = \int d^2\hat{\boldsymbol{\kappa}} \, Y_L^*(\hat{\boldsymbol{\kappa}}) Y_{L'}(\hat{\boldsymbol{\kappa}}) Y_{L''}(\hat{\boldsymbol{\kappa}}) \,. \tag{3.6.7}$$

Next we start from a second identity, namely, the expansion of Green's function [8, Sect. 16.1]

$$\frac{e^{i\kappa|\mathbf{r}'-\mathbf{r}|}}{|\mathbf{r}'-\mathbf{r}|} = 4\pi i\kappa \sum_L j_l(\kappa r') h_l^{(1)}(\kappa r) Y_L^*(\hat{\mathbf{r}}') Y_L(\hat{\mathbf{r}}) \,, \tag{3.6.8}$$

which is valid for $|\mathbf{r}'| < |\mathbf{r}|$; $h_l^{(1)}(\kappa r)$ denotes a spherical Hankel function as given by (3.1.6). Alternatively, the expansion can be written as

$$\frac{e^{i\kappa|\mathbf{r}'-\mathbf{r}|}}{|\mathbf{r}'-\mathbf{r}|} = 4\pi i\kappa \sum_{L''} j_{l''}(\kappa|\mathbf{r}'-\mathbf{r}+\mathbf{R}|) Y_{L''}^*(\widehat{\mathbf{r}'-\mathbf{r}+\mathbf{R}}) \, h_{l''}^{(1)}(\kappa R) Y_{L''}(\hat{\mathbf{R}}) \,, \tag{3.6.9}$$

which is valid for $|\mathbf{r}'-\mathbf{r}+\mathbf{R}| < |\mathbf{R}|$, but where the vector $\mathbf{R}$ is free otherwise. The same is true for the vectors $\mathbf{r}'$ and $\mathbf{r}$ as long as they fulfill the previous condition.

We now combine (3.6.9) with the expression (3.6.6) for the spherical Bessel function to get

$$\frac{e^{i\kappa|\mathbf{r}'-\mathbf{r}|}}{|\mathbf{r}'-\mathbf{r}|} = 4\pi i\kappa \sum_{LL'L''} 4\pi i^{l-l'-l''} j_l(\kappa r') j_{l'}(\kappa|\mathbf{r}-\mathbf{R}|)$$

$$Y_L^*(\hat{\mathbf{r}}') Y_{L'}(\widehat{\mathbf{r}-\mathbf{R}}) c_{LL'L''} h_{l''}^{(1)}(\kappa R) Y_{L''}(\hat{\mathbf{R}})$$

$$= 4\pi i\kappa \sum_L j_l(\kappa r') Y_L^*(\hat{\mathbf{r}}')$$

$$4\pi \sum_{L'L''} i^{l-l'-l''} j_{l'}(\kappa|\mathbf{r}-\mathbf{R}|) Y_{L'}(\widehat{\mathbf{r}-\mathbf{R}})$$

$$c_{LL'L''} h_{l''}^{(1)}(\kappa R) Y_{L''}(\hat{\mathbf{R}}) \,. \tag{3.6.10}$$

Comparing this to the identity (3.6.8) we can immediately read off

$$h_l^{(1)}(\kappa r) Y_L(\hat{\mathbf{r}}) = \sum_{L'} j_{l'}(\kappa|\mathbf{r}-\mathbf{R}|) Y_{L'}(\widehat{\mathbf{r}-\mathbf{R}})$$

$$4\pi \sum_{L''} i^{l-l'-l''} c_{LL'L''} h_{l''}^{(1)}(\kappa R) Y_{L''}(\hat{\mathbf{R}}) \,. \tag{3.6.11}$$

Finally, we note that the sum of the three angular momenta $l + l' + l''$ must be even since otherwise the Gaunt coefficient vanishes due to the parity of the spherical harmonics, (3.1.57). Hence we may exchange i by $-i$ in (3.6.11) and arrive at the following result

$$h_l^{(1)}(\kappa r) Y_L(\hat{\mathbf{r}}) = \sum_{L'} j_{l'}(\kappa|\mathbf{r}-\mathbf{R}|) Y_{L'}(\widehat{\mathbf{r}-\mathbf{R}})$$

$$4\pi \sum_{L''} i^{l'-l+l''} c_{LL'L''} h_{l''}^{(1)}(\kappa R) Y_{L''}(\hat{\mathbf{R}}) \,, \tag{3.6.12}$$

which is identical to (2.2.12) of [5]. It is particularly useful when combined with the traditional definition of the ASW envelope functions, which, in addition to the present choice, contains a factor i^l for each envelope function (see [5] for more details).

Still we have to fulfill the conditions $|\mathbf{r}'| < |\mathbf{r}|$ and $|\mathbf{r}' - \mathbf{r} + \mathbf{R}| < |\mathbf{R}|$ but since $\mathbf{r}'$ does not enter (3.6.12) any more we are free to fix it. The best choice seems to be $\mathbf{r}' = \mathbf{0}$ which leads to $0 < |\mathbf{r}|$ and $|\mathbf{r} - \mathbf{R}| < |\mathbf{R}|$. Pictorially speaking this means that the vector $\mathbf{r} - \mathbf{R}$ has to lie within the sphere centered at $\mathbf{R}$ and having a radius equal to the distance of the two sites where the spherical Hankel functions are centered.

In addition to the expansion theorem (3.6.11) for the spherical Hankel function we may easily derive an analogous theorem for the spherical Bessel function. To this end we start out from (3.6.6) and make the replacements

$$l \to l', \ l' \to l'', \ l'' \to l,$$

and

$$\mathbf{r} \to \mathbf{r} - \mathbf{R}, \ \mathbf{r}' \to -\mathbf{R}, \ \mathbf{r} - \mathbf{r}' \to \mathbf{r} .$$

This way we get

$$j_l(\kappa r)Y_L(\hat{\mathbf{r}}) = 4\pi \sum_{L'L''} i^{l'-l''-l} C_{L'L''L} j_{l'}(\kappa|\mathbf{r} - \mathbf{R}|)j_{l''}(\kappa R)Y_{L'}(\widehat{\mathbf{r} - \mathbf{R}})Y_{L''}(-\widehat{\mathbf{R}}) .$$

(3.6.13)

Here, we have additionally used the fact mentioned at the beginning of this section, namely, that Y_L and Y_L^* may be interchanged whenever they grow out of Legendre polynomials as is the case in (3.6.1). Taking into account the parity of the spherical harmonics, (3.1.57) we arrive at the result

$$j_l(\kappa r)Y_L(\hat{\mathbf{r}}) = \sum_{L'} j_{l'}(\kappa|\mathbf{r} - \mathbf{R}|)Y_{L'}(\widehat{\mathbf{r} - \mathbf{R}})$$
$$4\pi \sum_{L''} i^{l-l'-l''} C_{LL'L''} j_{l''}(\kappa R)Y_{L''}(\hat{\mathbf{R}}) ,$$

(3.6.14)

which is identical to (3.6.11) with the spherical Hankel functions replaced by spherical Bessel functions. In a final step we substract (3.6.14) from (3.6.12) in order to write down the corresponding identity for the spherical Neumann functions

$$n_l(\kappa r)Y_L(\hat{\mathbf{r}}) = \sum_{L'} j_{l'}(\kappa|\mathbf{r} - \mathbf{R}|)Y_{L'}(\widehat{\mathbf{r} - \mathbf{R}})$$
$$4\pi \sum_{L''} i^{l-l'-l''} C_{LL'L''} n_{l''}(\kappa R)Y_{L''}(\hat{\mathbf{R}}) ,$$

(3.6.15)

where n_l is the spherical Neumann function according to (3.1.2). As a consequence, the expansion theorems (3.6.11), (3.6.14) and (3.6.15) are valid for any linear combination f_l of the spherical Bessel and Neumann function. Hence, we may combine them into the identity

$$f_l(\kappa r)Y_L(\hat{\mathbf{r}}) = \sum_{L'} j_{l'}(\kappa|\mathbf{r} - \mathbf{R}|)Y_{L'}(\widehat{\mathbf{r} - \mathbf{R}})$$

$$4\pi \sum_{L''} i^{l-l'-l''} c_{LL'L''} f_{l''}(\kappa R)Y_{L''}(\hat{\mathbf{R}})$$

$$\text{for } |\mathbf{r} - \mathbf{R}| < |\mathbf{R}| . \tag{3.6.16}$$

Another set of identities, which is complementary to (3.6.16), may be derived from the formulation of (3.6.9) as

$$\frac{e^{i\kappa|\mathbf{r}'-\mathbf{r}|}}{|\mathbf{r}' - \mathbf{r}|} = 4\pi i\kappa \sum_{L''} j_{l''}(\kappa|\mathbf{r}' - \mathbf{R}|)Y_{L''}^*(\widehat{\mathbf{r}' - \mathbf{R}}) \, h_{l''}^{(1)}(\kappa|\mathbf{r} - \mathbf{R}|)Y_{L''}(\widehat{\mathbf{r} - \mathbf{R}}) . \tag{3.6.17}$$

It is valid for $|\mathbf{r}' - \mathbf{R}| < |\mathbf{r} - \mathbf{R}|$, but again the vector $\mathbf{R}$ is free otherwise. The same is true for the vectors $\mathbf{r}'$ and $\mathbf{r}$ as long as they fulfil the previous condition. Combining (3.6.17) with (3.6.6) for the spherical Bessel function we get

$$\frac{e^{i\kappa|\mathbf{r}'-\mathbf{r}|}}{|\mathbf{r}' - \mathbf{r}|} = 4\pi i\kappa \sum_{LL'L''} 4\pi i^{l-l'-l''} j_l(\kappa r')j_{l'}(\kappa R)$$

$$Y_L^*(\hat{\mathbf{r}'})Y_{L'}(\hat{\mathbf{R}})c_{LL'L''} h_{l''}^{(1)}(\kappa|\mathbf{r} - \mathbf{R}|)Y_{L''}(\widehat{\mathbf{r} - \mathbf{R}})$$

$$= 4\pi i\kappa \sum_{L} j_l(\kappa r')Y_L^*(\hat{\mathbf{r}'})$$

$$4\pi \sum_{L'L''} i^{l-l'-l''} j_{l'}(\kappa R)Y_{L'}(\hat{\mathbf{R}})c_{LL'L''}$$

$$h_{l''}^{(1)}(\kappa|\mathbf{r} - \mathbf{R}|)Y_{L''}(\widehat{\mathbf{r} - \mathbf{R}}) . \tag{3.6.18}$$

Comparison with the identity (3.6.8) leads to the result

$$h_l^{(1)}(\kappa r)Y_L(\hat{\mathbf{r}}) = \sum_{L'} h_{l'}^{(1)}(\kappa|\mathbf{r} - \mathbf{R}|)Y_{L'}(\widehat{\mathbf{r} - \mathbf{R}})$$

$$4\pi \sum_{L''} i^{l-l'-l''} c_{LL'L''} j_{l''}(\kappa R)Y_{L''}(\hat{\mathbf{R}}) , \tag{3.6.19}$$

where we have interchanged l' and l'' in order to emphasize the similarity to (3.6.11). Concerning the conditions $|\mathbf{r}'| < |\mathbf{r}|$ and $|\mathbf{r}' - \mathbf{R}| < |\mathbf{r} - \mathbf{R}|$ the vectors have to obey, we again opt to choose $\mathbf{r}' = \mathbf{0}$ this leading to $0 < |\mathbf{r}|$ and $|\mathbf{R}| < |\mathbf{r} - \mathbf{R}|$. Hence this time the vector $\mathbf{r} - \mathbf{R}$ has to lie outside the sphere centered at $\mathbf{R}$ and having a radius $|\mathbf{R}|$.

Again we subtract the identity (3.6.14) for the spherical Bessel functions this leading to the corresponding expression for the spherical Neumann functions

$$n_l(\kappa r)Y_L(\hat{\mathbf{r}}) = \sum_{L'} n_{l'}(\kappa|\mathbf{r} - \mathbf{R}|)Y_{L'}(\widehat{\mathbf{r} - \mathbf{R}})$$

$$4\pi \sum_{L''} i^{l-l'-l''} c_{LL'L''} j_{l''}(\kappa R)Y_{L''}(\hat{\mathbf{R}}) . \tag{3.6.20}$$

Finally, combining (3.6.19), (3.6.14) and (3.6.20) we arrive at the general result

$$f_l(\kappa r)Y_L(\hat{\mathbf{r}}) = \sum_{L'} f_{l'}(\kappa|\mathbf{r} - \mathbf{R}|)Y_{L'}(\widehat{\mathbf{r} - \mathbf{R}})$$
$$4\pi \sum_{L''} i^{l-l'-l''} c_{LL'L''} j_{l''}(\kappa R)Y_{L''}(\hat{\mathbf{R}})$$
$$\text{for } |\mathbf{r} - \mathbf{R}| > |\mathbf{R}| , \tag{3.6.21}$$

where f_l denotes any linear combination of j_l and n_l. Together with (3.6.16) it constitutes the set of fundamental expansion theorems.

In closing this section we point out again that in any of the above identities the spherical harmonics may be interchanged with their complex conjugate counterparts, since they all entered the formulas only as parts of Legendre polynomials, which are real by definition. For the same reason, according to (B.2.13) all equations may be identically formulated with the real-valued cubic harmonics as defined in App. B.2. Another important symmetry of the expansion theorems (3.6.16) and (3.6.21) stems from the fact, that the sum of the angular momenta must be even and, hence, all the i-factors in these equations are real. Finally, rewriting (3.6.9) with the exponential on the left-hand side replaced by the spherical Hankel function for $l = 0$ as given by (3.1.22)

$$h_0^{(1)}(\kappa r) = 4\pi \sum_{L'} (-)^{l'} j_{l'}(\kappa|\mathbf{r} - \mathbf{R}|)Y_{L'}^*(\widehat{\mathbf{r} - \mathbf{R}}) \, h_{l'}^{(1)}(\kappa R)Y_{L'}(\hat{\mathbf{R}}) , \tag{3.6.22}$$

we realize that this is simply the final result, (3.6.12), for the special case $l = 0$. Thus, just by inserting the identity (3.6.6) into an equation for $l = 0$ we were able to generalize this formula to the case of arbitrary l.

3.7 Structure Constants

In the present section we will combine the just derived expansion theorems with the definitions of the ASW envelope functions and their Bloch sums as given in Sects. 3.1 and 3.5. This will lead to the definition of structure constants, which comprise all information about the spatial structure. In addition, we will outline some of the symmetry properties of the structure constants.

As already mentioned in the preceding section the expansion theorems in their general form (3.6.16) and (3.6.21) are valid for any linear combination of the spherical Bessel functions. In the present context we concentrate on a few special situations. In particular, we need the theorem (3.6.16) for f_l being a spherical Hankel or Neumann function this corresponding to (3.6.11) and (3.6.15). Using the definitions (3.1.3) and (3.1.5) of the envelope functions we note

$$H_{L\kappa}(\mathbf{r}) = \sum_{L'} J_{L'\kappa}(\mathbf{r} - \mathbf{R})B_{L'L\kappa}(\mathbf{R}) \qquad \text{for } |\mathbf{r} - \mathbf{R}| < |\mathbf{R}| , \tag{3.7.1}$$

where the expansion coefficients $B_{L'L\kappa}(\mathbf{R})$ are the so-called structure constants. Comparing (3.6.11) and (3.7.1) we define

$$B_{L'L\kappa}(\mathbf{R}) := 4\pi \sum_{L''} i^{l-l'-l''} \kappa^{l+l'-l''} C_{LL'L''} H_{L''\kappa}(\mathbf{R}) \,. \tag{3.7.2}$$

In contrast, using (3.1.1) and (3.1.3) and the expansion theorem we get for the Neumann envelope function (3.6.15)

$$N_{L\kappa}(\mathbf{r}) = \sum_{L'} J_{L'\kappa}(\mathbf{r} - \mathbf{R}) B_{L'L\kappa}(\mathbf{R}) \qquad \text{for} \quad |\mathbf{r} - \mathbf{R}| < |\mathbf{R}| \,, \tag{3.7.3}$$

where the structure constants are now given by

$$B_{L'L\kappa}(\mathbf{R}) = 4\pi \sum_{L''} i^{l-l'-l''} \kappa^{l+l'-l''} C_{LL'L''} N_{L''\kappa}(\mathbf{R}) \,. \tag{3.7.4}$$

Again, the formulation with Neumann envelope functions is formally identical to the formulation based on Hankel envelope functions. For this reason, we used the same symbol B for the structure constants but will, as before, concentrate on the latter case, from which the corresponding expressions for the Neumann envelope functions can be readily deduced.

In passing, we point out that the definition (3.7.2) of the structure constant as well as definition (3.7.12) of the Bloch-summed structure constant as given below deviate from the original formulation as given in [14] or (2.2.15), (2.3.1), (2.3.3) and (2.3.5) of [5]. The difference is due to the i^l-factors, which were included in the original definition of the ASW basis functions and which we have omitted in the present treatment.

For later use we complement the definition (3.7.2) of the structure constants with their energy derivatives. Combining (3.7.2), (3.2.22), (3.1.17), and (3.2.26) we get

$$\dot{B}_{L'L\kappa}(\mathbf{R}) = \frac{\partial}{\partial\kappa^2} B_{L'L\kappa}(\mathbf{R})$$
$$= 2\pi \sum_{L''} i^{l-l'-l''} \kappa^{l+l'-l''} C_{LL'L''} Y_{L''}(\hat{\mathbf{R}})$$
$$\left[\frac{l+l'-l''}{\kappa^2} \bar{h}_{l''}^{(1)}(\kappa R) + R\bar{h}_{l''-1}^{(1)}(\kappa R)\right] \,. \tag{3.7.5}$$

From the definition (3.7.2) of the (real-space) structure constants we may read off several important properties. Obviously, they fulfill the symmetry relations

$$B_{LL'\kappa}(\mathbf{R}) = (-)^{l-l'} B_{L'L\kappa}(\mathbf{R}), \tag{3.7.6}$$

and

$$B_{L'L\kappa}(-\mathbf{R}) = (-)^{l-l'} B_{L'L\kappa}(\mathbf{R}) \,, \tag{3.7.7}$$

which latter follows from the parity of the spherical harmonics, (3.1.57) and the fact that the sum of all three angular momenta must be even. Taking (3.7.6) and (3.7.7) together we arrive at

$$B_{LL'\kappa}(-\mathbf{R}) = B_{L'L\kappa}(\mathbf{R}) \ . \tag{3.7.8}$$

In general, we prefer cubic harmonics rather than their complex counterparts and, hence, arrive at real-valued structure constants. This is due to the fact that the exponents of both the i-factor and κ in (3.7.2) are even and the envelope functions themselves were defined to be real quantities in their respective energy regions. We thus note

$$B^*_{L'L\kappa}(\mathbf{R}) = B_{L'L\kappa}(\mathbf{R}) \ . \tag{3.7.9}$$

Obviously, these symmetry relations likewise hold for the energy derivative (3.7.5). In addition, they simplify the numerical calculations and reduce the storage requirements substantially.

Next we turn to the case of crystalline translational symmetry and, in analogy to (3.5.1), rewrite (3.7.1) as

$$H_{L\kappa}(\boldsymbol{\tau} - \mathbf{R}_\mu) = \sum_{L'} J_{L'\kappa}(\boldsymbol{\tau}') B_{L'L\kappa}(\boldsymbol{\tau} - \boldsymbol{\tau}' - \mathbf{R}_\mu)$$

$$\text{for} \quad |\boldsymbol{\tau}'| < |\boldsymbol{\tau} - \boldsymbol{\tau}' - \mathbf{R}_\mu| \ , \tag{3.7.10}$$

in order to separate lattice vectors $\mathbf{R}_\mu$ from the vectors $\boldsymbol{\tau}$ and $\boldsymbol{\tau}'$ lying inside the unit cell. Performing the Bloch sum we arrive at

$$\sum_\mu (1 - \delta_{\mu 0}\delta(\boldsymbol{\tau})) e^{i\mathbf{k}\mathbf{R}_\mu} H_{L\kappa}(\boldsymbol{\tau} - \mathbf{R}_\mu)$$

$$= \sum_{L'} J_{L'\kappa}(\boldsymbol{\tau}') \sum_\mu (1 - \delta_{\mu 0}\delta(\boldsymbol{\tau})) e^{i\mathbf{k}\mathbf{R}_\mu} B_{L'L\kappa}(\boldsymbol{\tau} - \boldsymbol{\tau}' - \mathbf{R}_\mu) \ , \tag{3.7.11}$$

this leading directly to the definition of Bloch-summed structure constants by

$$B_{L'L\kappa}(\boldsymbol{\tau}, \mathbf{k}) := \sum_\mu (1 - \delta_{\mu 0}\delta(\boldsymbol{\tau})) e^{i\mathbf{k}\mathbf{R}_\mu} B_{L'L\kappa}(\boldsymbol{\tau} - \mathbf{R}_\mu)$$

$$= 4\pi \sum_{L''} i^{l-l'-l''} \kappa^{l+l'-l''} c_{LL'L''} \sum_\mu (1 - \delta_{\mu 0}\delta(\boldsymbol{\tau})) e^{i\mathbf{k}\mathbf{R}_\mu} H_{L''\kappa}(\boldsymbol{\tau} - \mathbf{R}_\mu)$$

$$= 4\pi \sum_{L''} i^{l-l'-l''} \kappa^{l+l'-l''} c_{LL'L''} D_{L''\kappa}(\boldsymbol{\tau}, \mathbf{k}) \ . \tag{3.7.12}$$

Here we have inserted the definition (3.7.2) of the real-space structure constants. In addition, we have benefited from the fact that the real-space structure constants themselves are given in terms of Hankel envelope functions, the Bloch sums of which were defined by (3.5.2). Note that we have used the same symbol for the real space structure constants and their Bloch sums. Finally, combining (3.7.10) to (3.7.12) and (3.5.2) we arrive at the expansion theorem for the Bloch sum of envelope functions

$$D_{L\kappa}(\boldsymbol{\tau}, \mathbf{k}) = \sum_{L'} J_{L'\kappa}(\boldsymbol{\tau}')B_{L'L\kappa}(\boldsymbol{\tau} - \boldsymbol{\tau}', \mathbf{k}) \,, \tag{3.7.13}$$

which is formally identical to the expansion theorem (3.7.1).

We complement the discussion of the Bloch-summed structure constants with the formulation of their derivatives with respect to the energy parameter κ^2. Combining (3.7.12) and (3.2.22) we get

$$\dot{B}_{L'L\kappa}(\boldsymbol{\tau}, \mathbf{k}) = \frac{\partial}{\partial \kappa^2} B_{L'L\kappa}(\boldsymbol{\tau}, \mathbf{k})$$

$$= 2\pi \sum_{L''} i^{l-l'-l''} \kappa^{l+l'-l''} c_{LL'L''}$$

$$\left[\frac{l+l'-l''}{\kappa^2} D_{L''\kappa}(\boldsymbol{\tau}, \mathbf{k}) + 2\dot{D}_{L''\kappa}(\boldsymbol{\tau}, \mathbf{k})\right] \,. \tag{3.7.14}$$

It may be readily combined with (3.5.32) to (3.5.35) and (3.5.39) in order to derive explicit formulas.

In passing, we discuss, as for the real-space structure constants, some properties of the Bloch-summed structure constant. First, we get from the definition (3.7.12)

$$B_{LL'\kappa}(\boldsymbol{\tau}, \mathbf{k}) = (-)^{l-l'} B_{L'L\kappa}(\boldsymbol{\tau}, \mathbf{k}) \,. \tag{3.7.15}$$

A second identity arise from the property (3.5.40) of the Bloch-summed envelope function as

$$B_{L'L\kappa}(-\boldsymbol{\tau}, \mathbf{k}) = (-)^{l-l'} B_{L'L\kappa}(\boldsymbol{\tau}, -\mathbf{k}) \,. \tag{3.7.16}$$

Finally, opting for cubic rather than spherical harmonics we combine (3.7.12) and (3.5.41) to get

$$B^*_{L'L\kappa}(\boldsymbol{\tau}, \mathbf{k}) = B_{L'L\kappa}(\boldsymbol{\tau}, -\mathbf{k}) \,. \tag{3.7.17}$$

In addition, combining the previous identities we arrive at the result

$$B_{LL'\kappa}(-\boldsymbol{\tau}, \mathbf{k}) = (-)^{l-l'} B^*_{LL'\kappa}(\boldsymbol{\tau}, \mathbf{k}) = B^*_{L'L\kappa}(\boldsymbol{\tau}, \mathbf{k}) \,. \tag{3.7.18}$$

In closing this section, we combine (3.2.47) to (3.2.49) for the gradients of the envelope functions with the expansion theorem (3.7.1) to derive some important identities for the structure constants. In particular, we note

$$\nabla H_{L\kappa}(\mathbf{r}) = \sum_{L''} \mathbf{G}_{L''L} \left[-\delta_{ll''-1} + \delta_{ll''+1}\kappa^2\right] H_{L''\kappa}(\mathbf{r})$$

$$= \sum_{L'} \sum_{L''} \mathbf{G}_{L''L} \left[-\delta_{ll''-1} + \delta_{ll''+1}\kappa^2\right] J_{L'\kappa}(\mathbf{r} - \mathbf{R})B_{L'L''\kappa}(\mathbf{R})$$

$$\tag{3.7.19}$$

as well as

$$\nabla H_{L\kappa}(\mathbf{r}) = \sum_{L''} \nabla J_{L''\kappa}(\mathbf{r} - \mathbf{R}) B_{L''L\kappa}(\mathbf{R})$$

$$= \sum_{L'} \sum_{L''} \mathbf{G}_{L'L''} \left[-\delta_{l''l'-1}\kappa^2 + \delta_{l''l'+1} \right] J_{L'\kappa}(\mathbf{r} - \mathbf{R}) B_{L''L\kappa}(\mathbf{R}) \ .$$

$$(3.7.20)$$

Comparing these formulas (and exchanging l and l') we arrive at

$$B_{LL''\kappa}(\mathbf{R})\mathbf{G}_{L''L'} \left[-\delta_{l'l''-1} + \delta_{l'l''+1}\kappa^2 \right]$$

$$= \mathbf{G}_{LL''} \left[-\delta_{l''l-1}\kappa^2 + \delta_{l''l+1} \right] B_{L''L'\kappa}(\mathbf{R})$$

$$= -\mathbf{G}_{LL''} \left[-\delta_{ll''-1} + \delta_{ll''+1}\kappa^2 \right] B_{L''L'\kappa}(\mathbf{R}) \ .$$

$$(3.7.21)$$

From this we obtain for the energy derivative

$$\dot{B}_{LL''\kappa}(\mathbf{R})\mathbf{G}_{L''L'} \left[-\delta_{l'l''-1} + \delta_{l'l''+1}\kappa^2 \right] + B_{LL''\kappa}(\mathbf{R})\mathbf{G}_{L''L'}\delta_{l'l''+1}$$

$$= \mathbf{G}_{LL''} \left[-\delta_{l''l-1}\kappa^2 + \delta_{l''l+1} \right] \dot{B}_{L''L'\kappa}(\mathbf{R}) - \mathbf{G}_{LL''}\delta_{l''l-1} B_{L''L'\kappa}(\mathbf{R}) \ ,$$

$$(3.7.22)$$

hence,

$$\dot{B}_{LL''\kappa}(\mathbf{R})\mathbf{G}_{L''L'} \left[-\delta_{l'l''-1} + \delta_{l'l''+1}\kappa^2 \right]$$

$$= \mathbf{G}_{LL''} \left[-\delta_{l''l-1}\kappa^2 + \delta_{l''l+1} \right] \dot{B}_{L''L'\kappa}(\mathbf{R})$$

$$- \mathbf{G}_{LL''}B_{L''L'\kappa}(\mathbf{R})\delta_{l''l-1} - B_{LL''\kappa}(\mathbf{R})\mathbf{G}_{L''L'}\delta_{l'l''+1} \ .$$

$$(3.7.23)$$

Again, we ask how the previous results would change if we had opted for the original definition of the radial part of the envelope functions with the additional i^l factor included, as it was proposed by Williams, Kübler, and Gelatt. According to the discussion at the end of Sect. 3.2, in this case the minus and plus signs before the first and second Kronecker-δ, respectively, in (3.7.19) to (3.7.21) would both turn into $+i$. As a consequence, the respective second terms on both the left- and right-hand side of (3.7.22) would also carry a $+i$ and in (3.7.23) the first and second term in the last line would be preceded by a $+i$ and $-i$, respectively.

Of course, since the matrices $\mathbf{G}_{LL'}$ are completely local and all the structural information of covered by the structure constants, all the previous derviations hold identically for the Bloch-summed counterparts of the respective real-space quantities. We thus complement (3.7.21) and (3.7.23) by the results

$$B_{LL''\kappa}(\boldsymbol{\tau}, \mathbf{k})\mathbf{G}_{L''L'} \left[-\delta_{l'l''-1} + \delta_{l'l''+1}\kappa^2 \right]$$

$$= \mathbf{G}_{LL''} \left[-\delta_{l''l-1}\kappa^2 + \delta_{l''l+1} \right] B_{L''L'\kappa}(\boldsymbol{\tau}, \mathbf{k})$$

$$= -\mathbf{G}_{LL''} \left[-\delta_{ll''-1} + \delta_{ll''+1}\kappa^2 \right] B_{L''L'\kappa}(\boldsymbol{\tau}, \mathbf{k}) \ ,$$

$$(3.7.24)$$

and

$$\dot{B}_{LL''\kappa}(\tau,\mathbf{k})\mathbf{G}_{L''L'}\left[-\delta_{l'l''-1}+\delta_{l'l''+1}\kappa^2\right]$$
$$= \mathbf{G}_{LL''}\left[-\delta_{l''l-1}\kappa^2+\delta_{l''l+1}\right]\dot{B}_{L''L'\kappa}(\tau,\mathbf{k})$$
$$-\mathbf{G}_{LL''}B_{L''L'\kappa}(\tau,\mathbf{k})\delta_{l''l-1}-B_{LL''\kappa}(\tau,\mathbf{k})\mathbf{G}_{L''L'}\delta_{l'l''+1} \, . \qquad (3.7.25)$$

3.8 Overlap Integrals of Envelope Functions

On the following pages we will derive closed expressions for the overlap integral of two Hankel or Neumann envelope functions. These integrals are needed for the overlap and Hamiltonian matrix when the combined correction is encountered as it is usual in the ASW method.

As before we want to treat both the "double κ" and the "single κ" case. In prinicple, the latter can be derived from the former by applying l'Hospitals rule to the final result. However, when we calculated radial integrals of the type (3.3.12) over products of envelope functions in Sect. 3.3 we realized that application of l'Hospitals rule was not straightforward. Instead we had to pass over to definite integrals first and only after that formal excursion we were able to properly perform the limit $\kappa_1 \rightarrow \kappa_2$ and arrive at the "single κ" case. As we have also learned in the context of (3.3.19) to (3.3.22) all these problems could be easily circumvented by starting rightaway from the "single κ" case and falling back on the energy derivative of Schrödinger's equation. We will follow this method also in the present section, which is especially devoted to the situation where the two functions entering the overlap integral are centered at different sites and where the advantages of the alternative method thus become more important. In passing we note that it was this case, for which the derivation as based on the energy derivative of Schrödinger's equation was first presented by Williams, Kübler, and Gelatt [14].

To be specific we first write down the overlap integral of two Hankel envelope functions centered at sites $\mathbf{R}_i$ and $\mathbf{R}_j$. It is defined as

$$\langle H_{L\kappa_1}(\mathbf{r}_i)|H_{L'\kappa_2}(\mathbf{r}_j)\rangle_\Omega = \int_\Omega d^3\mathbf{r}\, H^*_{L\kappa_1}(\mathbf{r}_i)H_{L'\kappa_2}(\mathbf{r}_j) \, , \qquad (3.8.1)$$

where

$$\mathbf{r}_i = \mathbf{r} - \mathbf{R}_i \, , \qquad (3.8.2)$$

as usual and Ω denotes the volume of integration to be specified lateron. As concerns the sites $\mathbf{R}_i$ and $\mathbf{R}_j$, we will keep things quite general and thus do not yet specify if they are equal or not. Actually, the case where $\mathbf{R}_i$ and $\mathbf{R}_j$ are equal has already been covered by the discussion starting with (3.3.12). As a consequence, the present section particularly aims at the situation, where $\mathbf{R}_i$ and $\mathbf{R}_j$ are different. However, in order to arrive at a uniform picture we want to treat all cases on the same footing and thus prefer to make any specifications as late as possible.

As already seen in Sect. 3.1, integrals of the type (3.8.1) with different values of the energy parameter κ usually are evaluated by starting from Schrödinger's equation for free particles or Helmholtz's equation

$$\left(\Delta + \kappa^2\right) H_{L\kappa}(\mathbf{r}) = 0 \ , \tag{3.8.3}$$

and applying Green's theorem in order to transform the volume integral to a surface integral. To be specific, we combine Schrödinger's equation (3.8.3) for both functions entering (3.8.1) and note

$$H^*_{L\kappa_1}(\mathbf{r}_i)\Delta H_{L'\kappa_2}(\mathbf{r}_j) - \Delta H^*_{L\kappa_1}(\mathbf{r}_i)H_{L'\kappa_2}(\mathbf{r}_j) = \left(\kappa_1^2 - \kappa_2^2\right) H^*_{L\kappa_1}(\mathbf{r}_i)H_{L'\kappa_2}(\mathbf{r}_j) \ . \tag{3.8.4}$$

Here we assumed operators to act only on the function following directly. On integrating both sides of (3.8.4) we get the intermediate result

$$\int_\Omega d^3\mathbf{r} \, H^*_{L\kappa_1}(\mathbf{r}_i)H_{L'\kappa_2}(\mathbf{r}_j)$$
$$= \frac{1}{\kappa_1^2 - \kappa_2^2} \int_\Omega d^3\mathbf{r} \left[H^*_{L\kappa_1}(\mathbf{r}_i)\Delta H_{L'\kappa_2}(\mathbf{r}_j) - \Delta H^*_{L\kappa_1}(\mathbf{r}_i)H_{L'\kappa_2}(\mathbf{r}_j)\right] \ . \tag{3.8.5}$$

For equal values of the energy parameter κ^2 we turn to the abovementioned alternative derivation originally proposed by Williams et al. [14]. Thus, as for the "double κ" case we start from Schrödinger's equation for free particles, (3.8.3), but perform the differentiation with respect to energy already at this early stage. This way we get from (3.8.3)

$$\left(\Delta + \kappa^2\right) \dot{H}_{L\kappa}(\mathbf{r}) + H_{L\kappa}(\mathbf{r}) = 0 \ , \tag{3.8.6}$$

where a dot as usual designates the derivative with respect to the energy parameter κ^2. From (3.8.6) and Schrödinger's equation (3.8.3) we are able to write down the identities

$$H^*_{L\kappa}(\mathbf{r}_i)H_{L'\kappa}(\mathbf{r}_j) = -\left(\Delta + \kappa^2\right) \dot{H}^*_{L\kappa}(\mathbf{r}_i)H_{L'\kappa}(\mathbf{r}_j) \ , \tag{3.8.7}$$

and

$$0 = \dot{H}^*_{L\kappa}(\mathbf{r}_i) \left(\Delta + \kappa^2\right) H_{L'\kappa}(\mathbf{r}_j) \ . \tag{3.8.8}$$

Adding (3.8.7) and (3.8.8) and integrating we arrive at the identity

$$\int_\Omega d^3\mathbf{r} \, H^*_{L\kappa}(\mathbf{r}_i)H_{L'\kappa}(\mathbf{r}_j)$$
$$= \int_\Omega d^3\mathbf{r} \left[\dot{H}^*_{L\kappa}(\mathbf{r}_i)\Delta H_{L'\kappa}(\mathbf{r}_j) - \Delta \dot{H}^*_{L\kappa}(\mathbf{r}_i)H_{L'\kappa}(\mathbf{r}_j)\right] \ , \tag{3.8.9}$$

which complements (3.8.5) for equal values of the energy parameters κ_1^2 and κ_2^2. Both (3.8.5) and (3.8.9) serve as a starting point for all subsequent evaluations.

Still we have to account for the singularities of the Hankel envelope functions at their respective origins and so we specify the volume of integration by

$$\Omega \longrightarrow \Omega_\delta = \begin{cases} \Omega_\infty - \Omega_i(\delta) & \text{for } i = j \\ \Omega_\infty - \Omega_i(\delta) - \Omega_j(\delta) & \text{for } i \neq j \end{cases}.$$

Here Ω_∞ denotes the volume of all space and $\Omega_i(\delta)$ is the volume of a sphere of radius δ centered at site $\mathbf{R}_i$. Thus we simply exclude the singularities for the time being but have to consider the limit $\delta \to 0$ lateron in case that we want to integrate over all space.

After this modification we apply Green's theorem to (3.8.5) or (3.8.9) and are then left with integrals over two or three surfaces, namely, the outer surface of the volume of integration and the one or two surfaces of the δ-spheres at the centers $\mathbf{R}_i$ and $\mathbf{R}_j$, respectively. At present, the outer surface is just the infinite surface. However, its contribution vanishes due to the asymptotic behaviour (3.1.48) of the spherical Hankel function for negative energies. Of course, for Neumann envelope functions we would have to introduce Bloch symmetry before. Then the infinite surface has to be replaced with the surface of the unit cell. Since we are integrating over the gradient of a periodic function this surface integral would also vanish.

As a consequence of the previous arguments we are left with only the surface integrals due to the δ-spheres excluded above. Concentrating first on the case where the two envelope functions are centered at the same site, i.e. $i = j$ we get for the "double κ" overlap integral

$$\int_{\Omega_\delta} d^3r\, H^*_{L\kappa_1}(\mathbf{r}_i) H_{L'\kappa_2}(\mathbf{r}_i)$$
$$= -\frac{1}{\kappa_1^2 - \kappa_2^2} \int_{\partial\Omega_i(\delta)} \hat{\mathbf{u}}_i d^2r_i \left[H^*_{L\kappa_1}(\mathbf{r}_i)\nabla H_{L'\kappa_2}(\mathbf{r}_i) - \nabla H^*_{L\kappa_1}(\mathbf{r}_i) H_{L'\kappa_2}(\mathbf{r}_i) \right],$$

$$(3.8.10)$$

where $\hat{\mathbf{u}}_i$ denotes the radial unit vector at sphere i pointing outward. With the help of

$$\hat{\mathbf{u}}_i \nabla = \frac{\partial}{\partial r_i}, \qquad (3.8.11)$$

the orthonormality of the spherical harmonics, and the definition (3.1.17) of the Hankel envelope function we then arrive at

$$\int_{\Omega_\delta} d^3r\, H^*_{L\kappa_1}(\mathbf{r}_i) H_{L'\kappa_2}(\mathbf{r}_i)$$
$$= -\frac{1}{\kappa_1^2 - \kappa_2^2} \left[r_i^2 \left(\bar{h}_l^{(1)}(\kappa_1 r_i)\frac{\partial}{\partial r_i}\bar{h}_l^{(1)}(\kappa_2 r_i) - \frac{\partial}{\partial r_i}\bar{h}_l^{(1)}(\kappa_1 r_i)\bar{h}_l^{(1)}(\kappa_2 r_i) \right) \right]_{r_i=\delta} \delta_{LL'},$$

$$(3.8.12)$$

which, as expected, is identical to (3.3.33).

For equal values of κ^2 we proceed in a similar way and apply Green's theorem to (3.8.9). Then we get in analogy to (3.8.10)

$$
\int_{\Omega_\delta} d^3\mathbf{r}\, H^*_{L\kappa}(\mathbf{r}_i) H_{L'\kappa}(\mathbf{r}_i)
$$

$$
= -\int_{\partial\Omega_i(\delta)} \hat{\mathbf{u}}_i d^2 r_i \left[\dot{H}^*_{L\kappa}(\mathbf{r}_i) \nabla H_{L'\kappa}(\mathbf{r}_i) - \nabla \dot{H}^*_{L\kappa}(\mathbf{r}_i) H_{L'\kappa}(\mathbf{r}_i) \right],
$$

$$(3.8.13)$$

and from this after performing the angular integration the result

$$
\int_{\Omega_\delta} d^3\mathbf{r}\, H^*_{L\kappa}(\mathbf{r}_i) H_{L'\kappa}(\mathbf{r}_i)
$$

$$
= - \left[r_i^2 \left(\frac{\partial}{\partial \kappa^2} \bar{h}^{(1)}_l(\kappa r_i) \frac{\partial}{\partial r_i} \bar{h}^{(1)}_l(\kappa r_i) - \frac{\partial}{\partial \kappa^2} \frac{\partial}{\partial r_i} \bar{h}^{(1)}_l(\kappa r_i) \bar{h}^{(1)}_l(\kappa r_i) \right) \right]_{r_i=\delta} \delta_{LL'},
$$

$$(3.8.14)$$

which is identical to (3.3.42).

As is apparent from both (3.8.12) and (3.8.14) it is not possible to perform the limit $\delta \to 0$. Actually, this situation never occurs in the ASW method since inside the atomic spheres the Hankel envelope functions are replaced by the augmented Hankel functions, which are regular by construction. For this reason, we will always need only the integral (3.8.12) or (3.8.14) extending over all space except for the atomic sphere, where the envelope functions are centered. In other words, we will only consider the case $\delta = S_i$. Just for completeness we rewrite (3.8.12) and (3.8.14) for this important case and arrive at the result

$$
\int_{\Omega_{S_i}} d^3\mathbf{r}\, H^*_{L\kappa_1}(\mathbf{r}_i) H_{L'\kappa_2}(\mathbf{r}_i)
$$

$$
= \int_{\Omega_\infty} d^3\mathbf{r}\, H^*_{L\kappa_1}(\mathbf{r}_i) H_{L'\kappa_2}(\mathbf{r}_i) - \int_{\Omega_i} d^3\mathbf{r}\, H^*_{L\kappa_1}(\mathbf{r}_i) H_{L'\kappa_2}(\mathbf{r}_i)
$$

$$
= \int_{S_i}^\infty dr\, r^2 \bar{h}^{(1)}_l(\kappa_1 r_i) \bar{h}^{(1)}_l(\kappa_2 r_i)\, \delta_{LL'}
$$

$$
= -\frac{1}{\kappa_1^2 - \kappa_2^2} \left[r_i^2 \left(\bar{h}^{(1)}_l(\kappa_1 r_i) \frac{\partial}{\partial r_i} \bar{h}^{(1)}_l(\kappa_2 r_i) - \frac{\partial}{\partial r_i} \bar{h}^{(1)}_l(\kappa_1 r_i) \bar{h}^{(1)}_l(\kappa_2 r_i) \right) \right]_{r_i=S_i} \delta_{LL'}
$$

$$
=: \langle H_{L\kappa_1}(\mathbf{r}_i) | H_{L'\kappa_2}(\mathbf{r}_j) \rangle_{\Omega_\infty} - \langle H_{L\kappa_1} | H_{L\kappa_2} \rangle_i \delta_{LL'}
$$

$$
=: \langle H_{L\kappa_1} | H_{L\kappa_2} \rangle'_i \delta_{LL'},
$$

$$(3.8.15)$$

for $\kappa_1 \neq \kappa_2$ and

$$
\int_{\Omega_{S_i}} d^3\mathbf{r}\, H^*_{L\kappa}(\mathbf{r}_i) H_{L'\kappa}(\mathbf{r}_i)
$$

$$
= \int_{\Omega_\infty} d^3\mathbf{r}\, H^*_{L\kappa}(\mathbf{r}_i) H_{L'\kappa}(\mathbf{r}_i) - \int_{\Omega_i} d^3\mathbf{r}\, H^*_{L\kappa}(\mathbf{r}_i) H_{L'\kappa}(\mathbf{r}_i)
$$

$$= \int_{S_i}^{\infty} dr\, r^2 \left[\bar{h}_l^{(1)}(\kappa r_i)\right]^2 \delta_{LL'}$$

$$= -\left[r_i^2 \left(\frac{\partial}{\partial \kappa^2} \bar{h}_l^{(1)}(\kappa r_i) \frac{\partial}{\partial r_i} \bar{h}_l^{(1)}(\kappa r_i) - \frac{\partial}{\partial \kappa^2} \frac{\partial}{\partial r_i} \bar{h}_l^{(1)}(\kappa r_i) \bar{h}_l^{(1)}(\kappa r_i) \right) \right]_{r_i=S_i} \delta_{LL'}$$

$$=: \langle H_{L\kappa}(\mathbf{r}_i)|H_{L'\kappa}(\mathbf{r}_j)\rangle_{\Omega_\infty} - \langle H_{L\kappa}|H_{L\kappa}\rangle_i \delta_{LL'}$$

$$=: \langle H_{L\kappa}|H_{L\kappa}\rangle_i' \delta_{LL'} , \tag{3.8.16}$$

for the "single κ" case. Here we have added the corresponding bra-ket notation of the integrals where the index i means integration over the atomic sphere at site $\mathbf{R}_i$. Note that in this notation we have omitted the arguments of the envelope functions whenever the integration extended over a single sphere.

Note that the volume of integration in (3.8.15) and (3.8.16) is *not* the interstitial region since we only excluded the atomic sphere at site $\mathbf{R}_i$. In order to arrive at the integral over the true interstitial region we still have to explicitly remove the contributions from the integrals over all other spheres. This we will come back to at the end of the section.

Next we turn to the situation, where both Hankel envelope functions entering (3.8.1) are centered at different sites, i.e. to the case $i \neq j$. According to the above specification of the volume of integration we then have to exclude the δ-spheres at both sites $\mathbf{R}_i$ and $\mathbf{R}_j$. Hence we write instead of (3.8.10)

$$\int_{\Omega_\delta} d^3r\, H_{L\kappa_1}^*(\mathbf{r}_i) H_{L'\kappa_2}(\mathbf{r}_j)$$

$$= -\frac{1}{\kappa_1^2 - \kappa_2^2} \int_{\partial\Omega_i(\delta)} \hat{\mathbf{u}}_i d^2r_i \left[H_{L\kappa_1}^*(\mathbf{r}_i) \nabla H_{L'\kappa_2}(\mathbf{r}_j) - \nabla H_{L\kappa_1}^*(\mathbf{r}_i) H_{L'\kappa_2}(\mathbf{r}_j) \right]$$

$$-\frac{1}{\kappa_1^2 - \kappa_2^2} \int_{\partial\Omega_j(\delta)} \hat{\mathbf{u}}_j d^2r_j \left[H_{L\kappa_1}^*(\mathbf{r}_i) \nabla H_{L'\kappa_2}(\mathbf{r}_j) - \nabla H_{L\kappa_1}^*(\mathbf{r}_i) H_{L'\kappa_2}(\mathbf{r}_j) \right] . \tag{3.8.17}$$

This way we are left with two surface integrals, each of which contains one function which is centered at the respective site while the second function is centered at the respective other site. For this reason we cannot directly split off the angular integration as we did for the case $i = j$ but first have to apply the expansion theorem. Then both functions entering the surface integral are centered at the site where the δ-sphere is located. Combining (3.8.17) with the expansion theorem (3.7.1) we thus get

$$\int_{\Omega_\delta} d^3r\, H_{L\kappa_1}^*(\mathbf{r}_i) H_{L'\kappa_2}(\mathbf{r}_j)$$

$$= -\frac{1}{\kappa_1^2 - \kappa_2^2} \int_{\partial\Omega_i(\delta)} \hat{\mathbf{u}}_i d^2r_i$$

$$\sum_{L''} \left[H_{L\kappa_1}^*(\mathbf{r}_i) \nabla J_{L''\kappa_2}(\mathbf{r}_i) - \nabla H_{L\kappa_1}^*(\mathbf{r}_i) J_{L''\kappa_2}(\mathbf{r}_i) \right] B_{L''L'\kappa_2}(\mathbf{R}_i - \mathbf{R}_j)$$

$$
-\frac{1}{\kappa_1^2 - \kappa_2^2} \int_{\partial\Omega_j(\delta)} \hat{\mathbf{u}}_j d^2 r_j
$$

$$
\sum_{L''} B^*_{L''L\kappa_1}(\mathbf{R}_j - \mathbf{R}_i) \left[J^*_{L''\kappa_1}(\mathbf{r}_j)\nabla H_{L'\kappa_2}(\mathbf{r}_j) - \nabla J^*_{L''\kappa_1}(\mathbf{r}_j)H_{L'\kappa_2}(\mathbf{r}_j) \right] .
$$

$$(3.8.18)$$

After this additional step we are able to use (3.8.11), the orthonormality of the spherical harmonics, and the definitions (3.1.16) and (3.1.17) of the envelope functions to write

$$
\int_{\Omega_\delta} d^3\mathbf{r} \, H^*_{L\kappa_1}(\mathbf{r}_i)H_{L'\kappa_2}(\mathbf{r}_j)
$$

$$
= -\frac{1}{\kappa_1^2 - \kappa_2^2} \left[r_i^2 \left(\bar{h}^{(1)}_l(\kappa_1 r_i)\frac{\partial}{\partial r_i}\bar{j}_l(\kappa_2 r_i) - \frac{\partial}{\partial r_i}\bar{h}^{(1)}_l(\kappa_1 r_i)\bar{j}_l(\kappa_2 r_i) \right) \right]_{r_i=\delta}
$$

$$
B_{LL'\kappa_2}(\mathbf{R}_i - \mathbf{R}_j)
$$

$$
-\frac{1}{\kappa_1^2 - \kappa_2^2} B^*_{L'L\kappa_1}(\mathbf{R}_j - \mathbf{R}_i)
$$

$$
\left[r_j^2 \left(\bar{j}_{l'}(\kappa_1 r_j)\frac{\partial}{\partial r_j}\bar{h}^{(1)}_{l'}(\kappa_2 r_j) - \frac{\partial}{\partial r_j}\bar{j}_{l'}(\kappa_1 r_j)\bar{h}^{(1)}_{l'}(\kappa_2 r_j) \right) \right]_{r_j=\delta} .
$$

$$(3.8.19)$$

For equal values of κ^2 we start from (3.8.9) and specify the volume of integration by all space except for the two δ-spheres at $\mathbf{R}_i$ and $\mathbf{R}_j$. Hence, we write

$$
\int_{\Omega_\delta} d^3\mathbf{r} \, H^*_{L\kappa}(\mathbf{r}_i)H_{L'\kappa}(\mathbf{r}_j)
$$

$$
= -\int_{\partial\Omega_i(\delta)} \hat{\mathbf{u}}_i d^2 r_i \left[H^*_{L\kappa}(\mathbf{r}_i)\nabla H_{L'\kappa}(\mathbf{r}_j) - \nabla H^*_{L\kappa}(\mathbf{r}_i)H_{L'\kappa}(\mathbf{r}_j) \right]
$$

$$
- \int_{\partial\Omega_j(\delta)} \hat{\mathbf{u}}_j d^2 r_j \left[H^*_{L\kappa}(\mathbf{r}_i)\nabla H_{L'\kappa}(\mathbf{r}_j) - \nabla H^*_{L\kappa}(\mathbf{r}_i)H_{L'\kappa}(\mathbf{r}_j) \right] ,
$$

$$(3.8.20)$$

which is the counterpart of (3.8.13). Again we have to include the expansion theorem in order to perform the angular integration at both sites $\mathbf{R}_i$ and $\mathbf{R}_j$. Insofar things are still the same as in the "double κ" case. There is however, a distinct difference. Due to the fact that in place of the energy denominator in (3.8.17) there is an energy derivative of one of the two functions in (3.8.20) application of the expansion theorem now leads to an extra term which contains the energy derivative of the structure constants. To be specific we write

$$
\int_{\Omega_\delta} d^3\mathbf{r} \, H^*_{L\kappa}(\mathbf{r}_i)H_{L'\kappa}(\mathbf{r}_j)
$$

$$
= -\int_{\partial\Omega_i(\delta)} \hat{\mathbf{u}}_i d^2 r_i
$$

$$\sum_{L''}\left[\dot{H}^*_{L\kappa}(\mathbf{r}_i)\nabla J_{L''\kappa}(\mathbf{r}_i) - \nabla\dot{H}^*_{L\kappa}(\mathbf{r}_i)J_{L''\kappa}(\mathbf{r}_i)\right]B_{L''L'\kappa}(\mathbf{R}_i - \mathbf{R}_j)$$

$$-\int_{\partial\Omega_j(\delta)}\hat{\mathbf{u}}_j d^2r_j$$

$$\sum_{L''}B^*_{L''L\kappa}(\mathbf{R}_j - \mathbf{R}_i)\left[\dot{J}^*_{L''\kappa}(\mathbf{r}_j)\nabla H_{L'\kappa}(\mathbf{r}_j) - \nabla\dot{J}^*_{L''\kappa}(\mathbf{r}_j)H_{L'\kappa}(\mathbf{r}_j)\right]$$

$$-\int_{\partial\Omega_j(\delta)}\hat{\mathbf{u}}_j d^2r_j$$

$$\sum_{L''}\dot{B}^*_{L''L\kappa}(\mathbf{R}_j - \mathbf{R}_i)\left[J^*_{L''\kappa}(\mathbf{r}_j)\nabla H_{L'\kappa}(\mathbf{r}_j) - \nabla J^*_{L''\kappa}(\mathbf{r}_j)H_{L'\kappa}(\mathbf{r}_j)\right]\ .$$

$$(3.8.21)$$

As before we are now able to perform the angular integration. Again using (3.8.11), the orthonormality of the spherical harmonics, and the definitions (3.1.16) and (3.1.17) of the envelope functions we write

$$\int_{\Omega_\delta}d^3\mathbf{r}\,H^*_{L\kappa}(\mathbf{r}_i)H_{L'\kappa}(\mathbf{r}_j)$$

$$= -\left[r_i^2\left(\frac{\partial}{\partial\kappa^2}\bar{h}^{(1)}_l(\kappa r_i)\frac{\partial}{\partial r_i}\bar{j}_l(\kappa r_i) - \frac{\partial}{\partial\kappa^2}\frac{\partial}{\partial r_i}\bar{h}^{(1)}_l(\kappa r_i)\bar{j}_l(\kappa r_i)\right)\right]_{r_i=\delta}$$

$$B_{LL'\kappa}(\mathbf{R}_i - \mathbf{R}_j)$$

$$-B^*_{L'L\kappa}(\mathbf{R}_j - \mathbf{R}_i)$$

$$\left[r_j^2\left(\frac{\partial}{\partial\kappa^2}\bar{j}_{l'}(\kappa r_j)\frac{\partial}{\partial r_j}\bar{h}^{(1)}_{l'}(\kappa r_j) - \frac{\partial}{\partial\kappa^2}\frac{\partial}{\partial r_j}\bar{j}_{l'}(\kappa r_j)\bar{h}^{(1)}_{l'}(\kappa r_j)\right)\right]_{r_j=\delta}$$

$$-\dot{B}^*_{L'L\kappa}(\mathbf{R}_j - \mathbf{R}_i)\left[r_j^2\left(\bar{j}_{l'}(\kappa r_j)\frac{\partial}{\partial r_j}\bar{h}^{(1)}_{l'}(\kappa r_j) - \frac{\partial}{\partial r_j}\bar{j}_{l'}(\kappa r_j)\bar{h}^{(1)}_{l'}(\kappa r_j)\right)\right]_{r_j=\delta}\ .$$

$$(3.8.22)$$

According to the definition (3.2.38) of the Wronskian and the identity (3.2.40) for the fundamental Wronskian of the barred Hankel and Bessel functions the last term in square brackets is (-1). In addition, the second term in square brackets can be rewritten with the help of the energy derivative of the fundamental Wronskian (3.2.40). For the latter we get from (3.2.38)

$$\frac{\partial}{\partial\kappa^2}W\{r\bar{h}^{(1)}_l(\kappa r), r\bar{j}_l(\kappa r); r\}$$

$$= \frac{\partial}{\partial\kappa^2}r\bar{h}^{(1)}_l(\kappa r)\frac{\partial}{\partial r}\left(r\bar{j}_l(\kappa r)\right) - \frac{\partial}{\partial\kappa^2}r\bar{j}_l(\kappa r)\frac{\partial}{\partial r}\left(r\bar{h}^{(1)}_l(\kappa r)\right)$$

$$+r\bar{h}^{(1)}_l(\kappa r)\frac{\partial}{\partial\kappa^2}\frac{\partial}{\partial r}\left(r\bar{j}_l(\kappa r)\right) - r\bar{j}_l(\kappa r)\frac{\partial}{\partial\kappa^2}\frac{\partial}{\partial r}\left(r\bar{h}^{(1)}_l(\kappa r)\right)\ . \quad (3.8.23)$$

Hence we arrive at the result

$$\int_{\Omega_\delta} d^3\mathbf{r}\, H^*_{L\kappa}(\mathbf{r}_i) H_{L'\kappa}(\mathbf{r}_j)$$

$$= -\left[r_i^2 \left(\frac{\partial}{\partial \kappa^2} \bar{h}_l^{(1)}(\kappa r_i) \frac{\partial}{\partial r_i} \bar{j}_l(\kappa r_i) - \frac{\partial}{\partial \kappa^2} \frac{\partial}{\partial r_i} \bar{h}_l^{(1)}(\kappa r_i) \bar{j}_l(\kappa r_i) \right) \right]_{r_i=\delta}$$

$$B_{LL'\kappa}(\mathbf{R}_i - \mathbf{R}_j)$$

$$- B^*_{L'L\kappa}(\mathbf{R}_j - \mathbf{R}_i)$$

$$\left[r_j^2 \left(\frac{\partial}{\partial \kappa^2} \bar{h}_{l'}^{(1)}(\kappa r_j) \frac{\partial}{\partial r_j} \bar{j}_{l'}(\kappa r_j) - \frac{\partial}{\partial \kappa^2} \frac{\partial}{\partial r_j} \bar{h}_{l'}^{(1)}(\kappa r_j) \bar{j}_{l'}(\kappa r_j) \right) \right]_{r_j=\delta}$$

$$+ \dot{B}^*_{L'L\kappa}(\mathbf{R}_j - \mathbf{R}_i)\,. \tag{3.8.24}$$

In contrast to the case $i = j$ we are now able to perform the limit $\delta \to 0$. To this end we use the identities (3.2.5), (3.2.11) as well as (3.1.43) to (3.1.45), which we combine with (3.8.19) to note

$$\langle H_{L\kappa_1}(\mathbf{r}_i)|H_{L'\kappa_2}(\mathbf{r}_j)\rangle_{\Omega_\infty}$$

$$= \int_{\Omega_\infty} d^3\mathbf{r}\, H^*_{L\kappa_1}(\mathbf{r}_i) H_{L'\kappa_2}(\mathbf{r}_j)$$

$$= \frac{1}{\kappa_2^2 - \kappa_1^2} \left[B_{LL'\kappa_2}(\mathbf{R}_i - \mathbf{R}_j) - B^*_{L'L\kappa_1}(\mathbf{R}_j - \mathbf{R}_i) \right], \tag{3.8.25}$$

for the "double κ" case. Again we have added the bra-ket notation of the integral for later use. For equal values of the energy parameter we combine the aforementioned identities with (3.8.24). There both square brackets on the right-hand side vanish for $\delta \to 0$ and so we are left with the result

$$\langle H_{L\kappa}(\mathbf{r}_i)|H_{L'\kappa}(\mathbf{r}_j)\rangle_{\Omega_\infty} = \int_{\Omega_\infty} d^3\mathbf{r}\, H^*_{L\kappa}(\mathbf{r}_i) H_{L'\kappa}(\mathbf{r}_j)$$

$$= \dot{B}^*_{L'L\kappa}(\mathbf{R}_j - \mathbf{R}_i)$$

$$= \dot{B}_{LL'\kappa}(\mathbf{R}_i - \mathbf{R}_j)\,, \tag{3.8.26}$$

where in the last step we have assumed to work with cubic rather than spherical harmonics and moreover used the identities (3.7.8) and (3.7.9).

Another important situation arises from equating the δ-spheres to the atomic spheres, i.e. from setting $\delta = S_i$ and $\delta = S_j$, respectively. We considered this case already for $i = j$, where we arrived at (3.8.15) and (3.8.16). However, due to the existence of the limit $\delta \to 0$ for $i \neq j$, we may now proceed in two different ways. We may either simply use $\delta = S_i$ and $\delta = S_j$ in (3.8.19) and (3.8.24) or else start from (3.8.25) and (3.8.26) and subtract the respective integrals over the atomic spheres. Preferring the first alternative we note

$$\int_{\Omega_\infty - \Omega_i - \Omega_j} d^3\mathbf{r}\, H^*_{L\kappa_1}(\mathbf{r}_i) H_{L'\kappa_2}(\mathbf{r}_j)$$

$$= \int_{\Omega_\infty} d^3\mathbf{r}\, H^*_{L\kappa_1}(\mathbf{r}_i) H_{L'\kappa_2}(\mathbf{r}_j)$$

$$-\int_{\Omega_i} d^3r\, H^*_{L\kappa_1}(\mathbf{r}_i) H_{L'\kappa_2}(\mathbf{r}_j) - \int_{\Omega_j} d^3r\, H^*_{L\kappa_1}(\mathbf{r}_i) H_{L'\kappa_2}(\mathbf{r}_j)$$

$$= -\frac{1}{\kappa_1^2 - \kappa_2^2} \left[r_i^2 \left(\bar{h}_l^{(1)}(\kappa_1 r_i) \frac{\partial}{\partial r_i} \bar{j}_l(\kappa_2 r_i) - \frac{\partial}{\partial r_i} \bar{h}_l^{(1)}(\kappa_1 r_i) \bar{j}_l(\kappa_2 r_i) \right) \right]_{r_i = S_i}$$

$$B_{LL'\kappa_2}(\mathbf{R}_i - \mathbf{R}_j)$$

$$-\frac{1}{\kappa_1^2 - \kappa_2^2} B^*_{L'L\kappa_1}(\mathbf{R}_j - \mathbf{R}_i)$$

$$\left[r_j^2 \left(\bar{j}_{l'}(\kappa_1 r_j) \frac{\partial}{\partial r_j} \bar{h}_{l'}^{(1)}(\kappa_2 r_j) - \frac{\partial}{\partial r_j} \bar{j}_{l'}(\kappa_1 r_j) \bar{h}_{l'}^{(1)}(\kappa_2 r_j) \right) \right]_{r_j = S_j}$$

$$= \langle H_{L\kappa_1}(\mathbf{r}_i) | H_{L'\kappa_2}(\mathbf{r}_j) \rangle_{\Omega_\infty}$$
$$- \langle H_{L\kappa_1} | J_{L\kappa_2} \rangle_i B_{LL'\kappa_2}(\mathbf{R}_i - \mathbf{R}_j) - B^*_{L'L\kappa_1}(\mathbf{R}_j - \mathbf{R}_i) \langle J_{L'\kappa_1} | H_{L'\kappa_2} \rangle_j ,$$

$$(3.8.27)$$

for different values of the energy parameter and

$$\int_{\Omega_\infty - \Omega_i - \Omega_j} d^3r\, H^*_{L\kappa}(\mathbf{r}_i) H_{L'\kappa}(\mathbf{r}_j)$$

$$= \int_{\Omega_\infty} d^3r\, H^*_{L\kappa}(\mathbf{r}_i) H_{L'\kappa}(\mathbf{r}_j)$$

$$- \int_{\Omega_i} d^3r\, H^*_{L\kappa}(\mathbf{r}_i) H_{L'\kappa}(\mathbf{r}_j) - \int_{\Omega_j} d^3r\, H^*_{L\kappa}(\mathbf{r}_i) H_{L'\kappa}(\mathbf{r}_j)$$

$$= - \left[r_i^2 \left(\frac{\partial}{\partial \kappa^2} \bar{h}_l^{(1)}(\kappa r_i) \frac{\partial}{\partial r_i} \bar{j}_l(\kappa r_i) - \frac{\partial}{\partial \kappa^2} \frac{\partial}{\partial r_i} \bar{h}_l^{(1)}(\kappa r_i) \bar{j}_l(\kappa r_i) \right) \right]_{r_i = S_i}$$

$$B_{LL'\kappa}(\mathbf{R}_i - \mathbf{R}_j)$$

$$- B^*_{L'L\kappa}(\mathbf{R}_j - \mathbf{R}_i)$$

$$\left[r_j^2 \left(\frac{\partial}{\partial \kappa^2} \bar{h}_{l'}^{(1)}(\kappa r_j) \frac{\partial}{\partial r_j} \bar{j}_{l'}(\kappa r_j) - \frac{\partial}{\partial \kappa^2} \frac{\partial}{\partial r_j} \bar{h}_{l'}^{(1)}(\kappa r_j) \bar{j}_{l'}(\kappa r_j) \right) \right]_{r_j = S_j}$$

$$+ \dot{B}^*_{L'L\kappa}(\mathbf{R}_j - \mathbf{R}_i)$$

$$= \langle H_{L\kappa}(\mathbf{r}_i) | H_{L'\kappa}(\mathbf{r}_j) \rangle_{\Omega_\infty}$$
$$- \langle H_{L\kappa} | J_{L\kappa} \rangle_i B_{LL'\kappa}(\mathbf{R}_i - \mathbf{R}_j) - B^*_{L'L\kappa}(\mathbf{R}_j - \mathbf{R}_i) \langle J_{L'\kappa} | H_{L'\kappa} \rangle_j , \quad (3.8.28)$$

for the "single κ" case. Note that the difference between the expressions (3.8.25) and (3.8.26) on the one hand and (3.8.27) and (3.8.28) on the other hand is identical to the radial integral as given by (3.3.32) and (3.3.41) times the structure constant.

Finally, we point out again that the region of integration considered so far still is *not* the true interstitial region. In order to calculate the integral over the interstitial we would have to remove not only the contribution from the sphere(s) where the two functions are centered but in addition the contributions from all other spheres. Since all these spheres lie outside the singularities of the Hankel envelope functions entering the integral the two cases $i = j$ and $i \neq j$ may be treated together again.

To be specific, we are thus aiming at the integral

$$\int_{\Omega_m} d^3\mathbf{r}\, H^*_{L\kappa_1}(\mathbf{r}_i)H_{L'\kappa_2}(\mathbf{r}_j)\,,$$

where Ω_m denotes the volume of a sphere of radius S_m centered at site $\mathbf{R}_m$ and $m \neq i$ and $m \neq j$. The calculation of this integral proceeds in exactly the same way as sketched at the beginning of this section. In other words, we make use of Schrödinger's equation and apply Green's theorem to transform the volume integral to a surface integral. Hence, we may rightaway start out from (3.8.5) and (3.8.9). Specifying the volume of integration by $\Omega \to \Omega_m$ we immediately arrive at

$$\int_{\Omega_m} d^3\mathbf{r}\, H^*_{L\kappa_1}(\mathbf{r}_i)H_{L'\kappa_2}(\mathbf{r}_i)$$
$$= \frac{1}{\kappa_1^2 - \kappa_2^2} \int_{\partial\Omega_m} \hat{\mathbf{u}}_m d^2 r_m \left[H^*_{L\kappa_1}(\mathbf{r}_i)\nabla H_{L'\kappa_2}(\mathbf{r}_j) - \nabla H^*_{L\kappa_1}(\mathbf{r}_i)H_{L'\kappa_2}(\mathbf{r}_j) \right]\,,$$
$$(3.8.29)$$

for $\kappa_1 \neq \kappa_2$. As already discussed for the above case $i \neq j$ we next have to apply the expansion theorem before the integral can be evaluated further. Actually, here we have to expand *both* envelope functions entering the integral. As a consequence we get from (3.8.29)

$$\int_{\Omega_m} d^3\mathbf{r}\, H^*_{L\kappa_1}(\mathbf{r}_i)H_{L'\kappa_2}(\mathbf{r}_j)$$
$$= \frac{1}{\kappa_1^2 - \kappa_2^2} \int_{\partial\Omega_m} \hat{\mathbf{u}}_m d^2 r_m \sum_{L''}\sum_{L'''} B^*_{L''L\kappa_1}(\mathbf{R}_m - \mathbf{R}_i)$$
$$\left[J^*_{L''\kappa_1}(\mathbf{r}_m)\nabla J_{L'''\kappa_2}(\mathbf{r}_m) - \nabla J^*_{L''\kappa_1}(\mathbf{r}_m)J_{L'''\kappa_2}(\mathbf{r}_m) \right]$$
$$B_{L'''L'\kappa_2}(\mathbf{R}_m - \mathbf{R}_j)\,.$$
$$(3.8.30)$$

Next, using (3.8.11), the orthonormality of the spherical harmonics, and the definition (3.1.17) of the Bessel envelope function we note

$$\int_{\Omega_m} d^3\mathbf{r}\, H^*_{L\kappa_1}(\mathbf{r}_i)H_{L'\kappa_2}(\mathbf{r}_j)$$
$$= \frac{1}{\kappa_1^2 - \kappa_2^2} \sum_{L''} B^*_{L''L\kappa_1}(\mathbf{R}_m - \mathbf{R}_i)$$
$$\left[r_m^2 \left(\bar{j}_{l''}(\kappa_1 r_m)\frac{\partial}{\partial r_m}\bar{j}_{l''}(\kappa_2 r_m) - \frac{\partial}{\partial r_m}\bar{j}_{l''}(\kappa_1 r_m)\bar{j}_{l''}(\kappa_2 r_m) \right) \right]_{r_m = S_m}$$
$$B_{L''L'\kappa_2}(\mathbf{R}_m - \mathbf{R}_j)$$
$$= \sum_{L''} B^*_{L''L\kappa_1}(\mathbf{R}_m - \mathbf{R}_i)\langle J_{L''\kappa_1}|J_{L''\kappa_2}\rangle_m B_{L''L'\kappa_2}(\mathbf{R}_m - \mathbf{R}_j)\,.$$
$$(3.8.31)$$

For equal values of κ we proceed in the same way and get

$$\int_{\Omega_m} d^3\mathbf{r}\, H^*_{L\kappa}(\mathbf{r}_i) H_{L'\kappa}(\mathbf{r}_j)$$
$$= \int_{\partial\Omega_m} \hat{\mathbf{u}}_m d^2 r_m \left[\dot{H}^*_{L\kappa}(\mathbf{r}_i)\nabla H_{L'\kappa}(\mathbf{r}_j) - \nabla \dot{H}^*_{L\kappa}(\mathbf{r}_i) H_{L'\kappa}(\mathbf{r}_j) \right] . \qquad (3.8.32)$$

Again we apply the expansion theorem (3.7.1) to both envelope functions this leading to

$$\int_{\Omega_m} d^3\mathbf{r}\, H^*_{L\kappa}(\mathbf{r}_i) H_{L'\kappa}(\mathbf{r}_j)$$
$$= \int_{\partial\Omega_m} \hat{\mathbf{u}}_m d^2 r_m \sum_{L''}\sum_{L'''} B^*_{L''L\kappa}(\mathbf{R}_m - \mathbf{R}_i)$$
$$\left[\dot{J}^*_{L''\kappa}(\mathbf{r}_m)\nabla J_{L'''\kappa}(\mathbf{r}_m) - \nabla \dot{J}^*_{L''\kappa}(\mathbf{r}_m) J_{L'''\kappa}(\mathbf{r}_m) \right] B_{L'''L'\kappa}(\mathbf{R}_m - \mathbf{R}_j)$$
$$+ \int_{\partial\Omega_m} \hat{\mathbf{u}}_m d^2 r_m \sum_{L''}\sum_{L'''} \dot{B}^*_{L''L\kappa}(\mathbf{R}_m - \mathbf{R}_i)$$
$$\left[J^*_{L''\kappa}(\mathbf{r}_m)\nabla J_{L'''\kappa}(\mathbf{r}_m) - \nabla J^*_{L''\kappa}(\mathbf{r}_m) J_{L'''\kappa}(\mathbf{r}_m) \right] B_{L'''L'\kappa}(\mathbf{R}_m - \mathbf{R}_j) . $$
$$(3.8.33)$$

As before, we perform the angular integration using (3.8.11), the orthonormality of the spherical harmonics and the definition (3.1.17) of the Bessel envelope function to note the result

$$\int_{\Omega_m} d^3\mathbf{r}\, H^*_{L\kappa}(\mathbf{r}_i) H_{L'\kappa}(\mathbf{r}_j)$$
$$= \sum_{L''} B^*_{L''L\kappa}(\mathbf{R}_m - \mathbf{R}_i)$$
$$\left[r_m^2 \left(\frac{\partial}{\partial\kappa^2}\bar{j}_l(\kappa r_m)\frac{\partial}{\partial r_m}\bar{j}_l(\kappa r_m) - \frac{\partial}{\partial\kappa^2}\frac{\partial}{\partial r_m}\bar{j}_l(\kappa r_m)\bar{j}_l(\kappa r_m) \right) \right]_{r_m = S_m}$$
$$B_{L''L'\kappa}(\mathbf{R}_m - \mathbf{R}_j)$$
$$+ \sum_{L''} \dot{B}^*_{L''L\kappa}(\mathbf{R}_m - \mathbf{R}_i)$$
$$\left[r_m^2 \left(\bar{j}_l(\kappa r_m)\frac{\partial}{\partial r_m}\bar{j}_l(\kappa r_m) - \frac{\partial}{\partial r_m}\bar{j}_l(\kappa r_m)\bar{j}_l(\kappa r_m) \right) \right]_{r_m = S_m}$$
$$B_{L''L'\kappa}(\mathbf{R}_m - \mathbf{R}_j)$$
$$= \sum_{L''} B^*_{L''L\kappa}(\mathbf{R}_m - \mathbf{R}_i)$$
$$\left[r_m^2 \left(\frac{\partial}{\partial\kappa^2}\bar{j}_l(\kappa r_m)\frac{\partial}{\partial r_m}\bar{j}_l(\kappa r_m) - \frac{\partial}{\partial\kappa^2}\frac{\partial}{\partial r_m}\bar{j}_l(\kappa r_m)\bar{j}_l(\kappa r_m) \right) \right]_{r_m = S_m}$$
$$B_{L''L'\kappa}(\mathbf{R}_m - \mathbf{R}_j)$$

$$= \sum_{L''} B^*_{L''L\kappa}(\mathbf{R}_m - \mathbf{R}_i)\langle J_{L''\kappa}|J_{L''\kappa}\rangle_m B_{L''L'\kappa}(\mathbf{R}_m - \mathbf{R}_j) \,. \tag{3.8.34}$$

With the previous identities at hand we are now in the position to calculate the integrals over the true interstitial region. To this end we combine (3.8.15), (3.8.16), (3.8.25) to (3.8.28), (3.8.31) as well as (3.8.34) and get

$$
\begin{aligned}
&\langle H_{L\kappa_1}(\mathbf{r}_i)|H_{L'\kappa_2}(\mathbf{r}_j)\rangle_{\Omega_{I\infty}} \\
&= \langle H_{L\kappa_1}|H_{L\kappa_2}\rangle'_i \delta_{LL'}\delta_{ij} \\
&\quad + \Bigg[\frac{1}{\kappa_2^2 - \kappa_1^2} \left[B_{LL'\kappa_2}(\mathbf{R}_i - \mathbf{R}_j) - B^*_{L'L\kappa_1}(\mathbf{R}_j - \mathbf{R}_i) \right] \left(1 - \delta(\kappa_1^2 - \kappa_2^2) \right) \\
&\quad\quad + \dot{B}_{LL'\kappa_1}(\mathbf{R}_i - \mathbf{R}_j)\delta(\kappa_1^2 - \kappa_2^2) - \langle H_{L\kappa_1}|J_{L\kappa_2}\rangle_i B_{LL'\kappa_2}(\mathbf{R}_i - \mathbf{R}_j) \\
&\quad\quad - B^*_{L'L\kappa_1}(\mathbf{R}_j - \mathbf{R}_i)\langle J_{L'\kappa_1}|H_{L'\kappa_2}\rangle_j \Bigg] (1 - \delta_{ij}) \\
&\quad - \sum_m \sum_{L''} (1 - \delta_{im})(1 - \delta_{mj}) B^*_{L''L\kappa_1}(\mathbf{R}_m - \mathbf{R}_i) \\
&\quad\quad\quad\quad\quad\quad \langle J_{L''\kappa_1}|J_{L''\kappa_2}\rangle_m B_{L''L'\kappa_2}(\mathbf{R}_m - \mathbf{R}_j) \,,
\end{aligned}
\tag{3.8.35}
$$

where $\Omega_{I\infty}$ denotes the interstitial region between atomic spheres in all space. In this context we point out that all integrals appearing on the right-hand side are radial integrals whereas all the structural information by now has condensed into the structure constants. We have thus cut the integrals over all space into local integrals, which contain only functions centered at the same site, and structure constants, which only depend on the relative positions of the atoms. Finally, this separation of intraatomic and interatomic quantities passes through to the matrix elements of the secular matrix as calculated in Sects. 2.2 and 4.2. Moreover, it will help in the calculation of such integrals for crystalline systems, to which we turn now.

According to the previous discussion introducing crystalline translational symmetry will not drastically change the integrals over products of envelope functions but only affect the form of the structure constants. In order to show this we first replace the Hankel envelope functions entering the integral (3.8.1) with the Bloch-summed envelope functions as defined by (3.5.2). Hence, we note

$$
\begin{aligned}
&\langle D_{L\kappa_1}(\mathbf{r}_i, \mathbf{k})|D_{L'\kappa_2}(\mathbf{r}_j, \mathbf{k})\rangle_\Omega \\
&= \sum_\mu \sum_\nu (1 - \delta_{\mu 0}\delta(\mathbf{r}_i))(1 - \delta_{\nu 0}\delta(\mathbf{r}_j)) \\
&\quad\quad\quad e^{-i\mathbf{k}\mathbf{R}_\mu}\langle H_{L\kappa_1}(\mathbf{r}_i - \mathbf{R}_\mu)|H_{L'\kappa_2}(\mathbf{r}_j - \mathbf{R}_\nu)\rangle_\Omega e^{i\mathbf{k}\mathbf{R}_\nu} \,,
\end{aligned}
\tag{3.8.36}
$$

where

$$\mathbf{r}_i = \mathbf{r} - \boldsymbol{\tau}_i \quad\text{and}\quad \mathbf{R}_{\mu i} := \mathbf{R}_\mu + \boldsymbol{\tau}_i \,, \tag{3.8.37}$$

and $\boldsymbol{\tau}_i$ is an atomic site within the unit cell at the origin. The changes due to the introduction of translational symmetry may be demonstrated best by the special case covered by (3.8.35). Inserting this into (3.8.36) we arrive at

$$
\langle D_{L\kappa_1}(\mathbf{r}_i,\mathbf{k})|D_{L'\kappa_2}(\mathbf{r}_j,\mathbf{k})\rangle_{\Omega_I}
$$
$$
= \frac{1}{N}\langle D_{L\kappa_1}(\mathbf{r}_i,\mathbf{k})|D_{L'\kappa_2}(\mathbf{r}_j,\mathbf{k})\rangle_{\Omega_{I\infty}}
$$
$$
= \langle H_{L\kappa_1}|H_{L\kappa_2}\rangle'_i\delta_{LL'}\delta_{ij}
$$
$$
+ \frac{1}{N}\sum_{\mu}\sum_{\nu}(1-\delta_{\mu0}\delta(\mathbf{r}_i))(1-\delta_{\nu0}\delta(\mathbf{r}_j))e^{-i\mathbf{k}\mathbf{R}_{\mu}}
$$
$$
\left[\left[\frac{1}{\kappa_2^2-\kappa_1^2}\left[B_{LL'\kappa_2}(\mathbf{R}_{\mu i}-\mathbf{R}_{\nu j})-B^*_{L'L\kappa_1}(\mathbf{R}_{\nu j}-\mathbf{R}_{\mu i})\right]\left(1-\delta(\kappa_1^2-\kappa_2^2)\right)\right.\right.
$$
$$
\left.+\dot{B}_{LL'\kappa_1}(\mathbf{R}_{\mu i}-\mathbf{R}_{\nu j})\delta(\kappa_1^2-\kappa_2^2)-\langle H_{L\kappa_1}|J_{L\kappa_2}\rangle_i B_{LL'\kappa_2}(\mathbf{R}_{\mu i}-\mathbf{R}_{\nu j})\right.
$$
$$
\left.-B^*_{L'L\kappa_1}(\mathbf{R}_{\nu j}-\mathbf{R}_{\mu i})\langle J_{L'\kappa_1}|H_{L'\kappa_2}\rangle_j\right](1-\delta_{\mu\nu}\delta_{ij})
$$
$$
-\sum_{\lambda m}\sum_{L''}(1-\delta_{\mu\lambda}\delta_{im})(1-\delta_{\lambda\nu}\delta_{mj})e^{i\mathbf{k}\mathbf{R}_{\lambda}}B^*_{L''L\kappa_1}(\mathbf{R}_{\lambda m}-\mathbf{R}_{\mu i})
$$
$$
\left.\langle J_{L''\kappa_1}|J_{L''\kappa_2}\rangle_m B_{L''L'\kappa_2}(\mathbf{R}_{\lambda m}-\mathbf{R}_{\nu j})e^{-i\mathbf{k}\mathbf{R}_{\lambda}}\right]e^{i\mathbf{k}\mathbf{R}_{\nu}}\,.
$$
$$
(3.8.38)
$$

At this stage we may use the definition (3.7.12) of the Bloch summed structure constants this leading to the final result

$$
\langle D_{L\kappa_1}(\mathbf{r}_i,\mathbf{k})|D_{L'\kappa_2}(\mathbf{r}_j,\mathbf{k})\rangle_{\Omega_I}
$$
$$
= \langle H_{L\kappa_1}|H_{L\kappa_2}\rangle'_i\delta_{LL'}\delta_{ij}
$$
$$
+\frac{1}{\kappa_2^2-\kappa_1^2}\left[B_{LL'\kappa_2}(\boldsymbol{\tau}_i-\boldsymbol{\tau}_j,\mathbf{k})-B^*_{L'L\kappa_1}(\boldsymbol{\tau}_j-\boldsymbol{\tau}_i,\mathbf{k})\right]\left(1-\delta(\kappa_1^2-\kappa_2^2)\right)
$$
$$
+\dot{B}_{LL'\kappa_1}(\boldsymbol{\tau}_i-\boldsymbol{\tau}_j,\mathbf{k})\delta(\kappa_1^2-\kappa_2^2)
$$
$$
-\langle H_{L\kappa_1}|J_{L\kappa_2}\rangle_i B_{LL'\kappa_2}(\boldsymbol{\tau}_i-\boldsymbol{\tau}_j,\mathbf{k})-B^*_{L'L\kappa_1}(\boldsymbol{\tau}_j-\boldsymbol{\tau}_i,\mathbf{k})\langle J_{L'\kappa_1}|H_{L'\kappa_2}\rangle_j
$$
$$
-\sum_{m}\sum_{L''}B^*_{L''L\kappa_1}(\boldsymbol{\tau}_m-\boldsymbol{\tau}_i,\mathbf{k})\langle J_{L''\kappa_1}|J_{L''\kappa_2}\rangle_m B_{L''L'\kappa_2}(\boldsymbol{\tau}_m-\boldsymbol{\tau}_j,\mathbf{k})\,.
$$
$$
(3.8.39)
$$

As mentioned before, this is identical to (3.8.35) with the structure constants replaced by their Bloch-summed counterparts. In a last step, we rewrite (3.8.39) for cubic rather than spherical harmonics. In that case the identity (3.7.18) holds and we obtain the result

$$
\langle D_{L\kappa_1}(\mathbf{r}_i,\mathbf{k})|D_{L'\kappa_2}(\mathbf{r}_j,\mathbf{k})\rangle_{\Omega_I}
$$
$$
= \langle H_{L\kappa_1}|H_{L\kappa_2}\rangle'_i\delta_{LL'}\delta_{ij}
$$

$$+ \frac{1}{\kappa_2^2 - \kappa_1^2} \left[B_{LL'\kappa_2}(\boldsymbol{\tau}_i - \boldsymbol{\tau}_j, \mathbf{k}) - B_{LL'\kappa_1}(\boldsymbol{\tau}_i - \boldsymbol{\tau}_j, \mathbf{k}) \right] \left(1 - \delta(\kappa_1^2 - \kappa_2^2) \right)$$

$$+ \dot{B}_{LL'\kappa_1}(\boldsymbol{\tau}_i - \boldsymbol{\tau}_j, \mathbf{k}) \delta(\kappa_1^2 - \kappa_2^2)$$

$$- \langle H_{L\kappa_1} | J_{L\kappa_2} \rangle_i B_{LL'\kappa_2}(\boldsymbol{\tau}_i - \boldsymbol{\tau}_j, \mathbf{k}) - B_{LL'\kappa_1}(\boldsymbol{\tau}_i - \boldsymbol{\tau}_j, \mathbf{k}) \langle J_{L'\kappa_1} | H_{L'\kappa_2} \rangle_j$$

$$- \sum_m \sum_{L''} B_{LL''\kappa_1}(\boldsymbol{\tau}_i - \boldsymbol{\tau}_m, \mathbf{k}) \langle J_{L''\kappa_1} | J_{L''\kappa_2} \rangle_m B_{L''L'\kappa_2}(\boldsymbol{\tau}_m - \boldsymbol{\tau}_j, \mathbf{k}) \; .$$

$$(3.8.40)$$

3.9 Pseudo Functions

In the previous sections of this chapter we have derived a lot of useful analytic expressions for the ASW envelope functions. Thereby we have learned about several appealing features of these functions. Nevertheless, as we also realized the singularities of the Hankel envelope functions at their origin, although cured by the concept of augmentation, are an obstacle for a more elegant formulation. Instead it would be desirable to have a set of functions at hand, which are regular and well behaving in all space.

The aforementioned disadvantage of the Hankel envelope functions can be superseeded in an elegant manner by the invention of pseudo functions. These functions are regular as well as smooth in all space. Moreover, they account for the many advantages offered by the envelope functions by changing the latter as little as possible. This is achieved by defining the pseudo function as the envelope function outside its on-center sphere while augmenting it with a smooth function inside this sphere. Note that there is no such augmentation in the off-center spheres, where the Hankel envelope functions are regular anyway. Eventually, with the definition of the pseudo functions at hand we will be able to give a very simple formulation of the ASW methods.

To be specific, we first turn to the region outside the atomic sphere at $\mathbf{R}_{\mu i}$ and require the pseudo function there to be identical to the envelope function. Hence, the pseudo function is defined by

$$H_{L\kappa}^0(\mathbf{r}_{\mu i})(1 - \Theta_{\mu i}) := H_{L\kappa}(\mathbf{r}_{\mu i})(1 - \Theta_{\mu i}) \, , \qquad (3.9.1)$$

where we denote the pseudo function by a superscript 0 and the Hankel envelope function is the same as given by (2.1.11). Note that from now on we will completely concentrate on the use of Hankel envelope functions. However, from the discussion at the beginning of Sect. 2.1 it is straightforward to reformulate the method in terms of Neumann envelope functions.

As mentioned before, both the envelope function and the just defined pseudo function are well-behaving smooth functions in all space outside the on-center atomic sphere. Hence, we did not distinguish between the true interstitial region and the region occupied by the off-center spheres.

Next we turn to the on-center sphere, where we are faced with the singularity of the Hankel envelope function at its origin. In this sphere, we arrive at the pseudo function by augmenting the Hankel envelope function by a smooth function. The simplest and most straighforward way to do this is the same as for the construction of the augmented functions in Sect. 2.1. To be concrete, we define

$$H^0_{L\kappa}(\mathbf{r}_{\mu i})\Theta_{\mu i} := \tilde{H}^0_{L\kappa}(\mathbf{r}_{\mu i}) := \tilde{h}^0_{l\kappa}(r_{\mu i})Y_L(\hat{\mathbf{r}}_{\mu i})\Theta_{\mu i} \ . \tag{3.9.2}$$

The radial part obeys the radial Schrödinger equation with the potential being the muffin-tin zero

$$\left[-\frac{1}{r_{\mu i}} \frac{\partial^2}{\partial r^2_{\mu i}} r_{\mu i} + \frac{l(l+1)}{r^2_{\mu i}} - v_{MTZ} - E^{(0)}_{l\kappa i} \right] \tilde{h}^0_{l\kappa}(r_{\mu i}) = 0 \ , \tag{3.9.3}$$

and is subject to the boundary conditions

$$\left[(\frac{\partial}{\partial r_{\mu i}})^n \left(\tilde{h}^0_{l\kappa}(r_{\mu i}) - \bar{h}^{(1)}_l(\kappa r_{\mu i}) \right) \right]_{r_{\mu i}=S_i} = 0 \ , \qquad n = 0, 1. \tag{3.9.4}$$

In addition, it must be a nodeless function as well as regular at the origin. Hence, the augmented pseudo function $\tilde{h}^0_{l\kappa}(r_{\mu i})$ is constructed in complete analogy to the augmented function $\tilde{h}_{l\kappa\sigma}(r_{\mu i})$ the only difference being the fact that a vanishing intraatomic potential is used here. As a consequence, we are able to take over many results derived for the augmented functions with the appropriate notational changes.

In contrast to the polynomial used in [5] as the radial part of the pseudo function within the on-center sphere, the present choice has the particular advantage to be a solution of the free radial Schrödinger equation. For this reason it will be identical to the augmented Hankel function for high values of the angular momentum, where the centrifugal term dominates the intraatomic potential. Moreover, we now have a pseudo function at hand, which obeys Schrödinger's equation for free particles in all space, is regular, continuous, and differentiable everywhere, and, finally, is quite smooth. The latter fact can be read off from the Fourier transform, which falls off as $\frac{1}{|\mathbf{q}|^4}$ for large $\mathbf{q}$ as has been shown in [5] and as we will also see in Sect. 3.11.

Next, combining (3.9.1) and (3.9.2) we obtain for the full pseudo function

$$H^0_{L\kappa}(\mathbf{r}_{\mu i}) = \tilde{H}^0_{L\kappa}(\mathbf{r}_{\mu i}) + H_{L\kappa}(\mathbf{r}_{\mu i})(1 - \Theta_{\mu i}) \ . \tag{3.9.5}$$

Still we have to take into account crystal translational symmetry and use Bloch sums of the pseudo functions, which we define as

$$D^0_{L\kappa}(\mathbf{r}_i, \mathbf{k}) := \sum_\mu e^{i\mathbf{k}\mathbf{R}_\mu} H^0_{L\kappa}(\mathbf{r}_{\mu i}) \ . \tag{3.9.6}$$

Combining this with the expression (3.9.5) for the pseudo function and the Bloch sum (3.5.2) of the Hankel envelope function we get

$$D^0_{L\kappa}(\mathbf{r}_i, \mathbf{k}) = D_{L\kappa}(\mathbf{r}_i, \mathbf{k}) + \hat{H}^0_{L\kappa}(\mathbf{r}_i) , \tag{3.9.7}$$

where we used the formal abbreviation

$$\hat{H}^0_{L\kappa}(\mathbf{r}_i) = \tilde{H}^0_{L\kappa}(\mathbf{r}_i) - H_{L\kappa}(\mathbf{r}_i)\Theta_i . \tag{3.9.8}$$

Having defined the pseudo functions as well as their Bloch sums we will in the following two sections turn to the calculations of overlap integrals of pseudo functions centered at different sites as well as to the calculation of their Fourier transform.

3.10 Overlap Integrals of Pseudo Functions

In the present section we turn to the calculation of overlap integrals of pseudo functions. Since the construction of the pseudo functions is much like augmentation we will thereby not only fall back on the results of Sect. 3.8 for the overlap integrals of the envelope functions but also benefit from the evaluation of the overlap matrix for the standard ASW method as outlined in Sect. 2.2. In doing so we skip the calculation of the overlap integrals of the real-space pseudo functions and turn directly to those of the corresponding Bloch sums.

To be concrete, we recall that the pseudo function was arrived at by augmenting the envelope function within the on-center sphere with the function $\tilde{H}^0_{L\kappa}(\mathbf{r}_{\mu i})$, whose radial part obeys the free-particle Schrödinger equation (3.9.3) with the boundary conditions (3.9.4). In contrast, in the off-center spheres no augmentation was performed. We are thus able to take over the expression (2.2.23) for the elements of the overlap matrix of the standard ASW method. With the appropriate notational changes accounting for the pseudo functions it reads as

$$
\begin{aligned}
&{}^0\langle L\kappa_1 i | L'\kappa_2 j \rangle^0_c \\
&= \left[\langle H^0_{L\kappa_1} | H^0_{L\kappa_2} \rangle_j + \langle H_{L\kappa_1} | H_{L\kappa_2} \rangle'_j \right] \delta_{LL'}\delta_{ij} \\
&\quad + \frac{1}{\kappa_2^2 - \kappa_1^2} \left[B_{LL'\kappa_2}(\boldsymbol{\tau}_i - \boldsymbol{\tau}_j, \mathbf{k}) - B^*_{L'L\kappa_1}(\boldsymbol{\tau}_j - \boldsymbol{\tau}_i, \mathbf{k}) \right] \left(1 - \delta(\kappa_1^2 - \kappa_2^2) \right) \\
&\quad + \dot{B}_{LL'\kappa_1}(\boldsymbol{\tau}_i - \boldsymbol{\tau}_j, \mathbf{k})\delta(\kappa_1^2 - \kappa_2^2) \\
&\quad + \left[\langle H^0_{L\kappa_1} | J_{L\kappa_2} \rangle_i - \langle H_{L\kappa_1} | J_{L\kappa_2} \rangle_i \right] B_{LL'\kappa_2}(\boldsymbol{\tau}_i - \boldsymbol{\tau}_j, \mathbf{k}) \\
&\quad + B^*_{L'L\kappa_1}(\boldsymbol{\tau}_j - \boldsymbol{\tau}_i, \mathbf{k}) \left[\langle J_{L'\kappa_1} | H^0_{L'\kappa_2} \rangle_j - \langle J_{L'\kappa_1} | H_{L'\kappa_2} \rangle_j \right] , \tag{3.10.1}
\end{aligned}
$$

where we have used the definition

$$|L\kappa i\rangle^0 := D^0_{L\kappa}(\mathbf{r}_i, \mathbf{k}). \tag{3.10.2}$$

Note that, in contrast to (2.2.23), the overlap matrix (3.10.1) does *not* contain three-center terms. This is due to the fact that in the off-center spheres the pseudo function is identical to the envelope function. Hence, in the language of the standard ASW method, the augmented Bessel function equals the Bessel envelope functions and thus terms corresponding to the last square brackets in (2.2.23) cancel out here. As a consequence, (3.10.1) is exact whereas in the overlap matrix (2.2.23) we had to truncate the L''-summation in the three-center contributions.

As in Sect. 2.2, it is useful to define the following abbreviations for the radial integrals

$$X_{L\kappa_1\kappa_2 i}^{0(S,1)} = \langle \tilde{H}_{L\kappa_1}^0 | \tilde{H}_{L\kappa_2}^0 \rangle_i + \langle H_{L\kappa_1} | H_{L\kappa_2} \rangle_i', \tag{3.10.3}$$

$$X_{L\kappa_1\kappa_2 i}^{0(S,2)} = \langle \tilde{H}_{L\kappa_1}^0 | J_{L\kappa_2} \rangle_i - \langle H_{L\kappa_1} | J_{L\kappa_2} \rangle_i + \frac{1}{\kappa_2^2 - \kappa_1^2} \left(1 - \delta(\kappa_1^2 - \kappa_2^2) \right) . \tag{3.10.4}$$

They allow to rewrite (3.10.1) as

$$
\begin{aligned}
{}^0\langle L\kappa_1 i | L'\kappa_2 j \rangle_c^0 \\
= X_{L\kappa_1\kappa_2 i}^{0(S,1)} \delta_{LL'} \delta_{ij} + \dot{B}_{LL'\kappa_1}(\boldsymbol{\tau}_i - \boldsymbol{\tau}_j, \mathbf{k}) \delta(\kappa_1^2 - \kappa_2^2) \\
+ X_{L\kappa_1\kappa_2 i}^{0(S,2)} B_{LL'\kappa_2}(\boldsymbol{\tau}_i - \boldsymbol{\tau}_j, \mathbf{k}) + B_{L'L\kappa_1}^*(\boldsymbol{\tau}_j - \boldsymbol{\tau}_i, \mathbf{k}) X_{L'\kappa_1\kappa_2 j}^{0(S,2)} .
\end{aligned}
\tag{3.10.5}
$$

Finally, we reformulate the previous results for cubic harmonics and write

$$
\begin{aligned}
{}^0\langle L\kappa_1 i | L'\kappa_2 j \rangle_c^0 \\
= X_{L\kappa_1\kappa_2 i}^{0(S,1)} \delta_{LL'} \delta_{ij} + \dot{B}_{LL'\kappa_1}(\boldsymbol{\tau}_i - \boldsymbol{\tau}_j, \mathbf{k}) \delta(\kappa_1^2 - \kappa_2^2) \\
+ X_{L\kappa_1\kappa_2 i}^{0(S,2)} B_{LL'\kappa_2}(\boldsymbol{\tau}_i - \boldsymbol{\tau}_j, \mathbf{k}) + B_{LL'\kappa_1}(\boldsymbol{\tau}_i - \boldsymbol{\tau}_j, \mathbf{k}) X_{L'\kappa_1\kappa_2 j}^{0(S,2)} .
\end{aligned}
\tag{3.10.6}
$$

Still, we have to look in more detail at those radial integrals entering (3.10.3) and (3.10.4), which contain the pseudo functions. Their calculation is performed in the same manner as that of the corresponding integrals entering the overlap matrix of the standard ASW method, which we discussed in the App. A.1. Here we leave the details of the calculations to the App. B.9.

3.11 Fourier Transform of Pseudo Functions

In the course of the development of the plane-wave based full-potential ASW method we will need the Fourier transform of the pseudo function as defined in Sect. 3.9. To this end we start from the definition (3.9.6) of the Bloch-summed pseudo function and define its Fourier transform by

$$D_{L\kappa}^0(\mathbf{r}_i, \mathbf{k}) =: \sum_n D_{L\kappa i}^0(\mathbf{K}_n + \mathbf{k}) e^{i(\mathbf{K}_n + \mathbf{k})\mathbf{r}_i}. \tag{3.11.1}$$

Note that we have used the same symbol for the Bloch-summed pseudo function and its Fourier transform. However, the correct meaning will always be clear from the different arguments and from the context.

The expansion coefficients entering (3.11.1) are given by

$$
\begin{aligned}
D^0_{L\kappa i}(\mathbf{K}_n + \mathbf{k}) &= \frac{1}{\Omega_c} \int_{\Omega_c} d^3\mathbf{r}_i \, D^0_{L\kappa}(\mathbf{r}_i, \mathbf{k}) e^{-i(\mathbf{K}_n+\mathbf{k})\mathbf{r}_i} \\
&= \frac{1}{\Omega_c} \sum_\mu \int_{\Omega_c} d^3\mathbf{r}_i \, H^0_{L\kappa}(\mathbf{r}_{\mu i}) e^{i\mathbf{k}\mathbf{R}_\mu} e^{-i(\mathbf{K}_n+\mathbf{k})\mathbf{r}_i} \\
&= \frac{1}{\Omega_c} \sum_\mu \int_{\Omega_c} d^3\mathbf{r}_i \, H^0_{L\kappa}(\mathbf{r}_{\mu i}) e^{-i(\mathbf{K}_n+\mathbf{k})\mathbf{r}_{\mu i}} \, .
\end{aligned}
\tag{3.11.2}
$$

Here we have, in the second step, used the definition (3.9.6) of the Bloch sum. Employing the translational invariance of the lattice we replace the lattice sum of integrals over the unit cell by a single integral extending over all space and write

$$
\begin{aligned}
&D^0_{L\kappa i}(\mathbf{K}_n + \mathbf{k}) \\
&= \frac{1}{\Omega_c} \int_\Omega d^3\mathbf{r}_i \, H^0_{L\kappa}(\mathbf{r}_i) e^{-i(\mathbf{K}_n+\mathbf{k})\mathbf{r}_i} \\
&= \frac{1}{\Omega_c} \int_{\Omega-\Omega_i} d^3\mathbf{r}_i \, H_{L\kappa}(\mathbf{r}_i) e^{-i(\mathbf{K}_n+\mathbf{k})\mathbf{r}_i} + \frac{1}{\Omega_c} \int_{\Omega_i} d^3\mathbf{r}_i \, \tilde{H}^0_{L\kappa}(\mathbf{r}_i) e^{-i(\mathbf{K}_n+\mathbf{k})\mathbf{r}_i} \, ,
\end{aligned}
\tag{3.11.3}
$$

where in the last step we have used (3.9.5) for the pseudo function. Inserting into (3.11.3) the expansion (3.6.1) of the plane waves in spherical Bessel functions and using the definition (3.1.3) of the Bessel envelope function as well as the orthonormality of the spherical harmonics we arrive at

$$
\begin{aligned}
&D^0_{L\kappa i}(\mathbf{K}_n + \mathbf{k}) \\
&= \frac{4\pi}{\Omega_c}(-i)^l |\mathbf{K}_n + \mathbf{k}|^l Y_L(\widehat{\mathbf{K}_n + \mathbf{k}}) \\
&\quad \left[\int_{\Omega-\Omega_i} d^3\mathbf{r}_i \, J^*_{L|\mathbf{K}_n+\mathbf{k}|}(\mathbf{r}_i) H_{L\kappa}(\mathbf{r}_i) + \int_{\Omega_i} d^3\mathbf{r}_i \, J^*_{L|\mathbf{K}_n+\mathbf{k}|}(\mathbf{r}_i) \tilde{H}^0_{L\kappa}(\mathbf{r}_i) \right] \, .
\end{aligned}
\tag{3.11.4}
$$

For the integrals in square brackets we fall back on the results (3.3.32) as well as (B.9.3)/(B.9.4) and get the result

$$
\begin{aligned}
&D^0_{L\kappa i}(\mathbf{K}_n + \mathbf{k}) \\
&= -\frac{4\pi}{\Omega_c}(-i)^l |\mathbf{K}_n + \mathbf{k}|^l Y_L(\widehat{\mathbf{K}_n + \mathbf{k}}) \\
&\quad \left[\frac{1}{\kappa^2 - (\mathbf{K}_n + \mathbf{k})^2} - \frac{1}{E^{(0)}_{l\kappa i} - (\mathbf{K}_n + \mathbf{k})^2} \right] \\
&\quad W\{r_i \bar{h}_l(\kappa r_i), r_i \bar{j}_l(|\mathbf{K}_n + \mathbf{k}|r_i); r_i\}|_{r_i = S_i}
\end{aligned}
$$

$$
= -\frac{4\pi}{\Omega_c}(-i)^l|\mathbf{K}_n + \mathbf{k}|^l Y_L(\widehat{\mathbf{K}_n + \mathbf{k}})
$$

$$
\left[\frac{1}{\kappa^2 - (\mathbf{K}_n + \mathbf{k})^2} - \frac{1}{E_{l\kappa i}^{(0)} - (\mathbf{K}_n + \mathbf{k})^2}\right]
$$

$$
S_i^2\left[\bar{h}_l^{(1)}(\kappa r_i)\frac{\partial}{\partial r_i}\bar{j}_l(|\mathbf{K}_n + \mathbf{k}|r_i) - \bar{j}_l(|\mathbf{K}_n + \mathbf{k}|r_i)\frac{\partial}{\partial r_i}\bar{h}_l^{(1)}(\kappa r_i)\right]_{r_i=S_i},
$$

$$\tag{3.11.5}$$

for $(\mathbf{K}_n + \mathbf{k})^2 \neq E_{l\kappa i}^{(0)}$ and

$$
D_{L\kappa i}^0(\mathbf{K}_n + \mathbf{k})
$$

$$
= -\frac{4\pi}{\Omega_c}(-i)^l|\mathbf{K}_n + \mathbf{k}|^l Y_L(\widehat{\mathbf{K}_n + \mathbf{k}})
$$

$$
\left[\frac{1}{\kappa^2 - (\mathbf{K}_n + \mathbf{k})^2}W\{r_i\bar{h}_l(\kappa r_i), r_i\bar{j}_l(|\mathbf{K}_n + \mathbf{k}|r_i); r_i\}|_{r_i=S_i} - S_{l\kappa i}^{(0)}\right]
$$

$$
= -\frac{4\pi}{\Omega_c}(-i)^l|\mathbf{K}_n + \mathbf{k}|^l Y_L(\widehat{\mathbf{K}_n + \mathbf{k}})
$$

$$
\left[\frac{S_i^2}{\kappa^2 - (\mathbf{K}_n + \mathbf{k})^2}\left[\bar{h}_l^{(1)}(\kappa r_i)\frac{\partial}{\partial r_i}\bar{j}_l(|\mathbf{K}_n + \mathbf{k}|r_i)\right.\right.
$$

$$
\left.\left. -\bar{j}_l(|\mathbf{K}_n + \mathbf{k}|r_i)\frac{\partial}{\partial r_i}\bar{h}_l^{(1)}(\kappa r_i)\right]_{r_i=S_i} - S_{l\kappa i}^{(0)}\right],
$$

$$\tag{3.11.6}$$

for $(\mathbf{K}_n + \mathbf{k})^2 = E_{l\kappa i}^{(0)}$. In case that $(\mathbf{K}_n + \mathbf{k})^2 = \kappa^2$ the Wronskian in both (3.11.5) and (3.11.6) becomes unity, hence, the difference of the Wronskians taken at the sphere radius and infinity vanishes. We thus apply l'Hospitals rule and get instead of (3.11.5) and (3.11.6) the results

$$
D_{L\kappa i}^0(\mathbf{K}_n + \mathbf{k})
$$

$$
= -\frac{4\pi}{\Omega_c}(-i)^l|\mathbf{K}_n + \mathbf{k}|^l Y_L(\widehat{\mathbf{K}_n + \mathbf{k}})
$$

$$
\left[S_i^2\left[\frac{\partial}{\partial \kappa^2}\left(\bar{h}_l^{(1)}(\kappa r_i)\right)\frac{\partial}{\partial r_i}(\bar{j}_l(\kappa r_i)) - \bar{j}_l(\kappa r_i)\frac{\partial}{\partial \kappa^2}\frac{\partial}{\partial r_i}\left(\bar{h}_l^{(1)}(\kappa r_i)\right)\right]_{r_i=S_i}\right.
$$

$$
\left. -\frac{W\{r_i\bar{h}_l(\kappa r_i), r_i\bar{j}_l(|\mathbf{K}_n + \mathbf{k}|r_i); r_i\}|_{r_i=S_i}}{E_{l\kappa i}^{(0)} - (\mathbf{K}_n + \mathbf{k})^2}\right],
$$

$$\tag{3.11.7}$$

for $(\mathbf{K}_n + \mathbf{k})^2 \neq E_{l\kappa i}^{(0)}$ and

$$
D_{L\kappa i}^0(\mathbf{K}_n + \mathbf{k})
$$

$$= -\frac{4\pi}{\Omega_c}(-i)^l |\mathbf{K}_n + \mathbf{k}|^l Y_L(\widehat{\mathbf{K}_n + \mathbf{k}})$$

$$\left[S_i^2 \left[\frac{\partial}{\partial \kappa^2} \left(\bar{h}_l^{(1)}(\kappa r_i) \right) \frac{\partial}{\partial r_i} \left(\bar{j}_l(\kappa r_i) \right) \right.\right.$$

$$\left.\left. - \bar{j}_l(\kappa r_i) \frac{\partial}{\partial \kappa^2} \frac{\partial}{\partial r_i} \left(\bar{h}_l^{(1)}(\kappa r_i) \right) \right]_{r_i = S_i} - S_{l\kappa i}^{(0)} \right] , \qquad (3.11.8)$$

for $(\mathbf{K}_n + \mathbf{k})^2 = E_{l\kappa i}^{(0)}$. Nevertheless, since the same energy denominator appears in the structure constants we assume that the case $(\mathbf{K}_n + \mathbf{k})^2 = \kappa^2$ has to be avoided anyway.

Special care needs the case $\mathbf{K}_n + \mathbf{k} \to \mathbf{0}$. According to (3.11.5) and (3.11.6) then all contributions except for those with $l = 0$ vanish. Using (3.1.44) we calculate

$$\bar{j}_l(|\mathbf{K}_n + \mathbf{k}|r_i) = \delta_{l0}, \qquad \text{and} \qquad \frac{\partial}{\partial r_i}\bar{j}_l(|\mathbf{K}_n + \mathbf{k}|r_i) = 0 , \qquad (3.11.9)$$

for $\mathbf{K}_n + \mathbf{k} \to \mathbf{0}$. With these identities at hand (3.11.5) to (3.11.6) reduce to

$$D_{L\kappa i}^0(\mathbf{0}) = \delta_{l0} \frac{\sqrt{4\pi}}{\Omega_c} \left[\frac{1}{\kappa^2} - \frac{1}{E_{0\kappa i}^{(0)}} \right] S_i^2 \frac{\partial}{\partial r_i} \bar{h}_0^{(1)}(\kappa r_i) \bigg|_{r_i = S_i}, \qquad (3.11.10)$$

for $E_{0\kappa i}^{(0)} \neq 0$ and

$$D_{L\kappa i}^0(\mathbf{0}) = \delta_{l0} \frac{\sqrt{4\pi}}{\Omega_c} \left[\frac{S_i^2}{\kappa^2} \frac{\partial}{\partial r_i} \bar{h}_0^{(1)}(\kappa r_i) \bigg|_{r_i = S_i} + S_{0\kappa i}^{(0)} \right] , \qquad (3.11.11)$$

for $E_{0\kappa i}^{(0)} = 0$. As before we consider, in addition, the case $(\mathbf{K}_n + \mathbf{k})^2 = \kappa^2 = 0$, where we use (3.1.45) for the barred Hankel function in the limit of small argument, i.e.

$$\bar{h}_l(\kappa r_i) = \delta_{l0} \frac{1}{r_i} , \qquad \text{and} \qquad \frac{\partial}{\partial r_i}\bar{h}_l(\kappa r_i) = -\delta_{l0} \frac{1}{r_i^2} . \qquad (3.11.12)$$

For $\kappa \to 0$ these expressions become independent of κ^2 and the derivatives with respect to κ^2 vanish. As a consequence, applying l'Hospitals rule gives zero for the corresponding terms and we are able to note instead of (3.11.9) and (3.11.10)

$$D_{L\kappa i}^0(\mathbf{0}) = \delta_{l0} \frac{\sqrt{4\pi}}{\Omega_c} \frac{1}{E_{0\kappa i}^{(0)}} , \qquad (3.11.13)$$

for $E_{0\kappa i}^{(0)} \neq 0$ and

$$D_{L\kappa i}^0(\mathbf{0}) = \delta_{l0} \frac{\sqrt{4\pi}}{\Omega_c} S_{0\kappa i}^{(0)} , \qquad (3.11.14)$$

for $E_{0\kappa i}^{(0)} = 0$.

As is obvious from the expression in square brackets in (3.11.5), the coefficients $D_{L\kappa i}^0(\mathbf{K}_n + \mathbf{k})$ entering the plane wave expansion (3.11.1) behave like $|\mathbf{K}_n + \mathbf{k}|^{-4}$ for large $\mathbf{K}_n$.

References

1. M. Abramowitz and I. A. Stegun, *Handbook of Mathematical Functions* (Dover, New York 1972)
2. P. E. Blöchl, Gesamtenergien, Kräfte und Metall-Halbleiter Grenzflächen. PhD thesis, Universität Stuttgart (1989)
3. M. Born and M. Bradburn, Proc. Cambridge Phil. Soc. **39**, 104 (1943)
4. P. P. Ewald, Ann. Phys. **64**, 253 (1921)
5. V. Eyert, Entwicklung und Implementation eines Full-Potential-ASW-Verfahrens. PhD thesis, Technische Hochschule Darmstadt (1991)
6. C. D. Gelatt, Jr., H. Ehrenreich, and R. E. Watson, Phys. Rev. B **15**, 1613 (1977)
7. F. S. Ham and B. Segall, Phys. Rev. **124**, 1786 (1961)
8. J. D. Jackson, *Classical Electrodynamics* (Wiley, New York 1975)
9. A. Messiah, *Quantum Mechanics*, vol 1 (North Holland, Amsterdam 1976)
10. R. D. Misra, Proc. Cambridge Phil. Soc. **36**, 173 (1940)
11. A. P. Smith and N. W. Ashcroft, Phys. Rev. B **38**, 12942 (1988)
12. M. P. Tosi, Cohesion of Ionic Solids in the Born Model. In: *Solid State Physics*, vol 16, ed by F. Seitz and D. Turnbull (Academic Press, New York 1964) pp 1–120
13. G. N. Watson, *Theory of Bessel Functions* (University Press, Cambridge 1966)
14. A. R. Williams, J. Kübler, and C. D. Gelatt, Jr., Phys. Rev. B **19**, 6094 (1979)

4

The Plane-Wave Based Full-Potential ASW Method

The standard ASW method as outlined in Chap. 2 has proven to be extremely fast, stable and reliable in a vast number of applications. At the same time, its basic formalism as well as its implementation are rather simple. However, the method suffers from sometimes too crude approximations and simplifications. One of these is the atomic-sphere approximation, which, as we have mentioned, hinders an accurate determination of the total energy and does not allow for the calculation, e.g. of electric field gradients and phonon frequencies. However, systematic developments towards a higher level of accuracy and inclusion of new features are not as straightforward as one might wish.

In the present chapter we will describe the plane-wave based full-potential ASW method, which was originally introduced in the 1980's and constitutes the first full-potential implementation of the ASW method [13]. In the present context, we will essentially refer to the original formulation but will also include more recent refinements.

The development of a plane-wave based full-potential ASW method is generally guided by two major requirements. First, the formulation should be in principle exact, which means that, within the framework of density-functional theory and the local-density approximation, there should be no further approximations. Second, the resulting code should allow for a computational load, which is still below that of the albeit even more accurate full-potential plane-wave methods.

These requirements lead to the following general guidelines. First, we have to abandon the atomic-sphere approximation and will use the muffin-tin approximation for the setup of the basis functions. In particular, we require that the spheres must not overlap. For this reason, we will use the terms muffin-tin sphere and atomic sphere synonymously throughout in this chapter unless otherwise noted. Second, we will use the same basis functions as in the standard ASW methods, which offer the great advantage of a minimal basis set and, hence, a small secular matrix. Finally, we will introduce the concept of additive augmentation for the basis functions and use it as well for all quantities of interest. To be specific, besides splitting each basis function into pseudo and local contributions, we will apply an analogous separation to the secular matrix, the electron density, the full potential, and the expression for

V. Eyert: *The Plane-Wave Based Full-Potential ASW Method*, Lect. Notes Phys. **719**, 117–173 (2007)
DOI 10.1007/978-3-540-71007-3_4

the total energy. For all these quantities, the respective pseudo part extends over all space and shows only small variations. As a consequence, it can be expanded in plane waves. Inside the non-overlapping muffin-tin spheres, the pseudo part is accompanied by local parts, which contain all the intrasphere details of the respective function and are well described by an expansion in spherical or cubic harmonics. In passing, we mention full-potential LMTO methods, which were developed along the same guidelines by Weyrich and by Blöchl [7, 32–36] as well as the full-charge LMTO scheme devised by Christensen [8].

The rather high computational demands of any full-potential method as compared to the standard ASW method motivates the search for an intermediate procedure, which would ideally offer the simplicity and speed of the latter and the increased accuracy of a full-potential method. In particular, such a method might give a first approach to the calculation of electric field gradients. Indeed, such a method is within reach and essentially based on the notion of reducing the pseudo contributions to a constant and using overlapping atomic spheres rather than touching muffin-tin spheres. Since this method thus appears as a limiting case of the full-potential approach an extra derivation is not justified. In contrast, we will mention the respective changes at several places within this chapter.

In order to sketch the formulation of the plane-wave based full-potential ASW method we follow the lines already taken in Chap. 2. To start with, we will reconsider the basis functions and outline the concept of additive augmentation. The resulting representation of the basis functions in terms of pseudo and local contributions will strongly influence the setup of the secular matrix as well as the representation of the electron density, the potential, and the total energy to be described in the subsequent sections.

4.1 Additive Augmentation

The construction of the ASW basis functions as outlined in Sect. 2.1 was based on Slater's muffin-tin approximation, which allowed for solving Schrödinger's equation separately in the interstitial region and the atomic spheres. It thus laid ground for a partial-wave approach, where the solutions in different portions of space are matched continuously and differentiably to form a single basis function. In the interstitial, we opted for spherical waves (for a fixed energy κ^2) as the so-called envelope functions, which, within the atomic spheres, were replaced by numerical solutions, the so-called augmented functions. In doing so, we arrived eventually at *augmented spherical waves*.

The so constructed basis functions have proven to offer a number of distinct advantages. In particular, due to the use of spherical waves, they form a minimal basis set of only few functions per atom, which fact lays ground for the very high efficiency of the ASW method. In addition, due to their atomic-like character, the ASW basis functions allow for a very intuitive analysis of the calculated results.

Finally, since the full potential deviates only little from the muffin-tin potential — a fact, which *a posteriori* justifies Slater's important step — partial waves built on the muffin-tin or atomic-sphere approximation still constitute a good starting point for a full-potential method. This is even more so since the wave function is determined variationally and, hence, may be well written in terms of basis functions, which solve Schrödinger's equation with only the muffin-tin potential in the respective portions of space. To summarize, using the basis functions of the standard ASW method as proposed in Sect. 2.1 we are well prepared for future developments. For this reason, there is no need for a restructuring of the ASW basis set.

Still, a slight reformulation of the basis functions will be necessary. This is for the following reason: In the standard ASW method we splitted in (2.2.11) and (2.2.22) each matrix element of the Hamiltonian and the overlap matrix into two parts. The first one was an integral extending over the whole unit cell and was built with the envelope functions. In contrast, the second integral carried all the additional contributions from inside the atomic spheres. This way the truncation of the L' sums in the one-center expansions (2.1.16) and (2.1.23) of Hankel functions in terms of Bessel functions centered at a different site could be justified. Moreover, the evaluation of the matrix elements became very simple. Nevertheless, that approach was quite *ad hoc* and not explicitly derived from a corresponding formulation of the basis functions. In the present context, we will make up for this weakness and start from a corresponding formulation of the basis functions. This will lead us to a different point of view of the augmentation process, which starts from the one-center expansion of the envelope function and replaces only the lower and intermediate partial waves by the respective augmented functions. In contrast, the higher partial waves are kept as they are [21, p. 41]. Note that only in the off-center spheres the envelope function and its one-center expansion differ, while in the on-center sphere they are identical.

To be specific, we will likewise want to write the basis functions as a sum of a rather slowly varying part extending through all space and intraatomic contributions, which contain all the atomic details and vanish outside the atomic spheres. These two parts we refer to as the pseudo and the local contributions, respectively. Note that, by definition, the pseudo part is identical to the envelope functions or, equivalently, the basis functions in the interstitial region and that the intraatomic contributions will allow for an expansion in spherical harmonics. The just outlined principle of separating a function has been already sketched in (2.1.2) and (2.1.4) for the muffin-tin potential and it has also been successfully employed in the original implementation of the plane-wave based full-potential ASW method [13].

To start with, we recognize the pseudo part of the basis function as the pseudo function defined in Sect. 3.9. Insofar we can fall back on the derivation given there. Nevertheless, in the present context it is useful to eplicitly distinguish between the interstitial region and the off-center spheres. Hence, we split off from (3.9.1) the contribution from the interstitial region,

$$H^0_{L\kappa}(\mathbf{r}_{\mu i})\Theta_I := H_{L\kappa}(\mathbf{r}_{\mu i})\Theta_I \, , \qquad (4.1.1)$$

where, as usual, we denote the pseudo function by a superscript 0 and the Hankel envelope function is the same as given by (2.1.11). Note that, as already mentioned in Sect. 3.9, we concentrate exclusively on the use of Hankel envelope functions. However, from the discussion at the beginning of Sect. 2.1 it is straightforward to reformulate the method in terms of Neumann envelope functions.

In the off-center spheres we likewise have from (3.9.1)

$$H_{L\kappa}^0(\mathbf{r}_{\mu i})\Theta_{\nu j}(1 - \delta_{\mu\nu}\delta_{ij}) := H_{L\kappa}(\mathbf{r}_{\mu i})\Theta_{\nu j}(1 - \delta_{\mu\nu}\delta_{ij}) . \tag{4.1.2}$$

We recall that in the off-center spheres this function can be expanded in Bessel envelope functions via the expansion theorem (2.1.16).

Finally, in the on-center sphere we employ (3.9.2) to (3.9.4) and arrive at the expression (3.9.5) for the full pseudo function,

$$H_{L\kappa}^0(\mathbf{r}_{\mu i}) = \tilde{H}_{L\kappa}^0(\mathbf{r}_{\mu i}) + H_{L\kappa}(\mathbf{r}_{\mu i})(1 - \Theta_{\mu i}) . \tag{4.1.3}$$

Note that the pseudo function does not depend on the spin index.

Having set up the pseudo function we are left with defining the local functions. According to the introductory remarks they are just given by the difference of the augmented and the pseudo function inside the atomic spheres.

Starting now with the on-center sphere we thus write down the following expression for the local function

$$\hat{H}_{L\kappa\sigma}(\mathbf{r}_{\mu i}) = \tilde{H}_{L\kappa\sigma}(\mathbf{r}_{\mu i}) - \tilde{H}_{L\kappa}^0(\mathbf{r}_{\mu i}) . \tag{4.1.4}$$

In contrast to the pseudo function the local function does depend on the spin index and reflects all the details of the intraatomic potential. Note that, due to the boundary conditions (2.1.15), the local function vanishes continuously and differentiably at the sphere boundary.

In the off-center spheres the construction of the local functions works in the same way as in the on-center spheres. However, as already outlined in Sect. 2.1, we first have to expand the Hankel envelope functions in terms of Bessel envelope functions centered in the off-center sphere and only after that we perform the augmentation, which leads to the augmented Bessel function. Now aiming at the local functions, i.e. at the difference of the augmented and the pseudo function, we proceed in exactly the same way in order to be consistent. With this in mind we define

$$\hat{J}_{L'\kappa\sigma}(\mathbf{r}_{\nu j}) = \tilde{J}_{L'\kappa\sigma}(\mathbf{r}_{\nu j}) - J_{L'\kappa}(\mathbf{r}_{\nu j})\Theta_{\nu j} . \tag{4.1.5}$$

Combining all the previous definitions we get for the basis functions, the *augmented spherical waves*, the result

$$H_{L\kappa\sigma}^\infty(\mathbf{r}_{\mu i}) = H_{L\kappa}^0(\mathbf{r}_{\mu i}) + \hat{H}_{L\kappa\sigma}(\mathbf{r}_{\mu i})$$
$$+ \sum_{L'\nu j}(1 - \delta_{\mu\nu}\delta_{ij})\hat{J}_{L'\kappa\sigma}(\mathbf{r}_{\nu j})B_{L'L\kappa}(\mathbf{R}_{\nu j} - \mathbf{R}_{\mu i}) , \tag{4.1.6}$$

which complements (2.1.24).

In a final step, we again take into account crystal translational symmetry and work with Bloch sums of the basis functions. Using the definition (3.9.6) of the

Bloch-summed pseudo function as well as the Bloch-summed structure constants (2.1.29) we get for the Bloch sum of the basis function

$$D_{L\kappa\sigma}^{\infty}(\mathbf{r}_i, \mathbf{k}) = D_{L\kappa}^{0}(\mathbf{r}_i, \mathbf{k}) + \hat{H}_{L\kappa\sigma}(\mathbf{r}_i) + \sum_{L'j} \hat{J}_{L'\kappa\sigma}(\mathbf{r}_j) B_{L'L\kappa}(\boldsymbol{\tau}_j - \boldsymbol{\tau}_i, \mathbf{k}) . \quad (4.1.7)$$

With this at hand we are able to write down the wave function in close analogy with (2.1.31) as

$$\begin{aligned}
\psi_{\mathbf{k}\sigma}(\mathbf{r}) = &\sum_{L\kappa i} c_{L\kappa i\sigma}(\mathbf{k}) D_{L\kappa}^{0}(\mathbf{r}_i, \mathbf{k}) \\
&+ \sum_{L\kappa i} c_{L\kappa i\sigma}(\mathbf{k}) \hat{H}_{L\kappa\sigma}(\mathbf{r}_i) \\
&+ \sum_{L'\kappa j} a_{L'\kappa j\sigma}(\mathbf{k}) \hat{J}_{L'\kappa\sigma}(\mathbf{r}_j) ,
\end{aligned} \quad (4.1.8)$$

where we have abbreviated

$$a_{L'\kappa j\sigma}(\mathbf{k}) = \sum_{Li} c_{L\kappa i\sigma}(\mathbf{k}) B_{L'L\kappa}(\boldsymbol{\tau}_j - \boldsymbol{\tau}_i, \mathbf{k}) . \quad (4.1.9)$$

Finally, we turn to the core states, which are treated in exactly the same manner as in the standard ASW method. Strictly speaking, this is true only as long as the overlapping atomic spheres are used. In contrast, as soon as we opt for the smaller muffin-tin spheres as in the present chapter, energetically higher lying core states might give finite contributions at the sphere boundary. These so-called semi-core states may then contribute to the bonding and must be rather attributed to the set of valence electrons and, hence, must be included in the secular matrix. Yet, since these states are not expected to display the same dispersion as the true valence states do, they are usually treated separately from the latter.

It was the purpose of the previous reformulation of the basis functions to prepare for a representation of the elements of the secular matrix as a sum of an integral extending over the unit cell, which contains only slowly varying functions, plus local, i.e. purely intraatomic contributions. Obviously, this will require the calculation of products of the basis functions. Since the same holds for the construction of the electron density to be performed lateron we will already now investigate such products in more detail and look for a proper representation.

To this end we will next evaluate products of the basis funcions in the form given by (4.1.7). However, when doing so we are faced with cross terms arising from the products of local and pseudo functions inside the atomic spheres. In order to allow for their efficient calculation it is useful to first write down the one-center expansion of the pseudo function within the atomic spheres, i.e. its representation in terms of the function $\tilde{H}_{L\kappa}^{0}(\mathbf{r}_{\mu i})$ as well as the Bessel functions times structure constants. Hence, we note

$$
\begin{aligned}
H^0_{L\kappa}(\mathbf{r}_{\mu i}) &= H^I_{L\kappa}(\mathbf{r}_{\mu i}) + \tilde{H}^0_{L\kappa}(\mathbf{r}_{\mu i}) \\
&\quad + \sum_{L'\nu j}(1 - \delta_{\mu\nu}\delta_{ij})J_{L'\kappa\sigma}(\mathbf{r}_{\nu j})B_{L'L\kappa}(\mathbf{R}_{\nu j} - \mathbf{R}_{\mu i})\Theta_{\nu j} \\
&= H^I_{L\kappa}(\mathbf{r}_{\mu i}) + \tilde{H}^0_{L\kappa}(\mathbf{r}_{\mu i}) \\
&\quad + \sum_{L'=0}^{L_{int}}\sum_{\nu j}(1 - \delta_{\mu\nu}\delta_{ij})J_{L'\kappa\sigma}(\mathbf{r}_{\nu j})B_{L'L\kappa}(\mathbf{R}_{\nu j} - \mathbf{R}_{\mu i})\Theta_{\nu j} \\
&\quad + \sum_{L'=L_{int}+1}^{\infty}\sum_{\nu j}(1 - \delta_{\mu\nu}\delta_{ij})J_{L'\kappa\sigma}(\mathbf{r}_{\nu j})B_{L'L\kappa}(\mathbf{R}_{\nu j} - \mathbf{R}_{\mu i})\Theta_{\nu j} ,
\end{aligned}
$$

$$(4.1.10)$$

where, in the last step, we have distinguished the lower and intermediate from the higher partial waves.

At this point it should be noted that we cannot simply neglect the high partial waves by truncating the L' summation. Instead, all partial waves must be considered in these one-center expansions of the pseudo functions. Note also that we did not simply omit the higher partial waves in (4.1.6) and (4.1.7), where the summation was likewise truncated after L_{int}. Actually, in that place we truncated the expansion of local functions, which could be interpreted as replacements of Bessel envelope functions by augmented Bessel functions. Since for high l values the augmented Bessel functions converge to the Bessel envelope functions it is well justified to truncate the summations in (4.1.6) and (4.1.7).

Keeping this in mind we multiply two basis functions of the type (4.1.7) and arrive at

$$
\begin{aligned}
&\left(D^{\infty}_{L\kappa_1\sigma}(\mathbf{r}_i,\mathbf{k})\right)^* D^{\infty}_{L'\kappa_2\sigma}(\mathbf{r}_j,\mathbf{k}) \\
&= \left(D^0_{L\kappa_1}(\mathbf{r}_i,\mathbf{k})\right)^* D^0_{L'\kappa_2}(\mathbf{r}_j,\mathbf{k}) \\
&\quad + \left[\hat{H}_{L\kappa_1\sigma}(\mathbf{r}_i) + \sum_{L''m}\hat{J}_{L''\kappa_1\sigma}(\mathbf{r}_m)B_{L''L\kappa_1}(\boldsymbol{\tau}_m - \boldsymbol{\tau}_i,\mathbf{k})\right]^* \\
&\quad \times \left[\hat{H}_{L'\kappa_2\sigma}(\mathbf{r}_j) + \sum_{L'''n}\hat{J}_{L'''\kappa_2\sigma}(\mathbf{r}_n)B_{L'''L'\kappa_2}(\boldsymbol{\tau}_n - \boldsymbol{\tau}_j,\mathbf{k})\right] \\
&\quad + \left[H^0_{L\kappa_1}(\mathbf{r}_i) + \sum_{L''=0}^{L_{int}}\sum_m J_{L''\kappa_1}(\mathbf{r}_m)\Theta_m B_{L''L\kappa_1}(\boldsymbol{\tau}_m - \boldsymbol{\tau}_i,\mathbf{k})\right]^* \\
&\quad \times \left[\hat{H}_{L'\kappa_2\sigma}(\mathbf{r}_j) + \sum_{L'''n}\hat{J}_{L'''\kappa_2\sigma}(\mathbf{r}_n)B_{L'''L'\kappa_2}(\boldsymbol{\tau}_n - \boldsymbol{\tau}_j,\mathbf{k})\right] \\
&\quad + \left[\hat{H}_{L\kappa_1\sigma}(\mathbf{r}_i) + \sum_{L''m}\hat{J}_{L''\kappa_1\sigma}(\mathbf{r}_m)B_{L''L\kappa_1}(\boldsymbol{\tau}_m - \boldsymbol{\tau}_i,\mathbf{k})\right]^*
\end{aligned}
$$

$$\times \left[H^0_{L'\kappa_2}(\mathbf{r}_j) + \sum_{L'''=0}^{L_{int}} \sum_n J_{L'''\kappa_2}(\mathbf{r}_n)\Theta_n B_{L'''L'\kappa_2}(\boldsymbol{\tau}_n - \boldsymbol{\tau}_j, \mathbf{k}) \right]$$

$$+ \left[\sum_{L''=L_{int}+1}^{\infty} \sum_m J_{L''\kappa_1}(\mathbf{r}_m)\Theta_m B_{L''L\kappa_1}(\boldsymbol{\tau}_m - \boldsymbol{\tau}_i, \mathbf{k}) \right]^*$$

$$\times \left[\hat{H}_{L'\kappa_2\sigma}(\mathbf{r}_j) + \sum_{L'''n} \hat{J}_{L'''\kappa_2\sigma}(\mathbf{r}_n) B_{L'''L'\kappa_2}(\boldsymbol{\tau}_n - \boldsymbol{\tau}_j, \mathbf{k}) \right]$$

$$+ \left[\hat{H}_{L\kappa_1\sigma}(\mathbf{r}_i) + \sum_{L''m} \hat{J}_{L''\kappa_1\sigma}(\mathbf{r}_m) B_{L''L\kappa_1}(\boldsymbol{\tau}_m - \boldsymbol{\tau}_i, \mathbf{k}) \right]^*$$

$$\times \left[\sum_{L'''=L_{int}+1}^{\infty} \sum_n J_{L'''\kappa_2}(\mathbf{r}_n)\Theta_n B_{L'''L'\kappa_2}(\boldsymbol{\tau}_n - \boldsymbol{\tau}_j, \mathbf{k}) \right] . \tag{4.1.11}$$

Note that, by definition, all the functions appearing in square brackets are restricted to their respective atomic spheres. Above equation may now be readily simplified by using the definitions (4.1.4) and (4.1.5) of the local functions and multiplying out. The cross terms arising then exactly cancel and we obtain the result

$$\left(D^\infty_{L\kappa_1\sigma}(\mathbf{r}_i, \mathbf{k}) \right)^* D^\infty_{L'\kappa_2\sigma}(\mathbf{r}_j, \mathbf{k})$$

$$= \left(D^0_{L\kappa_1}(\mathbf{r}_i, \mathbf{k}) \right)^* D^0_{L'\kappa_2}(\mathbf{r}_j, \mathbf{k})$$

$$+ \sum_m \left[\hat{H}_{L\kappa_1\sigma}(\mathbf{r}_m)\delta_{im} + \sum_{L''} \hat{J}_{L''\kappa_1\sigma}(\mathbf{r}_m) B_{L''L\kappa_1}(\boldsymbol{\tau}_m - \boldsymbol{\tau}_i, \mathbf{k}) \right]^*$$

$$\times \left[\tilde{H}_{L'\kappa_2\sigma}(\mathbf{r}_m)\delta_{mj} + \sum_{L'''} \tilde{J}_{L'''\kappa_2\sigma}(\mathbf{r}_m) B_{L'''L'\kappa_2}(\boldsymbol{\tau}_m - \boldsymbol{\tau}_j, \mathbf{k}) \right]$$

$$- \sum_m \left[H^0_{L\kappa_1}(\mathbf{r}_m)\delta_{im} + \sum_{L''=0}^{L_{int}} J_{L''\kappa_1}(\mathbf{r}_m)\Theta_m B_{L''L\kappa_1}(\boldsymbol{\tau}_m - \boldsymbol{\tau}_i, \mathbf{k}) \right]^*$$

$$\times \left[H^0_{L'\kappa_2}(\mathbf{r}_j)\delta_{mj} + \sum_{L'''}^{L_{int}} J_{L'''\kappa_2}(\mathbf{r}_m)\Theta_m B_{L'''L'\kappa_2}(\boldsymbol{\tau}_m - \boldsymbol{\tau}_j, \mathbf{k}) \right]$$

$$+ \sum_m \left[\sum_{L''=L_{int}+1}^{\infty} J_{L''\kappa_1}(\mathbf{r}_m)\Theta_m B_{L''L\kappa_1}(\boldsymbol{\tau}_m - \boldsymbol{\tau}_i, \mathbf{k}) \right]^*$$

$$\times \left[\hat{H}_{L'\kappa_2\sigma}(\mathbf{r}_m)\delta_{mj} + \sum_{L'''} \hat{J}_{L'''\kappa_2\sigma}(\mathbf{r}_m) B_{L'''L'\kappa_2}(\boldsymbol{\tau}_m - \boldsymbol{\tau}_j, \mathbf{k}) \right]$$

$$+ \sum_m \left[\hat{H}_{L\kappa_1\sigma}(\mathbf{r}_m)\delta_{im} + \sum_{L''} \hat{J}_{L''\kappa_1\sigma}(\mathbf{r}_m) B_{L''L\kappa_1}(\boldsymbol{\tau}_m - \boldsymbol{\tau}_i, \mathbf{k}) \right]^*$$

$$\times \left[\sum_{L'''=L_{int}+1}^{\infty} J_{L'''\kappa_2}(\mathbf{r}_m)\Theta_m B_{L'''L'\kappa_2}(\boldsymbol{\tau}_m - \boldsymbol{\tau}_j, \mathbf{k}) \right] . \tag{4.1.12}$$

Obviously the product of two basis functions now consists of three parts. The first one as made of the first term on the right-hand side is just the product of the smooth pseudo functions, which extends over all space. In contrast, the second and third terms are restricted to the atomic spheres. They represent the local part of the product of basis functions and they both include only lower and intermediate partial waves.

Finally, in the same manner as the local parts, the last two terms are restricted to the atomic spheres. However, they contain only products of pseudo functions belonging to the set of higher partial waves and local functions belonging to the set of lower and intermediate waves. As a consequence, due to the orthonormalization of the spherical harmonics, the two functions are strictly orthogonal. In addition, their radial parts hardly overlap. This is due to the fact that the higher partial waves vanish at least as $r^{l_{int}}$ at the origin of the respective atomic sphere. In contrast, the local functions are required to vanish continuously and differentiably at the sphere boundary.

To conclude, we can safely neglect these cross terms when evaluating products of the basis functions. In doing so, we do indeed arrive at a representation of the products of basis functions, which, in the same way as the basis functions themselves, allow for a separation into smooth pseudo parts and intraatomic local parts, where the latter arise as the difference of a term consisting only of augmented functions and another term, which derives exclusively from pseudo functions.

4.2 The Secular Matrix

With the representation of the basis functions in terms of pseudo and local contributions at hand we proceed to the setup of the secular matrix. In a full-potential method the latter contains additional contributions due to those terms of the potential, which go beyond the muffin-tin approximation. For simplicity, we follow the lines already taken in Sect. 2.2.

As with the standard ASW method we continue in assuming a global spin quantization axis. This allows to deal with two independent eigenvalue problems, one for each spin direction. As a consequence, we write for the effective spin-dependent Hamiltonian

$$
\begin{aligned}
H_\sigma &= -\Delta + v_\sigma(\mathbf{r}) - v_{MTZ} \\
&= -\Delta + v_\sigma^{MT}(\mathbf{r}) - v_{MTZ} + v_\sigma^{NMT}(\mathbf{r}) \,,
\end{aligned} \tag{4.2.1}
$$

which deviates from the expression used in the standard ASW method by the inclusion of the non-muffin-tin term.

Next, we separate, in the same manner as for the basis functions, the full potential into a slowly varying pseudo part, which extends over all space, and local contributions, which vanish continuously and differentiably at the boundaries of the atomic spheres. Thus, we write

$$v_\sigma(\mathbf{r}) = v_\sigma^0(\mathbf{r}) + \sum_i \hat{v}_\sigma(\mathbf{r}_i) \,, \tag{4.2.2}$$

where the superscript 0 and the hats again denote the pseudo and local parts, respectively. For the local contributions we use an expansion in spherical harmonics

$$\hat{v}_\sigma(\mathbf{r}_i) = \sum_K \hat{v}_{K\sigma}(r_i) Y_K(\hat{\mathbf{r}}_i)\Theta_i \,, \tag{4.2.3}$$

where the coefficients are readily calculated from

$$\hat{v}_{K\sigma}(r_i) = \int d\Omega_i \; \hat{v}_\sigma(\mathbf{r}_i) Y_K^*(\hat{\mathbf{r}}_i) \,. \tag{4.2.4}$$

While this representation of the local parts is rather unique it is not at all obvious, which representation to use for the pseudo part of the potential. Of course, due to the smoothness of the latter, it seems reasonable to use an expansion in plane waves as it has been originally proposed [13]. In addition, this choice allows for an in principle exact formulation of the method since the accuracy can be systematically controlled by the plane-wave cutoff. However, despite the smoothness of the pseudo part a lot of plane waves are needed for an accurate representation and, hence, the computational load is rather high in practice. For this reason, one might wish to have alternatives at hand.

To be specific, we write the pseudo part of the full potential as an expansion in plane waves

$$v_\sigma^0(\mathbf{r}) = \sum_t v_\sigma^0(\mathbf{K}_t)e^{i\mathbf{K}_t\mathbf{r}} \,, \tag{4.2.5}$$

where the expansion coefficients are given by

$$v_\sigma^0(\mathbf{K}_t) = \frac{1}{\Omega_c} \int_{\Omega_c} d^3\mathbf{r}\, v_\sigma^0(\mathbf{r}) e^{-i\mathbf{K}_t\mathbf{r}} \,. \tag{4.2.6}$$

Note that, as in Sect. 3.11, we used the same symbol for the real-space representation and the plane-wave expansion coefficients, which are distinguished only through their arguments.

At this point, we recall the possibility of proposing an intermediate approach, which would be half way between the standard ASW method and the full-potential methods. Motivated by the dispute about the best way to represent the interstitial or pseudo parts of the potential, the density and related quantities, we opt to reduce these interstitial quantities to their minimum. To be specific, we retain only the constant part of the pseudo potential, which corresponds to the $\mathbf{K}_t = \mathbf{0}$ term in (4.2.5). As we will see in Sect. 4.4, this term is just the muffin-tin zero, v_{MTZ}. In addition, in order to reduce the effect of this approximation, we use, following Andersen, space-filling atomic spheres rather than touching muffin-tin spheres. With these two means, we will be able to follow the treatment of the plane-wave based full-potential ASW method in the present and the subsequent sections and to obtain at each stage the above mentioned intermediate procedure. However, we will mark the important changes at each stage.

In addition to the previous representation of the pseudo part of the full potential we will also need its one-center representation, i.e. its representation inside the atomic spheres as an expansion in spherical harmonics. To this end we combine (4.2.5) with the expansion (3.6.1) of the plane waves in spherical Bessel functions and use the definition (3.1.3) of the Bessel envelope function as well as the orthonormality of the spherical harmonics to write

$$v_\sigma^0(\mathbf{r}) = \sum_n v_\sigma^0(\mathbf{K}_n) e^{i\mathbf{K}_n \boldsymbol{\tau}_i} e^{i\mathbf{K}_n \mathbf{r}_i}$$

$$= \sum_n v_\sigma^0(\mathbf{K}_n) e^{i\mathbf{K}_n \boldsymbol{\tau}_i} 4\pi \sum_K i^k |\mathbf{K}_n|^k J_{K|\mathbf{K}_n|}(\mathbf{r}_i) Y_K^*(\hat{\mathbf{K}}_n) \ . \qquad (4.2.7)$$

With this identity at hand we are able to define the one-center expansion of the pseudo part as

$$v_\sigma^0(\mathbf{r}_i) := \sum_K v_{K\sigma}^0(r_i) Y_K(\hat{\mathbf{r}}_i) \Theta_i$$

$$:= \sum_n v_\sigma^0(\mathbf{K}_n) e^{i\mathbf{K}_n \boldsymbol{\tau}_i} 4\pi \sum_K i^k |\mathbf{K}_n|^k J_{K|\mathbf{K}_n|}(\mathbf{r}_i) Y_K^*(\hat{\mathbf{K}}_n) \Theta_i \ , \qquad (4.2.8)$$

where the expansion coefficients are given by

$$v_{K\sigma}^0(r_i) = \int d\Omega_i \ v_\sigma^0(\mathbf{r}_i) Y_K^*(\hat{\mathbf{r}}_i)$$

$$= \sum_n v_\sigma^0(\mathbf{K}_n) e^{i\mathbf{K}_n \boldsymbol{\tau}_i} 4\pi i^k \bar{\jmath}_k(|\mathbf{K}_n|r_i) Y_K^*(\hat{\mathbf{K}}_n) \Theta_i \ . \qquad (4.2.9)$$

It is now rather trivial to set up the muffin-tin part of the full potential, which is needed for the construction of the basis functions. To this end, we just restrict the plane-wave representation (4.2.5) and the spherical-harmonics expansions (4.2.3) and (4.2.8) to the contributions from $\mathbf{K}_t = \mathbf{0}$ and $K = 0$, respectively. Note that, with only the $\mathbf{K}_t = \mathbf{0}$-contribution kept, the K-summation in (4.2.8) collapses to the $K = 0$-term, which, due to the limiting value of the spherical Bessel function as given by (3.1.41), reduces to $\frac{1}{4\pi}$. As a consequence, the one-center expansion of the pseudo part is just the constant $\mathbf{K}_t = \mathbf{0}$-term.

Separating the muffin-tin and non-muffin-tin parts of the full-potential Hamiltonian has the additional advantage that it allows to simplify the calculation of the matrix elements entering the secular matrix since all the basis functions in the respective regions of space are eigenfunctions of Schrödinger's equation formulated with the muffin-tin potential.

In order to simplify the subsequent calculation of the secular matrix we define, in accordance with (2.2.4) and (2.2.10), a bra-ket notation for all the functions involved by

$$|L\kappa i\rangle^\infty := D_{L\kappa\sigma}^\infty(\mathbf{r}_i, \mathbf{k}) \ , \qquad (4.2.10)$$

$$|L\kappa i\rangle^0 := D_{L\kappa}^0(\mathbf{r}_i, \mathbf{k}) \ , \qquad (4.2.11)$$

$$|\widetilde{L\kappa i}\rangle := \tilde{H}_{L\kappa\sigma}(\mathbf{r}_i) + \sum_{L'j} \tilde{J}_{L'\kappa\sigma}(\mathbf{r}_j) B_{L'L\kappa}(\boldsymbol{\tau}_j - \boldsymbol{\tau}_i, \mathbf{k}) , \tag{4.2.12}$$

$$|\widehat{L\kappa i}\rangle := \hat{H}_{L\kappa\sigma}(\mathbf{r}_i) + \sum_{L'j} \hat{J}_{L'\kappa\sigma}(\mathbf{r}_j) B_{L'L\kappa}(\boldsymbol{\tau}_j - \boldsymbol{\tau}_i, \mathbf{k}) , \tag{4.2.13}$$

$$|L\kappa i\rangle^{0^-} := H^0_{L\kappa}(\mathbf{r}_i) + \sum_{L'=0}^{L_{int}} \sum_{j} J_{L'\kappa}(\mathbf{r}_j) \Theta_j B_{L'L\kappa}(\boldsymbol{\tau}_j - \boldsymbol{\tau}_i, \mathbf{k}) , \tag{4.2.14}$$

$$|L\kappa i\rangle^{0^+} := \sum_{L'=L_{int}+1}^{\infty} \sum_{j} J_{L'\kappa}(\mathbf{r}_j) \Theta_j B_{L'L\kappa}(\boldsymbol{\tau}_j - \boldsymbol{\tau}_i, \mathbf{k}) . \tag{4.2.15}$$

In the last two equations we have explicitly distinguished the lower and intermediate partial waves of the pseudo part of the basis functions from their higher partial waves. Note that the set of (4.2.10) to (4.2.15) contains all the functions entering the product (4.1.12) of basis functions, which obviously also enter the secular matrix.

Next, using the representation (4.1.12) of products of basis functions as well as the two alternative representations (4.2.1) and (4.2.2) for the full potential and taking into account the respective regions, where the functions (4.2.10) to (4.2.15) are defined, we are able to write down the general matrix element of the Hamiltonian matrix as

$$^{\infty}\langle L\kappa_1 i | H_\sigma | L'\kappa_2 j\rangle^{\infty}_c$$
$$= {}^0\langle L\kappa_1 i | H_\sigma | L'\kappa_2 j\rangle^0_c + \langle \widetilde{L\kappa_1 i} | H_\sigma | \widetilde{L'\kappa_2 j}\rangle_c - {}^{0^-}\langle L\kappa_1 i | H_\sigma | L'\kappa_2 j\rangle^{0^-}_c$$
$$+ {}^{0^+}\langle L\kappa_1 i | H_\sigma | \widehat{L'\kappa_2 j}\rangle_c + \langle \widehat{L\kappa_1 i} | H_\sigma | L'\kappa_2 j\rangle^{0^+}_c$$
$$= {}^0\langle L\kappa_1 i | -\Delta + v^0_\sigma(\mathbf{r}) - v_{MTZ} | L'\kappa_2 j\rangle^0_c$$
$$+ \sum_m \Bigg[{}^0\langle L\kappa_1 i | \hat{v}_\sigma(\mathbf{r}_m) | L'\kappa_2 j\rangle^0_{M(m)}$$
$$+ \langle \widetilde{L\kappa_1 i} | -\Delta + v^{MT}_\sigma(\mathbf{r}) - v_{MTZ} + v^{NMT}_\sigma(\mathbf{r}) | \widetilde{L'\kappa_2 j}\rangle_{M(m)}$$
$$- {}^{0^-}\langle L\kappa_1 i | -\Delta + v^0_\sigma(\mathbf{r}_m) + \hat{v}_\sigma(\mathbf{r}_m) - v_{MTZ} | L'\kappa_2 j\rangle^{0^-}_{M(m)}$$
$$+ {}^{0^+}\langle L\kappa_1 i | -\Delta + v^{MT}_\sigma(\mathbf{r}) - v_{MTZ} + v^{NMT}_\sigma(\mathbf{r}) | \widehat{L'\kappa_2 j}\rangle_{M(m)}$$
$$+ \langle \widehat{L\kappa_1 i} | -\Delta + v^{MT}_\sigma(\mathbf{r}) - v_{MTZ} + v^{NMT}_\sigma(\mathbf{r}) | L'\kappa_2 j\rangle^{0^+}_{M(m)} \Bigg] ,$$
$$\tag{4.2.16}$$

where c and $M(m)$ denote integration over the unit cell and the muffin-tin sphere centered at site $\boldsymbol{\tau}_m$, respectively. In complete analogy to (4.1.12) for the products of basis functions, (4.2.16) contains the decomposition into smooth, local, and cross terms. Furthermore, it is the counterpart of the *ad hoc* representation (2.2.11) of the Hamiltonian matrix element of the standard ASW method. While discussing the general Hamiltonian matrix element of the full-potential ASW method we will also refer to the matrix element (2.2.11) and discuss, which approximations, if at all, are included in the latter.

First, we turn to the last two terms in square brackets in (4.2.16), which arise from the cross terms in the product (4.1.12). Since one of the factors in each of these two terms is a local function these integrals extend only over the atomic spheres. In addition, as already pointed out at the end of Sect. 4.1, in both terms the local functions comprise lower and intermediate partial waves whereas the respective other factor contains only higher partial waves. As a consequence, both factors are orthogonal due to the orthogonality of the spherical harmonics. Since the muffin-tin part of the Hamiltonian, i.e. the operator $-\Delta + v_\sigma^{MT}(\mathbf{r}) - v_{MTZ}$, conserves angular momentum, the respective integrals vanish exactly and we are left with only the non-muffin-tin parts, hence, the terms

$$
{}^{0^+}\langle L\kappa_1 i | H_\sigma | \widehat{L'\kappa_2 j} \rangle_c + \langle \widehat{L\kappa_1 i} | H_\sigma | L'\kappa_2 j \rangle_c^{0^+}
$$
$$
= {}^{0^+}\langle L\kappa_1 i | v_\sigma^{NMT}(\mathbf{r}) | \widehat{L'\kappa_2 j} \rangle_c + \langle \widehat{L\kappa_1 i} | v_\sigma^{NMT}(\mathbf{r}) | L'\kappa_2 j \rangle_c^{0^+} . \qquad (4.2.17)
$$

Yet, as has been also discussed at the end of Sect. 4.1, the radial functions entering these terms overlap only little. In addition, the non-muffin-tin parts of the potential inside the atomic spheres i.e. the non-spherical parts are very small as compared to the muffin-tin part. As a consequence, these integrals will give very small values and, although they could be calculated exactly, they can be safely neglected.

With the cross terms removed, the Hamiltonian matrix element assumes the form

$$
{}^{\infty}\langle L\kappa_1 i | H_\sigma | L'\kappa_2 j \rangle_c^{\infty}
$$
$$
= {}^{0}\langle L\kappa_1 i | - \Delta | L'\kappa_2 j \rangle_c^{0} + {}^{0}\langle L\kappa_1 i | v_\sigma^0(\mathbf{r}) - v_{MTZ} | L'\kappa_2 j \rangle_c^{0}
$$
$$
+ \sum_m \Big[{}^{0}\langle L\kappa_1 i | \hat{v}_\sigma(\mathbf{r}_m) | L'\kappa_2 j \rangle_{M(m)}^{0}
$$
$$
+ \langle \widehat{L\kappa_1 i} | - \Delta + v_\sigma^{MT}(\mathbf{r}) - v_{MTZ} | \widehat{L'\kappa_2 j} \rangle_{M(m)}
$$
$$
+ \langle \widehat{L\kappa_1 i} | v_\sigma^{NMT}(\mathbf{r}) | \widehat{L'\kappa_2 j} \rangle_{M(m)}
$$
$$
- {}^{0^-}\langle L\kappa_1 i | - \Delta | L'\kappa_2 j \rangle_{M(m)}^{0^-}
$$
$$
- {}^{0^-}\langle L\kappa_1 i | v_\sigma^0(\mathbf{r}_m) - v_{MTZ} | L'\kappa_2 j \rangle_{M(m)}^{0^-}
$$
$$
- {}^{0^-}\langle L\kappa_1 i | \hat{v}_\sigma(\mathbf{r}_m) | L'\kappa_2 j \rangle_{M(m)}^{0^-} \Big] , \qquad (4.2.18)
$$

which is now subject to further inspection.

To start with, we note that the first, fourth, and sixth term in (4.2.18) are identical to the matrix element of the Hamiltonian matrix as formulated for the standard ASW method. In particular, we write for the fourth term

$$
\langle \widehat{L\kappa_1 i} | - \Delta + v_\sigma^{MT}(\mathbf{r}) - v_{MTZ} | \widehat{L'\kappa_2 j} \rangle_{M(m)}
$$
$$
= \delta_{im} E_{l\kappa_2 m\sigma}^{(H)} \langle \tilde{H}_{L\kappa_1\sigma} | \tilde{H}_{L\kappa_2\sigma} \rangle_{M(m)} \delta_{LL'} \delta_{mj}
$$
$$
+ \delta_{im} E_{l\kappa_2 m\sigma}^{(J)} \langle \tilde{H}_{L\kappa_1\sigma} | \tilde{J}_{L\kappa_2\sigma} \rangle_{M(m)} B_{LL'\kappa_2}(\boldsymbol{\tau}_m - \boldsymbol{\tau}_j, \mathbf{k})
$$

$$+B^*_{L'L\kappa_1}(\boldsymbol{\tau}_m - \boldsymbol{\tau}_i, \mathbf{k})E^{(H)}_{l'\kappa_2 m\sigma}\langle \tilde{J}_{L'\kappa_1\sigma}|\tilde{H}_{L'\kappa_2\sigma}\rangle_{M(m)}\delta_{mj}$$

$$+\sum_{L''}B^*_{L''L\kappa_1}(\boldsymbol{\tau}_m - \boldsymbol{\tau}_i, \mathbf{k})E^{(J)}_{l''\kappa_2 m\sigma}\langle \tilde{J}_{L''\kappa_1\sigma}|\tilde{J}_{L''\kappa_2\sigma}\rangle_{M(m)}B_{L''L'\kappa_2}(\boldsymbol{\tau}_m - \boldsymbol{\tau}_j, \mathbf{k}) \,,$$

$$(4.2.19)$$

and for the sixth term

$$^{0^-}\langle L\kappa_1 i| - \Delta|L'\kappa_2 j\rangle^{0^-}_{M(m)}$$

$$= \delta_{im}E^{(0)}_{l\kappa_2 m}\langle \tilde{H}^0_{L\kappa_1}|\tilde{H}^0_{L\kappa_2}\rangle_{M(m)}\delta_{LL'}\delta_{mj}$$

$$+\delta_{im}\kappa_2^2\langle \tilde{H}^0_{L\kappa_1}|J_{L\kappa_2}\rangle_{M(m)}B_{LL'\kappa_2}(\boldsymbol{\tau}_m - \boldsymbol{\tau}_j, \mathbf{k})$$

$$+B^*_{L'L\kappa_1}(\boldsymbol{\tau}_m - \boldsymbol{\tau}_i, \mathbf{k})E^{(0)}_{l'\kappa_2 m}\langle J_{L'\kappa_1}|\tilde{H}^0_{L'\kappa_2}\rangle_{M(m)}\delta_{mj}$$

$$+\sum_{L''}B^*_{L''L\kappa_1}(\boldsymbol{\tau}_m - \boldsymbol{\tau}_i, \mathbf{k})\kappa_2^2\langle J_{L''\kappa_1}|J_{L''\kappa_2}\rangle_{M(m)}B_{L''L'\kappa_2}(\boldsymbol{\tau}_m - \boldsymbol{\tau}_j, \mathbf{k}) \,.$$

$$(4.2.20)$$

Here we have used the fact already discussed in connection with (3.9.2) to (3.9.4) that the pseudo function solves Helmholtz's equation in all regions of space including its own on-center sphere. As a consequence, we are able to reduce the integral over pseudo functions built with the Laplacian to overlap integrals times the respective eigenvalue.

For the first term on the right-hand side of (4.2.18) we proceed in the same way as in Sect. 2.2 and combine it with the sixth one. In doing so, we arrive at the overlap integral of two pseudo functions extending over the interstitial region. Since in the interstitial region the pseudo functions are identical to the envelope functions the special form of the pseudo function inside the on-center sphere does not enter at all and the integral is just the overlap integral of two envelope functions extending over the interstitial, which is given by (3.8.39). To conclude, the first, fourth, and sixth contribution to the Hamiltonian matrix element (4.2.18) reduce exactly to the result (2.2.15) obtained for the standard ASW method.

At this point it is very instructive to investigate how the general Hamiltonian matrix element (4.2.18) changes if we assume the potential to be of the muffin-tin form. In that case the pseudo part of the potential reduces to the constant muffin-tin zero and the non-muffin-tin contributions inside the spheres vanish. As a consequence, the second, fifth, and seventh term on the right-hand side of (4.2.18) cancel. In passing, we mention that the same is true for the cross terms in the original expression (4.2.16), which, according to the discussion preceding (4.2.17) vanish exactly, when the potential is of the muffin-tin form.

As a consequence, the only terms left of the Hamiltonian matrix element in the form (4.2.16) or (4.2.18), except those of the standard ASW method, are the third and eighth term of the latter expression. However, the difference of these two terms is rather small since the functions $|L\kappa i\rangle^0$ and $|L\kappa i\rangle^{0^-}$ differ only by higher partial waves, i.e. by Bessel envelope functions for angular momenta higher than l_{int}. These functions vanish at least as $r^{l_{int}}$ at the origin of the respective atomic sphere and have larger contributions only in the outer regions. In contrast, the local potential, which reduces to its spherical symmetric part within the muffin-tin approximation, goes to zero at the sphere boundary and shows larger contributions only in the inner regions of the spheres. To conclude, although being an approximation in both the standard and the full-potential ASW method, it is well justified to omit the third third and eighth term on the right-hand side of (4.2.18). We will thus follow the line taken already in the standard ASW method and do not include these two terms in the full-potential ASW method.

Still, we have to ask why these two terms did not appear in the Hamiltonian matrix element (2.2.11) of the standard ASW method, which should be identical to (4.2.18) within the muffin-tin approximation. The reason is quite simple. The difference is due to that fact that in (2.2.11) the integrals with the envelope functions were built with just the Laplacian and not with the Hamiltonian, as it should have been done. The Hamiltonian, within the muffin-tin approximation, reduces to the Laplacian in the interstitial region and gives rise to the envelope functions as solutions of Schrödinger's equation. However, inside the spheres, these solutions are subject to the local parts of the potential this leading to the third term in (4.2.18) with all partial waves included. For the same reason, the term to be subtracted inside the spheres is the matrix element of the envelope functions with the Hamiltonian, which again reduces to the Laplacian and the local potential, the latter of which leads to the eighth term in (4.2.18) with only the lower and intermediate partial waves. To conclude, the use of only the Laplacian in the integrals with the envelope functions in (2.2.11) of the standard ASW method is an approximation. It is superseded by the correct expressions (4.2.16) and (4.2.18). However, as we have just discussed, the error is very small and, eventually, the corresponding terms are neglected in both methods.

Next, we turn to the second term on the right-hand side of (4.2.18). Using (4.2.5), (3.11.1), and (4.2.11) for the plane-wave expansions of the pseudo parts of the full potential and the basis functions, respectively, we write

$$
{}^0\langle L\kappa_1 i | v_\sigma^0(\mathbf{r}) - v_{MTZ} | L'\kappa_2 j\rangle_c^0
$$

$$
= \int_{\Omega_c} d^3\mathbf{r}\, D_{L\kappa_1}^{0*}(\mathbf{r}_i, \mathbf{k}) \left(v_\sigma^0(\mathbf{r}) - v_{MTZ}\right) D_{L'\kappa_2}^0(\mathbf{r}_j, \mathbf{k})
$$

$$
= \int_{\Omega_c} d^3\mathbf{r} \left[\sum_n D_{L\kappa_1 i}^{0*}(\mathbf{K}_n + \mathbf{k})e^{-i(\mathbf{K}_n+\mathbf{k})\mathbf{r}_i}\right]
$$

$$
\left[\sum_t \left(v_\sigma^0(\mathbf{K}_t)e^{i\mathbf{K}_t\mathbf{r}} - v_{MTZ}\delta(\mathbf{K}_t)\right)\right]
$$

$$
\left[\sum_{n'} D_{L'\kappa_2 j}^0(\mathbf{K}_{n'} + \mathbf{k})e^{i(\mathbf{K}_{n'}+\mathbf{k})\mathbf{r}_j}\right]
$$

$$
= \Omega_c \sum_n \sum_{n'} D_{L\kappa_1 i}^{0*}(\mathbf{K}_n + \mathbf{k}) \left(v_\sigma^0(\mathbf{K}_n - \mathbf{K}_{n'}) - v_{MTZ}\delta(\mathbf{K}_n - \mathbf{K}_{n'})\right)
$$

$$
D_{L'\kappa_2 j}^0(\mathbf{K}_{n'} + \mathbf{k})e^{i(\mathbf{K}_n+\mathbf{k})\boldsymbol{\tau}_i} e^{-i(\mathbf{K}_{n'}+\mathbf{k})\boldsymbol{\tau}_j} . \qquad (4.2.21)
$$

As already indicated by the first line, these integrals are performed as a real-space integration. To this end, the plane-wave expansion of the pseudo functions has to be constructed, as described in Sect. 3.11. In addition, the Fourier transform to the real-space representation has to be performed. These two steps are the most time consuming parts of the present full-potential ASW method and, hence, any work to speed up the method should start here.

By now, we are left with only the fifth and seventh term on the right-hand side of (4.2.18). Both integrals extend over the atomic spheres and can be split into one-, two-, and three-center contributions in complete analogy to (4.2.19) and (4.2.20). We thus get for the integral with the augmented functions

$$
\begin{aligned}
&\langle \widetilde{L\kappa_1 i}|v_\sigma^{NMT}(\mathbf{r})|\widetilde{L'\kappa_2 j}\rangle_{M(m)} \\
&= \delta_{im}\langle \tilde{H}_{L\kappa_1\sigma}|v_\sigma^{NMT}(\mathbf{r})|\tilde{H}_{L'\kappa_2\sigma}\rangle_{M(m)}\delta_{mj} \\
&\quad + \sum_{L''}\delta_{im}\langle \tilde{H}_{L\kappa_1\sigma}|v_\sigma^{NMT}(\mathbf{r})|\tilde{J}_{L''\kappa_2\sigma}\rangle_{M(m)}B_{L''L'\kappa_2}(\boldsymbol{\tau}_m - \boldsymbol{\tau}_j,\mathbf{k}) \\
&\quad + \sum_{L''}B^*_{L''L\kappa_1}(\boldsymbol{\tau}_m - \boldsymbol{\tau}_i,\mathbf{k})\langle \tilde{J}_{L''\kappa_1\sigma}|v_\sigma^{NMT}(\mathbf{r})|\tilde{H}_{L'\kappa_2\sigma}\rangle_{M(m)}\delta_{mj} \\
&\quad + \sum_{L''}\sum_{L'''}B^*_{L''L\kappa_1}(\boldsymbol{\tau}_m - \boldsymbol{\tau}_i,\mathbf{k})\langle \tilde{J}_{L''\kappa_1\sigma}|v_\sigma^{NMT}(\mathbf{r})|\tilde{J}_{L'''\kappa_2\sigma}\rangle_{M(m)} \\
&\hspace{6cm} B_{L'''L'\kappa_2}(\boldsymbol{\tau}_m - \boldsymbol{\tau}_j,\mathbf{k}) . \quad (4.2.22)
\end{aligned}
$$

Note that, in contrast to (4.2.19) and (4.2.20), the intraatomic integrals are not diagonal in L since the non-spherical parts of the full potential couple states with different angular momentum. Still, we may decompose the non-muffin-tin potential into pseudo and local parts according to (4.2.2), (4.2.4), and (4.2.9) thereby arriving at

$$
\begin{aligned}
&\langle \tilde{F}_{L\kappa_1\sigma}|v_\sigma^{NMT}(\mathbf{r})|\tilde{G}_{L'\kappa_2\sigma}\rangle_{M(m)} \\
&= \langle \tilde{F}_{L\kappa_1\sigma}|\sum_K(1-\delta_{k0})\left(v^0_{K\sigma}(r_m)+\hat{v}_{K\sigma}(r_m)\right)Y_K(\hat{\mathbf{r}}_m)|\tilde{G}_{L'\kappa_2\sigma}\rangle_{M(m)} , \quad (4.2.23)
\end{aligned}
$$

where $\tilde{F}_{L\kappa_1\sigma}$ and $\tilde{G}_{L'\kappa_2\sigma}$ denote any of the augmented functions.

The seventh contribution to (4.2.18) is evaluated in the same manner the intraatomic integrals being

$$
\begin{aligned}
&{}^{0^-}\langle F_{L\kappa_1}|v_\sigma^0(\mathbf{r}) - v_{MTZ}|G_{L'\kappa_2}\rangle^{0^-}_{M(m)} \\
&= {}^{0^-}\langle F_{L\kappa_1}|\sum_K v^0_{K\sigma}(r_m)Y_K(\hat{\mathbf{r}}_m) - v_{MTZ}|G_{L'\kappa_2}\rangle^{0^-}_{M(m)} . \quad (4.2.24)
\end{aligned}
$$

Here the functions entering the integral are either an augmented pseudo function $\tilde{H}^0_{L\kappa}(\mathbf{r}_{\mu i})$ or a Bessel function $J_{L\kappa}(\mathbf{r}_{\mu i})$.

Equations (4.2.23) and (4.2.24) are evaluated by using the definitions (2.1.13), (2.1.17), (2.1.20), and (4.1.3) of the respective functions. Hence, we get, e.g. for the single terms on the right-hand side of (4.2.23)

$$\langle \tilde{F}_{L\kappa_1\sigma} | v_{K\sigma}^0(r_m) Y_K(\hat{\mathbf{r}}_m) | \tilde{G}_{L'\kappa_2\sigma} \rangle_{M(m)}$$

$$= c_{LL'K} \int_0^{s_m} dr_m r_m^2 \, \tilde{f}_{l\kappa_1\sigma}(r_m) v_{K\sigma}^0(r_m) \tilde{g}_{l'\kappa_2\sigma}(r_m) \,, \qquad (4.2.25)$$

where s_m denotes the muffin-tin radius and $\tilde{f}_{l\kappa_1\sigma}$ and $\tilde{g}_{l\kappa_1\sigma}$ are real valued augmented functions. An analogous relation holds for the terms on the right-hand side of (4.2.24). Thus, all these integrals can be decomposed into Gaunt coefficients and radial integrals, the latter of which are calculated numerically. In addition, owing to our choice of basis functions, which excludes the i^l factors from the basis functions, the integrals are real quantities.

Finally, we are able to combine (2.2.20) and (4.2.21) to (4.2.24) and write the general matrix element of the Hamiltonian as

$$^\infty \langle L\kappa_1 i | H_\sigma | L'\kappa_2 j \rangle_c^\infty$$

$$= \int_{\Omega_c} d^3\mathbf{r} \, D_{L\kappa_1}^{0*}(\mathbf{r}_i, \mathbf{k}) \left(v_\sigma^0(\mathbf{r}) - v_{MTZ} \right) D_{L'\kappa_2}^0(\mathbf{r}_j, \mathbf{k})$$

$$+ \left[X_{L\kappa_1\kappa_2 j\sigma}^{(H,1)} \delta_{LL'} + \langle \tilde{H}_{L\kappa_1\sigma} | v_\sigma^{NMT}(\mathbf{r}) | \tilde{H}_{L'\kappa_2\sigma} \rangle_{M(j)} \right.$$

$$\left. - {}^{0^-}\langle H_{L\kappa_1\sigma} | v_\sigma^0(\mathbf{r}) - v_{MTZ} | H_{L'\kappa_2\sigma} \rangle_{M(j)}^{0^-} \right] \delta_{ij}$$

$$+ \kappa_2^2 \dot{B}_{LL'\kappa_1}(\boldsymbol{\tau}_i - \boldsymbol{\tau}_j, \mathbf{k}) \delta(\kappa_1^2 - \kappa_2^2)$$

$$+ \sum_{L''} \left[X_{L\kappa_1\kappa_2 i\sigma}^{(H,2)} \delta_{LL''} + \langle \tilde{H}_{L\kappa_1\sigma} | v_\sigma^{NMT}(\mathbf{r}) | \tilde{J}_{L''\kappa_2\sigma} \rangle_{M(i)} \right.$$

$$\left. - {}^{0^-}\langle H_{L\kappa_1\sigma} | v_\sigma^0(\mathbf{r}) - v_{MTZ} | J_{L''\kappa_2\sigma} \rangle_{M(i)}^{0^-} \right] B_{L''L'\kappa_2}(\boldsymbol{\tau}_i - \boldsymbol{\tau}_j, \mathbf{k})$$

$$+ \sum_{L''} B_{L''L\kappa_1}^*(\boldsymbol{\tau}_j - \boldsymbol{\tau}_i, \mathbf{k})$$

$$\left[\left(X_{L'\kappa_1\kappa_2 j\sigma}^{(H,2)} + \delta(\kappa_1^2 - \kappa_2^2) \right) \delta_{L'L''} + \langle \tilde{J}_{L''\kappa_1\sigma} | v_\sigma^{NMT}(\mathbf{r}) | \tilde{H}_{L'\kappa_2\sigma} \rangle_{M(j)} \right.$$

$$\left. - {}^{0^-}\langle J_{L''\kappa_1\sigma} | v_\sigma^0(\mathbf{r}) - v_{MTZ} | H_{L'\kappa_2\sigma} \rangle_{M(j)}^{0^-} \right]$$

$$+ \sum_m \sum_{L''} \sum_{L'''} B_{L''L\kappa_1}^*(\boldsymbol{\tau}_m - \boldsymbol{\tau}_i, \mathbf{k})$$

$$\left[X_{L''\kappa_1\kappa_2 m\sigma}^{(H,3)} \delta_{L''L'''} + \langle \tilde{J}_{L''\kappa_1\sigma} | v_\sigma^{NMT}(\mathbf{r}) | \tilde{J}_{L'''\kappa_2\sigma} \rangle_{M(m)} \right.$$

$$\left. - {}^{0^-}\langle J_{L''\kappa_1\sigma} | v_\sigma^0(\mathbf{r}) - v_{MTZ} | J_{L'''\kappa_2\sigma} \rangle_{M(m)}^{0^-} \right]$$

$$B_{L'''L'\kappa_2}(\boldsymbol{\tau}_m - \boldsymbol{\tau}_j, \mathbf{k}) \,, \qquad (4.2.26)$$

where we have used the abbreviations (2.2.16) to (2.2.18) with the atomic-sphere radii replaced by the muffin-tin radii. In analogy to these abbreviations we define, in addition, the abbreviations

$$
\begin{aligned}
U^{(H,1)}_{L\kappa_1 L'\kappa_2 i\sigma} &= X^{(H,1)}_{L\kappa_1\kappa_2 i\sigma}\delta_{LL'} + \langle \tilde{H}_{L\kappa_1\sigma}|v^{NMT}_\sigma(\mathbf{r})|\tilde{H}_{L'\kappa_2\sigma}\rangle_{M(j)} \\
&\quad -{}^{0^-}\langle H_{L\kappa_1\sigma}|v^0_\sigma(\mathbf{r}) - v_{MTZ}|H_{L'\kappa_2\sigma}\rangle^{0^-}_{M(j)}, \quad (4.2.27)
\end{aligned}
$$

$$
\begin{aligned}
U^{(H,2)}_{L\kappa_1 L''\kappa_2 i\sigma} &= X^{(H,2)}_{L\kappa_1\kappa_2 i\sigma}\delta_{LL''} + \langle \tilde{H}_{L\kappa_1\sigma}|v^{NMT}_\sigma(\mathbf{r})|\tilde{J}_{L''\kappa_2\sigma}\rangle_{M(i)} \\
&\quad -{}^{0^-}\langle H_{L\kappa_1\sigma}|v^0_\sigma(\mathbf{r}) - v_{MTZ}|J_{L''\kappa_2\sigma}\rangle^{0^-}_{M(i)}, \quad (4.2.28)
\end{aligned}
$$

$$
\begin{aligned}
U^{(H,3)}_{L''\kappa_1 L'''\kappa_2 i\sigma} &= X^{(H,3)}_{L''\kappa_1\kappa_2 m\sigma}\delta_{L''L'''} + \langle \tilde{J}_{L''\kappa_1\sigma}|v^{NMT}_\sigma(\mathbf{r})|\tilde{J}_{L'''\kappa_2\sigma}\rangle_{M(m)} \\
&\quad -{}^{0^-}\langle J_{L''\kappa_1\sigma}|v^0_\sigma(\mathbf{r}) - v_{MTZ}|J_{L'''\kappa_2\sigma}\rangle^{0^-}_{M(m)}, \quad (4.2.29)
\end{aligned}
$$

which allow to rewrite (4.2.26) as

$$
\begin{aligned}
&{}^\infty\langle L\kappa_1 i|H_\sigma|L'\kappa_2 j\rangle^\infty_c \\
&= \int_{\Omega_c} d^3\mathbf{r}\, D^{0*}_{L\kappa_1}(\mathbf{r}_i,\mathbf{k})\left(v^0_\sigma(\mathbf{r}) - v_{MTZ}\right)D^0_{L'\kappa_2}(\mathbf{r}_j,\mathbf{k}) \\
&\quad + U^{(H,1)}_{L\kappa_1 L'\kappa_2 i\sigma}\delta_{ij} + \kappa_2^2\dot{B}_{LL'\kappa_1}(\boldsymbol{\tau}_i - \boldsymbol{\tau}_j,\mathbf{k})\delta(\kappa_1^2 - \kappa_2^2) \\
&\quad + \sum_{L''}U^{(H,2)}_{L\kappa_1 L''\kappa_2 i\sigma}B_{L''L'\kappa_2}(\boldsymbol{\tau}_i - \boldsymbol{\tau}_j,\mathbf{k}) \\
&\quad + \sum_{L''}B^*_{L''L\kappa_1}(\boldsymbol{\tau}_j - \boldsymbol{\tau}_i,\mathbf{k})\left(U^{(H,2)*}_{L'\kappa_2 L''\kappa_1 j\sigma} + \delta(\kappa_1^2 - \kappa_2^2)\delta_{L'L''}\right) \\
&\quad + \sum_m\sum_{L''}\sum_{L'''}B^*_{L''L\kappa_1}(\boldsymbol{\tau}_m - \boldsymbol{\tau}_i,\mathbf{k})U^{(H,3)}_{L''\kappa_1 L'''\kappa_2 m\sigma}B_{L'''L'\kappa_2}(\boldsymbol{\tau}_m - \boldsymbol{\tau}_j,\mathbf{k}).
\end{aligned}
$$

$$(4.2.30)$$

Here we have also used the identity (A.1.8). Finally, since we prefer to work with cubic rather than spherical harmonics, we use (3.7.18) and obtain for the Hamiltonian matrix element

$$
\begin{aligned}
&{}^\infty\langle L\kappa_1 i|H_\sigma|L'\kappa_2 j\rangle^\infty_c \\
&= \int_{\Omega_c} d^3\mathbf{r}\, D^{0*}_{L\kappa_1}(\mathbf{r}_i,\mathbf{k})\left(v^0_\sigma(\mathbf{r}) - v_{MTZ}\right)D^0_{L'\kappa_2}(\mathbf{r}_j,\mathbf{k}) \\
&\quad + U^{(H,1)}_{L\kappa_1 L'\kappa_2 i\sigma}\delta_{ij} + \kappa_2^2\dot{B}_{LL'\kappa_1}(\boldsymbol{\tau}_i - \boldsymbol{\tau}_j,\mathbf{k})\delta(\kappa_1^2 - \kappa_2^2) \\
&\quad + \sum_{L''}U^{(H,2)}_{L\kappa_1 L''\kappa_2 i\sigma}B_{L''L'\kappa_2}(\boldsymbol{\tau}_i - \boldsymbol{\tau}_j,\mathbf{k}) \\
&\quad + \sum_{L''}B_{LL''\kappa_1}(\boldsymbol{\tau}_i - \boldsymbol{\tau}_j,\mathbf{k})\left(U^{(H,2)*}_{L'\kappa_2 L''\kappa_1 j\sigma} + \delta(\kappa_1^2 - \kappa_2^2)\delta_{L'L''}\right) \\
&\quad + \sum_m\sum_{L''}\sum_{L'''}B_{LL''\kappa_1}(\boldsymbol{\tau}_i - \boldsymbol{\tau}_m,\mathbf{k})U^{(H,3)}_{L''\kappa_1 L'''\kappa_2 m\sigma}B_{L'''L'\kappa_2}(\boldsymbol{\tau}_m - \boldsymbol{\tau}_j,\mathbf{k}).
\end{aligned}
$$

$$(4.2.31)$$

At this stage it is very simple to specify the general Hamiltonian matrix elements of the intermediate approach mentioned at the beginning of this chapter. Within this approach the respective first terms on the right-hand sides of (4.2.30) and (4.2.31), i.e. the integrals extending over the unit cell would vanish due to the fact that the pseudo part of the full potential would reduce to the muffin-tin zero. The same would happen to the respective last terms of (4.2.27) to (4.2.29). In contrast, we would keep the respective second terms of these latter equations, which were to be build with the muffin-tin radii replaced by the atomic-sphere radii and are the only terms beyond the standard ASW method.

Next we turn to the calculation of the matrix elements of the overlap matrix. Since we still use the same basis functions as for the standard ASW method the overlap matrix is also identical to that given in Sect. 2.2. This can be proven by starting from (4.2.31) but setting all energies in numerators to one and all additional potentials to zero. However, we should have actually started from the representation of products of the basis functions as covered by (4.1.12) and constructed the overlap matrix in analogy to the initial expression (4.2.16) for the Hamiltonian matrix. We thus write for the overlap matrix

$$
\begin{aligned}
{}^{\infty}\langle L\kappa_1 i | L'\kappa_2 j\rangle_c^{\infty} \\
= {}^{0}\langle L\kappa_1 i | L'\kappa_2 j\rangle_c^{0} + \langle \widetilde{L\kappa_1 i} | \widetilde{L'\kappa_2 j}\rangle_c - {}^{0^-}\langle L\kappa_1 i | L'\kappa_2 j\rangle_c^{0^-} \\
+ {}^{0^+}\langle L\kappa_1 i | \widehat{L'\kappa_2 j}\rangle_c + \langle \widehat{L\kappa_1 i} | L'\kappa_2 j\rangle_c^{0^+} \\
= {}^{0}\langle L\kappa_1 i | L'\kappa_2 j\rangle_c^{0} \\
+ \sum_m \Big[\langle \widetilde{L\kappa_1 i} | \widetilde{L'\kappa_2 j}\rangle_{M(m)} - {}^{0^-}\langle L\kappa_1 i | L'\kappa_2 j\rangle_{M(m)}^{0^-} \\
+ {}^{0^+}\langle L\kappa_1 i | \widehat{L'\kappa_2 j}\rangle_{M(m)} + \langle \widehat{L\kappa_1 i} | L'\kappa_2 j\rangle_{M(m)}^{0^+} \Big] .
\end{aligned} \tag{4.2.32}
$$

According to the discussion at the end of Sect. 4.1 the cross terms, i.e. the last two terms in the square brackets, cancel due to the orthonormality of the spherical harmonics and, apart from the differences in the definitions of the bra and ket states, we are indeed left with the same expression as in the standard ASW method, (2.2.22). We may thus fall back on (2.2.27) and note the result

$$
\begin{aligned}
{}^{\infty}\langle L\kappa_1 i | L'\kappa_2 j\rangle_c^{\infty} \\
= X^{(S,1)}_{L\kappa_1\kappa_2 i\sigma}\delta_{LL'}\delta_{ij} + \dot{B}_{LL'\kappa_1}(\boldsymbol{\tau}_i - \boldsymbol{\tau}_j, \mathbf{k})\delta(\kappa_1^2 - \kappa_2^2) \\
+ X^{(S,2)}_{L\kappa_1\kappa_2 i\sigma}B_{LL'\kappa_2}(\boldsymbol{\tau}_i - \boldsymbol{\tau}_j, \mathbf{k}) + B^*_{L'L\kappa_1}(\boldsymbol{\tau}_j - \boldsymbol{\tau}_i, \mathbf{k})X^{(S,2)}_{L'\kappa_2\kappa_1 j\sigma} \\
+ \sum_m \sum_{L''} B^*_{L''L\kappa_1}(\boldsymbol{\tau}_m - \boldsymbol{\tau}_i, \mathbf{k})X^{(S,3)}_{L''\kappa_1\kappa_2 m\sigma}B_{L''L'\kappa_2}(\boldsymbol{\tau}_m - \boldsymbol{\tau}_j, \mathbf{k}) ,
\end{aligned} \tag{4.2.33}
$$

where we have used the abbreviations (2.2.24) to (2.2.26) with the atomic-sphere radii replaced by the muffin-tin radii. Finally, we prefer again to work with cubic harmonics and write

$$
\begin{aligned}
{}^{\infty}\langle L\kappa_1 i | L'\kappa_2 j\rangle_c^{\infty} \\
= X^{(S,1)}_{L\kappa_1\kappa_2 i\sigma}\delta_{LL'}\delta_{ij} + \dot{B}_{LL'\kappa_1}(\boldsymbol{\tau}_i - \boldsymbol{\tau}_j, \mathbf{k})\delta(\kappa_1^2 - \kappa_2^2)
\end{aligned}
$$

$$+X^{(S,2)}_{L\kappa_1\kappa_2 i\sigma}B_{LL'\kappa_2}(\boldsymbol{\tau}_i-\boldsymbol{\tau}_j,\mathbf{k})+B_{LL'\kappa_1}(\boldsymbol{\tau}_i-\boldsymbol{\tau}_j,\mathbf{k})X^{(S,2)}_{L'\kappa_2\kappa_1 j\sigma}$$

$$+\sum_m\sum_{L''}B_{LL''\kappa_1}(\boldsymbol{\tau}_i-\boldsymbol{\tau}_m,\mathbf{k})X^{(S,3)}_{L''\kappa_1\kappa_2 m\sigma}B_{L''L'\kappa_2}(\boldsymbol{\tau}_m-\boldsymbol{\tau}_j,\mathbf{k})\ . \tag{4.2.34}$$

To conclude, in the same manner as for the standard ASW method we have eventually arrived at a rather clear formulation of the general matrix element. Except for the integral over pseudo functions extending over all space again structural information is separated from intraatomic information this fact reducing the work to be performed at every $\mathbf{k}$-point to a minimum. Yet, in the full-potential ASW there are more intraatomic quantities entering the Hamiltonian matrix. This is due to the non-spherical contributions to the potential inside the spheres, which lead to a higher number of intraatomic integrals. In contrast, the contributions from the pseudo potential have a rather simple structure and allow for a high degree of vectorization.

In passing, we mention common practice of the standard ASW method to extract phase factors $e^{i\mathbf{k}\boldsymbol{\tau}_i}$ from the secular matrix in order to simplify computations. To be specific, instead of using the expression (4.2.31) we evaluate the Hamiltonian matrix

$$e^{-i\mathbf{k}\boldsymbol{\tau}_i}{}^\infty\langle L\kappa_1 i|H_\sigma|L'\kappa_2 j\rangle_c^\infty\,e^{i\mathbf{k}\boldsymbol{\tau}_j}$$

$$=\int_{\Omega_c}d^3\mathbf{r}\,e^{-i\mathbf{k}\boldsymbol{\tau}_i}D^{0*}_{L\kappa_1}(\mathbf{r}_i,\mathbf{k})\left(v^0_\sigma(\mathbf{r})-v_{MTZ}\right)D^0_{L'\kappa_2}(\mathbf{r}_j,\mathbf{k})e^{i\mathbf{k}\boldsymbol{\tau}_j}$$

$$+U^{(H,1)}_{L\kappa_1 L'\kappa_2 i\sigma}\delta_{ij}+\kappa_2^2\dot{B}_{LL'\kappa_1}(\boldsymbol{\tau}_i-\boldsymbol{\tau}_j,\mathbf{k})e^{-i\mathbf{k}(\boldsymbol{\tau}_i-\boldsymbol{\tau}_j)}\delta(\kappa_1^2-\kappa_2^2)$$

$$+\sum_{L''}U^{(H,2)}_{L\kappa_1 L''\kappa_2 i\sigma}B_{L''L'\kappa_2}(\boldsymbol{\tau}_i-\boldsymbol{\tau}_j,\mathbf{k})e^{-i\mathbf{k}(\boldsymbol{\tau}_i-\boldsymbol{\tau}_j)}$$

$$+\sum_{L''}e^{-i\mathbf{k}(\boldsymbol{\tau}_i-\boldsymbol{\tau}_j)}B_{LL''\kappa_1}(\boldsymbol{\tau}_i-\boldsymbol{\tau}_j,\mathbf{k})\left(U^{(H,2)*}_{L'\kappa_2 L''\kappa_1 j\sigma}+\delta(\kappa_1^2-\kappa_2^2)\delta_{L'L''}\right)$$

$$+\sum_m\sum_{L''}\sum_{L'''}e^{-i\mathbf{k}(\boldsymbol{\tau}_i-\boldsymbol{\tau}_m)}B_{LL''\kappa_1}(\boldsymbol{\tau}_i-\boldsymbol{\tau}_m,\mathbf{k})U^{(H,3)}_{L''\kappa_1 L'''\kappa_2 m\sigma}$$

$$B_{L'''L'\kappa_2}(\boldsymbol{\tau}_m-\boldsymbol{\tau}_j,\mathbf{k})e^{-i\mathbf{k}(\boldsymbol{\tau}_m-\boldsymbol{\tau}_j)}\ , \tag{4.2.35}$$

which grows out of the original one by a unitary transformation. For this reason, the norm of the wave function, which is given by (2.3.12), must be unaltered. Hence, the Hamiltonian matrix (4.2.35) will give rise to modified coefficients

$$c_{L\kappa j\sigma}(\mathbf{k})\,e^{-i\mathbf{k}\boldsymbol{\tau}_j}\ ,$$

such that the norm does not change. Since the wave function likewise remains unchanged the phase factors do not affect the derivations presented in the following sections.

As concerns evaluation of the modified Hamiltonian matrix we would have to calculate modified structure constants

$$B_{LL'\kappa_1}(\boldsymbol{\tau}_i-\boldsymbol{\tau}_j,\mathbf{k})e^{-i\mathbf{k}(\boldsymbol{\tau}_i-\boldsymbol{\tau}_j)}\ ,$$

instead of the structure constants themselves and, due to the definition (3.7.12), this means to calculate Bloch-summed envelope functions times the phase factor,

$$D_{L''\kappa_1}(\boldsymbol{\tau}_i - \boldsymbol{\tau}_j, \mathbf{k})e^{-i\mathbf{k}(\boldsymbol{\tau}_i-\boldsymbol{\tau}_j)}.$$

The same holds for the energy derivative of the structure constant. Eventually, the previous expression has to be combined with the results (3.5.18) to (3.5.21) for the Bloch-summed envelope function, where extracting the phase factor $e^{i\mathbf{k}\boldsymbol{\tau}}$ leads to substantial reduction in computer time.

While the previous considerations hold for both the standard and the full-potential ASW method, the first term on the right-hand side of (4.2.35) is unique to the latter. Combining it with the definition (3.11.1) of the pseudo function in terms of its Fourier expansion we write

$$
\begin{aligned}
D^0_{L\kappa}(\mathbf{r}_i, \mathbf{k})e^{i\mathbf{k}\boldsymbol{\tau}_i} &= \sum_n D^0_{L\kappa i}(\mathbf{K}_n + \mathbf{k})e^{i(\mathbf{K}_n+\mathbf{k})\mathbf{r}_i}e^{i\mathbf{k}\boldsymbol{\tau}_i} \\
&= \sum_n D^0_{L\kappa i}(\mathbf{K}_n + \mathbf{k})e^{i\mathbf{K}_n\mathbf{r}_i}e^{i\mathbf{k}\mathbf{r}} ,
\end{aligned}
\tag{4.2.36}
$$

where, in the second step, we have used the definition (2.1.26). Finally, since the pseudo functions always appear in products of the type (4.1.12) entering either the secular matrix or the electron density, the second exponential on the right-hand side of (4.2.36) likewise cancels out. We are thus left only with exponentials of reciprocal lattice vectors $\mathbf{K}_n$.

4.3 Electron Density

Having described the setup of the secular matrix of the full-potential ASW method we turn to the calculation of the spin-dependent electron density. Actually, the evaluation of the full electron density is independent of the way the expansion coefficients $c_{L\kappa i\sigma}(\mathbf{k})$ of the wave function in terms of the basis functions have been calculated. In other words, for the density it does not matter if the secular matrix were built with only the muffin-tin potential or with the full potential. As already discussed in Sect. 2.3, even the muffin-tin potential allows for the calculation of the full electron density without any shape approximation. The only reason for not evaluating the full electron density at that place was the restriction to the muffin-tin potential in the subsequent calculations. To conclude, the construction of the electron density from the wave function is a good starting point for a full-potential treatment whenever this starts from a previous muffin-tin calculation.

Since the construction of the valence electron density as based on the density of states as described in Sect. 2.3 can be done only within the shape approximation, we will now use the wave function directly. Hence, we write, as in (2.3.1), for the spin-dependent valence electron density

$$\rho_{val,\sigma}(\mathbf{r}) = \sum_{\mathbf{k}n} |\psi_{\mathbf{k}n\sigma}(\mathbf{r})|^2 \Theta(E_F - \varepsilon_{\mathbf{k}n\sigma}) . \tag{4.3.1}$$

As before, we have implied to use Fermi statistics by summing over the occupied states up to the Fermi energy E_F; all energies are referred to the muffin-tin zero. n labels the different eigenstates. For simplicity in writing, this band index will be absorbed into the $\mathbf{k}$-point label. In contrast to constructing the valence electron density from the (partial) densities of states, the direct use of the wave function has the distinct advantage of being fully in the spirit of the variational principle coming with density-functional theory [35].

Obviously, the construction of the valence electron density from (4.3.1) requires calculation of the product of the wave functions. To this end, we again use the representation of products of basis functions as outlined at the end of Sect. 4.1. Thus, we write the valence electron density as

$$\rho_{val,\sigma}(\mathbf{r}) = \rho^0_{val,\sigma}(\mathbf{r}) + \hat{\rho}_{val,\sigma}(\mathbf{r}) + \rho^{mix}_{val,\sigma}(\mathbf{r}) , \qquad (4.3.2)$$

where the three parts are readily calculated by combining (2.1.31), (4.1.7) to (4.1.9), and (4.1.12) to give

$$\rho^0_{val,\sigma}(\mathbf{r}) = \sum_{\mathbf{k}} \left| \sum_{L\kappa i} c_{L\kappa i\sigma}(\mathbf{k}) D^0_{L\kappa}(\mathbf{r}_i, \mathbf{k}) \right|^2 \Theta(E_F - \varepsilon_{\mathbf{k}\sigma}) , \qquad (4.3.3)$$

$$\hat{\rho}_{val,\sigma}(\mathbf{r}) = \sum_{\mathbf{k}} \left[\left| \sum_{L\kappa i}^{L_{int}} \left(c_{L\kappa i\sigma}(\mathbf{k}) \tilde{H}_{L\kappa\sigma}(\mathbf{r}_i) + a_{L\kappa i\sigma}(\mathbf{k}) \tilde{J}_{L\kappa\sigma}(\mathbf{r}_i) \right) \right|^2 \right.$$
$$\left. - \left| \sum_{L\kappa i}^{L_{int}} \left(c_{L\kappa i\sigma}(\mathbf{k}) H^0_{L\kappa}(\mathbf{r}_i) + a_{L\kappa i\sigma}(\mathbf{k}) J_{L\kappa}(\mathbf{r}_i)\Theta_i \right) \right|^2 \right]$$
$$\Theta(E_F - \varepsilon_{\mathbf{k}\sigma}) , \qquad (4.3.4)$$

and

$$\rho^{mix}_{val,\sigma}(\mathbf{r}) = \sum_{\mathbf{k}} \left[\left(\sum_{L=L_{int}+1}^{\infty} \sum_{\kappa i} a_{L\kappa i\sigma}(\mathbf{k}) J_{L\kappa}(\mathbf{r}_i)\Theta_m \right)^* \right.$$
$$\left. \left(\sum_{L\kappa i}^{L_{int}} \left(c_{L\kappa i\sigma}(\mathbf{k}) \hat{H}_{L\kappa\sigma}(\mathbf{r}_i) + a_{L\kappa i\sigma}(\mathbf{k}) \hat{J}_{L\kappa\sigma}(\mathbf{r}_i) \right) \right) + h.c. \right]$$
$$\Theta(E_F - \varepsilon_{\mathbf{k}\sigma}) . \qquad (4.3.5)$$

Here, we recognize the division of the products in three parts as outlined at the end of Sect. 4.1. As before, we will not take into account the last contribution, which arises from products of lower/intermediate and higher partial waves. The first and second term behave like pseudo and local functions, i.e. the first term is smooth and extends over all space while the second term contains all the intraatomic details and vanishes outside the atomic spheres.

It is the objective of this section to represent the remaining contributions to the valence electron density as an expansion in the appropriate sets of functions. With respect to the pseudo part this is achieved by an expansion in plane waves as has been done already in (4.2.5) for the pseudo part of the full potential. Hence, we write for the pseudo part of the full valence electron density

$$\rho_{val,\sigma}^0(\mathbf{r}) = \sum_t \rho_{val,\sigma}^0(\mathbf{K}_t) e^{i\mathbf{K}_t \mathbf{r}} , \tag{4.3.6}$$

where the expansion coefficients are given by

$$\rho_{val,\sigma}^0(\mathbf{K}_t) = \frac{1}{\Omega_c} \int_{\Omega_c} d^3\mathbf{r} \, \rho_{val,\sigma}^0(\mathbf{r}) e^{-i\mathbf{K}_t \mathbf{r}} . \tag{4.3.7}$$

In order to explicitly determine the pseudo part of the valence electron density $\rho_{val,\sigma}^0(\mathbf{r})$ we combine (4.3.3) with the plane-wave expansion (3.11.1) of the pseudo functions to

$$\rho_{val,\sigma}^0(\mathbf{r}) = \sum_{L\kappa_1 i} \sum_{L'\kappa_2 j} \sum_{\mathbf{k}} c_{L\kappa_1 i\sigma}^*(\mathbf{k}) c_{L'\kappa_2 j\sigma}(\mathbf{k})$$
$$D_{L\kappa_1}^{0*}(\mathbf{r}_i, \mathbf{k}) D_{L'\kappa_2}^0(\mathbf{r}_j, \mathbf{k}) \Theta(E_F - \varepsilon_{\mathbf{k}\sigma})$$
$$= \sum_{L\kappa_1 i} \sum_{L'\kappa_2 j} \sum_{\mathbf{k}} c_{L\kappa_1 i\sigma}^*(\mathbf{k}) c_{L'\kappa_2 j\sigma}(\mathbf{k}) \Theta(E_F - \varepsilon_{\mathbf{k}\sigma})$$
$$\left[\sum_n D_{L\kappa_1 i}^{0*}(\mathbf{K}_n + \mathbf{k}) e^{-i(\mathbf{K}_n + \mathbf{k})\mathbf{r}_i} \right]$$
$$\left[\sum_{n'} D_{L'\kappa_2 j}^0(\mathbf{K}_{n'} + \mathbf{k}) e^{i(\mathbf{K}_{n'} + \mathbf{k})\mathbf{r}_j} \right] . \tag{4.3.8}$$

From this we get for the Fourier coefficients

$$\rho_{val,\sigma}^0(\mathbf{K}_t)$$
$$= \sum_{L\kappa_1 i} \sum_{L'\kappa_2 j} \sum_{\mathbf{k}} c_{L\kappa_1 i\sigma}^*(\mathbf{k}) c_{L'\kappa_2 j\sigma}(\mathbf{k}) \Theta(E_F - \varepsilon_{\mathbf{k}\sigma})$$
$$\left[\sum_n D_{L\kappa_1 i}^{0*}(\mathbf{K}_n + \mathbf{k}) e^{i(\mathbf{K}_n + \mathbf{k})\boldsymbol{\tau}_i} \right]$$
$$\left[\sum_{n'} D_{L'\kappa_2 j}^0(\mathbf{K}_{n'} + \mathbf{k}) e^{-i(\mathbf{K}_{n'} + \mathbf{k})\boldsymbol{\tau}_j} \right]$$
$$\frac{1}{\Omega_c} \int_{\Omega_c} d^3\mathbf{r} \, e^{-i(\mathbf{K}_n + \mathbf{K}_t - \mathbf{K}_{n'})\mathbf{r}}$$
$$= \sum_{L\kappa_1 i} \sum_{L'\kappa_2 j} \sum_{\mathbf{k}} c_{L\kappa_1 i\sigma}^*(\mathbf{k}) c_{L'\kappa_2 j\sigma}(\mathbf{k}) \Theta(E_F - \varepsilon_{\mathbf{k}\sigma})$$

$$\sum_n D^{0*}_{L\kappa_1 i}(\mathbf{K}_n + \mathbf{k}) D^{0}_{L'\kappa_2 j}(\mathbf{K}_n + \mathbf{K}_t + \mathbf{k})$$

$$e^{i(\mathbf{K}_n+\mathbf{k})\boldsymbol{\mathcal{T}}_i} e^{-i(\mathbf{K}_n+\mathbf{K}_t+\mathbf{k})\boldsymbol{\mathcal{T}}_j} \,. \tag{4.3.9}$$

In passing, we note that, in the same manner as the calculation of the integrals (4.2.21) over the product of pseudo basis functions and the pseudo part of the potential, the evaluation of the pseudo part of the valence electron density in practice is performed in real space using (4.3.8).

Next, turning to the local contributions of the electron density as given by (4.3.4) we proceed in a similar manner and exchange the summations over $\mathbf{k}$ and L. This is useful since the eigenvectors $c_{L\kappa i\sigma}(\mathbf{k})$ as well as the $a_{L\kappa i\sigma}(\mathbf{k})$ do not depend on coordinates in space. Employing the definitions (2.1.13), (2.1.17), (2.1.20), and (3.9.2) of the augmented as well as the pseudo functions we get from (4.3.4)

$$\hat{\rho}_{val,\sigma}(\mathbf{r})$$
$$= \sum_i \hat{\rho}_{val,\sigma}(\mathbf{r}_i)$$
$$= \sum_i \sum_L^{L'_{max}} \sum_{L'}^{L'_{max}} \sum_{\kappa_1} \sum_{\kappa_2} Y_L^*(\hat{\mathbf{r}}_i) Y_{L'}(\hat{\mathbf{r}}_i)$$
$$\Bigg[\sum_{\mathbf{k}} c^*_{L\kappa_1 i\sigma}(\mathbf{k}) c_{L'\kappa_2 i\sigma}(\mathbf{k}) \left(\tilde{h}_{l\kappa_1\sigma}(r_i)\tilde{h}_{l'\kappa_2\sigma}(r_i) - \tilde{h}^0_{l\kappa_1}(r_i)\tilde{h}^0_{l'\kappa_2}(r_i) \right)$$
$$+ \sum_{\mathbf{k}} c^*_{L\kappa_1 i\sigma}(\mathbf{k}) a_{L'\kappa_2 i\sigma}(\mathbf{k}) \left(\tilde{h}_{l\kappa_1\sigma}(r_i)\tilde{j}_{l'\kappa_2\sigma}(r_i) - \tilde{h}^0_{l\kappa_1}(r_i)\tilde{j}_{l'}(\kappa_2 r_i) \right)$$
$$+ \sum_{\mathbf{k}} a^*_{L\kappa_1 i\sigma}(\mathbf{k}) c_{L'\kappa_2 i\sigma}(\mathbf{k}) \left(\tilde{j}_{l\kappa_1\sigma}(r_i)\tilde{h}_{l'\kappa_2\sigma}(r_i) - \bar{j}_l(\kappa_1 r_i)\tilde{h}^0_{l'\kappa_2}(r_i) \right)$$
$$+ \sum_{\mathbf{k}} a^*_{L\kappa_1 i\sigma}(\mathbf{k}) a_{L'\kappa_2 i\sigma}(\mathbf{k}) \left(\tilde{j}_{l\kappa_1\sigma}(r_i)\tilde{j}_{l'\kappa_2\sigma}(r_i) - \bar{j}_l(\kappa_1 r_i)\bar{j}_{l'}(\kappa_2 r_i) \right) \Bigg]$$
$$\Theta(E_F - \varepsilon_{\mathbf{k}\sigma}) \,. \tag{4.3.10}$$

As in Sect. 2.3, the angular momentum summations include both the lower and intermediate waves. Next we split the local electron density into different angular momentum contributions

$$\hat{\rho}_{val,\sigma}(\mathbf{r}_i) = \sum_K \hat{\rho}_{val,K\sigma}(r_i) Y_K(\hat{\mathbf{r}}_i)\Theta_i \,, \tag{4.3.11}$$

with the coefficients given by

$$\hat{\rho}_{val,K\sigma}(r_i) = \int d\Omega_i \, \hat{\rho}_{val,\sigma}(\mathbf{r}_i) Y_K^*(\hat{\mathbf{r}}_i) \,. \tag{4.3.12}$$

Defining local density matrices by

$$W^{(1)}_{L\kappa_1 L'\kappa_2 i\sigma} = \sum_{\mathbf{k}} c^*_{L\kappa_1 i\sigma}(\mathbf{k}) c_{L'\kappa_2 i\sigma}(\mathbf{k}) \Theta(E_F - \varepsilon_{\mathbf{k}\sigma}) , \qquad (4.3.13)$$

$$W^{(2)}_{L\kappa_1 L'\kappa_2 i\sigma} = \sum_{\mathbf{k}} c^*_{L\kappa_1 i\sigma}(\mathbf{k}) a_{L'\kappa_2 i\sigma}(\mathbf{k}) \Theta(E_F - \varepsilon_{\mathbf{k}\sigma}) , \qquad (4.3.14)$$

$$W^{(3)}_{L\kappa_1 L'\kappa_2 i\sigma} = \sum_{\mathbf{k}} a^*_{L\kappa_1 i\sigma}(\mathbf{k}) a_{L'\kappa_2 i\sigma}(\mathbf{k}) \Theta(E_F - \varepsilon_{\mathbf{k}\sigma}) , \qquad (4.3.15)$$

and inserting (4.3.10) into (4.3.12) we arrive at the result

$$
\begin{aligned}
&\hat{\rho}_{val,K\sigma}(r_i) \\
&= \tilde{\rho}_{val,K\sigma}(r_i) - \rho^0_{val,K\sigma}(r_i) \\
&= \sum_L^{L'_{max}} \sum_{L'}^{L'_{max}} \sum_{\kappa_1} \sum_{\kappa_2} c_{LL'K} \\
&\quad \left[W^{(1)}_{L\kappa_1 L'\kappa_2 i\sigma} \left(\tilde{h}_{l\kappa_1\sigma}(r_i)\tilde{h}_{l'\kappa_2\sigma}(r_i) - \tilde{h}^0_{l\kappa_1}(r_i)\tilde{h}^0_{l'\kappa_2}(r_i) \right) \right. \\
&\quad + W^{(2)}_{L\kappa_1 L'\kappa_2 i\sigma} \left(\tilde{h}_{l\kappa_1\sigma}(r_i)\tilde{j}_{l'\kappa_2\sigma}(r_i) - \tilde{h}^0_{l\kappa_1}(r_i)\bar{j}_{l'}(\kappa_2 r_i) \right) \\
&\quad + W^{(2)*}_{L'\kappa_2 L\kappa_1 i\sigma} \left(\tilde{j}_{l\kappa_1\sigma}(r_i)\tilde{h}_{l'\kappa_2\sigma}(r_i) - \bar{j}_l(\kappa_1 r_i)\tilde{h}^0_{l'\kappa_2}(r_i) \right) \\
&\quad \left. + W^{(3)}_{L\kappa_1 L'\kappa_2 i\sigma} \left(\tilde{j}_{l\kappa_1\sigma}(r_i)\tilde{j}_{l'\kappa_2\sigma}(r_i) - \bar{j}_l(\kappa_1 r_i)\bar{j}_{l'}(\kappa_2 r_i) \right) \right] . \qquad (4.3.16)
\end{aligned}
$$

Note that the coefficients result from a summation over all elements of a Hermitian matrix of dimension L_{int} times the number of interstitial energies κ^2 and, hence, turn out to be real quantities. As a consequence, we have to store only the real parts of the local density matrices (4.3.13) to (4.3.15). Again, this simplification is due to our new choice of basis functions, which excludes the i^l factors from the basis functions.

As concerns the norm of the wave function (4.1.8) we point to the fact that the eigenvectors resulting from the solution of the generalized eigenproblem (2.2.7) are correctly normalized and so is the wave function. Since we do no longer employ the shape approximation there is no renormalization as in Sect. 2.3 and so we stay with the correct normalization without any approximation.

Finally, the valence electron density has to be complemented by the core electron density. The latter is taken from the corresponding expression (2.3.20) given within the standard ASW method. Combining the densities of the valence and the core electrons we arrive at the total spin-dependent electron density

$$
\begin{aligned}
\rho_{el,\sigma}(\mathbf{r}) &= \rho^0_{val,\sigma}(\mathbf{r}) + \sum_i \left(\hat{\rho}_{val,\sigma}(\mathbf{r}_i) + \rho_{core,\sigma}(\mathbf{r}_i) \right) \\
&= \rho^0_{val,\sigma}(\mathbf{r}) + \sum_i \sum_K \left(\hat{\rho}_{val,K\sigma}(r_i) + \sqrt{4\pi}\delta_{k0}\rho_{core,\sigma}(r_i) \right) Y_K(\hat{\mathbf{r}}_i)
\end{aligned}
$$

$$=: \rho_{el,\sigma}^0(\mathbf{r}) + \sum_i \sum_K \hat{\rho}_{el,K\sigma}(r_i) Y_K(\hat{\mathbf{r}}_i)$$

$$=: \rho_{el,\sigma}^0(\mathbf{r}) + \sum_i \hat{\rho}_{el,\sigma}(\mathbf{r}_i) \ .$$ (4.3.17)

In closing this section, we turn again to the intermediate approach and aim at reducing the effort to calculate the full-potential contributions to their minimum. First of all, this would mean to stay only with the $\mathbf{K}_t = \mathbf{0}$ term of the pseudo part of the density as given by (4.3.7). This term is most easily calculated by combining (4.3.7) and (4.3.8) to write

$$\rho_{val,\sigma}^0(\mathbf{0}) = \frac{1}{\Omega_c} \int_{\Omega_c} d^3\mathbf{r} \, \rho_{val,\sigma}^0(\mathbf{r})$$

$$= \frac{1}{\Omega_c} \sum_{L\kappa_1 i} \sum_{L'\kappa_2 j} \sum_{\mathbf{k}} c_{L\kappa_1 i\sigma}^*(\mathbf{k}) c_{L'\kappa_2 j\sigma}(\mathbf{k})$$

$$\int_{\Omega_c} d^3\mathbf{r} \, D_{L\kappa_1}^{0*}(\mathbf{r}_i, \mathbf{k}) D_{L'\kappa_2}^0(\mathbf{r}_j, \mathbf{k}) \Theta(E_F - \varepsilon_{\mathbf{k}\sigma}) \ .$$

(4.3.18)

The integral in the last line is just the overlap integral of the pseudo functions as discussed in Sect. 3.10. Plugging the results gained in that context into (4.3.18) we arrive at the result

$$\rho_{val,\sigma}^0(\mathbf{0}) = \frac{1}{\Omega_c} \sum_{L\kappa_1 i} \sum_{L'\kappa_2 j} \sum_{\mathbf{k}} c_{L\kappa_1 i\sigma}^*(\mathbf{k}) c_{L'\kappa_2 j\sigma}(\mathbf{k})$$

$$\left[X_{L\kappa_1\kappa_2 i}^{0(S,1)} \delta_{LL'} \delta_{ij} + \dot{B}_{LL'\kappa_1}(\boldsymbol{\tau}_i - \boldsymbol{\tau}_j, \mathbf{k}) \delta(\kappa_1^2 - \kappa_2^2) \right.$$

$$+ X_{L\kappa_1\kappa_2 i}^{0(S,2)} B_{LL'\kappa_2}(\boldsymbol{\tau}_i - \boldsymbol{\tau}_j, \mathbf{k}) + B_{LL'\kappa_1}(\boldsymbol{\tau}_i - \boldsymbol{\tau}_j, \mathbf{k}) X_{L'\kappa_1\kappa_2 j}^{0(S,2)} \right]$$

$$\Theta(E_F - \varepsilon_{\mathbf{k}\sigma}) \ ,$$ (4.3.19)

which allows for a calculation of the constant pseudo density at a minimal effort.

Of course, we would also have to modify the local contributions. In particular, we will have to replace the contributions from the pseudo functions to (4.3.16) by the just calculated constant density. We would thus write for the local parts

$$\hat{\rho}_{val,K\sigma}(r_i)$$

$$= \tilde{\rho}_{val,K\sigma}(r_i) - \rho_{val,K\sigma}^0(r_i)$$

$$= \sum_L^{L'_{max}} \sum_{L'}^{L'_{max}} \sum_{\kappa_1} \sum_{\kappa_2} c_{LL'K}$$

$$\left[W_{L\kappa_1 L'\kappa_2 i\sigma}^{(1)} \tilde{h}_{l\kappa_1\sigma}(r_i) \tilde{h}_{l'\kappa_2\sigma}(r_i) + W_{L\kappa_1 L'\kappa_2 i\sigma}^{(2)} \tilde{h}_{l\kappa_1\sigma}(r_i) \tilde{j}_{l'\kappa_2\sigma}(r_i) \right.$$

$$+W^{(2)*}_{L'\kappa_2 L\kappa_1 i\sigma}\tilde{\jmath}_{l\kappa_1\sigma}(r_i)\tilde{h}_{l'\kappa_2\sigma}(r_i) + W^{(3)}_{L\kappa_1 L'\kappa_2 i\sigma}\tilde{\jmath}_{l\kappa_1\sigma}(r_i)\tilde{\jmath}_{l'\kappa_2\sigma}(r_i)\Bigg]$$

$$-\sqrt{4\pi}\rho^0_{val,\sigma}(\mathbf{K}_t = 0)\delta_{K0} , \tag{4.3.20}$$

which now extend through the atomic-sphere rather than the muffin-tin sphere.

4.4 The Effective Potential

With the full electron density at hand we are in a position to perform the last step of the self-consistency cycle and calculate the full potential. In doing so we require, of course, the full potential to be described in terms of pseudo and local contributions as formulated in (4.2.2), (4.2.3), and (4.2.5).

As already outlined in Sect. 2.4, the effective potential consists, within the framework of density-functional theory and the local-density approximation, of the external, Hartree, and exchange-correlation potential [15]. While the latter comprises the non-classical parts of the electron-electron interaction, the former two parts represent the classical electrostatic potentials arising from the nuclear and the electronic charges, respectively. For the calculation of the total electrostatic potential it is thus useful to combine the electron density as given by (4.3.17) with the point density due to the nuclei as given by (2.4.1). The resulting total density reads as

$$\rho(\mathbf{r}) = \sum_\sigma \rho_{el,\sigma}(\mathbf{r}) + \rho_{nucl}(\mathbf{r}) . \tag{4.4.1}$$

For the purpose of calculating the classical electrostatic potential we split the total density (4.4.1), like all other quantities before, into pseudo and local contributions

$$\rho(\mathbf{r}) = \rho^0(\mathbf{r}) + \sum_i \hat{\rho}(\mathbf{r}_i) , \tag{4.4.2}$$

which are defined by

$$\rho^0(\mathbf{r}) = \sum_\sigma \rho^0_{val,\sigma}(\mathbf{r}) , \tag{4.4.3}$$

and

$$\hat{\rho}(\mathbf{r}_i) = \sum_\sigma \left(\hat{\rho}_{val,\sigma}(\mathbf{r}_i) + \rho_{core,\sigma}(\mathbf{r}_i)\right) + \rho_{nucl}(\mathbf{r}_i) . \tag{4.4.4}$$

The classical electrostatic potential is then given as a linear functional of the full density

$$v_{es}(\mathbf{r}) = 2 \int d^3\mathbf{r}' \frac{\rho(\mathbf{r}')}{|\mathbf{r} - \mathbf{r}'|}$$

$$= 2 \int d^3\mathbf{r}' \frac{\rho^0(\mathbf{r}')}{|\mathbf{r} - \mathbf{r}'|} + 2 \sum_{\mu i} \int_{\Omega_i} d^3\mathbf{r}'_{\mu i} \frac{\hat{\rho}(\mathbf{r}'_{\mu i})}{|\mathbf{r}_{\mu i} - \mathbf{r}'_{\mu i}|} . \tag{4.4.5}$$

Note the factor 2 entering (4.4.5), which results from our choice of atomic units ($e^2 = 2$) taken in Sect. 1.2. As already mentioned in Sect. 2.4, it reflects the fact that we have calculated densities rather than charge densities and, hence, the extra e factor making a charge density from a particle density has to be included here.

In evaluating the electrostatic potential as well as separating it into pseudo and local parts we have to be aware of the fact that the local parts of the density contribute to the electrostatic potential outside their own sphere. For this reason, the pseudo part of this potential does not arise from the pseudo part of the density alone but comprises also contributions due to the local parts of the density. To start with, we set up the potential arising from the pseudo density as

$$v_{es}^{00}(\mathbf{r}) := 2 \int d^3r' \, \frac{\rho^0(\mathbf{r}')}{|\mathbf{r} - \mathbf{r}'|} \, . \tag{4.4.6}$$

It obliges Poisson's equation

$$\Delta v_{es}^{00}(\mathbf{r}) = -8\pi\rho^0(\mathbf{r}) \, , \tag{4.4.7}$$

which is most easily solved by employing the plane-wave expansions of the pseudo parts of both the density

$$\rho^0(\mathbf{r}) = \sum_t \rho^0(\mathbf{K}_t)e^{i\mathbf{K}_t\mathbf{r}}$$

$$= \sum_t \sum_\sigma \rho_{val,\sigma}^0(\mathbf{K}_t)e^{i\mathbf{K}_t\mathbf{r}} \, , \tag{4.4.8}$$

and the classical electrostatic potential. In analogy to (4.2.5) the latter reads as

$$v_{es}^{00}(\mathbf{r}) = \sum_t v_{es}^{00}(\mathbf{K}_t)e^{i\mathbf{K}_t\mathbf{r}} \, . \tag{4.4.9}$$

Application of the Laplacian reduces to multiplying the t'th term in the series with $-\mathbf{K}_t^2$ and we note the result

$$v_{es}^{00}(\mathbf{K}_t) = \frac{8\pi}{|\mathbf{K}_t|^2}\rho^0(\mathbf{K}_t) \, . \tag{4.4.10}$$

Here, the term with $\mathbf{K}_t = 0$ needs special attention and will be discussed in more detail below.

In evaluating the second contribution to (4.4.5), the electrostatic potential arising from the local parts of the density, we first consider the situation, where the position $\mathbf{r}$ lies outside all atomic spheres, i.e. in the interstitial region. In this case the resulting electrostatic potential trivially belongs to the pseudo part. Furthermore, it can be simply written in terms of the multipole moments of the local densities, which are defined by

$$M_{Ki} := \int_{\Omega_i} d^3\mathbf{r}_i \ (r_i)^k Y_K^*(\hat{\mathbf{r}}_i)\hat{\rho}(\mathbf{r}_i)$$

$$= \int_0^{S_i} dr_i \ r_i^{k+2} \hat{\rho}_K(r_i)$$

$$= \int_0^{S_i} dr_i \ r_i^{k+2} \left[\sum_\sigma \left(\hat{\rho}_{val,K\sigma}(r_i) + \sqrt{4\pi}\delta_{k0}\rho_{core,\sigma}(r_i) \right) + \sqrt{4\pi}\delta_{k0}\rho_{nucl}(r_i) \right]$$

$$= \int_0^{S_i} dr_i \ r_i^{k+2} \left[\sum_\sigma \left(\hat{\rho}_{val,K\sigma}(r_i) + \sqrt{4\pi}\delta_{k0}\rho_{core,\sigma}(r_i) \right) \right] - \frac{1}{\sqrt{4\pi}}\delta_{k0} Z_i \ .$$

$$(4.4.11)$$

With the multipole moments at hand, evaluation of the pseudo part of the electrostatic potential in the interstitial region is straightforward. Using the spherical-harmonics expansion (4.3.11) of the local densities as well as the identity (2.4.3) we write

$$v_{es}^0(\mathbf{r})\Theta_I = v_{es}^{00}(\mathbf{r}) + 2\sum_{\mu i} \int_{\Omega_i} d^3\mathbf{r}'_{\mu i} \ \frac{\hat{\rho}(\mathbf{r}'_{\mu i})}{|\mathbf{r}_{\mu i} - \mathbf{r}'_{\mu i}|}$$

$$= v_{es}^{00}(\mathbf{r}) + 2\sum_{\mu i} \int_{\Omega_i} d^3\mathbf{r}'_{\mu i} \sum_K \frac{4\pi}{2k+1} \frac{(r'_{\mu i})^k}{r_{\mu i}^{k+1}} Y_K^*(\hat{\mathbf{r}}'_{\mu i}) Y_K(\hat{\mathbf{r}}_{\mu i})\hat{\rho}(\mathbf{r}'_{\mu i})$$

$$= v_{es}^{00}(\mathbf{r}) + \sum_K \frac{8\pi}{2k+1} \sum_{\mu i} \frac{1}{r_{\mu i}^{k+1}} Y_K(\hat{\mathbf{r}}_{\mu i})$$

$$\int_{\Omega_i} d^3\mathbf{r}'_{\mu i} \ (r'_{\mu i})^k Y_K^*(\hat{\mathbf{r}}'_{\mu i})\hat{\rho}(\mathbf{r}'_{\mu i})$$

$$= v_{es}^{00}(\mathbf{r}) + \sum_K \frac{8\pi}{(2k+1)!!} \sum_i M_{Ki} \sum_\mu H_{K\kappa=0}(\mathbf{r}_{\mu i})$$

$$= v_{es}^{00}(\mathbf{r}) + \sum_K \frac{8\pi}{(2k+1)!!} \sum_i M_{Ki} D_{K\kappa=0}(\mathbf{r}_i, \mathbf{k} = \mathbf{0}) \ , \qquad (4.4.12)$$

where in the last steps we employed the definition (3.1.17) of the Hankel function, its asymptotic behaviour (3.1.45) for small arguments as well as the definition (3.5.2) of the Bloch-summed Hankel function. Next, taking advantage of the representation (3.5.18) to (3.5.22) of the latter, we obtain

$$v_{es}^0(\mathbf{r})\Theta_I = v_{es}^{00}(\mathbf{r}) + \sum_K \frac{8\pi}{(2k+1)!!} \sum_i M_{Ki}$$

$$\left[D_{K\kappa=0}^{(1)}(\mathbf{r}_i, \mathbf{k} = \mathbf{0}) + D_{K\kappa=0}^{(2)}(\mathbf{r}_i, \mathbf{k} = \mathbf{0}) + D_{K\kappa=0}^{(3)}(\mathbf{r}_i) \right]$$

$$= v_{es}^{00}(\mathbf{r}) - \frac{4\pi}{\Omega_c} \sum_n e^{i\mathbf{K}_n\mathbf{r}} \sum_i e^{-i\mathbf{K}_n\boldsymbol{\tau}_i}$$

$$\sum_K \frac{8\pi}{(2k+1)!!} (-i)^k M_{Ki} |\mathbf{K}_n|^{k-2} Y_K(\hat{\mathbf{K}}_n) e^{-\frac{\mathbf{K}_n^2}{\eta_M}}$$

$$+\frac{2}{\sqrt{\pi}}\sum_K\frac{8\pi}{(2k+1)!!}\sum_i M_{Ki}2^k$$

$$\sum_\mu(1-\delta_{\mu 0}\delta(\mathbf{r}_i))|\mathbf{r}_{\mu i}|^k Y_K(\hat{\mathbf{r}}_{\mu i})\int_{\frac{1}{2}\eta_M^{1/2}}^\infty \xi^{2k}e^{-\mathbf{r}_{\mu i}^2\xi^2}\,d\xi$$

$$-4\eta_M^{1/2}\sum_i M_{0i}\delta(\mathbf{r}_i)\delta_{k0}\,. \tag{4.4.13}$$

Note that inclusion of the term $D^{(3)}_{K\kappa=0}(\mathbf{r}_i)$ is only for formal reasons since the δ-distribution coming with it vanishes in the interstitial region.

In passing, we point to the formal similarity of the second term in the last line of (4.4.12) to the Madelung potential calculated within the standard ASW method in (2.4.9). In particular, for $\mathbf{r}=\boldsymbol{\tau}_j$ and with the multipole moment of the local charge density replaced by that of the total intraatomic charge the $K=0$-term in the K-summation of (4.4.12) reduces exactly to the Madelung potential given by (2.4.9).

The previous evaluation of the real-space lattice sum with the help of the Ewald method allows for a deeper discussion of the potential arising from the local multipoles. To start with, we conclude from comparing (4.4.13) to (4.4.10) that the electrostatic potential corresponding to the term $D^{(1)}_{K\kappa=0}(\mathbf{r}_i,\mathbf{k}=\mathbf{0})$ is that of a density $\rho_{aux}(\mathbf{r})$ with Fourier coefficients

$$\rho_{aux}(\mathbf{K}_n)=\frac{4\pi}{\Omega_c}\sum_i e^{-i\mathbf{K}_n\boldsymbol{\tau}_i}\sum_K(-i)^k\frac{M_{Ki}}{(2k+1)!!}|\mathbf{K}_n|^k Y_K(\hat{\mathbf{K}}_n)e^{-\frac{\mathbf{K}_n^2}{\eta_M}}\,. \tag{4.4.14}$$

As we will see immediately, this density can be written as a superposition of localized densities of the form

$$\rho_{aux}(\mathbf{r})=\sum_i\rho_{aux}(\mathbf{r}_i)=\sum_i\sum_K\rho_{aux,K}(r_i)Y_K(\hat{\mathbf{r}}_i)\,, \tag{4.4.15}$$

where each coefficient in the spherical-harmonics expansion consists of a Gaussian of width $4/\sqrt{\eta_M}$ scaled to the multipole moment M_{Ki},

$$\rho_{aux,K}(r_i)=g_{Ki}\frac{2}{\sqrt{\pi}}2^k\left(\frac{1}{2}\eta_M^{1/2}\right)^{2k}r_i^k e^{-\frac{1}{4}\eta_M r_i^2}$$

$$=\frac{4}{\sqrt{\pi}}2^k\frac{M_{Ki}}{(2k+1)!!}\left(\frac{1}{2}\eta_M^{1/2}\right)^{2k+3}r_i^k e^{-\frac{1}{4}\eta_M r_i^2}\,. \tag{4.4.16}$$

Here, the prefactors were chosen in accordance with the convention used in the integral representation of the Hankel envelope function as given by (3.4.5). The factor g_{Ki}, which guarantees the correct scaling,

$$g_{Ki}\frac{2}{\sqrt{\pi}}2^k\left(\frac{1}{2}\eta_M^{1/2}\right)^{2k}\int_0^\infty r_i^{2k+2}e^{-\frac{1}{4}\eta_M r_i^2}\,dr_i\stackrel{!}{=}M_{Ki}\,, \tag{4.4.17}$$

follows from (C.4.12) as

$$g_{Ki} = \frac{M_{Ki}\,\eta_M^{3/2}}{4(2k+1)!!}.$$

(4.4.18)

Note that the integration in (4.4.17) should have extended only to the sphere radius rather than to infinity. However, the resulting error can be systematically reduced by adapting the Ewald parameter η_M. We will turn to this issue below.

With the definitions (4.4.5) and (4.4.11) as well as the Poisson transform equation (B.4.8) at hand, the plane-wave expansion of the auxiliary density (4.4.15) is calculated as

$$
\begin{aligned}
\rho_{aux}(\mathbf{K}_t) &= \frac{1}{\Omega_c}\int_{\Omega_c} d^3\mathbf{r}\,\rho_{aux}(\mathbf{r})e^{-i\mathbf{K}_t\mathbf{r}} \\
&= \frac{1}{\Omega_c}\sum_i\int_{\Omega_c} d^3\mathbf{r}\,\rho_{aux}(\mathbf{r}_{\mu i})e^{-i\mathbf{K}_t\boldsymbol{\tau}_i}e^{-i\mathbf{K}_t\mathbf{r}_i} \\
&= \frac{1}{\Omega_c}\sum_i e^{-i\mathbf{K}_t\boldsymbol{\tau}_i}\int_{\Omega_c} d^3\mathbf{r}\,\sum_K \rho_{aux,K}(r_{\mu i})Y_K(\hat{\mathbf{r}}_{\mu i})e^{-i\mathbf{K}_t\mathbf{r}_i} \\
&= \frac{1}{\Omega_c}\sum_i e^{-i\mathbf{K}_t\boldsymbol{\tau}_i}\int_{\Omega_c} d^3\mathbf{r}\,\sum_K \frac{4}{\sqrt{\pi}}2^k\frac{M_{Ki}}{(2k+1)!!}\left(\frac{1}{2}\eta_M^{1/2}\right)^{2k+3} \\
&\qquad \sum_\mu |\mathbf{r}_{\mu i}|^k e^{-\frac{1}{4}\eta_M\mathbf{r}_{\mu i}^2}Y_K(\hat{\mathbf{r}}_{\mu i})e^{-i\mathbf{K}_t\mathbf{r}_i} \\
&= \frac{1}{\Omega_c}\sum_i e^{-i\mathbf{K}_t\boldsymbol{\tau}_i}\int_{\Omega_c} d^3\mathbf{r}\,\sum_K \frac{4}{\sqrt{\pi}}2^k\frac{M_{Ki}}{(2k+1)!!}\left(\frac{1}{2}\eta_M^{1/2}\right)^{2k+3} \\
&\qquad \frac{\pi^{\frac{3}{2}}}{\Omega_c}(-i)^k 2^{-k}\left(\frac{1}{2}\eta_M^{1/2}\right)^{-(2k+3)} \\
&\qquad \sum_n e^{i\mathbf{K}_n\mathbf{r}_i}|\mathbf{K}_n|^k Y_K(\widehat{\mathbf{K}_n})e^{-\frac{\mathbf{K}_n^2}{\eta_M}}e^{-i\mathbf{K}_t\mathbf{r}_i} \\
&= \frac{4\pi}{\Omega_c}\sum_i e^{-i\mathbf{K}_t\boldsymbol{\tau}_i}\sum_K(-i)^k\frac{M_{Ki}}{(2k+1)!!}|\mathbf{K}_t|^k Y_K(\widehat{\mathbf{K}_t})e^{-\frac{\mathbf{K}_t^2}{\eta_M}},
\end{aligned}
$$

(4.4.19)

which is indeed identical to (4.4.14). We have thus shown that the $D^{(1)}_{K\kappa=0}$-contribution to the interstitial electrostatic potential arises from Gaussian densities of width $4/\sqrt{\eta_M}$, which are localized at the atomic positions and scaled to the multipole moments M_{Ki}. However, we are still left with the $D^{(2)}_{K\kappa=0}$-term entering (4.4.13). Its role becomes clearer from taking the limit $\eta_M \to \infty$, in which case this term vanishes. At the same time, the Gaussians turn into δ-distributions localized at the atomic sites.

The important point to notice is that in the present context we are interested only in the electrostatic potential in the interstitial region, while the shape of this potential inside the atomic spheres has no relevance. Hence, details of the potential on a length scale smaller than the smallest atomic sphere radius need not be calculated. We may thus safely omit the $D^{(2)}_{K\kappa=0}$-contribution to (4.4.13) as long as a large enough η_M is used. Following the discussion in Sect. 3.5, we choose

$$\eta_M = \frac{6.5}{R^2_{min}} , \qquad (4.4.20)$$

where R_{min} designates the smallest atomic sphere radius. Note that, with this choice, the integral entering (4.4.17) can be safely limited to the sphere radius rather than to infinity.

With the previous considerations we have arrived at a completely new method of representing the electrostatic potential, which offers several advantages and, in a slightly different way, was first outlined by Weinert [31]. It starts from the separation (4.4.2), but adds and subtracts the auxiliary density (4.4.15) to the pseudo and local part, respectively. We thus replace (4.4.3) and (4.4.4) by

$$\rho^0(\mathbf{r}) = \sum_\sigma \rho^0_{val,\sigma}(\mathbf{r}) + \rho_{aux}(\mathbf{r}) , \qquad (4.4.21)$$

and

$$\hat{\rho}(\mathbf{r}_i) = \sum_\sigma \left(\hat{\rho}_{val,\sigma}(\mathbf{r}_i) + \rho_{core,\sigma}(\mathbf{r}_i) \right) + \rho_{nucl}(\mathbf{r}_i) - \rho_{aux}(\mathbf{r}_i) . \qquad (4.4.22)$$

Since, by construction, the auxiliary density inside each atomic sphere has the same multipole moments as the sum of valence, core and nuclear densities the local density (4.4.22) produces no electrostatic field outside its atomic sphere. Instead, the influence of the intraatomic physical charges is completely covered by the contribution of the auxiliary density to the pseudo density (4.4.21). We are thus in a position to define the pseudo and local parts of the electrostatic potential by

$$v_{es}(\mathbf{r}) = v^0_{es}(\mathbf{r}) + \sum_i \hat{v}_{es}(\mathbf{r}_i) , \qquad (4.4.23)$$

where

$$v^0_{es}(\mathbf{r}) = 2 \int d^3r' \, \frac{\rho^0(\mathbf{r})}{|\mathbf{r} - \mathbf{r}'|} = 2 \int d^3r' \, \frac{\sum_\sigma \rho^0_{val,\sigma}(\mathbf{r}) + \rho_{aux}(\mathbf{r})}{|\mathbf{r} - \mathbf{r}'|} , \qquad (4.4.24)$$

and

$$\hat{v}_{es}(\mathbf{r}_i) := 2 \int_{\Omega_i} d^3r'_i \, \frac{\hat{\rho}(\mathbf{r}'_i)}{|\mathbf{r}_i - \mathbf{r}'_i|} . \qquad (4.4.25)$$

Due to the above construction, the local part indeed vanishes at and beyond the respective atomic sphere radius while the pseudo part extends smoothly throughout all space. Equation (4.4.24) thus offers the additional advantage over the representation (4.4.12) of being valid not only in the interstitial region but also inside the atomic spheres. Using the plane-wave expansion (4.4.14) of the auxiliary density we obtain the Fourier coefficients of the pseudo part of the electrostatic potential as

$$v_{es}^0(\mathbf{K}_t) = \frac{8\pi}{|\mathbf{K}_t|^2} \left(\sum_\sigma \rho_{val,\sigma}^0(\mathbf{K}_t) + \rho_{aux}(\mathbf{K}_t) \right) . \qquad (4.4.26)$$

In particular, for $\mathbf{K}_t = \mathbf{0}$ we recall (4.3.7) and (4.4.19) for the Fourier coefficients of the pseudo and auxiliary density, respectively, and note

$$\sum_\sigma \rho_{val,\sigma}^0(\mathbf{K}_t = \mathbf{0}) + \rho_{aux}(\mathbf{K}_t = \mathbf{0})$$

$$= \frac{1}{\Omega_c} \int_{\Omega_c} d^3\mathbf{r} \left(\sum_\sigma \rho_{val,\sigma}^0(\mathbf{r}) + \rho_{aux}(\mathbf{r}) \right)$$

$$= \frac{1}{\Omega_c} \int_{\Omega_c} d^3\mathbf{r} \sum_\sigma \rho_{val,\sigma}^0(\mathbf{r}) + \frac{\sqrt{4\pi}}{\Omega_c} \sum_i M_{0i}$$

$$= \frac{1}{\Omega_c} \int_{\Omega_c} d^3\mathbf{r} \sum_\sigma \rho_{val,\sigma}^0(\mathbf{r})$$

$$+ \frac{1}{\Omega_c} \sum_i \int_{\Omega_i} d^3\mathbf{r}_i \left[\sum_\sigma (\hat{\rho}_{val,\sigma}(\mathbf{r}_i) + \rho_{core,\sigma}(\mathbf{r}_i)) \right] - \frac{1}{\Omega_c} \sum_i Z_i . \qquad (4.4.27)$$

The $\mathbf{K}_t = \mathbf{0}$-coefficient thus corresponds to the total, electronic plus nuclear, charge and is zero for charge neutrality reasons. As a consequence, also the $\mathbf{K}_t = \mathbf{0}$-term of the electrostatic potential vanishes.

For the local parts we employ the spherical-harmonics expansions (4.3.17) and (4.4.15) of the local parts of the electronic and auxiliary density, respectively, as well as the identity (2.4.3) and obtain

$$\hat{v}_{es}(\mathbf{r}_i) = 2 \int d\Omega_i' \int_0^{S_i} dr_i'(r_i')^2 \sum_K \hat{\rho}_K(r_i') Y_K(\hat{\mathbf{r}}_i')$$

$$\sum_{K'} \frac{4\pi}{2k'+1} \frac{r_<^{k'}}{r_>^{k'+1}} Y_{K'}^*(\hat{\mathbf{r}}_i') Y_{K'}(\hat{\mathbf{r}}_i)$$

$$= \sum_K \frac{8\pi}{2k+1} \int_0^{S_i} dr_i'(r_i')^2 \frac{r_<^k}{r_>^{k+1}} \hat{\rho}_K(r_i') Y_K(\hat{\mathbf{r}}_i)$$

$$= \sum_K \hat{v}_{es,K}(r_i) Y_K(\hat{\mathbf{r}}_i) , \qquad (4.4.28)$$

where we have used the orthonormality of the spherical harmonics and where

$$\hat{v}_{es,K}(r_i) = \frac{8\pi}{2k+1} \left[\frac{1}{r_i^{k+1}} \int_0^{r_i} dr_i' \, r_i'^{k+2} \hat{\rho}_K(r_i') + r_i^k \int_{r_i}^{S_i} dr_i' \, r_i'^{1-k} \hat{\rho}_K(r_i') \right] .$$

$$(4.4.29)$$

Finally, we take advantage of the fact that the electrostatic potential is a linear functional of the density and write the local part as the difference of the potentials arising from the true and the pseudo density as

$$\hat{v}_{es}(\mathbf{r}_i) = \tilde{v}_{es}(\mathbf{r}_i) - v_{es}^{0-}(\mathbf{r}_i) = 2 \int_{\Omega_i} d^3 \mathbf{r}'_i \frac{\tilde{\rho}(\mathbf{r}'_i) - \rho^{0-}(\mathbf{r}'_i)}{|\mathbf{r}_i - \mathbf{r}'_i|} , \qquad (4.4.30)$$

where

$$\tilde{\rho}(\mathbf{r}_i) = \tilde{\rho}_{el}(\mathbf{r}_i) + \rho_{nucl}(\mathbf{r}_i)$$
$$= \sum_{\sigma} (\tilde{\rho}_{val,\sigma}(\mathbf{r}_i) + \rho_{core,\sigma}(\mathbf{r}_i)) + \rho_{nucl}(\mathbf{r}_i) \qquad (4.4.31)$$

and

$$\rho^{0-}(\mathbf{r}_i) = \rho_{el}^{0-}(\mathbf{r}_i) = \sum_{\sigma} \rho_{val,\sigma}^{0-}(\mathbf{r}_i) + \rho_{aux}(\mathbf{r}_i) . \qquad (4.4.32)$$

From these definitions the two components of the local part of the electrostatic potential are calculated along the lines given by (4.4.28) and (4.4.29).

Next we turn to the non-classical and spin-dependent exchange-correlation potential. Its calculation turns out to be somewhat more complicated since, within density-functional theory and the local-density approximation, the exchange-correlation potential and energy density are local but at the same time non-linear functions of the electron density

$$v_{xc,\sigma}(\mathbf{r}) := v_{xc,\sigma} \left[\rho_{el,\sigma}(\mathbf{r}), \rho_{el,-\sigma}(\mathbf{r}) \right] . \qquad (4.4.33)$$

As a consequence, the separation of the electron density into pseudo and local parts, which can be expanded in plane waves and spherical harmonics, respectively, is not preserved. However, in order to make a connection to the prescription (4.2.2) to (4.2.6) of the full potential in terms of plane waves and spherical harmonics we likewise need the corresponding representation of the exchange-correlation potential.

To do so, we start enforcing the separation of the exchange-correlation potential in pseudo and local parts by writing

$$v_{xc,\sigma}(\mathbf{r}) = v_{xc,\sigma}^{0}(\mathbf{r}) + \sum_{i} \hat{v}_{xc,\sigma}(\mathbf{r}_i) . \qquad (4.4.34)$$

Since the pseudo part of the electron density is given on a regular mesh extending over the real-space unit cell we are able to calculate the exchange-correlation potential arising from the pseudo electron density at each mesh point and define the result as the pseudo part of the exchange-correlation potential as

$$v_{xc,\sigma}^{0}(\mathbf{r}) := v_{xc,\sigma} \left[\rho_{el,\sigma}^{0}(\mathbf{r}), \rho_{el,-\sigma}^{0}(\mathbf{r}) \right] . \qquad (4.4.35)$$

Like the pseudo part of the electron density this potential is rather smooth and thus will allow for a plane-wave expansion

$$v_{xc,\sigma}^{0}(\mathbf{r}) = \sum_{t} v_{xc,\sigma}^{0}(\mathbf{K}_t) e^{i\mathbf{K}_t \cdot \mathbf{r}} , \qquad (4.4.36)$$

with the expansion coefficients given by

$$v_{xc,\sigma}^{0}(\mathbf{K}_t) = \frac{1}{\Omega_c} \int_{\Omega_c} d^3\mathbf{r}\, v_{xc,\sigma}^{0}(\mathbf{r}) e^{-i\mathbf{K}_t\mathbf{r}}$$

$$= \frac{1}{\Omega_c} \int_{\Omega_c} d^3\mathbf{r}\, v_{xc,\sigma} \left[\rho_{el,\sigma}^{0}(\mathbf{r}), \rho_{el,-\sigma}^{0}(\mathbf{r}) \right] e^{-i\mathbf{K}_t\mathbf{r}} . \qquad (4.4.37)$$

Inside the atomic spheres we start from the one-center expansions of the true electronic and the pseudo electron density

$$\rho_{el,\sigma}(\mathbf{r}_i) = \sum_K \rho_{el,K\sigma}(r_i) Y_K(\hat{\mathbf{r}}_i) , \qquad (4.4.38)$$

and

$$\rho_{el,\sigma}^{0-}(\mathbf{r}_i) = \sum_K \rho_{el,K\sigma}^{0-}(r_i) Y_K(\hat{\mathbf{r}}_i) . \qquad (4.4.39)$$

Each of these densities again allows to calculate a corresponding exchange-correlation potential via (4.4.33). While the potential due to the true electronic density is the true exchange-correlation potential inside the atomic sphere, the potential growing out of the pseudo density is used to compensate for the exchange-correlation potential of the pseudo density as calculated on the real-space mesh. The local part of the exchange-correlation potential is thus given by

$$\hat{v}_{xc,\sigma}(\mathbf{r}_i) = \tilde{v}_{xc,\sigma}(\mathbf{r}_i) - v_{xc,\sigma}^{0-}(\mathbf{r}_i)$$

$$= v_{xc,\sigma} \left[\rho_{el,\sigma}(\mathbf{r}_i), \rho_{el,-\sigma}(\mathbf{r}_i) \right] - v_{xc,\sigma} \left[\rho_{el,\sigma}^{0-}(\mathbf{r}_i), \rho_{el,-\sigma}^{0-}(\mathbf{r}_i) \right] . \qquad (4.4.40)$$

Since the true electronic and the pseudo density are identical at the sphere boundary the local part of the exchange-correlation potential is also confined to the region of the atomic spheres and goes to zero continuously and differentiably at the sphere boundary.

Still, we are seeking for a spherical-harmonics expansion of the potentials entering (4.4.40). It cannot be derived directly from the corresponding expansion of the electron density due to the nonlinearity of the exchange-correlation potential. Yet, we may define the spherical-harmonics expansion by

$$\hat{v}_{xc,\sigma}(\mathbf{r}_i) = \sum_K \hat{v}_{xc,K\sigma}(r_i) Y_K(\hat{\mathbf{r}}_i) , \qquad (4.4.41)$$

where the coefficients are calculated from

$$\hat{v}_{xc,K\sigma}(r_i) = \int d\Omega_i\, \hat{v}_{xc,\sigma}(\mathbf{r}_i) Y_K^*(\hat{\mathbf{r}}_i)$$

$$= \int d\Omega_i\, \left\{ \tilde{v}_{xc,\sigma}(\mathbf{r}_i) - v_{xc,\sigma}^{0-}(\mathbf{r}_i) \right\} Y_K^*(\hat{\mathbf{r}}_i)$$

$$= \int d\Omega_i\, \left\{ v_{xc,\sigma} \left[\rho_{el,\sigma}(\mathbf{r}_i), \rho_{el,-\sigma}(\mathbf{r}_i) \right] \right.$$

$$\left. - v_{xc,\sigma} \left[\rho_{el,\sigma}^{0-}(\mathbf{r}_i), \rho_{el,-\sigma}^{0-}(\mathbf{r}_i) \right] \right\} Y_K^*(\hat{\mathbf{r}}_i) . \qquad (4.4.42)$$

In contrast to the procedure based on Taylor expansions of the non-spherical parts of the electron densities about the spherical part as employed in [13], we here opt for a numerical calculation of the integrals entering (4.4.42) and replace the angular integration by a weighted sum over a set of selected independent directions $\hat{\mathbf{r}}_{i,\alpha}$,

$$\hat{v}_{xc,K\sigma}(r_i) = \sum_{\alpha} w_\alpha \left\{ \tilde{v}_{xc,\sigma}(\mathbf{r}_{i,\alpha}) - v_{xc,\sigma}^{0-}(\mathbf{r}_{i,\alpha}) \right\} Y_K^*(\hat{\mathbf{r}}_{i,\alpha})$$

$$= \sum_{\alpha} w_\alpha \left\{ v_{xc,\sigma} \left[\rho_{el,\sigma}(\mathbf{r}_{i,\alpha}), \rho_{el,-\sigma}(\mathbf{r}_{i,\alpha}) \right] \right.$$

$$\left. - v_{xc,\sigma} \left[\rho_{el,\sigma}^{0-}(\mathbf{r}_{i,\alpha}), \rho_{el,-\sigma}^{0-}(\mathbf{r}_{i,\alpha}) \right] \right\} Y_K^*(\hat{\mathbf{r}}_{i,\alpha}) . \quad (4.4.43)$$

Optimal schemes to perform this integration over the surface of a sphere have been widely discussed in the literature [3, 6, 16, 17, 20, 22, 26, 28, 29].

In a very last step, we have to fix the muffin-tin zero. At variance with the procedure outlined at the end of Sect. 2.4 for the standard ASW method we choose the muffin-tin zero as the $\mathbf{K}_t = \mathbf{0}$-coefficient of the pseudo part of the potential. Since, due to charge neutrality, the corresponding contribution to the electrostatic potential vanishes, we are left with only the $\mathbf{K}_t = \mathbf{0}$-coefficient of the pseudo part of the exchange-correlation potential.

As in the previous section, we explore the possibility of reinventing the overlapping atomic spheres and reducing the pseudo contributions to their $\mathbf{K}_t = \mathbf{0}$-contribution. As we have already discussed in connection with (4.4.27), the $\mathbf{K}_t = \mathbf{0}$-term of the total density vanishes for charge neutrality reasons and, as a consequence, the constant term of the pseudo part of the electrostatic potential vanishes. However, in this case the auxiliary density, being reduced to its $\mathbf{K}_t = \mathbf{0}$-term, does not properly describe the multipole moments of the local charge density. As a consequence, their influence on the potential outside the respective atomic sphere must be taken into account by the Madelung potential just in the same manner as in Sect. 2.4. As has become clear from the discussion following (4.4.13), the Madelung potential may be regarded as a special case of the potential generated by the auxiliary density.

Since the pseudo parts of both the valence electron density and the auxiliary density cancel we will likewise have to remove their respective one-center expansions from the formalism. As a consequence, we set both $\rho^{0-}(\mathbf{r}_i)$ and $v_{es}^{0-}(\mathbf{r}_i)$ to zero.

For the exchange-correlation potential the situation is somewhat simpler due to its locality. Here, we would first calculate the exchange-correlation potential of the constant density $\rho_{val,\sigma}^0(\mathbf{K}_t = \mathbf{0})$ as given by (4.4.19), the result being $v_{xc,\sigma}^0(\mathbf{K}_t = \mathbf{0})$. After that, we calculate the local part of the exchange-correlation potential as described above with the second term in (4.4.40) replaced by the constant value $v_{xc,\sigma}^{0-}(\mathbf{r}_i) = v_{xc,\sigma}^0(\mathbf{K}_t = \mathbf{0})$.

By now we have arrived at a complete determination of the full potential from the different contributions to the density. In order to keep things clear we summarize the sequence of all the calculations, which are needed to construct the full potential and the matrix elements of the secular matrix, in the flow diagram displayed in

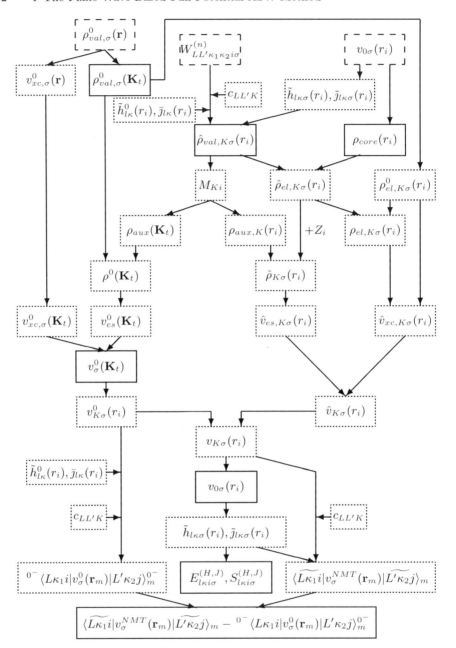

Fig. 4.1. Flow diagram of the real-space calculations of the plane-wave based full-potential ASW method. Input and output variables are highlighted by *dashed* and *solid boxes*, respectively

Fig. 4.1. Eventually, using one of the schemes for accelerating the iterations towards self-consistency described in [14] we are in a position to feed the just calculated potential back into the secular matrix and, hence, to close the self-consistency cycle of the full-potential ASW method.

4.5 Total Energy

In the second but last section of this chapter we turn to a refined calculation of the total energy. In particular, with the present full-potential formalism at hand, we aim at going beyond the errors introduced by the atomic-sphere approximation. Nevertheless, the general basis for calculating the total energy is still the same and so we start from the expression given by density-functional theory. There, the total energy is written as the sum of kinetic, electrostatic, and exchange-correlation energy [4, 5, 10, 12, 15, 18, 25]. This expression was also the starting point of the calculations in Sect. 2.5, see (2.5.1). However, as is well known and as was also outlined in Sect. 2.5, the kinetic energy can be rewritten as the sum over all single-particle energies minus the so-called "double counting" terms by starting from the Kohn-Sham equations, multiplying them with the complex conjugate of the respective wave functions, summing over all eigenstates and integrating over all space [15, 18]. After these manipulations the total energy reads as

$$
\begin{aligned}
E_T &= E_T\left[\rho_{el,\uparrow}(\mathbf{r}), \rho_{el,\downarrow}(\mathbf{r})\right] \\
&= \sum_{\mathbf{k}\sigma} E_\sigma(\mathbf{k})\Theta(E_F - E_\sigma(\mathbf{k})) + \sum_{nlmi\sigma} E_{nli\sigma} + v_{MTZ}\sum_i Q_i \\
&\quad - \frac{1}{N}\int\int d^3\mathbf{r}\, d^3\mathbf{r}'\, \frac{\rho_{el}(\mathbf{r})\rho_{el}(\mathbf{r}')}{|\mathbf{r} - \mathbf{r}'|} + \frac{1}{N}\sum_{\mu\nu}\sum_{ij}(1 - \delta_{\mu\nu}\delta_{ij})\frac{Z_i Z_j}{|\mathbf{R}_{\nu j} - \mathbf{R}_{\mu i}|} \\
&\quad + E_{xc}\left[\rho_{el,\uparrow}(\mathbf{r}), \rho_{el,\downarrow}(\mathbf{r})\right] - \frac{1}{N}\sum_\sigma \int d^3\mathbf{r}\, v_{xc,\sigma}(\mathbf{r})\rho_{el,\sigma}(\mathbf{r})\,,
\end{aligned}
\tag{4.5.1}
$$

where all integrations extend over the whole crystal and the third term takes care of the fact that the single-particle energies were referred to the muffin-tin zero whereas the potential is not. Note that (4.5.1) is identical to the last step of (2.5.5). Differences will merely grow out of the specific form of the full potential.

As for the construction of the exchange-correlation potential in Sect. 4.4, we use the local-density approximation for the exchange-correlation energy functional [10, 12, 15, 18, 25] and note

$$
E_{xc}\left[\rho_{el,\uparrow}(\mathbf{r}), \rho_{el,\downarrow}(\mathbf{r})\right] = \frac{1}{N}\sum_\sigma \int d^3\mathbf{r}\, \varepsilon_{xc,\sigma}(\mathbf{r})\rho_{el}(\mathbf{r})\,,
\tag{4.5.2}
$$

where the function $\varepsilon_{xc,\sigma}$ is a local function of the spin-dependent electron densities, which, in complete analogy to (4.4.33), reads as

$$
\varepsilon_{xc,\sigma}(\mathbf{r}) := \varepsilon_{xc,\sigma}\left[\rho_{el,\uparrow}(\mathbf{r}), \rho_{el,\downarrow}(\mathbf{r})\right]\,.
\tag{4.5.3}
$$

Again, this expression is evaluated using one of the standard parametrizations for the local-density approximation. Combining (4.5.2) with the last term on the right-hand side of (4.5.1) we thus write

$$E_{xc}\left[\rho_{el,\uparrow}(\mathbf{r}),\rho_{el,\downarrow}(\mathbf{r})\right] - \frac{1}{N}\sum_{\sigma}\int d^3\mathbf{r}\, v_{xc,\sigma}(\mathbf{r})\rho_{el,\sigma}(\mathbf{r})$$

$$= \sum_{\sigma}\int_{\Omega_c} d^3\mathbf{r}\,\left[\varepsilon_{xc,\sigma}(\mathbf{r})\rho_{el}(\mathbf{r}) - v_{xc,\sigma}(\mathbf{r})\rho_{el,\sigma}(\mathbf{r})\right]\,. \tag{4.5.4}$$

While, within the standard ASW method, we were able to express the total energy as a sum of atomic contributions plus the Madelung energy, we will obtain a more elaborate result for the full-potential ASW methods. To start with, we turn, in the same way as in Sect. 2.5, to the band energy contribution. However, since we abandoned the moment analysis (2.3.20), we can no longer simply rewrite it in terms of electron numbers and energies but have to stay with the energy weighted integral over the density of states. Adding the core state energies we note

$$\sum_{\mathbf{k}\sigma} E_\sigma(\mathbf{k})\Theta(E_F - E_\sigma(\mathbf{k})) + \sum_{nlmi\sigma} E_{nli\sigma}$$

$$= \sum_{\sigma}\int_{-\infty}^{E_F} dE \sum_{\mathbf{k}} E_\sigma(\mathbf{k})\delta(E - E_\sigma(\mathbf{k})) + \sum_{nli\sigma}(2l+1)E_{nli\sigma}$$

$$= \sum_{\sigma}\int_{-\infty}^{E_F} dE\, E\rho_\sigma(E) + \sum_{nli\sigma}(2l+1)E_{nli\sigma}\,. \tag{4.5.5}$$

Next we turn to the "double counting" terms and look especially at the fourth and fifth term on the right-hand side of (4.5.1). Using (2.4.1) and (4.4.1) for the nuclear and total density, respectively, we write

$$-\frac{1}{N}\int\int d^3\mathbf{r}d^3\mathbf{r}'\, \frac{\rho_{el}(\mathbf{r})\rho_{el}(\mathbf{r}')}{|\mathbf{r}-\mathbf{r}'|} + \frac{1}{N}\sum_{\mu\nu}\sum_{ij}(1-\delta_{\mu\nu}\delta_{ij})\frac{Z_iZ_j}{|\mathbf{R}_{\nu j}-\mathbf{R}_{\mu i}|}$$

$$= -\frac{1}{N}\int\int d^3\mathbf{r}d^3\mathbf{r}'\, \frac{(\rho_{el}(\mathbf{r})+\rho_{nucl}(\mathbf{r}))\,(\rho_{el}(\mathbf{r}')-\rho_{nucl}(\mathbf{r}'))}{|\mathbf{r}-\mathbf{r}'|}$$

$$\quad -\frac{1}{N}\sum_{\mu\nu}\sum_{ij}\delta_{\mu\nu}\delta_{ij}\frac{Z_iZ_j}{|\mathbf{R}_{\nu j}-\mathbf{R}_{\mu i}|}$$

$$= -\frac{1}{N}\int\int d^3\mathbf{r}d^3\mathbf{r}'\, \frac{\rho(\mathbf{r})\,(\rho_{el}(\mathbf{r}')-\rho_{nucl}(\mathbf{r}'))}{|\mathbf{r}-\mathbf{r}'|} - \frac{1}{N}\sum_{\mu\nu}\sum_{ij}\delta_{\mu\nu}\delta_{ij}\frac{Z_iZ_j}{|\mathbf{R}_{\nu j}-\mathbf{R}_{\mu i}|}$$

$$= -\frac{1}{2}\int_{\Omega_c} d^3\mathbf{r}\, v_{es}(\mathbf{r})\,(\rho_{el}(\mathbf{r})-\rho_{nucl}(\mathbf{r})) - \sum_{i}\lim_{\mathbf{r}\to 0}\frac{Z_iZ_i}{|\mathbf{r}|}$$

$$= -\frac{1}{2}\int_{\Omega_c} d^3\mathbf{r}\, v_{es}(\mathbf{r})\rho_{el}(\mathbf{r}) - \frac{1}{2}\sum_{i}\lim_{\mathbf{r}\to\tau_i}\left(v_{es}(\mathbf{r})+\frac{2Z_i}{|\mathbf{r}-\tau_i|}\right)Z_i\,, \tag{4.5.6}$$

where, we have explicitly singled out the nuclear self-interaction in the intermediate steps and used (4.4.5) for the classical Coulomb potential.

Inserting (4.5.4) to (4.5.6) into (4.5.1) we thus arrive at the intermediate result

$$
E_T = \sum_\sigma \int_{-\infty}^{E_F} dE \, E\rho_\sigma(E) + \sum_{nli\sigma}(2l+1)E_{nli\sigma} + v_{MTZ}\sum_i Q_i
$$
$$
- \frac{1}{2}\int_{\Omega_c} d^3\mathbf{r} \, v_{es}(\mathbf{r})\rho_{el}(\mathbf{r}) - \frac{1}{2}\sum_i \lim_{\mathbf{r}\to\boldsymbol{\tau}_i}\left(v_{es}(\mathbf{r}) + \frac{2Z_i}{|\mathbf{r}-\boldsymbol{\tau}_i|}\right)Z_i
$$
$$
+ \sum_\sigma \int_{\Omega_c} d^3\mathbf{r} \, [\varepsilon_{xc,\sigma}(\mathbf{r})\rho_{el}(\mathbf{r}) - v_{xc,\sigma}(\mathbf{r})\rho_{el,\sigma}(\mathbf{r})] \,, \tag{4.5.7}
$$

which is still quite general and, in particular, does not refer to the specifics of the plane-wave based full-potential ASW method. As a consequence, it may serve as a starting point for the evaluation of the total energy within any full-potential method. In passing, we point out that the integrals over the unit cell entering on the right-hand side of (4.5.7) may be regarded as overlap integrals of the respective two quantities involved.

Being more specific now and concentrating on the electrostatic contribution first, we fall back on (4.4.23) and (4.3.17) for the classical Coulomb potential and the electronic density in terms of pseudo and local contributions. In addition, we use the decomposition of the latter into local augmented and pseudo parts according to (4.3.16) and (4.4.30) to (4.4.32) and write the first term in the second line of (4.5.7) as

$$
-\frac{1}{2}\int_{\Omega_c} d^3\mathbf{r} \, v_{es}(\mathbf{r})\rho_{el}(\mathbf{r})
$$
$$
= -\frac{1}{2}\int_{\Omega_c} d^3\mathbf{r} \, v_{es}^0(\mathbf{r})\rho_{el}^0(\mathbf{r})
$$
$$
-\frac{1}{2}\sum_i \int_{\Omega_i} d^3\mathbf{r}_i \, [\hat{v}_{es}(\mathbf{r}_i)\hat{\rho}_{el}(\mathbf{r}_i) + \hat{v}_{es}(\mathbf{r}_i)\rho_{el}^0(\mathbf{r}_i) + v_{es}^0(\mathbf{r}_i)\hat{\rho}_{el}(\mathbf{r}_i)]
$$
$$
= -\frac{1}{2}\int_{\Omega_c} d^3\mathbf{r} \, v_{es}^0(\mathbf{r})\rho_{el}^0(\mathbf{r}) - \frac{1}{2}\sum_i \int_{\Omega_i} d^3\mathbf{r}_i \, [\tilde{v}_{es}(\mathbf{r}_i)\tilde{\rho}_{el}(\mathbf{r}_i) - v_{es}^{0-}(\mathbf{r}_i)\rho_{el}^{0-}(\mathbf{r}_i)] \,.
$$
$$
\tag{4.5.8}
$$

In contrast, for the second term in that line we note

$$
-\frac{1}{2}\sum_i \lim_{\mathbf{r}\to\boldsymbol{\tau}_i}\left(v_{es}(\mathbf{r}) + \frac{2Z_i}{|\mathbf{r}-\boldsymbol{\tau}_i|}\right)Z_i
$$
$$
= -\frac{1}{2}\sum_i Z_i v_{es}^0(\boldsymbol{\tau}_i) + \frac{1}{2}\sum_i Z_i v_{es}^{0-}(\mathbf{r}_i=0) - \frac{1}{2}\sum_i \lim_{\mathbf{r}_i\to 0}\left(\tilde{v}_{es}(\mathbf{r}_i) + \frac{2Z_i}{|\mathbf{r}_i|}\right)Z_i \,.
$$
$$
\tag{4.5.9}
$$

In further evaluating these expressions we turn to the respective first terms on the right-hand sides of (4.5.8) and (4.5.9). Using the Fourier expansions (4.3.6) and (4.4.8) of the pseudo parts of the electron density and the classical Coulomb potential, respectively, we obtain

$$-\frac{1}{2}\int_{\Omega_c}d^3r\,v_{es}^0(\mathbf{r})\rho_{el}^0(\mathbf{r}) = -\frac{1}{2}\sum_{mn}v_{es}^0(\mathbf{K}_m)\rho_{el}^0(\mathbf{K}_n)\int_{\Omega_c}d^3r\,e^{i(\mathbf{K}_m+\mathbf{K}_n)\mathbf{r}}$$

$$= -\frac{1}{2}\Omega_c\sum_n(v_{es}^0(\mathbf{K}_n))^*\rho_{el}^0(\mathbf{K}_n)\ . \tag{4.5.10}$$

In the last step we have used the fact that both the density and the classical Coulomb potential are real quantities and thus their Fourier coefficients have to fulfill the condition

$$v_{es}^0(-\mathbf{K}_n) \overset{!}{=} (v_{es}^0(\mathbf{K}_n))^* \quad \forall n\ . \tag{4.5.11}$$

In a very similar manner we rewrite the first term on the right-hand side of (4.5.9) and obtain

$$-\frac{1}{2}\sum_i Z_i v_{es}^0(\boldsymbol{\tau}_i) = -\frac{1}{2}\sum_i Z_i\sum_n v_{es}^0(\mathbf{K}_n)\int_{\Omega_c}d^3r\,e^{i\mathbf{K}_n\mathbf{r}}\delta(\mathbf{r}_i)$$

$$= -\frac{1}{2}\sum_i Z_i\sum_n v_{es}^0(\mathbf{K}_n)e^{i\mathbf{K}_n\boldsymbol{\tau}_i}\ . \tag{4.5.12}$$

In order to simplify the last term on the right-hand side of (4.5.9) we define the electrostatic potential generated by the electronic density inside the sphere by

$$\tilde{v}_{es}^{(el)}(\mathbf{r}_i) = \tilde{v}_{es}(\mathbf{r}_i) + \frac{2Z_i}{|\mathbf{r}_i|} = 2\int_{\Omega_i}d^3r_i'\,\frac{\tilde{\rho}_{el}(\mathbf{r}_i')}{|\mathbf{r}_i-\mathbf{r}_i'|}\ , \tag{4.5.13}$$

and note

$$-\frac{1}{2}\sum_i\lim_{\mathbf{r}_i\to 0}\left(\tilde{v}_{es}(\mathbf{r}_i)+\frac{2Z_i}{|\mathbf{r}_i|}\right)Z_i = -\frac{1}{2}\sum_i Z_i\lim_{\mathbf{r}_i\to 0}\tilde{v}_{es}^{(el)}(\mathbf{r}_i)$$

$$= -\sum_i Z_i\int_{\Omega_i}d^3r_i'\,\frac{\tilde{\rho}_{el}(\mathbf{r}_i')}{|\mathbf{r}_i'|}$$

$$= -\frac{1}{2}\sum_i\int_{\Omega_i}d^3r_i\,\frac{2Z_i}{|\mathbf{r}_i|}\tilde{\rho}_{el}(\mathbf{r}_i)\ . \tag{4.5.14}$$

Combining (4.5.8) to (4.5.14) we thus arrive at

$$-\frac{1}{2}\int_{\Omega_c}d^3r\,v_{es}(\mathbf{r})\rho_{el}(\mathbf{r}) - \frac{1}{2}\sum_i\lim_{\mathbf{r}\to\boldsymbol{\tau}_i}\left(v_{es}(\mathbf{r})+\frac{2Z_i}{|\mathbf{r}-\boldsymbol{\tau}_i|}\right)Z_i$$

$$= -\frac{1}{2}\Omega_c\sum_n(v_{es}^0(\mathbf{K}_n))^*\rho_{el}^0(\mathbf{K}_n) - \frac{1}{2}\sum_i Z_i\sum_n v_{es}^0(\mathbf{K}_n)e^{i\mathbf{K}_n\boldsymbol{\tau}_i}$$

$$-\frac{1}{2}\sum_i\int_{\Omega_i}d^3r_i\left[\left(\tilde{v}_{es}(\mathbf{r}_i)+\frac{2Z_i}{|\mathbf{r}_i|}\right)\tilde{\rho}_{el}(\mathbf{r}_i)-v_{es}^{0-}(\mathbf{r}_i)\rho_{el}^{0-}(\mathbf{r}_i)\right]$$

$$+\frac{1}{2}\sum_i Z_i v_{es}^{0-}(\mathbf{r}_i = \mathbf{0})$$

$$= -\frac{1}{2}\Omega_c \sum_n (v_{es}^0(\mathbf{K}_n))^* \rho_{el}^0(\mathbf{K}_n) - \frac{1}{2}\sum_i Z_i \sum_n v_{es}^0(\mathbf{K}_n)e^{i\mathbf{K}_n\boldsymbol{\tau}_i}$$

$$-\frac{1}{2}\sum_{Ki}\int_0^{S_i} dr_i r_i^2 \left[\left(\tilde{v}_{es,K}(r_i) + 2\sqrt{4\pi}\delta_{k0}\frac{Z_i}{r_i}\right)\tilde{\rho}_{el,K}(r_i) - v_{es,K}^{0-}(r_i)\rho_{el,K}^{0-}(r_i)\right]$$

$$+\frac{1}{2\sqrt{4\pi}}\sum_i Z_i v_{es,K=0}^{0-}(r_i = 0) , \tag{4.5.15}$$

where, in the last step, we have used the spherical-harmonics expansions (4.4.28) and (4.3.17) of the local contributions.

For the exchange-correlation contribution to the total energy as covered by the last line of (4.5.7) we proceed along the same lines as for the exchange-correlation potential described in Sect. 4.4 and divide the exchange-correlation energy density into pseudo and local parts according to

$$\varepsilon_{xc,\sigma}(\mathbf{r}) = \varepsilon_{xc,\sigma}^0(\mathbf{r}) + \sum_i \hat{\varepsilon}_{xc,\sigma}(\mathbf{r}_i) . \tag{4.5.16}$$

The pseudo part is just the exchange-correlation energy density of the pseudo density given on the real-space mesh

$$\varepsilon_{xc,\sigma}^0(\mathbf{r}) := \varepsilon_{xc,\sigma}\left[\rho_{el,\uparrow}^0(\mathbf{r}), \rho_{el,\downarrow}^0(\mathbf{r})\right] , \tag{4.5.17}$$

with the functional on the right-hand side being evaluated using the notions of density-functional theory and the local-density approximation and a suitable parametrization of the latter [15]. In addition, we require that the pseudo part can be written as a plane-wave expansion

$$\varepsilon_{xc,\sigma}^0(\mathbf{r}) = \sum_t \varepsilon_{xc,\sigma}^0(\mathbf{K}_t)e^{i\mathbf{K}_t\mathbf{r}} , \tag{4.5.18}$$

with the expansion coefficients given by

$$\varepsilon_{xc,\sigma}^0(\mathbf{K}_t) = \frac{1}{\Omega_c}\int_{\Omega_c} d^3\mathbf{r}\, \varepsilon_{xc,\sigma}^0(\mathbf{r})e^{-i\mathbf{K}_t\mathbf{r}}$$

$$= \frac{1}{\Omega_c}\int_{\Omega_c} d^3\mathbf{r}\, \varepsilon_{xc,\sigma}\left[\rho_{el,\uparrow}^0(\mathbf{r}), \rho_{el,\downarrow}^0(\mathbf{r})\right]e^{-i\mathbf{K}_t\mathbf{r}} . \tag{4.5.19}$$

In contrast, inside the atomic spheres the local part of the exchange-correlation energy density is defined as the difference between the true energy density and the energy density calculated from the local pseudo density

$$\hat{\varepsilon}_{xc,\sigma}(\mathbf{r}_i) = \varepsilon_{xc,\sigma}\left[\rho_{el,\uparrow}(\mathbf{r}_i), \rho_{el,\downarrow}(\mathbf{r}_i)\right] - \varepsilon_{xc,\sigma}\left[\rho_{el,\uparrow}^{0-}(\mathbf{r}_i), \rho_{el,\downarrow}^{0-}(\mathbf{r}_i)\right] , \tag{4.5.20}$$

where the densities are given by (4.4.38) and (4.4.39). In the same manner as the exchange-correlation potential the local part of the energy density is confined to the region of the atomic spheres and goes to zero continuously and differentiably at the sphere boundary. Although this conforms to our requirements on the representation of the energy density we are, nevertheless, not able to calculate the local part of the exchange-correlation energy density directly from the local contributions of the pseudo and local parts of the density. Still, we are able to note the expansion into spherical harmonics

$$\hat{\varepsilon}_{xc,\sigma}(\mathbf{r}_i) = \sum_K \hat{\varepsilon}_{xc,K\sigma}(r_i) Y_K(\hat{\mathbf{r}}_i) , \tag{4.5.21}$$

where the coefficients are calculated as

$$
\begin{aligned}
&\hat{\varepsilon}_{xc,K\sigma}(r_i) \\
&= \int d\Omega_i \, \hat{\varepsilon}_{xc,\sigma}(\mathbf{r}_i) Y_K^*(\hat{\mathbf{r}}_i) \\
&= \int d\Omega_i \, \left\{ \varepsilon_{xc,\sigma} \left[\rho_{el,\uparrow}(\mathbf{r}_i), \rho_{el,\downarrow}(\mathbf{r}_i) \right] - \varepsilon_{xc,\sigma} \left[\rho_{el,\uparrow}^{0-}(\mathbf{r}_i), \rho_{el,\downarrow}^{0-}(\mathbf{r}_i) \right] \right\} Y_K^*(\hat{\mathbf{r}}_i) .
\end{aligned}
\tag{4.5.22}
$$

Again we opt for a numerical calculation of the integrals entering (4.5.23) in close analogy to that of the calculations of the exchange-correlation potential as sketched in (4.4.43).

Finally, inserting the (4.4.34) and (4.5.17) for the exchange-correlation potential and energy density into the last line of (4.5.7) and using the respective plane-wave and spherical-harmonics expansions we arrive at the intermediate result

$$
\begin{aligned}
&\sum_\sigma \int_{\Omega_c} d^3\mathbf{r} \, \left[\varepsilon_{xc,\sigma}(\mathbf{r})\rho_{el}(\mathbf{r}) - v_{xc,\sigma}(\mathbf{r})\rho_{cl,\sigma}(\mathbf{r}) \right] \\
&= \sum_\sigma \int_{\Omega_c} d^3\mathbf{r} \, \left[\varepsilon_{xc,\sigma}^0(\mathbf{r})\rho_{el}^0(\mathbf{r}) - v_{xc,\sigma}^0(\mathbf{r})\rho_{el,\sigma}^0(\mathbf{r}) \right] \\
&\quad + \sum_{i\sigma} \int_{\Omega_i} d^3\mathbf{r}_i \, \left[\varepsilon_{xc,\sigma}^0(\mathbf{r}_i)\hat{\rho}_{el}(\mathbf{r}_i) - v_{xc,\sigma}^0(\mathbf{r}_i)\hat{\rho}_{el,\sigma}(\mathbf{r}_i) \right. \\
&\qquad\qquad + \hat{\varepsilon}_{xc,\sigma}(\mathbf{r}_i)\rho_{el}^0(\mathbf{r}_i) - \hat{v}_{xc,\sigma}(\mathbf{r}_i)\rho_{el,\sigma}^0(\mathbf{r}_i) \\
&\qquad\qquad \left. + \hat{\varepsilon}_{xc,\sigma}(\mathbf{r}_i)\hat{\rho}_{el}^0(\mathbf{r}_i) - \hat{v}_{xc,\sigma}(\mathbf{r}_i)\hat{\rho}_{el,\sigma}^0(\mathbf{r}_i) \right] \\
&= \sum_\sigma \int_{\Omega_c} d^3\mathbf{r} \, \left[\varepsilon_{xc,\sigma}^0(\mathbf{r})\rho_{el}^0(\mathbf{r}) - v_{xc,\sigma}^0(\mathbf{r})\rho_{el,\sigma}^0(\mathbf{r}) \right] \\
&\quad + \sum_{i\sigma} \int_{\Omega_i} d^3\mathbf{r}_i \, \left[\tilde{\varepsilon}_{xc,\sigma}(\mathbf{r}_i)\tilde{\rho}_{el}(\mathbf{r}_i) - \tilde{v}_{xc,\sigma}(\mathbf{r}_i)\tilde{\rho}_{el,\sigma}(\mathbf{r}_i) \right. \\
&\qquad\qquad \left. - \varepsilon_{xc,\sigma}^{0-}(\mathbf{r}_i)\rho_{el}^{0-}(\mathbf{r}_i) + v_{xc,\sigma}^{0-}(\mathbf{r}_i)\rho_{el,\sigma}^{0-}(\mathbf{r}_i) \right]
\end{aligned}
$$

$$= \Omega_c \sum_\sigma \sum_n \left[\left(\varepsilon^0_{xc,\sigma}(\mathbf{K}_n) \right)^* \rho^0_{el}(\mathbf{K}_n) - \left(v^0_{xc,\sigma}(\mathbf{K}_n) \right)^* \rho^0_{el,\sigma}(\mathbf{K}_n) \right]$$

$$+ \sum_{Ki\sigma} \int_0^{S_i} dr_i r_i^2 \left[\tilde{\varepsilon}_{xc,K\sigma}(r_i) \tilde{\rho}_{el,K}(r_i) - \tilde{v}_{xc,K\sigma}(r_i) \tilde{\rho}_{el,K\sigma}(r_i) \right.$$

$$\left. - \varepsilon^{0-}_{xc,K\sigma}(r_i) \rho^{0-}_{el,K}(r_i) + v^{0-}_{xc,K\sigma}(r_i) \rho^{0-}_{el,K\sigma}(r_i) \right] . \qquad (4.5.23)$$

In the last step, we have used the following identity, which is based on the Fourier expansions (4.3.6), (4.4.36), and (4.5.18) of the pseudo parts of the charge density as well as the exchange-correlation potential and energy density, respectively

$$\sum_\sigma \int_{\Omega_c} d^3r \left[\varepsilon^0_{xc,\sigma}(\mathbf{r}) \rho^0_{el}(\mathbf{r}) - v^0_{xc,\sigma}(\mathbf{r}) \rho^0_{el,\sigma}(\mathbf{r}) \right]$$

$$= \sum_\sigma \sum_{mn} \left[\varepsilon^0_{xc,\sigma}(\mathbf{K}_m) \rho^0_{el}(\mathbf{K}_n) - v^0_{xc,\sigma}(\mathbf{K}_m) \rho^0_{el,\sigma}(\mathbf{K}_n) \right] \int_{\Omega_c} d^3r \, e^{i(\mathbf{K}_m + \mathbf{K}_n)\mathbf{r}}$$

$$= \Omega_c \sum_\sigma \sum_n \left[\left(\varepsilon^0_{xc,\sigma}(\mathbf{K}_n) \right)^* \rho^0_{el}(\mathbf{K}_n) - \left(v^0_{xc,\sigma}(\mathbf{K}_n) \right)^* \rho^0_{el,\sigma}(\mathbf{K}_n) \right] . \qquad (4.5.24)$$

Again, we have used the fact that the electron density as well as the exchange-correlation potential and energy density are real quantities.

We are now able to insert the intermediate results (4.5.15) and (4.5.23) into (4.5.7) and arrive at the final result

$$E_T = E_T \left[\rho_{el,\uparrow}(\mathbf{r}), \rho_{el,\downarrow}(\mathbf{r}) \right]$$

$$+ \sum_\sigma \int_{-\infty}^{E_F} dE \, E\rho_\sigma(E) + \sum_{nli\sigma} (2l+1) E_{nli\sigma} + v_{MTZ} \sum_i Q_i$$

$$- \frac{1}{2} \Omega_c \sum_n (v^0_{es}(\mathbf{K}_n))^* \rho^0_{el}(\mathbf{K}_n) - \frac{1}{2} \sum_i Z_i \sum_n v^0_{es}(\mathbf{K}_n) e^{i\mathbf{K}_n \boldsymbol{\tau}_i}$$

$$- \frac{1}{2} \sum_{Ki} \int_0^{S_i} dr_i r_i^2 \left[\left(\tilde{v}_{es,K}(r_i) + 2\sqrt{4\pi}\delta_{k0} \frac{Z_i}{r_i} \right) \tilde{\rho}_{el,K}(r_i) - v^{0-}_{es,K}(r_i) \rho^{0-}_{el,K}(r_i) \right]$$

$$+ \frac{1}{2\sqrt{4\pi}} \sum_i Z_i \, v^{0-}_{es,K=0}(r_i = 0)$$

$$+ \Omega_c \sum_\sigma \sum_n \left[\left(\varepsilon^0_{xc,\sigma}(\mathbf{K}_n) \right)^* \rho^0_{el}(\mathbf{K}_n) - \left(v^0_{xc,\sigma}(\mathbf{K}_n) \right)^* \rho^0_{el,\sigma}(\mathbf{K}_n) \right]$$

$$+ \sum_{Ki\sigma} \int_0^{S_i} dr_i r_i^2 \left[\tilde{\varepsilon}_{xc,K\sigma}(r_i) \tilde{\rho}_{el,K}(r_i) - \tilde{v}_{xc,K\sigma}(r_i) \tilde{\rho}_{el,K\sigma}(r_i) \right.$$

$$\left. - \varepsilon^{0-}_{xc,K\sigma}(r_i) \rho^{0-}_{el,K}(r_i) + v^{0-}_{xc,K\sigma}(r_i) \rho^{0-}_{el,K\sigma}(r_i) \right] , \qquad (4.5.25)$$

which closes the present outline of the full-potential ASW method.

Nevertheless, once again we consider the intermediate approach, where all pseudo quantities are reduced to their $\mathbf{K}_n = \mathbf{0}$-terms and the atomic-sphere approximation is reinstalled. However, as outlined in Sect. 4.4, the $\mathbf{K}_n = \mathbf{0}$-term of the electrostatic potential vanishes due to charge neutrality. As a consequence, we also had to remove the one-center expansion of the potential $v_{es}^{0-}(\mathbf{r}_i)$. For the final expression (4.5.25) of the total energy this means that all the pseudo contributions in the third, fourth and fifth line cancel and the double-counting terms are reduced to the first term in the fourth line.

In contrast, the pseudo parts arising from the exchange-correlation potential and energy density, if reduced to their $\mathbf{K}_n = \mathbf{0}$-terms, give rise to a constant contribution, which, according to the discussion at the end of Sect. 4.4, must be compensated by the terms in the last line of (4.5.25) by setting

$$\rho_{el}^{0-}(r_i) = \rho_{el}^{0}(\mathbf{K}_n = \mathbf{0}),$$

as well as

$$v_{xc,\sigma}^{0-}(\mathbf{r}_i) = v_{xc,\sigma}^{0}(\mathbf{K}_n = \mathbf{0}) \qquad \text{and} \qquad \varepsilon_{xc,\sigma}^{0-}(\mathbf{r}_i) = \varepsilon_{xc,\sigma}^{0}(\mathbf{K}_n = \mathbf{0}).$$

Finally, the electrostatic contributions to the total energy have to be repaired by reintroducing the Madelung energy as given by (2.5.17). To conclude, the total energy expression of the intermediate approach is rather similar to result arising within the standard ASW method except for the inclusion of the non-spherical contributions inside the atomic spheres.

4.6 Optical Properties

Within the framework of the Kubo linear-response formalism the optical conductivity tensor is given by [9, 11, 19, 23, 24, 30]

$$\sigma_{\lambda\lambda'}(E) = \frac{ie^2\hbar}{m^2\Omega_c} \sum_{\mathbf{k}} \sum_{n\sigma} \sum_{n'\sigma'} \frac{f(\varepsilon_{\mathbf{k}n\sigma}) - f(\varepsilon_{\mathbf{k}n'\sigma'})}{\varepsilon_{\mathbf{k}n'\sigma'} - \varepsilon_{\mathbf{k}n\sigma}} \frac{\Pi_{\mathbf{k}n\sigma n'\sigma'}^{\lambda} \Pi_{\mathbf{k}n'\sigma' n\sigma}^{\lambda'}}{E - (\varepsilon_{\mathbf{k}n'\sigma'} - \varepsilon_{\mathbf{k}n\sigma}) + i0^+},$$

$$(4.6.1)$$

where λ and λ' denote the Cartesian components, $\varepsilon_{\mathbf{k}n\sigma}$ are the band energies, and $f(\varepsilon_{\mathbf{k}n\sigma})$ is the Fermi function. The factors in the numerators are the Cartesian components of matrix elements of the momentum operator taken between single-particle states

$$\Pi_{\mathbf{k}n\sigma n'\sigma'} = \langle \psi_{\mathbf{k}n\sigma} | \left[\mathbf{p} + \frac{\alpha}{8}\boldsymbol{\sigma} \times \nabla V(\mathbf{r}) \right] | \psi_{\mathbf{k}n'\sigma'} \rangle . \qquad (4.6.2)$$

Here, α denotes the fine structure constant, which we specified in (1.2.2). The spin-orbit term entering these matrix elements is known to be much smaller than the canonical momentum operator and, hence, can be neglected. As a consequence, spin-flip transitions are suppressed and the spin indices of the two wave functions involved become identical.

As it stands, (4.6.1) includes both interband and intraband transitions. At zero temperature only bands at the Fermi energy contribute to the latter, which can be cast into the form

$$\bar{\sigma}_{\lambda\lambda'}(E) = \frac{\omega_{p;\lambda\lambda'}^2}{4\pi} \frac{i}{\omega + i\gamma_D} \,, \tag{4.6.3}$$

where $\gamma_D = 1/\tau_D$ and τ_D is the phenomenological Drude electron relaxation time [1]. In most applications, we adopt the perfect crystal approximation and let $\gamma_D \to 0$. Finally,

$$\omega_{p;\lambda\lambda'}^2 = \frac{4\pi e^2 \hbar^2}{m^2 \Omega_c} \sum_{\mathbf{k}} \sum_{n\sigma} \Pi_{\mathbf{k}n\sigma n\sigma}^{\lambda} \Pi_{\mathbf{k}n\sigma n\sigma}^{\lambda'} \delta(\varepsilon_n(\mathbf{k}) - E_F) = \omega_{p;\lambda'\lambda}^2 \,, \tag{4.6.4}$$

denotes the symmetric plasma frequency tensor. The interband optical conductivity tensor can be rewritten as

$$\hat{\sigma}_{\lambda\lambda'}(E) = \frac{ie^2\hbar}{m^2 \Omega_c} \sum_{\mathbf{k}} \sum_{\substack{n\sigma \\ occ}} \sum_{\substack{n'\sigma' \\ unocc}} \frac{1}{\varepsilon_{\mathbf{k}n'\sigma'} - \varepsilon_{\mathbf{k}n\sigma}} \left[\frac{\Pi_{\mathbf{k}n\sigma n'\sigma'}^{\lambda} \Pi_{\mathbf{k}n'\sigma' n\sigma}^{\lambda'}}{E - (\varepsilon_{\mathbf{k}n'\sigma'} - \varepsilon_{\mathbf{k}n\sigma}) + i0^+} \right.$$
$$\left. + \frac{(\Pi_{\mathbf{k}n\sigma n'\sigma'}^{\lambda} \Pi_{\mathbf{k}n'\sigma' n\sigma}^{\lambda'})^*}{E + (\varepsilon_{\mathbf{k}n'\sigma'} - \varepsilon_{\mathbf{k}n\sigma}) + i0^+} \right] \,. \tag{4.6.5}$$

This contribution may in turn be expressed in terms of the oscillator strength tensor, which is defined as [27]

$$f_{n\sigma n'\sigma'}^{\lambda\lambda'} = \frac{\Pi_{\mathbf{k}n\sigma n'\sigma'}^{\lambda} \Pi_{\mathbf{k}n'\sigma' n\sigma}^{\lambda'}}{\varepsilon_{\mathbf{k}n'\sigma'} - \varepsilon_{\mathbf{k}n\sigma}} \,, \tag{4.6.6}$$

as

$$\hat{\sigma}_{\lambda\lambda'}(E) = \frac{ie^2\hbar}{m^2 \Omega_c} \sum_{\mathbf{k}} \sum_{\substack{n\sigma \\ occ}} \sum_{\substack{n'\sigma' \\ unocc}} \left[\frac{f_{n\sigma n'\sigma'}^{\lambda\lambda'}}{E - (\varepsilon_{\mathbf{k}n'\sigma'} - \varepsilon_{\mathbf{k}n\sigma}) + i0^+} \right.$$
$$\left. + \frac{(f_{n\sigma n'\sigma'}^{\lambda\lambda'})^*}{E + (\varepsilon_{\mathbf{k}n'\sigma'} - \varepsilon_{\mathbf{k}n\sigma}) + i0^+} \right] \,. \tag{4.6.7}$$

With the help of Dirac's identity

$$\frac{1}{E - \varepsilon_n(\mathbf{k}) \pm i0^+} = \mathcal{P} \frac{1}{E - \varepsilon_n(\mathbf{k})} \mp i\pi\delta(E - \varepsilon_n(\mathbf{k})) \,, \tag{4.6.8}$$

where $\mathcal{P}$ denotes Cauchy's principal value, we may readily rewrite the interband optical conductivity as

$$\hat{\sigma}_{\lambda\lambda'}(E)$$

$$= \frac{ie^2\hbar}{m^2\Omega_c} \sum_{\mathbf{k}} \sum_{\substack{n\sigma \\ occ}} \sum_{\substack{n'\sigma' \\ unocc}} \left[\mathcal{P}\frac{f_{n\sigma n'\sigma'}^{\lambda\lambda'}}{E-(\varepsilon_{\mathbf{k}n'\sigma'}-\varepsilon_{\mathbf{k}n\sigma})} + \mathcal{P}\frac{(f_{n\sigma n'\sigma'}^{\lambda\lambda'})^*}{E+(\varepsilon_{\mathbf{k}n'\sigma'}-\varepsilon_{\mathbf{k}n\sigma})} \right]$$

$$+ \frac{e^2\hbar\pi}{m^2\Omega_c} \sum_{\mathbf{k}} \sum_{\substack{n\sigma \\ occ}} \sum_{\substack{n'\sigma' \\ unocc}} \left[f_{n\sigma n'\sigma'}^{\lambda\lambda'}\delta\left(E-(\varepsilon_{\mathbf{k}n'\sigma'}-\varepsilon_{\mathbf{k}n\sigma})\right) \right.$$

$$\left. + (f_{n\sigma n'\sigma'}^{\lambda\lambda'})^*\delta\left(E+(\varepsilon_{\mathbf{k}n'\sigma'}-\varepsilon_{\mathbf{k}n\sigma})\right) \right]$$

$$= \frac{ie^2\hbar}{m^2\Omega_c} \sum_{\mathbf{k}} \sum_{\substack{n\sigma \\ occ}} \sum_{\substack{n'\sigma' \\ unocc}} \left[\mathcal{P}\frac{f_{n\sigma n'\sigma'}^{\lambda\lambda'}}{E-(\varepsilon_{\mathbf{k}n'\sigma'}-\varepsilon_{\mathbf{k}n\sigma})} + \mathcal{P}\frac{(f_{n\sigma n'\sigma'}^{\lambda\lambda'})^*}{E+(\varepsilon_{\mathbf{k}n'\sigma'}-\varepsilon_{\mathbf{k}n\sigma})} \right]$$

$$+ \frac{e^2\hbar\pi}{m^2\Omega_c} \sum_{\mathbf{k}} \sum_{\substack{n\sigma \\ occ}} \sum_{\substack{n'\sigma' \\ unocc}} f_{n\sigma n'\sigma'}^{\lambda\lambda'}\delta\left(E-(\varepsilon_{\mathbf{k}n'\sigma'}-\varepsilon_{\mathbf{k}n\sigma})\right) , \qquad (4.6.9)$$

where in the last step we have used the fact that the second term in the second square brackets does not contribute for positive energies E. For the same reason, the denominator in the second term in the first square brackets does not vanish and the sum over states reduces to a standard integral. Usually, the first and second contribution on the right-hand side are designated as the dispersive and absorptive part of the optical conductivity. In general these are not identical to the real and imaginary parts of $\hat{\sigma}_{\lambda\lambda'}$ since the oscillator strength tensor likewise may be a complex quantity. We thus split the latter into its real an imaginary part and arrive at

$$\hat{\sigma}_{\lambda\lambda'}(E) = \hat{\sigma}_{\lambda\lambda'}^{(1)}(E) + i\hat{\sigma}_{\lambda\lambda'}^{(2)}(E)$$

$$= -\frac{e^2\hbar}{m^2\Omega_c} \sum_{\mathbf{k}} \sum_{\substack{n\sigma \\ occ}} \sum_{\substack{n'\sigma' \\ unocc}} \left[\mathcal{P}\frac{\Im f_{n\sigma n'\sigma'}^{\lambda\lambda'}}{E-(\varepsilon_{\mathbf{k}n'\sigma'}-\varepsilon_{\mathbf{k}n\sigma})} - \mathcal{P}\frac{\Im f_{n\sigma n'\sigma'}^{\lambda\lambda'}}{E+(\varepsilon_{\mathbf{k}n'\sigma'}-\varepsilon_{\mathbf{k}n\sigma})} \right]$$

$$+ i\frac{e^2\hbar}{m^2\Omega_c} \sum_{\mathbf{k}} \sum_{\substack{n\sigma \\ occ}} \sum_{\substack{n'\sigma' \\ unocc}} \left[\mathcal{P}\frac{\Re f_{n\sigma n'\sigma'}^{\lambda\lambda'}}{E-(\varepsilon_{\mathbf{k}n'\sigma'}-\varepsilon_{\mathbf{k}n\sigma})} + \mathcal{P}\frac{\Re f_{n\sigma n'\sigma'}^{\lambda\lambda'}}{E+(\varepsilon_{\mathbf{k}n'\sigma'}-\varepsilon_{\mathbf{k}n\sigma})} \right]$$

$$+ \frac{e^2\hbar\pi}{m^2\Omega_c} \sum_{\mathbf{k}} \sum_{\substack{n\sigma \\ occ}} \sum_{\substack{n'\sigma' \\ unocc}} \Re f_{n\sigma n'\sigma'}^{\lambda\lambda'}\delta\left(E-(\varepsilon_{\mathbf{k}n'\sigma'}-\varepsilon_{\mathbf{k}n\sigma})\right)$$

$$+ i\frac{e^2\hbar\pi}{m^2\Omega_c} \sum_{\mathbf{k}} \sum_{\substack{n\sigma \\ occ}} \sum_{\substack{n'\sigma' \\ unocc}} \Im f_{n\sigma n'\sigma'}^{\lambda\lambda'}\delta\left(E-(\varepsilon_{\mathbf{k}n'\sigma'}-\varepsilon_{\mathbf{k}n\sigma})\right) . \qquad (4.6.10)$$

Hence, both the real and imaginary parts of the optical conductivity consist of dispersive and absorptive contributions.

With the optical conductivity at hand we are able to derive several related quantities. In particular, the dielectric function tensor is given by

$$\varepsilon_{\lambda\lambda'}(E) = 4\pi\epsilon_0 \left(\delta_{\lambda\lambda'} + i\hbar\frac{4\pi\sigma_{\lambda\lambda'}(E)}{E}\right) .\tag{4.6.11}$$

In the same manner as the optical conductivity, the dielectric function can be splitted into interband and intraband contributions. For the latter we obtain from combining (4.6.3) and (4.6.11)

$$\bar\varepsilon_{\lambda\lambda'}(E) = 4\pi\epsilon_0 \left(\delta_{\lambda\lambda'} - \frac{\omega^2_{p;\lambda\lambda'}}{\omega^2 + \gamma_D^2} + i\frac{\gamma_D\omega^2_{p;\lambda\lambda'}}{\omega(\omega^2 + \gamma_D^2)}\right) .\tag{4.6.12}$$

Having the dielectric function at hand, we arrive at the complex refraction index

$$\tilde n_{\lambda\lambda'}(E) = n_{\lambda\lambda'}(E) + i\kappa_{\lambda\lambda'}(E) = \sqrt{\varepsilon_{\lambda\lambda'}(E)} ,\tag{4.6.13}$$

where n is the real refraction index and κ the extinction coefficient, as well as at the energy-loss function

$$\tilde e_{\lambda\lambda'}(E) = -\Im\frac{1}{\varepsilon_{\lambda\lambda'}(E)} .\tag{4.6.14}$$

Finally, we may wish to calculate the effective-mass tensor for each band and at each **k**-point from inverting the matrix [9, 19, 27, 30]

$$\left(\frac{m}{m^*_{n\sigma}(\mathbf{k})}\right)_{\lambda\lambda'} = \delta_{\lambda\lambda'} - \frac{2}{m}\sum_{n'\neq n}\frac{\Pi^\lambda_{\mathbf{k}n\sigma n'\sigma}\Pi^{\lambda'}_{\mathbf{k}n'\sigma n\sigma}}{\varepsilon_{\mathbf{k}n'\sigma} - \varepsilon_{\mathbf{k}n\sigma}}$$
$$= \delta_{\lambda\lambda'} - \frac{2}{m}\sum_{n'\neq n}f^{\lambda\lambda'}_{n\sigma n'\sigma} .\tag{4.6.15}$$

This result can be easily derived from $\mathbf{k}\cdot\mathbf{p}$ perturbation theory [2, App. E]. Note that Wang and Callaway give the same formula with a plus sign in front of the sum, which, however, according to Maurer violates the sum rule and must be a typo [19].

In order to evaluate the oscillator strength tensor in the framework of the ASW method we use $\mathbf{p} = -i\nabla$ as well as the representation (4.1.12) of products of basis functions and, taking into account the respective regions, where the functions (4.2.10) to (4.2.15) are defined, we are essentially left with the calculation of the matrix elements built with the basis functions, which, in complete analogy to the expression (4.2.16) for the Hamiltonian matrix, read as

$$^\infty\langle L\kappa_1 i| - i\nabla|L'\kappa_2 j\rangle^\infty_c$$
$$= {}^0\langle L\kappa_1 i| - i\nabla|L'\kappa_2 j\rangle^0_c + \langle\widetilde{L\kappa_1 i}| - i\nabla|\widetilde{L'\kappa_2 j}\rangle_c - {}^{0^-}\langle L\kappa_1 i| - i\nabla|L'\kappa_2 j\rangle^{0^-}_c$$
$$+ {}^{0^+}\langle L\kappa_1 i| - i\nabla|\widehat{L'\kappa_2 j}\rangle_c + \langle\widehat{L\kappa_1 i}| - i\nabla|L'\kappa_2 j\rangle^{0^+}_c$$
$$= {}^0\langle L\kappa_1 i| - i\nabla|L'\kappa_2 j\rangle^0_c$$
$$- i\sum_m\left[\langle\widetilde{L\kappa_1 i}|\nabla|\widetilde{L'\kappa_2 j}\rangle_{M(m)} - {}^{0^-}\langle L\kappa_1 i|\nabla|L'\kappa_2 j\rangle^{0^-}_{M(m)}\right.$$
$$\left. + {}^{0^+}\langle L\kappa_1 i|\nabla|\widehat{L'\kappa_2 j}\rangle_{M(m)} + \langle\widehat{L\kappa_1 i}|\nabla|L'\kappa_2 j\rangle^{0^+}_{M(m)}\right] .\tag{4.6.16}$$

Here, as before, c and $M(m)$ denote integration over the unit cell and the muffin-tin sphere centered at site $\boldsymbol{\tau}_m$, respectively. Note that we have suppressed the spin index since spin-flip transitions were ruled out. Like (4.1.12) and (4.2.16) for the products of basis functions and the Hamiltonian matrix, respectively, (4.6.16) contains the decomposition into smooth, local, and cross terms. As already outlined in the context of the construction of the Hamiltonian matrix the cross terms, i.e. the last two terms in the square brackets, vanish due to the orthonormality of the spherical harmonics.

In evaluating the remaining terms we start with the first term on the right-hand side of (4.6.16), namely, the integral over the unit cell, which has to be built with the pseudo functions. However, in the same manner as in Sect. 4.2 for the secular matrix, we formally replace the pseudo functions by the original envelope functions. Yet, since the envelope and the pseudo functions differ only inside the on-center sphere and the contributions from inside this sphere will be compensated by the second term in the square brackets of (4.6.16), the final result will be the same. Next, defining in complete analogy to (2.2.10)

$$|L\kappa\mu i\rangle := H_{L\kappa}(\mathbf{r}_{\mu i}) , \tag{4.6.17}$$

we rewrite the integral over the unit cell as an integral over all space as

$$
\begin{aligned}
&\langle L\kappa_1 i| - i\nabla | L'\kappa_2 j \rangle_c \\
&= \sum_{\mu}\sum_{\nu} e^{-i\mathbf{k}\mathbf{R}_{\mu}} \langle L\kappa_1\mu i| - i\nabla | L'\kappa_2 \nu j\rangle_c \, e^{i\mathbf{k}\mathbf{R}_{\nu}} \\
&= \sum_{\mu} e^{-i\mathbf{k}\mathbf{R}_{\mu}} \langle L\kappa_1\mu i| - i\nabla | L'\kappa_2 0 j\rangle \\
&= -i \sum_{\mu} e^{-i\mathbf{k}\mathbf{R}_{\mu}} \int_{\Omega_\infty} d^3\mathbf{r}\, H^*_{L\kappa_1}(\mathbf{r}_{\mu i})\nabla H_{L'\kappa_2}(\mathbf{r}_{0j}) , \tag{4.6.18}
\end{aligned}
$$

where the index 0 marks an arbitrarily chosen unit cell of the crystal. Inserting into this expression (3.2.47) for the gradient of the Hankel envelope functions we arrive at the intermediate result

$$
\begin{aligned}
&\int_{\Omega_\infty} d^3\mathbf{r}\, H^*_{L\kappa_1}(\mathbf{r}_{\mu i})\nabla H_{L'\kappa_2}(\mathbf{r}_{0j}) \\
&= \sum_{L''} \mathbf{G}_{L''L'} \left[-\delta_{l'l''-1} + \delta_{l'l''+1}\kappa_2^2 \right] \int_{\Omega_\infty} d^3\mathbf{r}\, H^*_{L\kappa_1}(\mathbf{r}_{\mu i})H_{L''\kappa_2}(\mathbf{r}_{0j}) \\
&= -\sum_{L''} \mathbf{G}_{L''L'} (i\kappa_2)^{l'-l''+1} \int_{\Omega_\infty} d^3\mathbf{r}\, H^*_{L\kappa_1}(\mathbf{r}_{\mu i})H_{L''\kappa_2}(\mathbf{r}_{0j}) , \tag{4.6.19}
\end{aligned}
$$

with the matrices $\mathbf{G}_{L''L'}$ given by (3.2.44). We have thus replaced the integral built with the gradient by the product of two simpler matrices. One of the factors is, essentially, one of the matrices $\mathbf{G}_{L''L'}$ and the other factor is the standard overlap integral of two Hankel envelope functions, which is part of the overlap matrix (4.2.34) and, hence, well known. Note that the matrices $\mathbf{G}_{L''L'}$ contain only local quantities. As a consequence, all the structural information is contained exclusively in the overlap integral. For this reason, we can easily reintroduce Bloch symmetry and combine (4.6.18) and (4.6.19) to the result

$$\langle L\kappa_1 i| -i\nabla |L'\kappa_2 j\rangle_c = -i \sum_{L''} \langle L\kappa_1 i|L''\kappa_2 j\rangle_c \mathbf{G}_{L''L'} \left[-\delta_{l'l''-1} + \delta_{l'l''+1}\kappa_2^2 \right] . \quad (4.6.20)$$

For the special case $i \neq j$ we fall back on (3.8.39) and, ignoring the subtracted intrasphere terms as well as the term for the case $i = j$ entering that equation, we note

$$\langle L\kappa_1 i| - i\nabla |L'\kappa_2 j\rangle_c$$

$$= -\frac{i}{\kappa_2^2 - \kappa_1^2} \sum_{L''} \left[B_{LL''\kappa_2}(\boldsymbol{\tau}_i - \boldsymbol{\tau}_j, \mathbf{k}) - B^*_{L''L\kappa_1}(\boldsymbol{\tau}_j - \boldsymbol{\tau}_i, \mathbf{k}) \right]$$

$$\left(1 - \delta(\kappa_1^2 - \kappa_2^2) \right) \mathbf{G}_{L''L'} \left[-\delta_{l'l''-1} + \delta_{l'l''+1}\kappa_2^2 \right]$$

$$-i \sum_{L''} \dot{B}_{LL''\kappa_1}(\boldsymbol{\tau}_i - \boldsymbol{\tau}_j, \mathbf{k})\delta(\kappa_1^2 - \kappa_2^2)\mathbf{G}_{L''L'} \left[-\delta_{l'l''-1} + \delta_{l'l''+1}\kappa_2^2 \right]$$

$$= -\frac{i}{\kappa_2^2 - \kappa_1^2} \sum_{L''} \mathbf{G}_{LL''} \left[-\delta_{l''l-1}\kappa_2^2 + \delta_{l''l+1} \right] B_{L''L'\kappa_2}(\boldsymbol{\tau}_i - \boldsymbol{\tau}_j, \mathbf{k}) \left(1 - \delta(\kappa_1^2 - \kappa_2^2) \right)$$

$$-\frac{i}{\kappa_1^2 - \kappa_2^2} \sum_{L''} B^*_{L''L\kappa_1}(\boldsymbol{\tau}_j - \boldsymbol{\tau}_i, \mathbf{k})\mathbf{G}_{L''L'} \left[-\delta_{l'l''-1} + \delta_{l'l''+1}\kappa_2^2 \right] \left(1 - \delta(\kappa_1^2 - \kappa_2^2) \right)$$

$$-i \sum_{L''} \dot{B}_{LL''\kappa_1}(\boldsymbol{\tau}_i - \boldsymbol{\tau}_j, \mathbf{k})\mathbf{G}_{L''L'} \left[-\delta_{l'l''-1} + \delta_{l'l''+1}\kappa_2^2 \right] \delta(\kappa_1^2 - \kappa_2^2) . \quad (4.6.21)$$

Here, we have in the second step used the identity (3.7.24) in order to prepare for the derivations below.

Next we turn to the first two terms in the square brackets of (4.6.16) and, using the expansion theorems (2.1.16) and (2.1.23) for the envelope and the augmented functions, we arrive at the usual separation into one-, two-, and three-center terms

$$\langle \widetilde{L\kappa_1 i}|\nabla|\widetilde{L'\kappa_2 j}\rangle_{M(m)}$$

$$= \delta_{im}\langle \tilde{H}_{L\kappa_1\sigma}|\nabla|\tilde{H}_{L'\kappa_2\sigma}\rangle_{M(m)}\delta_{mj}$$

$$+ \sum_{L''} \delta_{im}\langle \tilde{H}_{L\kappa_1\sigma}|\nabla|\tilde{J}_{L''\kappa_2\sigma}\rangle_{M(m)} B_{L''L'\kappa_2}(\boldsymbol{\tau}_m - \boldsymbol{\tau}_j, \mathbf{k})$$

$$+ \sum_{L''} B^*_{L''L\kappa_1}(\boldsymbol{\tau}_m - \boldsymbol{\tau}_i, \mathbf{k})\langle \tilde{J}_{L''\kappa_1\sigma}|\nabla|\tilde{H}_{L'\kappa_2\sigma}\rangle_{M(m)}\delta_{mj}$$

$$+ \sum_{L''} \sum_{L'''} B^*_{L''L\kappa_1}(\boldsymbol{\tau}_m - \boldsymbol{\tau}_i, \mathbf{k})\langle \tilde{J}_{L''\kappa_1\sigma}|\nabla|\tilde{J}_{L'''\kappa_2\sigma}\rangle_{M(m)} B_{L'''L'\kappa_2}(\boldsymbol{\tau}_m - \boldsymbol{\tau}_j, \mathbf{k})$$

$$(4.6.22)$$

for the first term. Note that the intraatomic integrals are not diagonal in L since the gradient couples states with different angular momenta. In general, the integrals entering (4.6.22) assume the form

$$\langle \tilde{F}_{L\kappa_1\sigma} | \nabla | \tilde{G}_{L'\kappa_2\sigma} \rangle_{M(m)}$$

$$= \int_0^{s_m} dr_m r_m \, \tilde{f}_{l\kappa_1\sigma}(r_m) \int d^2\hat{\mathbf{r}}_m \, \mathcal{Y}_L(\hat{\mathbf{r}}_m) \, (r_m \nabla) \, \tilde{g}_{l'\kappa_2\sigma}(r_m) \mathcal{Y}_{L'}(\hat{\mathbf{r}}_m)$$

$$= \int_0^{s_m} dr_m r_m \, \tilde{f}_{l\kappa_1\sigma}(r_m) \int d^2\hat{\mathbf{r}}_m \, \mathcal{Y}_L(\hat{\mathbf{r}}_m) \, (r_m \nabla_r + r_m \nabla_{\vartheta,\varphi}) \, \tilde{g}_{l'\kappa_2\sigma}(r_m) \mathcal{Y}_{L'}(\hat{\mathbf{r}}_m)$$

$$= \mathbf{G}_{LL'} \int_0^{s_m} dr_m r_m^2 \, \tilde{f}_{l\kappa_1\sigma}(r_m) \frac{\partial}{\partial r_m} \tilde{g}_{l'\kappa_2\sigma}(r_m)$$

$$+ \mathbf{D}_{LL'} \int_0^{s_m} dr_m r_m \, \tilde{f}_{l\kappa_1\sigma}(r_m) \tilde{g}_{l'\kappa_2\sigma}(r_m)$$

$$= \mathbf{G}_{LL'} \int_0^{s_m} dr_m r_m^2 \, \tilde{f}_{l\kappa_1\sigma}(r_m) \frac{\partial}{\partial r_m} \tilde{g}_{l'\kappa_2\sigma}(r_m)$$

$$+ \mathbf{G}_{LL'} \left[\delta_{l'l-1}(-l') + \delta_{l'l+1}(l'+1) \right] \int_0^{s_m} dr_m r_m \, \tilde{f}_{l\kappa_1\sigma}(r_m) \tilde{g}_{l'\kappa_2\sigma}(r_m) \, , \quad (4.6.23)$$

with the matrices $\mathbf{D}_{LL'}$ given by (3.2.45) and (3.2.46), respectively. $\tilde{F}_{L\kappa_1\sigma}$ and $\tilde{G}_{L'\kappa_2\sigma}$ denote any of the augmented functions and $\tilde{f}_{l\kappa_1\sigma}$ and $\tilde{g}_{l\kappa_1\sigma}$ are their real valued radial parts. The radial integrals are evaluated numerically.

The second term in the square brackets of (4.6.16) is likewise evaluated using the expansion theorems (2.1.16) and then reads as

$$^{0^-}\langle L\kappa_1 i | \nabla | L'\kappa_2 j \rangle^{0^-}_{M(m)}$$

$$= \delta_{im}\langle H_{L\kappa_1} | \nabla | H_{L'\kappa_2} \rangle_{M(m)} \delta_{mj}$$

$$+ \sum_{L''} \delta_{im}\langle H_{L\kappa_1} | \nabla | J_{L''\kappa_2} \rangle_{M(m)} B_{L''L'\kappa_2}(\boldsymbol{\tau}_m - \boldsymbol{\tau}_j, \mathbf{k})$$

$$+ \sum_{L''} B^*_{L''L\kappa_1}(\boldsymbol{\tau}_m - \boldsymbol{\tau}_i, \mathbf{k})\langle J_{L''\kappa_1} | \nabla | H_{L'\kappa_2} \rangle_{M(m)} \delta_{mj}$$

$$+ \sum_{L''} \sum_{L'''} B^*_{L''L\kappa_1}(\boldsymbol{\tau}_m - \boldsymbol{\tau}_i, \mathbf{k})\langle J_{L''\kappa_1} | \nabla | J_{L'''\kappa_2} \rangle_{M(m)} B_{L'''L'\kappa_2}(\boldsymbol{\tau}_m - \boldsymbol{\tau}_j, \mathbf{k}) \, .$$

$$(4.6.24)$$

As for the integral over the unit cell we have here formally replaced the pseudo functions by the original envelope functions in order to allow for a cancellation of the integrals involving Hankel envelope functions within the on-center spheres. Next, using (3.2.47) and (3.2.49) for the gradients of the Hankel and Bessel envelope functions as well as the fact that the intraatomic overlap integrals of these functions are diagonal in L we arrive at

$$^{0^-}\langle L\kappa_1 i | \nabla | L'\kappa_2 j \rangle^{0^-}_{M(m)}$$

$$= \delta_{im}\langle H_{L\kappa_1} | H_{L\kappa_2} \rangle_{M(m)} \mathbf{G}_{LL'} \left[-\delta_{l'l-1} + \delta_{l'l+1}\kappa_2^2 \right] \delta_{mj}$$

$$+ \sum_{L''} \delta_{im}\langle H_{L\kappa_1} | J_{L\kappa_2} \rangle_{M(m)} \mathbf{G}_{LL''} \left[-\delta_{l''l-1}\kappa_2^2 + \delta_{l''l+1} \right] B_{L''L'\kappa_2}(\boldsymbol{\tau}_m - \boldsymbol{\tau}_j, \mathbf{k})$$

$$+ \sum_{L''} B^*_{L''L\kappa_1}(\boldsymbol{\tau}_m - \boldsymbol{\tau}_i, \mathbf{k})\langle J_{L''\kappa_1} | H_{L''\kappa_2} \rangle_{M(m)} \mathbf{G}_{L''L'} \left[-\delta_{l'l''-1} + \delta_{l'l''+1}\kappa_2^2 \right] \delta_{mj}$$

$$+ \sum_{L''} \sum_{L'''} B^*_{L''L\kappa_1}(\boldsymbol{\tau}_m - \boldsymbol{\tau}_i, \mathbf{k}) \langle J_{L''\kappa_1} | J_{L''\kappa_2} \rangle_{M(m)} \mathbf{G}_{L''L'''} \left[-\delta_{l'''l''-1}\kappa_2^2 + \delta_{l'''l''+1} \right]$$

$$B_{L'''L'\kappa_2}(\boldsymbol{\tau}_m - \boldsymbol{\tau}_j, \mathbf{k}) \,. \tag{4.6.25}$$

It is interesting to discuss the angular momentum cutoffs used with the previous formula. As usual, Hankel functions in (4.6.22) and (4.6.23) are included only for the lower partial waves, whereas Bessel functions are taken into account for both the lower and intermediate partial waves. However, the same angular momentum cutoffs are used in (4.6.23) and (4.6.25) despite the fact that application of the gradient produces functions with the angular momentum l lowered and increased by one, respectively. In particular, the function with the increased l value is ignored for the highest partial waves. Yet, this error occurs for both the augmented and the envelope functions and thus is compensated to large parts.

Finally, we insert the intermediate results (4.6.20) to (4.6.25) into the initial formula for the momentum operator matrix elements, (4.6.16) and note

$$^{\infty}\langle L\kappa_1 i | - i\nabla | L'\kappa_2 j \rangle^{\infty}_c$$

$$= -i \sum_{L''} \langle L\kappa_1 i | L''\kappa_2 j \rangle_c \mathbf{G}_{L''L'} \left[-\delta_{l'l''-1} + \delta_{l'l''+1}\kappa_2^2 \right]$$

$$-i \left[\langle \tilde{H}_{L\kappa_1\sigma} | \nabla | \tilde{H}_{L'\kappa_2\sigma} \rangle_{M(i)} \right.$$

$$\left. - \langle H_{L\kappa_1} | H_{L\kappa_2} \rangle_{M(i)} \mathbf{G}_{LL'} \left[-\delta_{l'l-1} + \delta_{l'l+1}\kappa_2^2 \right] \right] \delta_{ij}$$

$$-i \sum_{L''} \left[\langle \tilde{H}_{L\kappa_1\sigma} | \nabla | \tilde{J}_{L''\kappa_2\sigma} \rangle_{M(i)} \right.$$

$$\left. - \langle H_{L\kappa_1} | J_{L\kappa_2} \rangle_{M(i)} \mathbf{G}_{LL''} \left[-\delta_{l''l-1}\kappa_2^2 + \delta_{l''l+1} \right] \right]$$

$$B_{L''L'\kappa_2}(\boldsymbol{\tau}_i - \boldsymbol{\tau}_j, \mathbf{k})$$

$$-i \sum_{L''} B^*_{L''L\kappa_1}(\boldsymbol{\tau}_j - \boldsymbol{\tau}_i, \mathbf{k})$$

$$\left[\langle \tilde{J}_{L''\kappa_1\sigma} | \nabla | \tilde{H}_{L'\kappa_2\sigma} \rangle_{M(j)} \right.$$

$$\left. - \langle J_{L''\kappa_1} | H_{L''\kappa_2} \rangle_{M(j)} \mathbf{G}_{L''L'} \left[-\delta_{l'l''-1} + \delta_{l'l''+1}\kappa_2^2 \right] \right]$$

$$-i \sum_m \sum_{L''} \sum_{L'''} B^*_{L''L\kappa_1}(\boldsymbol{\tau}_m - \boldsymbol{\tau}_i, \mathbf{k})$$

$$\left[\langle \tilde{J}_{L''\kappa_1\sigma} | \nabla | \tilde{J}_{L'''\kappa_2\sigma} \rangle_{M(m)} \right.$$

$$\left. - \langle J_{L''\kappa_1} | J_{L''\kappa_2} \rangle_{M(m)} \mathbf{G}_{L''L'''} \left[-\delta_{l'''l''-1}\kappa_2^2 + \delta_{l'''l''+1} \right] \right]$$

$$B_{L'''L'\kappa_2}(\boldsymbol{\tau}_m - \boldsymbol{\tau}_j, \mathbf{k}) \,. \tag{4.6.26}$$

For the overlap integral entering the first term on the right-hand side we use the second formulation of (4.6.21) for the case $i \neq j$ and combine the first two terms of that formulation with the intraatomic integrals arising from (4.6.25). In contrast, for $i = j$, we obtain, apart from the $\mathbf{G}_{L''L}$ factor, the overlap integral of two Hankel envelope functions extending over all space. As usual, we combine this overlap integral with the second term in the first square brackets of (4.6.26). This removes the singularity of the Hankel envelope function at the origin and leads to an integral extending from the sphere radius to infinity. We thus arrive at the result

$$
{}^{\infty}\langle L\kappa_1 i| - i\nabla|L'\kappa_2 j\rangle_c^{\infty}
$$

$$
= -i\left[\langle \tilde{H}_{L\kappa_1\sigma}|\nabla|\tilde{H}_{L'\kappa_2\sigma}\rangle_{M(i)}\right.
$$

$$
\left. + \langle H_{L\kappa_1}|H_{L\kappa_2}\rangle'_{M(i)}\mathbf{G}_{LL'}\left[-\delta_{l'l-1} + \delta_{l'l+1}\kappa_2^2\right]\right]\delta_{ij}
$$

$$
- \frac{i}{\kappa_2^2 - \kappa_1^2}\sum_{L''}\mathbf{G}_{LL''}\left[-\delta_{l''l-1}\kappa_2^2 + \delta_{l''l+1}\right]B_{L''L'\kappa_2}(\boldsymbol{\tau}_i - \boldsymbol{\tau}_j, \mathbf{k})\left(1 - \delta(\kappa_1^2 - \kappa_2^2)\right)
$$

$$
- \frac{i}{\kappa_1^2 - \kappa_2^2}\sum_{L''}B^*_{L''L\kappa_1}(\boldsymbol{\tau}_j - \boldsymbol{\tau}_i, \mathbf{k})\mathbf{G}_{L''L'}\left[-\delta_{l'l''-1} + \delta_{l'l''+1}\kappa_2^2\right]\left(1 - \delta(\kappa_1^2 - \kappa_2^2)\right)
$$

$$
- i\sum_{L''}\dot{B}_{LL''\kappa_1}(\boldsymbol{\tau}_i - \boldsymbol{\tau}_j, \mathbf{k})\mathbf{G}_{L''L'}\left[-\delta_{l'l''-1} + \delta_{l'l''+1}\kappa_2^2\right]\delta(\kappa_1^2 - \kappa_2^2)
$$

$$
- i\sum_{L''}\left[\langle \tilde{H}_{L\kappa_1\sigma}|\nabla|\tilde{J}_{L''\kappa_2\sigma}\rangle_{M(i)}\right.
$$

$$
\left. - \langle H_{L\kappa_1}|J_{L\kappa_2}\rangle_{M(i)}\mathbf{G}_{LL''}\left[-\delta_{l''l-1}\kappa_2^2 + \delta_{l''l+1}\right]\right]
$$

$$
B_{L''L'\kappa_2}(\boldsymbol{\tau}_i - \boldsymbol{\tau}_j, \mathbf{k})
$$

$$
- i\sum_{L''}B^*_{L''L\kappa_1}(\boldsymbol{\tau}_j - \boldsymbol{\tau}_i, \mathbf{k})
$$

$$
\left[\langle \tilde{J}_{L''\kappa_1\sigma}|\nabla|\tilde{H}_{L'\kappa_2\sigma}\rangle_{M(j)}\right.
$$

$$
\left. - \langle J_{L''\kappa_1}|H_{L''\kappa_2}\rangle_{M(j)}\mathbf{G}_{L''L'}\left[-\delta_{l'l''-1} + \delta_{l'l''+1}\kappa_2^2\right]\right]
$$

$$
- i\sum_{m}\sum_{L''}\sum_{L'''}B^*_{L''L\kappa_1}(\boldsymbol{\tau}_m - \boldsymbol{\tau}_i, \mathbf{k})
$$

$$
\left[\langle \tilde{J}_{L''\kappa_1\sigma}|\nabla|\tilde{J}_{L'''\kappa_2\sigma}\rangle_{M(m)}\right.
$$

$$
\left. - \langle J_{L''\kappa_1}|J_{L''\kappa_2}\rangle_{M(m)}\mathbf{G}_{L''L'''}\left[-\delta_{l'''l''-1}\kappa_2^2 + \delta_{l'''l''+1}\right]\right]
$$

$$
B_{L'''L'\kappa_2}(\boldsymbol{\tau}_m - \boldsymbol{\tau}_j, \mathbf{k})
$$

$$= -i\Bigg[\langle\tilde{H}_{L\kappa_1\sigma}|\nabla|\tilde{H}_{L'\kappa_2\sigma}\rangle_{M(i)}$$

$$+\langle H_{L\kappa_1}|H_{L\kappa_2}\rangle'_{M(i)}\mathbf{G}_{LL'}\left[-\delta_{l'l-1}+\delta_{l'l+1}\kappa_2^2\right]\Bigg]\delta_{ij}$$

$$-i\sum_{L''}\dot{B}_{LL''\kappa_1}(\boldsymbol{\tau}_i-\boldsymbol{\tau}_j,\mathbf{k})\mathbf{G}_{L''L'}\left[-\delta_{l'l''-1}+\delta_{l'l''+1}\kappa_2^2\right]\delta(\kappa_1^2-\kappa_2^2)$$

$$-i\sum_{L''}\Bigg[\langle\tilde{H}_{L\kappa_1\sigma}|\nabla|\tilde{J}_{L''\kappa_2\sigma}\rangle_{M(i)}$$

$$-\left(\langle H_{L\kappa_1}|J_{L\kappa_2}\rangle_{M(i)}-\frac{1}{\kappa_2^2-\kappa_1^2}\left(1-\delta(\kappa_1^2-\kappa_2^2)\right)\right)$$

$$\mathbf{G}_{LL''}\left[-\delta_{l''l-1}\kappa_2^2+\delta_{l''l+1}\right]\Bigg]B_{L''L'\kappa_2}(\boldsymbol{\tau}_i-\boldsymbol{\tau}_j,\mathbf{k})$$

$$-i\sum_{L''}B^*_{L''L\kappa_1}(\boldsymbol{\tau}_j-\boldsymbol{\tau}_i,\mathbf{k})$$

$$\Bigg[\langle\tilde{J}_{L''\kappa_1\sigma}|\nabla|\tilde{H}_{L'\kappa_2\sigma}\rangle_{M(j)}$$

$$-\left(\langle J_{L''\kappa_1}|H_{L''\kappa_2}\rangle_{M(j)}-\frac{1}{\kappa_1^2-\kappa_2^2}\left(1-\delta(\kappa_1^2-\kappa_2^2)\right)\right)$$

$$\mathbf{G}_{L''L'}\left[-\delta_{l'l''-1}+\delta_{l'l''+1}\kappa_2^2\right]\Bigg]$$

$$-i\sum_m\sum_{L''}\sum_{L'''}B^*_{L''L\kappa_1}(\boldsymbol{\tau}_m-\boldsymbol{\tau}_i,\mathbf{k})$$

$$\Bigg[\langle\tilde{J}_{L''\kappa_1\sigma}|\nabla|\tilde{J}_{L'''\kappa_2\sigma}\rangle_{M(m)}$$

$$-\langle J_{L''\kappa_1}|J_{L''\kappa_2}\rangle_{M(m)}\mathbf{G}_{L''L'''}\left[-\delta_{l'''l''-1}\kappa_2^2+\delta_{l'''l''+1}\right]\Bigg]$$

$$B_{L'''L'\kappa_2}(\boldsymbol{\tau}_m-\boldsymbol{\tau}_j,\mathbf{k})\ . \tag{4.6.27}$$

As for the Hamiltonian and overlap matrices in Sect. 2.2 it is useful to define the following abbreviations for the one-, two-, and three-center contributions

$$\mathbf{U}^{(M,1)}_{L\kappa_1L'\kappa_2i\sigma}=\langle\tilde{H}_{L\kappa_1\sigma}|\nabla|\tilde{H}_{L'\kappa_2\sigma}\rangle_{M(i)}$$

$$+\langle H_{L\kappa_1}|H_{L\kappa_2}\rangle'_{M(i)}\mathbf{G}_{LL'}\left[-\delta_{l'l-1}+\delta_{l'l+1}\kappa_2^2\right]\ , \tag{4.6.28}$$

$$\mathbf{U}^{(M,2)}_{L\kappa_1L''\kappa_2i\sigma}=\langle\tilde{H}_{L\kappa_1\sigma}|\nabla|\tilde{J}_{L''\kappa_2\sigma}\rangle_{M(i)}$$

$$-\left(\langle H_{L\kappa_1}|J_{L\kappa_2}\rangle_{M(i)}-\frac{1}{\kappa_2^2-\kappa_1^2}\left(1-\delta(\kappa_1^2-\kappa_2^2)\right)\right)$$

$$\mathbf{G}_{LL''}\left[-\delta_{l''l-1}\kappa_2^2+\delta_{l''l+1}\right]\ , \tag{4.6.29}$$

$$\mathbf{U}^{(M,2')}_{L''\kappa_1 L'\kappa_2 i\sigma} = \langle \tilde{J}_{L''\kappa_1\sigma}|\nabla|\tilde{H}_{L'\kappa_2\sigma}\rangle_{M(i)}$$
$$- \left(\langle J_{L''\kappa_1}|H_{L''\kappa_2}\rangle_{M(i)} - \frac{1}{\kappa_1^2 - \kappa_2^2}\left(1 - \delta(\kappa_1^2 - \kappa_2^2)\right)\right)$$
$$\mathbf{G}_{L''L'}\left[-\delta_{l'l''-1} + \delta_{l'l''+1}\kappa_2^2\right]\,, \qquad (4.6.30)$$

$$\mathbf{U}^{(M,3)}_{L''\kappa_1 L'''\kappa_2 i\sigma} = \langle \tilde{J}_{L''\kappa_1\sigma}|\nabla|\tilde{J}_{L'''\kappa_2\sigma}\rangle_{M(i)}$$
$$- \langle J_{L''\kappa_1}|J_{L''\kappa_2}\rangle_{M(i)}\mathbf{G}_{L''L'''}\left[-\delta_{l'''l''-1}\kappa_2^2 + \delta_{l'''l''+1}\right]\,.$$
$$(4.6.31)$$

With these identities at hand we note for the momentum operator matrix element

$$^{\infty}\langle L\kappa_1 i| - i\nabla|L'\kappa_2 j\rangle_c^{\infty}$$
$$= -i\mathbf{U}^{(M,1)}_{L\kappa_1 L'\kappa_2 i\sigma}\delta_{ij}$$
$$-i\sum_{L''}\dot{B}_{LL''\kappa_1}(\boldsymbol{\tau}_i - \boldsymbol{\tau}_j, \mathbf{k})\mathbf{G}_{L''L'}\left[-\delta_{l'l''-1} + \delta_{l'l''+1}\kappa_2^2\right]\delta(\kappa_1^2 - \kappa_2^2)$$
$$-i\sum_{L''}\mathbf{U}^{(M,2)}_{L\kappa_1 L''\kappa_2 i\sigma}B_{L''L'\kappa_2}(\boldsymbol{\tau}_i - \boldsymbol{\tau}_j, \mathbf{k})$$
$$-i\sum_{L''}B^*_{L''L\kappa_1}(\boldsymbol{\tau}_j - \boldsymbol{\tau}_i, \mathbf{k})\mathbf{U}^{(M,2')}_{L''\kappa_1 L'\kappa_2 j\sigma}$$
$$-i\sum_{m}\sum_{L''}\sum_{L'''}B^*_{L''L\kappa_1}(\boldsymbol{\tau}_m - \boldsymbol{\tau}_i, \mathbf{k})\mathbf{U}^{(M,3)}_{L''\kappa_1 L'''\kappa_2 m\sigma}B_{L'''L'\kappa_2}(\boldsymbol{\tau}_m - \boldsymbol{\tau}_j, \mathbf{k})\,.$$
$$(4.6.32)$$

Finally, since we opted for the real cubic rather than the complex spherical harmonics, we use (3.7.18) and write the momentum operator matrix element as

$$^{\infty}\langle L\kappa_1 i| - i\nabla|L'\kappa_2 j\rangle_c^{\infty}$$
$$= -i\mathbf{U}^{(M,1)}_{L\kappa_1 L'\kappa_2 i\sigma}\delta_{ij}$$
$$-i\sum_{L''}\dot{B}_{LL''\kappa_1}(\boldsymbol{\tau}_i - \boldsymbol{\tau}_j, \mathbf{k})\mathbf{G}_{L''L'}\left[-\delta_{l'l''-1} + \delta_{l'l''+1}\kappa_2^2\right]\delta(\kappa_1^2 - \kappa_2^2)$$
$$-i\sum_{L''}\mathbf{U}^{(M,2)}_{L\kappa_1 L''\kappa_2 i\sigma}B_{L''L'\kappa_2}(\boldsymbol{\tau}_i - \boldsymbol{\tau}_j, \mathbf{k})$$
$$-i\sum_{L''}B_{LL''\kappa_1}(\boldsymbol{\tau}_i - \boldsymbol{\tau}_j, \mathbf{k})\mathbf{U}^{(M,2')}_{L''\kappa_1 L'\kappa_2 j\sigma}$$
$$-i\sum_{m}\sum_{L''}\sum_{L'''}B_{LL''\kappa_1}(\boldsymbol{\tau}_i - \boldsymbol{\tau}_m, \mathbf{k})\mathbf{U}^{(M,3)}_{L''\kappa_1 L'''\kappa_2 m\sigma}B_{L'''L'\kappa_2}(\boldsymbol{\tau}_m - \boldsymbol{\tau}_j, \mathbf{k})\,.$$
$$(4.6.33)$$

Note that (4.6.32) and (4.6.33) have a very similar algebraic structure as (4.2.30) and (4.2.31) for the Hamiltonian matrix and thus can be easily evaluated along the same lines. However, we point out that the matrices $\mathbf{U}^{(M,n)}_{L\kappa_1 L'\kappa_2 i\sigma}$ have non-vanishing elements for $l' = l \pm 1$ only and, hence, the setup of the momentum operator matrices needs much less operations as compared to the calculation of the Hamiltonian matrix.

In a final step, we combine the result (4.6.33) with the representation (4.1.8) of the wave function in terms of the basis functions and obtain

$$
{}^\infty\langle\psi_{\mathbf{k}\sigma}| - i\nabla|\psi_{\mathbf{k}\sigma}\rangle_c^\infty
$$

$$
= -i\sum_{L\kappa_1 i}\sum_{L'\kappa_2} c^*_{L\kappa_1 i\sigma}(\mathbf{k})\mathbf{U}^{(M,1)}_{L\kappa_1 L'\kappa_2 i\sigma}c_{L'\kappa_2 i\sigma}(\mathbf{k})
$$

$$
-i\sum_{L\kappa_1 i}\sum_{L'\kappa_2 j}\sum_{L''} c^*_{L\kappa_1 i\sigma}(\mathbf{k})\dot{B}^*_{L''L\kappa_1}(\boldsymbol{\tau}_j - \boldsymbol{\tau}_i, \mathbf{k})
$$

$$
\mathbf{G}_{L''L'}\left[-\delta_{l'l''-1} + \delta_{l'l''+1}\kappa_2^2\right]\delta(\kappa_1^2 - \kappa_2^2)c_{L'\kappa_2 i\sigma}(\mathbf{k})
$$

$$
-i\sum_{L\kappa_1 i}\sum_{L'\kappa_2 j}\sum_{L''} c^*_{L\kappa_1 i\sigma}(\mathbf{k})\mathbf{U}^{(M,2)}_{L\kappa_1 L''\kappa_2 i\sigma}B_{L''L'\kappa_2}(\boldsymbol{\tau}_i - \boldsymbol{\tau}_j, \mathbf{k})c_{L'\kappa_2 j\sigma}(\mathbf{k})
$$

$$
-i\sum_{L\kappa_1 i}\sum_{L'\kappa_2 j}\sum_{L''} c^*_{L\kappa_1 i\sigma}(\mathbf{k})B^*_{L''L\kappa_1}(\boldsymbol{\tau}_j - \boldsymbol{\tau}_i, \mathbf{k})\mathbf{U}^{(M,2')}_{L''\kappa_1 L'\kappa_2 j\sigma}c_{L'\kappa_2 j\sigma}(\mathbf{k})
$$

$$
-i\sum_{L\kappa_1 i}\sum_{L'\kappa_2 j}\sum_{m}\sum_{L''}\sum_{L'''} c^*_{L\kappa_1 i\sigma}(\mathbf{k})B^*_{L''L\kappa_1}(\boldsymbol{\tau}_m - \boldsymbol{\tau}_i, \mathbf{k})\mathbf{U}^{(M,3)}_{L''\kappa_1 L'''\kappa_2 m\sigma}
$$

$$
B_{L'''L'\kappa_2}(\boldsymbol{\tau}_m - \boldsymbol{\tau}_j, \mathbf{k})c_{L'\kappa_2 j\sigma}(\mathbf{k}) . \tag{4.6.34}
$$

Using the abbreviation (4.1.9) as well as the analogous formula for the energy derivative,

$$
\dot{a}_{L'\kappa j\sigma}(\mathbf{k}) = \sum_{Li} c_{L\kappa i\sigma}(\mathbf{k})\dot{B}_{L'L\kappa}(\boldsymbol{\tau}_j - \boldsymbol{\tau}_i, \mathbf{k}) , \tag{4.6.35}
$$

we arrive at the compact representation

$$
{}^\infty\langle\psi_{\mathbf{k}\sigma}| - i\nabla|\psi_{\mathbf{k}\sigma}\rangle_c^\infty
$$

$$
= -i\sum_i\sum_{L\kappa_1}\sum_{L'\kappa_2} c^*_{L\kappa_1 i\sigma}(\mathbf{k})\mathbf{U}^{(M,1)}_{L\kappa_1 L'\kappa_2 i\sigma}c_{L'\kappa_2 i\sigma}(\mathbf{k})
$$

$$
-i\sum_j\sum_{L''\kappa_1}\sum_{L'\kappa_2} \dot{a}^*_{L''\kappa_1 j\sigma}(\mathbf{k})\mathbf{G}_{L''L'}\left[-\delta_{l'l''-1} + \delta_{l'l''+1}\kappa_2^2\right]\delta(\kappa_1^2 - \kappa_2^2)c_{L'\kappa_2 i\sigma}(\mathbf{k})
$$

$$
-i\sum_i\sum_{L\kappa_1}\sum_{L''\kappa_2} c^*_{L\kappa_1 i\sigma}(\mathbf{k})\mathbf{U}^{(M,2)}_{L\kappa_1 L''\kappa_2 i\sigma}a_{L''\kappa_2 i\sigma}(\mathbf{k})
$$

$$
-i\sum_j\sum_{L''\kappa_1}\sum_{L'\kappa_2} a^*_{L''\kappa_1 j\sigma}(\mathbf{k})\mathbf{U}^{(M,2')}_{L''\kappa_1 L'\kappa_2 j\sigma}c_{L'\kappa_2 i\sigma}(\mathbf{k})
$$

$$
-i\sum_m\sum_{L''\kappa_1}\sum_{L'''\kappa_2} a^*_{L''\kappa_1 m\sigma}(\mathbf{k})\mathbf{U}^{(M,3)}_{L''\kappa_1 L'''\kappa_2 m\sigma}a_{L'''\kappa_2 m\sigma}(\mathbf{k})
$$

$$
= -i\sum_i\sum_{L\kappa_1}\sum_{L'\kappa_2}\Bigg\{ c^*_{L\kappa_1 i\sigma}(\mathbf{k})\mathbf{U}^{(M,1)}_{L\kappa_1 L'\kappa_2 i\sigma}c_{L'\kappa_2 i\sigma}(\mathbf{k})
$$

$$
\dot{a}^*_{L\kappa_1 i\sigma}(\mathbf{k})\mathbf{G}_{LL'}\left[-\delta_{l'l-1} + \delta_{l'l+1}\kappa_2^2\right]\delta(\kappa_1^2 - \kappa_2^2)c_{L'\kappa_2 i\sigma}(\mathbf{k})
$$

$$
c^*_{L\kappa_1 i\sigma}(\mathbf{k})\mathbf{U}^{(M,2)}_{L\kappa_1 L'\kappa_2 i\sigma}a_{L'\kappa_2 i\sigma}(\mathbf{k})
$$

$$a^*_{L\kappa_1 i\sigma}(\mathbf{k})\mathbf{U}^{(M,2')}_{L\kappa_1 L'\kappa_2 i\sigma}c_{L'\kappa_2 i\sigma}(\mathbf{k})$$

$$\left. a^*_{L\kappa_1 i\sigma}(\mathbf{k})\mathbf{U}^{(M,3)}_{L\kappa_1 L'\kappa_2 i\sigma}a_{L'\kappa_2 i\sigma}(\mathbf{k})\right\} \ . \tag{4.6.36}$$

In closing this section, we point to the afore mentioned intermediate approach, which omits the pseudo quantities in the interstitial region and uses overlapping atomic spheres. In the present context, this approach is easily adopted by replacing in (4.6.16) and all following equations the integrations over the muffin-tin spheres by integrations over the atomic spheres.

References

1. V. N. Antonov, P. M. Oppeneer, A. N. Yaresko, A. Y. Perlov, and T. Kraft, Phys. Rev. B **56**, 13012 (1997)
2. N. W. Ashcroft and N. D. Mermin, *Solid State Physics* (Holt-Saunders, Philadelphia 1976)
3. K. Atkinson, J. Austral. Math. Soc. B **23**, 332 (1982)
4. U. von Barth, Density-Functional Theory for Solids. In: *The Electronic Structure of Complex Systems*, ed by P. Phariseau and W. Temmerman (Plenum Press, New York 1984) pp 67–140
5. U. von Barth, An Overview of Density-Functional Theory. In: *Many-Body Phenomena at Surfaces*, ed by D. Langreth and H. Suhl (Academic Press, Orlando 1984) pp 3–50
6. Z. P. Bažant and B. H. Oh, Z. Angew. Math. Mech. **66**, 37 (1986)
7. P. E. Blöchl, Gesamtenergien, Kräfte und Metall-Halbleiter Grenzflächen. PhD thesis, Universität Stuttgart (1989)
8. N. E. Christensen, Phys. Rev. B **29**, 5547 (1984)
9. G. Czycholl, *Theoretische Festkörperphysik*, (Springer, Berlin 2004)
10. R. M. Dreizler and E. K. U. Gross, *Density-Functional Theory* (Springer, Berlin 1990)
11. H. Ebert, Rep. Prog. Phys. **59**, 1665 (1996)
12. H. Eschrig, *The Fundamentals of Density-Functional Theory* (Edition am Gutenbergplatz, Leipzig 2003)
13. V. Eyert, Entwicklung und Implementation eines Full-Potential-ASW-Verfahrens. PhD thesis, Technische Hochschule Darmstadt (1991)
14. V. Eyert, J. Comput. Phys. **124**, 271 (1996)
15. V. Eyert, *Electronic Structure of Crystalline Materials*, 2nd edn (University of Augsburg, Augsburg 2005)
16. P. Keast, J. Comput. Appl. Math. **17**, 151 (1987)
17. P. Keast and J. C. Diaz, SIAM J. Numer. Anal. **20**, 406 (1983)
18. J. Kübler and V. Eyert, Electronic structure calculations. In: *Electronic and Magnetic Properties of Metals and Ceramics*, ed by K. H. J. Buschow (VCH Verlagsgesellschaft, Weinheim 1992) pp 1–145; vol 3A of *Materials Science and Technology*, ed by R. W. Cahn, P. Haasen, and E. J. Kramer (VCH Verlagsgesellschaft, Weinheim 1991–1996)
19. T. Maurer, Berechnung des magneto-optischen Kerr-Effekts. Diploma thesis, Technische Hochschule Darmstadt (1991)
20. A. D. McLaren, Math. Comput. **17**, 361 (1963)

21. M. S. Methfessel, Multipole Green Functions for Electronic Structure Calculations. PhD thesis, University of Nijmegen (1986)
22. I. P. Mysovskih, Sov. Math. Dokl. **18**, 925 (1977)
23. P. M. Oppeneer, T. Maurer, J. Sticht, and J. Kübler, Phys. Rev. B **45**, 10924 (1992)
24. P. M. Oppeneer, J. Sticht, T. Maurer, and J. Kübler, Z. Phys. B **88**, 309 (1992)
25. R. G. Parr and W. Yang, *Density-Functional Theory of Atoms and Molecules* (Oxford University Press, Oxford 1989)
26. A. S. Popov, Comput. Math. Math. Phys. **35**, 369 (1995)
27. H. W. A. M. Rompa, R. Eppenga, and M. F. H. Schuurmans, Physica **145B**, 5 (1987)
28. A. H. Stroud, *Approximate Calculation of Multiple Integrals* (Prentice-Hall, Englewood Cliffs 1971)
29. A. H. Stroud, SIAM J. Numer. Anal. **10**, 559 (1973)
30. C. S. Wang and J. Callaway, Phys. Rev. **B9**, 4897 (1974)
31. M. Weinert, J. Math. Phys. **22**, 2433 (1981)
32. K.-H. Weyrich, Solid State Commun. **54**, 975 (1985)
33. K.-H. Weyrich and R. Siems, Jap. J. Appl. Phys., Suppl. **24**, 201 (1985)
34. K.-H. Weyrich and R. Siems, Z. Phys. **61**, 63 (1985)
35. K.-H. Weyrich, Phys. Rev. B **37**, 10269 (1988)
36. K.-H. Weyrich, L. Brey, and N. E. Christensen, Phys. Rev. B **38**, 1392 (1988)

A

Details of the Standard ASW Method

A.1 Integrals over Augmented Functions

In the course of the construction of ASW basis functions in Sect. 2.1 we have defined the augmented Hankel and Bessel functions in (2.1.13) and (2.1.20), respectively. Both types of augmented functions have to fulfill the radial Schrödinger equations (2.1.14) and (2.1.21) subject to the boundary conditions (2.1.15) and (2.1.22) and the regularity at the origin. As already mentioned in connection with (2.1.22), the radial equations (2.1.14) and (2.1.21) are identical and the difference between the augmented functions $\tilde{h}_{l\kappa\sigma}$ and $\tilde{j}_{l\kappa\sigma}$ as well as the energies $E_{l\kappa i\sigma}^{(H)}$ and $E_{l\kappa i\sigma}^{(J)}$ thus arises exclusively from the different boundary conditions (2.1.15) and (2.1.22).

It is the subject of the present section to define the radial integrals over the augmented functions and to set up interrelations between them. Using (2.1.13) to (2.1.15) and (2.1.20) to (2.1.22) we thus define the Hankel integral

$$
\begin{aligned}
S_{l\kappa i\sigma}^{(H)} &:= \langle \tilde{H}_{L\kappa\sigma} | \tilde{H}_{L'\kappa\sigma} \rangle_i \delta_{LL'} \\
&= \int_{\Omega_i} d^3\mathbf{r}_i \, \tilde{H}_{L\kappa\sigma}^*(\mathbf{r}_i) \tilde{H}_{L\kappa\sigma}(\mathbf{r}_i) \\
&= \int_0^{S_i} dr_i \, r_i^2 \tilde{h}_{l\kappa\sigma}(r_i) \tilde{h}_{l\kappa\sigma}(r_i) \,,
\end{aligned}
\tag{A.1.1}
$$

as well as the Bessel integral

$$
\begin{aligned}
S_{l\kappa i\sigma}^{(J)} &:= \langle \tilde{J}_{L\kappa\sigma} | \tilde{J}_{L'\kappa\sigma} \rangle_i \delta_{LL'} \\
&= \int_{\Omega_i} d^3\mathbf{r}_i \, \tilde{J}_{L\kappa\sigma}^*(\mathbf{r}_i) \tilde{J}_{L\kappa\sigma}(\mathbf{r}_i) \\
&= \int_0^{S_i} dr_i \, r_i^2 \tilde{j}_{l\kappa\sigma}(r_i) \tilde{j}_{l\kappa\sigma}(r_i) \,.
\end{aligned}
\tag{A.1.2}
$$

V. Eyert: *Details of the Standard ASW Method*, Lect. Notes Phys. **719**, 175–197 (2007)
DOI 10.1007/978-3-540-71007-3_5   © Springer-Verlag Berlin Heidelberg 2007

In addition, we define the mixed integrals

$$\langle \tilde{H}_{L\kappa_1\sigma} | \tilde{H}_{L'\kappa_2\sigma} \rangle_i \delta_{LL'} = \int_{\Omega_i} d^3 r_i \, \tilde{H}^*_{L\kappa_1\sigma}(\mathbf{r}_i) \tilde{H}_{L\kappa_2\sigma}(\mathbf{r}_i)$$

$$= \int_0^{S_i} dr_i \, r_i^2 \tilde{h}_{l\kappa_1\sigma}(r_i) \tilde{h}_{l\kappa_2\sigma}(r_i) \,, \qquad \text{(A.1.3)}$$

$$\langle \tilde{J}_{L\kappa_1\sigma} | \tilde{J}_{L'\kappa_2\sigma} \rangle_i \delta_{LL'} = \int_{\Omega_i} d^3 r_i \, \tilde{J}^*_{L\kappa_1\sigma}(\mathbf{r}_i) \tilde{J}_{L\kappa_2\sigma}(\mathbf{r}_i)$$

$$= \int_0^{S_i} dr_i \, r_i^2 \tilde{j}_{l\kappa_1\sigma}(r_i) \tilde{j}_{l\kappa_2\sigma}(r_i) \,, \qquad \text{(A.1.4)}$$

and

$$\langle \tilde{H}_{L\kappa_1\sigma} | \tilde{J}_{L'\kappa_2\sigma} \rangle_i \delta_{LL'} = \langle \tilde{J}_{L'\kappa_2\sigma} | \tilde{H}_{L\kappa_1\sigma} \rangle_i \delta_{LL'}$$

$$= \int_{\Omega_i} d^3 r_i \, \tilde{H}^*_{L\kappa_1\sigma}(\mathbf{r}_i) \tilde{J}_{L\kappa_2\sigma}(\mathbf{r}_i)$$

$$= \int_0^{S_i} dr_i \, r_i^2 \tilde{h}_{l\kappa_1\sigma}(r_i) \tilde{j}_{l\kappa_2\sigma}(r_i) \,. \qquad \text{(A.1.5)}$$

Whereas the last integral contains two different kinds of functions the integrals (A.1.3) and (A.1.4) are just extensions of the Hankel and Bessel integrals (A.1.1) and (A.1.2) to different values of κ^2. However, as we will see all these mixed integrals can be treated on the same footing and, in particular, they can all be expressed in terms of the Hankel and Bessel energies $E^{(H)}_{l\kappa i\sigma}$ and $E^{(J)}_{l\kappa i\sigma}$ as well as the Hankel and Bessel integrals $S^{(H)}_{l\kappa i\sigma}$ and $S^{(J)}_{l\kappa i\sigma}$. This is a consequence of the fact that all the augmented functions fulfill the same radial Schrödinger equation the only difference being the particular boundary conditions. Insofar the difference between augmented Hankel or Bessel functions for different values of κ^2 is equivalent to the difference between an augmented Hankel and an augmented Bessel function for the same κ^2.

To be concrete, we turn to the integral (A.1.5) and recall the radial Schrödinger equations (2.1.14) and (2.1.21) for the augmented functions. The evaluation of the integral is then done along the same lines as for the corresponding integrals of the spherical Hankel and Bessel functions as described in the context of (3.3.12) to (3.3.22). Thus we first combine the radial equations (2.1.14) and (2.1.21) to note

$$r_i \tilde{h}_{l\kappa_1\sigma}(r_i) \frac{\partial^2}{\partial r_i^2} r_i \tilde{j}_{l\kappa_2\sigma}(r_i) - r_i \tilde{j}_{l\kappa_2\sigma}(r_i) \frac{\partial^2}{\partial r_i^2} r_i \tilde{h}_{l\kappa_1\sigma}(r_i)$$

$$= (E^{(H)}_{l\kappa_1 i\sigma} - E^{(J)}_{l\kappa_2 i\sigma}) \, r_i^2 \tilde{h}_{l\kappa_1\sigma}(r_i) \tilde{j}_{l\kappa_2\sigma}(r_i) \,. \qquad \text{(A.1.6)}$$

Integrating the left-hand side by parts we get

$$(E_{l\kappa_1 i\sigma}^{(H)} - E_{l\kappa_2 i\sigma}^{(J)}) \int_0^{S_i} dr_i \, r_i^2 \tilde{h}_{l\kappa_1\sigma}(r_i) \tilde{j}_{l\kappa_2\sigma}(r_i)$$

$$= \left[r_i \tilde{h}_{l\kappa_1\sigma}(r_i) \frac{\partial}{\partial r_i} r_i \tilde{j}_{l\kappa_2\sigma}(r_i) - r_i \tilde{j}_{l\kappa_2\sigma}(r_i) \frac{\partial}{\partial r_i} r_i \tilde{h}_{l\kappa_1\sigma}(r_i) \right]_0^{S_i}$$

$$= W\{r_i \tilde{h}_{l\kappa_1\sigma}(r_i), r_i \tilde{j}_{l\kappa_2\sigma}(r_i); r_i\}|_{r_i=S_i} - W\{r_i \tilde{h}_{l\kappa_1\sigma}(r_i), -r_i \tilde{j}_{l\kappa_2\sigma}(r_i); r_i\}|_{r_i=0} \, .$$

$$(A.1.7)$$

Due to the regularity of the augmented functions at the origin the second Wronskian on the right-hand side vanishes and, due to the boundary conditions (2.1.15) and (2.1.22), the first Wronskian can be formulated identically with the corresponding envelope functions. Hence we arrive at the result

$$\langle \tilde{H}_{L\kappa_1\sigma} | \tilde{J}_{L'\kappa_2\sigma} \rangle_i \delta_{LL'}$$

$$= \langle \tilde{J}_{L'\kappa_2\sigma} | \tilde{H}_{L\kappa_1\sigma} \rangle_i \delta_{LL'}$$

$$= \int_0^{S_i} dr_i \, r_i^2 \tilde{h}_{l\kappa_1\sigma}(r_i) \tilde{j}_{l\kappa_2\sigma}(r_i)$$

$$= \frac{1}{E_{l\kappa_1 i\sigma}^{(H)} - E_{l\kappa_2 i\sigma}^{(J)}} W\{r_i \bar{h}_l(\kappa_1 r_i), r_i \bar{j}_l(\kappa_2 r_i); r_i\}|_{r_i=S_i} \, , \qquad (A.1.8)$$

which can be interpreted as the counterpart of (3.3.32) for the augmented functions.

By now, we have reduced the number of quantities characterizing each partial wave to four, namely the Hankel and Bessel energies $E_{l\kappa i\sigma}^{(H)}$ and $E_{l\kappa i\sigma}^{(J)}$ as well as the Hankel and Bessel integrals $S_{l\kappa i\sigma}^{(H)}$ and $S_{l\kappa i\sigma}^{(J)}$.

For completeness we note the identity

$$E_{l\kappa_1 i\sigma}^{(H)} \langle \tilde{J}_{L'\kappa_2\sigma} | \tilde{H}_{L\kappa_1\sigma} \rangle_i \delta_{LL'}$$

$$= E_{l\kappa_1 i\sigma}^{(H)} \langle \tilde{H}_{L\kappa_1\sigma} | \tilde{J}_{L'\kappa_2\sigma} \rangle_i \delta_{LL'}$$

$$= E_{l\kappa_1 i\sigma}^{(H)} \frac{1}{E_{l\kappa_1 i\sigma}^{(H)} - E_{l\kappa_2 i\sigma}^{(J)}} W\{r_i \bar{h}_l(\kappa_1 r_i), r_i \bar{j}_l(\kappa_2 r_i); r_i\}|_{r_i=S_i}$$

$$= \left[E_{l\kappa_2 i\sigma}^{(J)} \frac{1}{E_{l\kappa_1 i\sigma}^{(H)} - E_{l\kappa_2 i\sigma}^{(J)}} + 1 \right] W\{r_i \bar{h}_l(\kappa_1 r_i), r_i \bar{j}_l(\kappa_2 r_i); r_i\}|_{r_i=S_i}$$

$$= E_{l\kappa_2 i\sigma}^{(J)} \langle \tilde{H}_{L\kappa_1\sigma} | \tilde{J}_{L'\kappa_2\sigma} \rangle_i \delta_{LL'} + W\{r_i \bar{h}_l(\kappa_1 r_i), r_i \bar{j}_l(\kappa_2 r_i); r_i\}|_{r_i=S_i}$$

$$= E_{l\kappa_2 i\sigma}^{(J)} \langle \tilde{J}_{L'\kappa_2\sigma} | \tilde{H}_{L\kappa_1\sigma} \rangle_i \delta_{LL'} + W\{r_i \bar{h}_l(\kappa_1 r_i), r_i \bar{j}_l(\kappa_2 r_i); r_i\}|_{r_i=S_i} \, ,$$

$$(A.1.9)$$

which follows directly from (A.1.8) and which will be useful for calculating the Hamiltonian matrix elements in Sect. 2.2.

Of particular importance in this context is the case $\kappa_1^2 = \kappa_2^2 = \kappa^2$, where, according to (3.2.40), the Wronskian entering (A.1.8) and (A.1.9) reduces to unity.

For different values of the energy parameter it might still happen that the Hankel energy $E_{l\kappa_1 i\sigma}^{(H)}$ and the Bessel energy $E_{l\kappa_2 i\sigma}^{(J)}$ corresponding to a different value of the energy parameter are identical. In this case the augmented Hankel function $\tilde{h}_{l\kappa_1\sigma}$ and the augmented Bessel function $\tilde{j}_{l\kappa_2\sigma}$ by construction would also be identical and we were able to note

$$
\int_0^{S_i} dr_i \, r_i^2 \tilde{h}_{l\kappa_1\sigma}(r_i) \tilde{j}_{l\kappa_2\sigma}(r_i)
$$

$$
= \int_0^{S_i} dr_i \, r_i^2 \tilde{h}_{l\kappa_1\sigma}(r_i) \tilde{h}_{l\kappa_1\sigma}(r_i) = S_{l\kappa_1 i\sigma}^{(H)}
$$

$$
= \int_0^{S_i} dr_i \, r_i^2 \tilde{j}_{l\kappa_2\sigma}(r_i) \tilde{j}_{l\kappa_2\sigma}(r_i) = S_{l\kappa_2 i\sigma}^{(J)} \ , \tag{A.1.10}
$$

for $E_{l\kappa_1 i\sigma}^{(H)} = E_{l\kappa_2 i\sigma}^{(J)}$.

The evaluation of the other two mixed integrals, (A.1.3) and (A.1.4), is performed in complete analogy with the previous derivation. Thus we are able to directly note the results

$$
\langle \tilde{H}_{L\kappa_1\sigma} | \tilde{H}_{L'\kappa_2\sigma} \rangle_i \delta_{LL'}
$$

$$
= \langle \tilde{H}_{L'\kappa_2\sigma} | \tilde{H}_{L\kappa_1\sigma} \rangle_i \delta_{LL'}
$$

$$
= \int_0^{S_i} dr_i \, r_i^2 \tilde{h}_{l\kappa_1\sigma}(r_i) \tilde{h}_{l\kappa_2\sigma}(r_i)
$$

$$
= \frac{1}{E_{l\kappa_1 i\sigma}^{(H)} - E_{l\kappa_2 i\sigma}^{(H)}} W\{r_i\bar{h}_l(\kappa_1 r_i), r_i\bar{h}_l(\kappa_2 r_i); r_i\}|_{r_i=S_i} \ , \tag{A.1.11}
$$

and

$$
\langle \tilde{J}_{L\kappa_1\sigma} | \tilde{J}_{L'\kappa_2\sigma} \rangle_i \delta_{LL'}
$$

$$
= \langle \tilde{J}_{L'\kappa_2\sigma} | \tilde{J}_{L\kappa_1\sigma} \rangle_i \delta_{LL'}
$$

$$
= \int_0^{S_i} dr_i \, r_i^2 \tilde{j}_{l\kappa_1\sigma}(r_i) \tilde{j}_{l\kappa_2\sigma}(r_i)
$$

$$
= \frac{1}{E_{l\kappa_1 i\sigma}^{(J)} - E_{l\kappa_2 i\sigma}^{(J)}} W\{r_i\bar{j}_l(\kappa_1 r_i), r_i\bar{j}_l(\kappa_2 r_i); r_i\}|_{r_i=S_i} \ . \tag{A.1.12}
$$

Again, we point to the formal similarity with (3.3.30), which holds for the Bessel envelope functions.

Finally, we write down the identities corresponding to (A.1.9) which read as

$$
E_{l\kappa_1 i\sigma}^{(H)} \langle \tilde{H}_{L\kappa_1\sigma} | \tilde{H}_{L'\kappa_2\sigma} \rangle_i \delta_{LL'}
$$

$$
= E_{l\kappa_1 i\sigma}^{(H)} \langle \tilde{H}_{L'\kappa_2\sigma} | \tilde{H}_{L\kappa_1\sigma} \rangle_i \delta_{LL'}
$$

$$
= E_{l\kappa_1 i\sigma}^{(H)} \frac{1}{E_{l\kappa_1 i\sigma}^{(H)} - E_{l\kappa_2 i\sigma}^{(H)}} W\{r_i\bar{h}_l(\kappa_1 r_i), r_i\bar{h}_l(\kappa_2 r_i); r_i\}|_{r_i=S_i}
$$

$$= \left[E^{(H)}_{l\kappa_2 i\sigma} \frac{1}{E^{(H)}_{l\kappa_1 i\sigma} - E^{(H)}_{l\kappa_2 i\sigma}} + 1 \right] W\{r_i \bar{h}_l(\kappa_1 r_i), r_i \bar{h}_l(\kappa_2 r_i); r_i\}|_{r_i=S_i}$$

$$= E^{(H)}_{l\kappa_2 i\sigma} \langle \tilde{H}_{L\kappa_1\sigma} | \tilde{H}_{L'\kappa_2\sigma} \rangle_i \delta_{LL'} + W\{r_i \bar{h}_l(\kappa_1 r_i), r_i \bar{h}_l(\kappa_2 r_i); r_i\}|_{r_i=S_i}$$

$$= E^{(H)}_{l\kappa_2 i\sigma} \langle \tilde{H}_{L\kappa_2\sigma} | \tilde{H}_{L'\kappa_1\sigma} \rangle_i \delta_{LL'} + W\{r_i \bar{h}_l(\kappa_1 r_i), r_i \bar{h}_l(\kappa_2 r_i); r_i\}|_{r_i=S_i} \ ,$$

$$(\text{A.1.13})$$

and

$$E^{(J)}_{l\kappa_1 i\sigma} \langle \tilde{J}_{L\kappa_1\sigma} | \tilde{J}_{L'\kappa_2\sigma} \rangle_i \delta_{LL'}$$

$$= E^{(J)}_{l\kappa_1 i\sigma} \langle \tilde{J}_{L'\kappa_2\sigma} | \tilde{J}_{L\kappa_1\sigma} \rangle_i \delta_{LL'}$$

$$= E^{(J)}_{l\kappa_1 i\sigma} \frac{1}{E^{(J)}_{l\kappa_1 i\sigma} - E^{(J)}_{l\kappa_2 i\sigma}} W\{r_i \bar{j}_l(\kappa_1 r_i), r_i \bar{j}_l(\kappa_2 r_i); r_i\}|_{r_i=S_i}$$

$$= \left[E^{(J)}_{l\kappa_2 i\sigma} \frac{1}{E^{(J)}_{l\kappa_1 i\sigma} - E^{(J)}_{l\kappa_2 i\sigma}} + 1 \right] W\{r_i \bar{j}_l(\kappa_1 r_i), r_i \bar{j}_l(\kappa_2 r_i); r_i\}|_{r_i=S_i}$$

$$= E^{(J)}_{l\kappa_2 i\sigma} \langle \tilde{J}_{L\kappa_1\sigma} | \tilde{J}_{L'\kappa_2\sigma} \rangle_i \delta_{LL'} + W\{r_i \bar{j}_l(\kappa_1 r_i), r_i \bar{j}_l(\kappa_2 r_i); r_i\}|_{r_i=S_i}$$

$$= E^{(J)}_{l\kappa_2 i\sigma} \langle \tilde{J}_{L\kappa_2\sigma} | \tilde{J}_{L'\kappa_1\sigma} \rangle_i \delta_{LL'} + W\{r_i \bar{j}_l(\kappa_1 r_i), r_i \bar{j}_l(\kappa_2 r_i); r_i\}|_{r_i=S_i} \ .$$

$$(\text{A.1.14})$$

A.2 Moments of the Partial Densities of States

In this section we describe the calculation of the moments of the partial densities of states (DOS). These moments are needed for the subsequent moment analysis and, in addition, the lowest moment is connected to the calculation of the Fermi energy from the total density of states. The moments are defined by (2.3.16) as

$$M^{(k)} = \int_{-\infty}^{E_F} dE \ E^k \rho(E) \ . \tag{A.2.1}$$

Here we have omitted the indices referring to a specific partial density of states; in the following the symbol ρ will refer to the total or any of the partial DOS.

Since the electronic energies are bounded from below it is sufficient to calculate the integral and, hence, the partial densities of states only in a limited energy range between suitably chosen boundaries E_{min} and E_{max}. While the linear tetrahedron method to be described in App. D.5 allows for a direct calculation of the integrals (A.2.1), the sampling methods, which will be dealt with in more detail in Apps. D.3 and D.4, supply the densities of states on a discrete set of points between E_{min} and E_{max}. In the present section, we concentrate on this latter case, which requires a numerical calculation of the above integrals. In doing so, we use an equidistant mesh by dividing the whole energy range into N intervals of length

$$\Delta E = \frac{1}{N}(E_{max} - E_{min}) ,$$ (A.2.2)

and define the mesh points by

$$E_i = E_{min} + i\Delta E, \qquad i = 0, \dots, N .$$ (A.2.3)

The density of states at these points is given by

$$\rho_i = \rho(E_i) .$$ (A.2.4)

Next we approximate the density of states in each interval by its linear interpolation between neighbouring meshpoints,

$$\bar{\rho}_i(E) = \rho_i + \frac{\Delta\rho_i}{\Delta E}(E - E_i) ,$$ (A.2.5)

where

$$\Delta\rho_i = \rho_{i+1} - \rho_i .$$ (A.2.6)

The integral appearing in (A.2.1) may now be written as a sum of integrals over the intervals as

$$M^{(k)} = \sum_{i=0}^{N_F-1} \int_{E_i}^{E_{i+1}} dE \, E^k \bar{\rho}_i(E) + \int_{E_{N_F}}^{E_F} dE \, E^k \bar{\rho}_i(E)$$

$$=: \sum_{i=0}^{N_F-1} \hat{M}_i^{(k)}(E_{i+1}) + \hat{M}_{N_F}^{(k)}(E_F) .$$ (A.2.7)

Here we have denoted the highest meshpoint below the Fermi energy as N_F and defined the integral

$$\hat{M}_i^{(k)}(E') = \int_{E_i}^{E'} dE \, E^k \bar{\rho}(E) .$$ (A.2.8)

Although the integration can be performed in a straightforward manner it is useful to first rewrite the energy power with the help of Binomi's formula as

$$E^k = (E - E_i + E_i)^k = \sum_{m=0}^{k} \binom{k}{m}(E - E_i)^{k-m} E_i^m .$$ (A.2.9)

Using this as well as the linear approximation (A.2.5) to the density of states within each interval we get for the integral (A.2.8)

$$\hat{M}_i^{(k)}(E') = \sum_{m=0}^{k} \binom{k}{m} E_i^m \int_{E_i}^{E'} dE \, (E - E_i)^{k-m} \left[\rho_i + \frac{\Delta\rho_i}{\Delta E}(E - E_i) \right]$$

$$= \sum_{m=0}^{k} \binom{k}{m} E_i^m \left[\frac{1}{k-m+1}\rho_i(E' - E_i)^{k-m+1} \right.$$

$$+\frac{1}{k-m+2}\frac{\Delta\rho_i}{\Delta E}(E'-E_i)^{k-m+2}\Bigg]$$

$$=\sum_{m=0}^{k}\binom{k}{m}E_i^m(E'-E_i)^{k-m+1}\frac{1}{(k-m+1)(k-m+2)}$$

$$\Bigg[(k-m+2)\rho_i+(k-m+1)\Delta\rho_i\frac{E'-E_i}{\Delta E}\Bigg]\,.$$

$$(A.2.10)$$

For all the integrals extending over a full intervall the previous result reduces to

$$\hat{M}_i^{(k)}(E_{i+1})=\sum_{m=0}^{k}\binom{k}{m}E_i^m(\Delta E)^{k-m+1}\frac{[\rho_i+(k-m+1)\rho_{i+1}]}{(k-m+1)(k-m+2)}\,.$$

$$(A.2.11)$$

In contrast, for the last interval, which contains the Fermi energy, the final result reads as

$$\hat{M}_{N_F}^{(k)}(E_F)=\sum_{m=0}^{k}\binom{k}{m}E_{N_F}^m(E_F-E_{N_F})^{k-m+1}\frac{1}{(k-m+1)(k-m+2)}$$

$$\Bigg[(k-m+2)\rho_{N_F}+(k-m+1)\Delta\rho_{N_F}\frac{E_F-E_{N_F}}{\Delta E}\Bigg]\,.$$

$$(A.2.12)$$

Finally, by combining (A.2.7), (A.2.11) and (A.2.12) we note, in particular, the results for $k=0$

$$M^{(0)}=\frac{1}{2}\Delta E\sum_{i=0}^{N_F-1}(\rho_i+\rho_{i+1})+(E_F-E_{N_F})\Bigg[\rho_{N_F}+\frac{1}{2}\Delta\rho_{N_F}\frac{E_F-E_{N_F}}{\Delta E}\Bigg]\,,$$

$$(A.2.13)$$

and $k=1$

$$M^{(1)}=\frac{1}{2}\Delta E\sum_{i=0}^{N_F-1}E_i(\rho_i+\rho_{i+1})$$

$$+E_{N_F}(E_F-E_{N_F})\Bigg[\rho_{N_F}+\frac{1}{2}\Delta\rho_{N_F}\frac{E_F-E_{N_F}}{\Delta E}\Bigg]$$

$$+\frac{1}{6}(\Delta E)^2\sum_{i=0}^{N_F-1}(\rho_i+2\rho_{i+1})$$

$$+(E_F-E_{N_F})^2\Bigg[\frac{1}{2}\rho_{N_F}+\frac{1}{3}\Delta\rho_{N_F}\frac{E_F-E_{N_F}}{\Delta E}\Bigg]\,,\quad(A.2.14)$$

which denote the total valence electron number Q_{tot} and the total band energy, respectively.

A.3 Determination of the Fermi Energy

Having outlined the practical calculation of the moments of the partial densities of states we turn to the evaluation of the Fermi energy. It is fixed by the condition that the zeroth moment of the total density of states as given by (A.2.13) equals the total charge Q_{tot} of the system. As in Sect. A.2 we concentrate on the situation, where the densities of states have been calculated with the help of one of the sampling methods. In the formulation (A.2.13) we assumed that the highest meshpoint N_F below the Fermi energy has already been identified. Hence, the Fermi energy must be somewhere in the interval between the meshpoints N_F and $N_F + 1$. It is useful to define the charge left for this interval by

$$Q'_{tot} = Q_{tot} - \frac{1}{2}\Delta E \sum_{i=0}^{N_F-1} (\rho_i + \rho_{i+1}) . \tag{A.3.1}$$

With this definition we get from (A.2.13)

$$Q'_{tot} = \rho_{N_F}(E_F - E_{N_F}) + \frac{1}{2}\frac{\Delta\rho_{N_F}}{\Delta E}(E_F - E_{N_F})^2 . \tag{A.3.2}$$

Thus, within the linear approximation for the density of states inside each energy interval as proposed in App. A.2, the Fermi energy arises as the solution of a quadratic equation. Yet, care has to be taken in choosing the correct solution. In order to sidestep any difficulty we rewrite (A.3.2) as

$$\left(\frac{\Delta\rho_{N_F}}{\Delta E}\right)^2 (E_F - E_{N_F})^2 + 2\rho_{N_F}\frac{\Delta\rho_{N_F}}{\Delta E}(E_F - E_{N_F}) - 2Q'_{tot}\frac{\Delta\rho_{N_F}}{\Delta E} = 0 . \tag{A.3.3}$$

From this we get the solutions

$$\frac{\Delta\rho_{N_F}}{\Delta E}(E_F - E_{N_F}) = -\rho_{N_F} \pm \sqrt{\rho_{N_F}^2 + 2Q'_{tot}\frac{\Delta\rho_{N_F}}{\Delta E}} . \tag{A.3.4}$$

In order to make a choice between these two solutions we combine (A.3.4) with the expression (A.2.5) for the density of states at the Fermi energy

$$\bar{\rho}_{N_F}(E_F) = \rho_{N_F} + \frac{\Delta\rho_{N_F}}{\Delta E}(E_F - E_{N_F})$$

$$= \pm\sqrt{\rho_{N_F}^2 + 2Q'_{tot}\frac{\Delta\rho_{N_F}}{\Delta E}} . \tag{A.3.5}$$

Since the density of states must be a positive number we can rule out the solution with the negative sign and arrive at a unique solution for the Fermi energy

$$\frac{E_F - E_{N_F}}{\Delta E} = \frac{\bar{\rho}_{N_F}(E_F) - \rho_{N_F}}{\Delta\rho_{N_F}} . \tag{A.3.6}$$

Still, our previous considerations hold only for zero temperature, where the Fermi-Dirac function is identical to the step function. In contrast, for finite temperatures the chemical potential μ, which turns into the Fermi energy for $T = 0$, arises from solving the equation

$$Q_{tot} = \int_{-\infty}^{+\infty} dE \, \rho(E) \frac{1}{e^{\beta(E-\mu)} + 1} , \tag{A.3.7}$$

where $\beta = 1/k_B T$. Finite temperatures are of special interest for semiconductors, where they lead to the excitation of charge carriers across the optical band gap. In the following we adopt the usual strategy to determine the number of excited carriers from the electronic structure calculated at zero temperature. Concentrating on nondegenerate semiconductors, which are defined by the conditions [1, Chap. 28]

$$E_c - \mu \gg k_B T \quad \text{and} \quad \mu - E_v \gg k_B T$$

for the chemical potential, the valence band maximum E_v and the conduction band minimum E_c, we approximate the Fermi function by [1, Chap. 28]

$$\frac{1}{e^{\beta(E-\mu)} + 1} = 1 - \frac{1}{e^{-\beta(E-\mu)} + 1} \approx 1 - e^{\beta(E-\mu)} \quad \text{for } E < E_v , \tag{A.3.8}$$

$$\frac{1}{e^{\beta(E-\mu)} + 1} \approx e^{-\beta(E-\mu)} \quad \text{for } E > E_c , \tag{A.3.9}$$

and rewrite the integral (A.3.7) as

$$\begin{aligned} Q_{tot} &= \int_{-\infty}^{E_v} dE \, \rho(E)(1 - e^{\beta(E-\mu)}) + \int_{E_c}^{+\infty} dE \, \rho(E) e^{-\beta(E-\mu)} \\ &= Q_{tot}(T = 0) - e^{-\beta\mu} \int_{-\infty}^{E_v} dE \, \rho(E) e^{\beta E} + e^{\beta\mu} \int_{E_c}^{+\infty} dE \, \rho(E) e^{-\beta E} \\ &=: Q_{tot}(T = 0) - Q_p(T) + Q_n(T) . \end{aligned} \tag{A.3.10}$$

In the last line we have defined the number of excited holes in the valence band, $Q_p(T)$, as well as the number of electrons in the conduction band, $Q_n(T)$. Furthermore, we have extracted the exponential with the chemical potential from the integrals. Next, we apply again the linear approximation to the density of states within each energy interval and combine (A.3.10) with (A.2.5)

$$\begin{aligned} Q_p(T) &= e^{-\beta\mu} \sum_{i=0}^{N_v} \int_{E_i}^{E_{i+1}} dE \, \bar{\rho}_i(E) \, e^{\beta E} \\ &= e^{-\beta\mu} \sum_{i=0}^{N_v} \left[\left(\rho_i - \frac{\Delta\rho_i}{\Delta E} E_i\right) \int_{E_i}^{E_{i+1}} dE \, e^{\beta E} + \frac{\Delta\rho_i}{\Delta E} \int_{E_i}^{E_{i+1}} dE \, E \, e^{\beta E} \right] , \end{aligned} \tag{A.3.11}$$

$$Q_n(T) = e^{\beta\mu} \sum_{i=N_c}^{N-1} \int_{E_i}^{E_{i+1}} dE \, \bar{\rho}_i(E) \, e^{-\beta E}$$

$$= e^{\beta\mu} \sum_{i=N_c}^{N-1} \left[(\rho_i - \frac{\Delta\rho_i}{\Delta E} E_i) \int_{E_i}^{E_{i+1}} dE\, e^{-\beta E} + \frac{\Delta\rho_i}{\Delta E} \int_{E_i}^{E_{i+1}} dE\, E\, e^{-\beta E} \right] .$$

$$(A.3.12)$$

N_v and N_c denote the meshpoints just below the valence band maximum E_v and the conduction band minimum E_c, respectively. We have assumed that E_v, μ, and E_c all are in different energy intervals, which is reasonable for nondegenerate semiconductors and the fine mesh usually chosen for sampling the density of states; for standard calculations we recommend $\Delta E = 1\,\mathrm{mRyd}$. Evaluating the integrals we obtain

$$Q_p(T) = e^{-\beta\mu} \sum_{i=0}^{N_v} \left[\frac{1}{\beta}(\rho_i - \frac{\Delta\rho_i}{\Delta E} E_i)e^{\beta E} + \frac{1}{\beta}\frac{\Delta\rho_i}{\Delta E} E e^{\beta E} - \frac{1}{\beta^2}\frac{\Delta\rho_i}{\Delta E} e^{\beta E} \right]_{E_i}^{E_{i+1}}$$

$$= \frac{1}{\beta} e^{-\beta\mu} \sum_{i=0}^{N_v} \left[\bar{\rho}_i(E)e^{\beta E} \right]_{E_i}^{E_{i+1}} - \frac{1}{\beta^2} e^{-\beta\mu} \sum_{i=0}^{N_v} \frac{\Delta\rho_i}{\Delta E} \left[e^{\beta E} \right]_{E_i}^{E_{i+1}} ,$$

$$(A.3.13)$$

$$Q_n(T) = -e^{\beta\mu} \sum_{i=N_c}^{N-1} \left[\frac{1}{\beta}(\rho_i - \frac{\Delta\rho_i}{\Delta E} E_i)e^{-\beta E} + \frac{1}{\beta}\frac{\Delta\rho_i}{\Delta E} E e^{-\beta E} + \frac{1}{\beta^2}\frac{\Delta\rho_i}{\Delta E} e^{-\beta E} \right]_{E_i}^{E_{i+1}}$$

$$= -\frac{1}{\beta} e^{\beta\mu} \sum_{i=N_c}^{N-1} \left[\bar{\rho}_i(E)e^{-\beta E} \right]_{E_i}^{E_{i+1}} - \frac{1}{\beta^2} e^{\beta\mu} \sum_{i=N_c}^{N-1} \frac{\Delta\rho_i}{\Delta E} \left[e^{-\beta E} \right]_{E_i}^{E_{i+1}} ,$$

$$(A.3.14)$$

where we have used the definition (A.2.5) of the interpolated density of states in the last step. In (A.3.12) the sums containing $\bar{\rho}_i(E)$ cancel because successive terms add $\bar{\rho}_i(E_i)e^{-\beta E_i}$ with positive and negative sign and the density of states vanishes for $i = 0$ and $i = N$ as well as within the gap. We are thus left with the result

$$Q_p(T) = -\frac{1}{\beta^2} e^{-\beta(\mu-E_v)} \sum_{i=0}^{N_v} \frac{\Delta\rho_i}{\Delta E} \left[e^{\beta(E-E_v)} \right]_{E_i}^{E_{i+1}}$$

$$= -\frac{1}{\beta^2} e^{-\beta(\mu-E_v)} \left(e^{\beta\Delta E} - 1 \right) \sum_{i=0}^{N_v} \frac{\Delta\rho_i}{\Delta E} e^{\beta(E_i-E_v)}$$

$$= -\frac{1}{\beta^2} e^{-\beta(\mu-E_v)} \left(e^{\beta\Delta E} - 1 \right) \frac{1}{\Delta E}$$

$$\left[\sum_{i=0}^{N_v-1} \Delta\rho_i e^{\beta(E_i-E_v)} - \frac{\Delta E}{E_v - E_{N_v}} \rho_{N_v} \frac{1 - e^{\beta(E_{N_v}-E_v)}}{e^{\beta\Delta E} - 1} \right] , \qquad (A.3.15)$$

$$Q_n(T) = -\frac{1}{\beta^2} e^{\beta(\mu-E_c)} \sum_{i=N_c}^{N-1} \frac{\Delta\rho_i}{\Delta E} \left[e^{-\beta(E-E_c)} \right]_{E_i}^{E_{i+1}}$$

$$= -\frac{1}{\beta^2}e^{\beta(\mu-E_c)}\left(e^{-\beta\Delta E}-1\right)\sum_{i=N_c}^{N-1}\frac{\Delta\rho_i}{\Delta E}e^{-\beta(E_i-E_c)}$$

$$= -\frac{1}{\beta^2}e^{\beta(\mu-E_c)}\left(e^{-\beta\Delta E}-1\right)\frac{1}{\Delta E}$$

$$\left[\frac{\Delta E}{E_{N_c+1}-E_c}\rho_{N_c+1}\frac{e^{-\beta(E_{N_c+1}-E_c)}-1}{e^{-\beta\Delta E}-1}+\sum_{i=N_c+1}^{N-1}\Delta\rho_i e^{-\beta(E_i-E_c)}\right] ,$$

$$(A.3.16)$$

respectively, where we have extracted an additional exponential containing the band edge, such that the exponents appearing in the sums vanish at these points. As a consequence, the temperature dependence of the integrals is rather small, which fact facilitates their numerical evaluation. In addition, we have in the respective last steps treated the intervals containing the valence band maximum and the conduction band minimum separately in order to avoid numerical inaccuracies. In the case of an intrinsic semiconductor the number of electrons and holes must be identical and we arrive at the following condition for the chemical potential

$$e^{2\beta\left[\mu-\frac{1}{2}(E_c+E_v)\right]}$$

$$= -e^{\beta\Delta E}\frac{\sum_{i=0}^{N_v}\Delta\rho_i e^{\beta(E_i-E_v)}}{\sum_{i=N_c}^{N-1}\Delta\rho_i e^{-\beta(E_i-E_c)}}$$

$$= -e^{\beta\Delta E}\frac{\left[\sum_{i=0}^{N_v-1}\Delta\rho_i e^{\beta(E_i-E_v)}-\frac{\Delta E}{E_v-E_{N_v}}\rho_{N_v}\frac{1-e^{\beta(E_{N_v}-E_v)}}{e^{\beta\Delta E}-1}\right]}{\left[\frac{\Delta E}{E_{N_c+1}-E_c}\rho_{N_c+1}\frac{e^{-\beta(E_{N_c+1}-E_c)}-1}{e^{-\beta\Delta E}-1}+\sum_{i=N_c+1}^{N-1}\Delta\rho_i e^{-\beta(E_i-E_c)}\right]} ,$$

$$(A.3.17)$$

where again we have singled out the terms containing the band edges in the second step.

In case we do not want to work with the density of states but prefer to use the band energies directly as in the context of the linear tetrahedron method, we start from the definitions of $Q_p(T)$ and $Q_n(T)$ underlying (A.3.10) and write

$$Q_p(T) = e^{-\beta\mu}\int_{-\infty}^{E_v}dE\,\frac{1}{\Omega_{BZ}}\sum_n\int_{\Omega_{BZ}}d^3\mathbf{k}\,\delta(E-\varepsilon_n(\mathbf{k}))e^{\beta E}$$

$$= \frac{1}{\Omega_{BZ}}e^{-\beta\mu}\sum_n\int_{\Omega_{BZ}}d^3\mathbf{k}\,\Theta(E_v-\varepsilon_n(\mathbf{k}))e^{\beta\varepsilon_n(\mathbf{k})}$$

$$= \frac{1}{\Omega_{BZ}}e^{-\beta(\mu-E_v)}\sum_n\int_{\Omega_{BZ}}d^3\mathbf{k}\,\Theta(E_v-\varepsilon_n(\mathbf{k}))e^{\beta(\varepsilon_n(\mathbf{k})-E_v)} ,\quad (A.3.18)$$

$$Q_n(T) = e^{\beta\mu}\int_{E_c}^{+\infty}dE\,\frac{1}{\Omega_{BZ}}\sum_n\int_{\Omega_{BZ}}d^3\mathbf{k}\,\delta(E-\varepsilon_n(\mathbf{k}))e^{-\beta E}$$

$$= \frac{1}{\Omega_{BZ}}e^{\beta\mu}\sum_n\int_{\Omega_{BZ}}d^3\mathbf{k}\,\Theta(\varepsilon_n(\mathbf{k})-E_c)e^{-\beta\varepsilon_n(\mathbf{k})}$$

$$= \frac{1}{\Omega_{BZ}} e^{\beta(\mu - E_c)} \sum_n \int_{\Omega_{BZ}} d^3k\, \Theta(\varepsilon_n(\mathbf{k}) - E_c) e^{-\beta(\varepsilon_n(\mathbf{k}) - E_c)} . \qquad (A.3.19)$$

Here we have inserted the standard definition of the density of states; n labels the band index. Again using the fact that for an intrinsic semiconductor the number of electrons and holes must be identical we obtain for the chemical potential

$$e^{2\beta[\mu - \frac{1}{2}(E_c + E_v)]} = \frac{\sum_n \int_{\Omega_{BZ}} d^3k\, \Theta(E_v - \varepsilon_n(\mathbf{k})) e^{\beta(\varepsilon_n(\mathbf{k}) - E_v)}}{\sum_n \int_{\Omega_{BZ}} d^3k\, \Theta(\varepsilon_n(\mathbf{k}) - E_c) e^{-\beta(\varepsilon_n(\mathbf{k}) - E_c)}} . \qquad (A.3.20)$$

To summarize, with the formulas (A.3.15)/(A.3.16) or (A.3.18)/(A.3.19) at hand, we may for a given temperature first calculate the chemical potential and then the number of excited carriers in the valence and conduction band.

A.4 Moment Analysis of the Partial Densities of States

With the moments $M^{(k)}$, $(k = 0, \ldots, 3)$ of the partial densities of states at hand we are in a position to perform the moment analysis. It aims at calculating two energies $E^{(\alpha)}$ and weights $Q^{(\alpha)}$, such that the auxiliary density of states

$$\tilde{\rho}(E) := \sum_{\alpha=1}^{2} \delta(E - E^{(\alpha)}) Q^{(\alpha)}, \qquad (A.4.1)$$

has the same first four moments as the true partial density of states. To be specific, we combine (2.3.16) and (2.3.18) and obtain

$$M^{(0)} = Q^{(1)} + Q^{(2)} , \qquad (A.4.2)$$
$$M^{(1)} = E^{(1)} Q^{(1)} + E^{(2)} Q^{(2)} , \qquad (A.4.3)$$
$$M^{(2)} = (E^{(1)})^2 Q^{(1)} + (E^{(2)})^2 Q^{(2)} , \qquad (A.4.4)$$
$$M^{(3)} = (E^{(1)})^3 Q^{(1)} + (E^{(2)})^3 Q^{(2)} . \qquad (A.4.5)$$

To solve this set of equations we first rewrite (A.4.4) and (A.4.5) with the help of the first pair of equations as

$$(E^{(1)})^2 Q^{(1)} + (E^{(2)})^2 Q^{(2)} = E^{(1)} \left(M^{(1)} - E^{(2)} Q^{(2)} \right)$$
$$= +E^{(2)} \left(M^{(1)} - E^{(1)} Q^{(1)} \right)$$
$$= \left(E^{(1)} + E^{(2)} \right) M^{(1)} - E^{(1)} E^{(2)} M^{(0)}$$
$$\overset{!}{=} M^{(2)} , \qquad (A.4.6)$$
$$(E^{(1)})^3 Q^{(1)} + (E^{(2)})^3 Q^{(2)} = E^{(1)} \left(M^{(2)} - (E^{(2)})^2 Q^{(2)} \right)$$
$$= +E^{(2)} \left(M^{(2)} - (E^{(1)})^2 Q^{(1)} \right)$$

$$= \left(E^{(1)} + E^{(2)} \right) M^{(2)} - E^{(1)} E^{(2)} M^{(1)}$$

$$\stackrel{!}{=} M^{(3)} . \tag{A.4.7}$$

From these two equations we get immediately the result

$$E^{(1)} + E^{(2)} = \frac{1}{D_E} \left(M^{(1)} M^{(2)} - M^{(0)} M^{(3)} \right) =: 2E_0 , \tag{A.4.8}$$

$$E^{(1)} E^{(2)} = \frac{1}{D_E} \left((M^{(2)})^2 - M^{(1)} M^{(3)} \right) =: E_p^2 , \tag{A.4.9}$$

where we have abbreviated the coefficient determinant by

$$D_E = (M^{(1)})^2 - M^{(0)} M^{(2)} . \tag{A.4.10}$$

We are now able to calculate the energies $E^{(1)}$ and $E^{(2)}$ by using the identity

$$\Delta^2 := \left(E^{(1)} - E^{(2)} \right)^2 = \left(E^{(1)} + E^{(2)} \right)^2 - 4 E^{(1)} E^{(2)} = 4 \left(E_0^2 - E_p^2 \right) , \tag{A.4.11}$$

hence,

$$\Delta =: + \left(E^{(2)} - E^{(1)} \right) . \tag{A.4.12}$$

Combining (A.4.8) and (A.4.11) we get the result

$$E^{(1)} = E_0 - \sqrt{E_0^2 - E_p^2} \quad \text{and} \quad E^{(2)} = E_0 + \sqrt{E_0^2 - E_p^2} , \tag{A.4.13}$$

where, by labelling the solutions, we strictly enforced the condition $E^{(1)} \leq E^{(2)}$, i.e. $\Delta \geq 0$.

In order to evaluate the weights we solve the set of (A.4.2) and (A.4.3) and arrive at

$$Q^{(1)} = \frac{1}{\Delta} \left(E^{(2)} M^{(0)} - M^{(1)} \right) , \tag{A.4.14}$$

$$Q^{(2)} = \frac{1}{\Delta} \left(M^{(1)} - E^{(1)} M^{(0)} \right) . \tag{A.4.15}$$

Still we have to discuss those situations, where one or both of D_E and Δ vanishes. Of course, Δ vanishes for $E^{(1)} = E^{(2)}$. For D_E we rewrite (A.4.10) with the help of (A.4.2) to (A.4.4) as

$$D_E = - \left(E^{(2)} - E^{(1)} \right)^2 Q^{(1)} Q^{(2)} = -\Delta^2 Q^{(1)} Q^{(2)} . \tag{A.4.16}$$

Hence, both quantities vanish at the same time. However, in this case the solution of the problem is rather obvious from (A.4.2) to (A.4.5) and we have the trivial result

$$Q^{(1)} = Q^{(2)} = \frac{M^{(0)}}{2} \, , \tag{A.4.17}$$

$$E^{(1)} = E^{(2)} = E_0 = \frac{M^{(1)}}{M^{(0)}} = \frac{M^{(2)}}{M^{(1)}} = \frac{M^{(3)}}{M^{(2)}} \, . \tag{A.4.18}$$

However, if $M^{(0)} = M^{(1)} = M^{(2)} = 0$ we make the choice

$$E^{(1)} = E^{(2)} = E_0 = 0 \, . \tag{A.4.19}$$

Nevertheless, in order to avoid numerical instabilities, we will at any rate enforce a finite energy splitting of at least 2×10^{-6} Ryd.

To sum up, the calculations of the energies and weights proceeds along the following lines: First we evaluate D_E from (A.4.10). In case it vanishes we fix E_0 according to (A.4.18)/(A.4.19) and set $\frac{\Delta^2}{4}$ to 10^{-12}. In contrast for $D_E \neq 0$ the quantities E_0 and $\frac{\Delta^2}{4} = (E_0^2 - E_p^2)$ are calculated from (A.4.8)/(A.4.9). Again, we require $\frac{\Delta^2}{4}$ to be at least 10^{-12}. With these prepositions we are eventually able to calculate the energies and weights from (A.4.14) to (A.4.16).

A.5 Intraatomic Radial Mesh

As widely discussed in Chaps. 2 and 4, inside the atomic spheres all quantities of interest are represented by spherical-harmonics expansions, which, for the standard ASW method, includes only the $l = 0$-term. The calculation of the respectice radial functions is usually performed on an exponential radial mesh given by

$$r_i = B \left(e^{A(i-1)} - 1 \right) \qquad \text{for } i = 1, \ldots, N \, . \tag{A.5.1}$$

This mesh is fully characterized by the parameters A and B as well as the total number of mesh points. We use 500 to 1000 points depending on the size of the atomic sphere and the atomic number, while the parameter A usually is set to 0.02. From this the parameter B is uniquely determined by the condition that the outermost mesh point is identical to the radius of the atomic sphere, $S = r_N$ or the muffin-tin sphere, $s = r_N$. The so specified intraatomic radial mesh is used in all ASW methods.

In order to facilitate the evaluation of integrals or the solution of differential equations on the radial mesh we define, in general, the function

$$r(x) = B \left(e^{Ax} - 1 \right) \qquad \text{for } 0 \leq x \leq S \, . \tag{A.5.2}$$

From this we have

$$\frac{dr}{dx} = BAe^{Ax} = A \left(r(x) + B \right) \, , \tag{A.5.3}$$

and

$$x(r) = \frac{1}{A} \ln \left(1 + \frac{r}{B} \right) \, . \tag{A.5.4}$$

We thus obtain, e.g. for the radial integral of a function $f(r)$ the expression

$$\int_0^S dr\, r^2 f(r) = \int_0^{x(S)} dx\, \frac{dr}{dx} r^2(x) f(r(x)) \;, \tag{A.5.5}$$

which is accessible to standard numerical integration techniques. Using, in particular, Simpson's rule for single intervals we arrive at

$$\int_0^{x(S)} dx\, \frac{dr}{dx} r^2(x) f(r(x))$$

$$= \frac{1}{3} \sum_{i=1}^{(N-1)/2} \left[f_{2i-1} r_{2i-1}^2 \frac{dr}{dx}\bigg|_{x=2i-2} + 4 f_{2i} r_{2i}^2 \frac{dr}{dx}\bigg|_{x=2i-1} + f_{2i+1} r_{2i+1}^2 \frac{dr}{dx}\bigg|_{x=2i} \right].$$

$$\tag{A.5.6}$$

Another example arises from second order differential equations as, e.g. Poisson's or Schrödinger's equation. In this case we define

$$rf(r) = \sqrt{\frac{dr}{dx}} F(x) \;, \tag{A.5.7}$$

which gives rise to

$$\frac{d}{dr} rf(r) = \frac{dx}{dr} \frac{d}{dx} \sqrt{\frac{dr}{dx}} F(x)$$

$$= \frac{dx}{dr} \frac{1}{2} \sqrt{\frac{dx}{dr}} \left(\frac{d}{dx} \frac{dr}{dx} \right) F(x) + \frac{dx}{dr} \sqrt{\frac{dr}{dx}} F'(x)$$

$$= \sqrt{\frac{dx}{dr}} \frac{1}{2} A F(x) + \sqrt{\frac{dx}{dr}} F'(x) \;, \tag{A.5.8}$$

and

$$\frac{d^2}{dr^2} rf(r) = \frac{dx}{dr} \frac{d}{dx} \left[\sqrt{\frac{dx}{dr}} \frac{1}{2} A F(x) + \sqrt{\frac{dx}{dr}} F'(x) \right]$$

$$= \frac{dx}{dr} \left[-\frac{1}{2} \left(\frac{dr}{dx} \right)^{-\frac{3}{2}} \left(\frac{d}{dx} \frac{dr}{dx} \right) \right] \left(\frac{1}{2} A F(x) + F'(x) \right)$$

$$+ \frac{dx}{dr} \sqrt{\frac{dx}{dr}} \left(\frac{1}{2} A F'(x) + F''(x) \right)$$

$$= \left(-\frac{1}{2} A \left(\frac{dr}{dx} \right)^{-\frac{3}{2}} \right) \left(\frac{1}{2} A F(x) + F'(x) \right)$$

$$+ \left(\frac{dr}{dx} \right)^{-\frac{3}{2}} \left(\frac{1}{2} A F'(x) + F''(x) \right)$$

$$= \left(\frac{dr}{dx} \right)^{-\frac{3}{2}} \left[F''(x) - \frac{A^2}{4} F(x) \right] \;. \tag{A.5.9}$$

Note that (A.5.8) and (A.5.9) do not contain the first derivative of the function $F(x)$. The radial Schrödinger equation

$$\left[-\frac{d^2}{dr^2} + \frac{l(l+1)}{r^2} + v(r) - E \right] r f(r) = 0, \tag{A.5.10}$$

may be thus cast into the form

$$F''(x) = \left[\left(\frac{dr}{dx} \right)^2 \left(\frac{l(l+1)}{r^2} + v(r) - E \right) + \frac{A^2}{4} \right] F(x) . \tag{A.5.11}$$

A.6 Solving Poisson's Equation Inside the Atomic Spheres

The charge density inside an atomic sphere gives rise to an electrostatic potential

$$v_{es}(\mathbf{r}) := 2 \int_\Omega d^3\mathbf{r}' \, \frac{\rho(\mathbf{r}')}{|\mathbf{r} - \mathbf{r}'|} , \tag{A.6.1}$$

where the integral extends over the atomic sphere region. Of course, this expression is equivalent to Poisson's equation

$$\Delta v_{es}(\mathbf{r}) = -8\pi\rho(\mathbf{r}) . \tag{A.6.2}$$

In general, both the intraatomic electron density and the electrostatic potential can be represented by their spherical-harmonics expansions

$$\rho(\mathbf{r}) = \sum_L \rho_L(r) Y_L(\hat{\mathbf{r}}), \tag{A.6.3}$$

and

$$v_{es}(\mathbf{r}) = \sum_L v_{es,L}(r) Y_L(\hat{\mathbf{r}}) , \tag{A.6.4}$$

where the coefficients are determined by

$$\rho_L(r) = \int d\Omega \, \rho(\mathbf{r}) Y_L^*(\hat{\mathbf{r}}) , \tag{A.6.5}$$

and

$$v_{es,L}(r) = \int d\Omega \, v_{es}(\mathbf{r}) Y_L^*(\hat{\mathbf{r}}) . \tag{A.6.6}$$

For the standard ASW method the previous expressions reduce to the $l = 0$-terms. Next, using the identity (2.4.3)

$$\frac{1}{|\mathbf{r} - \mathbf{r}'|} = \sum_L \frac{4\pi}{2l+1} \frac{r_<^l}{r_>^{l+1}} Y_L(\hat{\mathbf{r}}) Y_L^*(\hat{\mathbf{r}}') , \tag{A.6.7}$$

we rewrite (A.6.1) as

$$v_{es}(\mathbf{r}) = 2 \int d\Omega' \int_0^S dr' r'^2 \sum_L \rho_L(r') Y_L(\hat{\mathbf{r}}')$$

$$\sum_{L'} \frac{4\pi}{2l'+1} \frac{r_<^{l'}}{r_>^{l'+1}} Y_{L'}^*(\hat{\mathbf{r}}') Y_{L'}(\hat{\mathbf{r}})$$

$$= \sum_L \frac{8\pi}{2l+1} \int_0^S dr' r'^2 \frac{r_<^l}{r_>^{l+1}} \rho_L(r') Y_L(\hat{\mathbf{r}}) . \tag{A.6.8}$$

Here we have used the orthonormality of the spherical harmonics. From (A.6.8) we obtain

$$v_{es,L}(r) = \frac{8\pi}{2l+1} \left[\frac{1}{r^{l+1}} \int_0^r dr'\, r'^{l+2} \rho_L(r') + r^l \int_r^S dr'\, r'^{1-l} \rho_L(r') \right] . \tag{A.6.9}$$

In particular, we get for the potential at the origin

$$v_{es,L}(r = 0) = 8\pi \delta_{l0} \int_0^S dr'\, r' \rho_0(r') , \tag{A.6.10}$$

and at the sphere boundary

$$v_{es,L}(r = S) = \frac{8\pi}{2l+1} \frac{1}{S^{l+1}} \int_0^S dr'\, r'^{l+2} \rho_L(r') . \tag{A.6.11}$$

In order to solve the radial problem (A.6.9), we turn back to Poisson's equation (A.6.2) and combine it with the representation of the Laplace operator in spherical coordinates,

$$\Delta = \frac{1}{r} \frac{\partial^2}{\partial r^2} r - \frac{\mathbf{L}^2}{r^2} , \tag{A.6.12}$$

to

$$\left[\frac{1}{r} \frac{d^2}{dr^2} r - \frac{l(l+1)}{r^2} \right] v_{es,L}(r) = -8\pi \rho_L(r) , \tag{A.6.13}$$

or, equivalently,

$$\left[\frac{d^2}{dr^2} - \frac{l(l+1)}{r^2} \right] r v_{es,L}(r) + 8\pi r \rho_L(r) = 0 . \tag{A.6.14}$$

In the rest of this section we will prepare this differential equation for its numerical integration. In doing so, we will to large parts follow the outline given by Loucks [2]. To be specific, we start using the approach (A.5.7),

$$r v_{es,L}(r) = \sqrt{\frac{dr}{dx}} W(x) , \tag{A.6.15}$$

where r and x are related by definition (A.5.2). This leads to the identity (A.5.9) for the second derivative,

$$\frac{d^2}{dr^2} r v_{es,L}(r) = \left(\frac{dr}{dx}\right)^{-\frac{3}{2}} \left[W''(x) - \frac{A^2}{4}W(x)\right] . \tag{A.6.16}$$

Complementing (A.6.15) by the definition

$$8\pi \left(\frac{dr}{dx}\right)^{\frac{3}{2}} r\rho_L(r) = F(x) , \tag{A.6.17}$$

we cast the differential (A.6.14) into the form

$$W''(x) - \tilde{A}(x)W(x) + F(x) = 0 , \tag{A.6.18}$$

where we have used the abbreviation

$$\tilde{A}(x) = \frac{A^2}{4} + \frac{l(l+1)}{r^2}\left(\frac{dr}{dx}\right)^2 . \tag{A.6.19}$$

In order to discretize this equation we start from Taylor's expansion of the function $W(x)$ about a point x

$$W(x \pm \Delta) = W(x) \pm \Delta W'(x) + \frac{\Delta^2}{2!}W''(x) \pm \frac{\Delta^3}{3!}W'''(x) + \frac{\Delta^4}{4!}W^{(iv)}(x) + \dots . \tag{A.6.20}$$

Adding both series we arrive at

$$W(x + \Delta) + W(x - \Delta) - 2W(x) = \Delta^2 W''(x) + \frac{\Delta^4}{12}W^{(iv)}(x) + \dots . \tag{A.6.21}$$

Taking the second derivative of this latter expression and neglecting terms of order 6 and higher we obtain

$$W''(x + \Delta) + W''(x - \Delta) - 2W''(x) = \Delta^2 W^{(iv)}(x) . \tag{A.6.22}$$

Finally, we combine these last two identities and obtain the result

$$\frac{\Delta^2}{12}\left(W''(x + \Delta) + W''(x - \Delta) - 2W''(x)\right) + \Delta^2 W''(x)$$
$$= W(x + \Delta) + W(x - \Delta) - 2W(x) . \tag{A.6.23}$$

Next we insert into this latter result the differential (A.6.18) and note

$$\left(1 - \frac{\Delta^2}{12}\tilde{A}(x + \Delta)\right) W(x + \Delta) + \left(1 - \frac{\Delta^2}{12}\tilde{A}(x - \Delta)\right) W(x - \Delta)$$
$$- 2\left(1 - \frac{\Delta^2}{12}\tilde{A}(x)\right) W(x)$$
$$= \Delta^2 \tilde{A}(x)W(x) - \frac{\Delta^2}{12}\left(F(x + \Delta) + F(x - \Delta) + 10F(x)\right) . \tag{A.6.24}$$

Discretization with $\Delta = 1$ yields the final expression

$$\left(1 - \frac{1}{12}\tilde{A}_{j+1}\right)W_{j+1} + \left(1 - \frac{1}{12}\tilde{A}_{j-1}\right)W_{j-1} - 2\left(1 - \frac{1}{12}\tilde{A}_j\right)W_j$$

$$= \tilde{A}_j W_j - \frac{1}{12}\left(F_{j+1} + F_{j-1} + 10F_j\right) . \tag{A.6.25}$$

Further evaluation of this result is done by recursion using the definitions

$$X_j = X_{j-1} + \tilde{A}_j W_j - \frac{1}{12}\left(F_{j+1} + F_{j-1} + 10F_j\right) \tag{A.6.26}$$

and

$$Y_j = \left(1 - \frac{1}{12}\tilde{A}_j\right)W_j , \tag{A.6.27}$$

where

$$X_1 = 0, \tag{A.6.28}$$

and

$$Y_1 = 0 . \tag{A.6.29}$$

Inserting these into (A.6.25) we obtain

$$Y_{j+1} - Y_j - (Y_j - Y_{j-1}) = X_j - X_{j-1} , \tag{A.6.30}$$

hence,

$$Y_{j+1} = Y_j + X_j , \tag{A.6.31}$$

which allows for a rapid determination of the electrostatic potential,

$$v_{es,L}(r_j) = \frac{1}{r_j}\sqrt{\left(\frac{dr}{dx}\right)_j}\left(1 - \frac{1}{12}\tilde{A}_j\right)^{-1}Y_j$$

$$= \frac{r_j}{r_j^2 - \frac{1}{48}A^2 r_j^2 - \frac{1}{12}l(l+1)\left(\frac{dr}{dx}\right)_j^2}\sqrt{\left(\frac{dr}{dx}\right)_j}Y_j . \tag{A.6.32}$$

Special care has to be taken when the denonimator on the right hand side vanishes, i.e. when

$$1 - \frac{1}{12}\tilde{A}_j = 0 , \tag{A.6.33}$$

Using the identities (A.6.19), (A.5.2), (A.5.3) as well as the abbreviation

$$\alpha = 12 - \frac{A^2}{4} , \tag{A.6.34}$$

we rewrite (A.6.33) as

$$\alpha - l(l+1)\frac{A^2}{(1 - e^{-Ax})^2} = 0 , \tag{A.6.35}$$

and, solving for x, we get the result

$$x = -\frac{1}{A}\ln\left[1 - A\sqrt{\frac{l(l+1)}{\alpha}}\right] \approx \sqrt{\frac{l(l+1)}{\alpha}} \ . \tag{A.6.36}$$

It leads to $x \approx 0.4$ for $l = 1$ and $x \approx 3.1$ for $l = 10$. The divergence of the electrostatic potential as calculated from (A.6.32) thus arises near the nucleus. In order to avoid it, we start out from the approach

$$v_{es,L}(r_j) = r_j^l \tag{A.6.37}$$

for the first few points for $l \neq 0$ and note

$$W_1 = 0 \quad \text{and} \quad W_j = r_j^{l+1}/\sqrt{\left(\frac{dr}{dx}\right)_j} \quad \text{for } j = 2, 3 \ . \tag{A.6.38}$$

From this we have

$$X_2 = Y_3 = Y_2 \ , \tag{A.6.39}$$

and can evaluate all X and Y for higher values of j via the above recursion relations.

At the very end, we still have to add to the above solution of the inhomogeneous (A.6.14) that of the homogeneous equation,

$$v_{es,L}(r_j) = A_L r_j^l \ , \tag{A.6.40}$$

where the constant A_L is determined by the boundary conditions, i.e. by the electrostatic potential at the sphere radius.

A.7 Gradients of the Intraatomic Electron Density

In the course of constructing the exchange-correlation potential from the local part of the electronic density we want to include the option to go beyond the local density approximation and to use, in particular, the generalized gradient approximation. This requires knowledge of various combinations of gradients of the local part of the electron density, which, inside the atomic spheres, is generally given by a spherical-harmonics expansion

$$\rho(\mathbf{r}) = \sum_L \rho_L(r)Y_L(\hat{\mathbf{r}}) =: P \ . \tag{A.7.1}$$

Here we have introduced the symbol P for the electron density in order to facilitate the following derivations. As before, (A.7.1) is valid for both a full-potential scheme and the standard ASW method, in which latter case it reduces to the $l = 0$-term.

Due to the symmetry of the problem we will need the gradient in spherical coordinates, i.e.

$$\nabla = \nabla_r + \nabla_{\vartheta,\varphi} = \hat{\mathbf{u}}_r \frac{\partial}{\partial r} + \hat{\mathbf{u}}_\vartheta \frac{1}{r}\frac{\partial}{\partial \vartheta} + \hat{\mathbf{u}}_\varphi \frac{1}{r\sin\vartheta}\frac{\partial}{\partial \varphi} \ , \tag{A.7.2}$$

where

$$\hat{\mathbf{u}}_r = \begin{pmatrix} \sin\vartheta\cos\varphi \\ \sin\vartheta\sin\varphi \\ \cos\vartheta \end{pmatrix}, \quad \hat{\mathbf{u}}_\vartheta = \begin{pmatrix} \cos\vartheta\cos\varphi \\ \cos\vartheta\sin\varphi \\ -\sin\vartheta \end{pmatrix}, \quad \hat{\mathbf{u}}_\varphi = \begin{pmatrix} -\sin\varphi \\ \cos\varphi \\ 0 \end{pmatrix}, \quad (A.7.3)$$

are the three standard unit vectors.

With these notions at hand we are able to write the gradient of the density (A.7.1) as

$$\nabla\rho(\mathbf{r}) = \hat{\mathbf{u}}_r \left[\sum_L \frac{\partial}{\partial r}\rho_L(r)Y_L(\hat{\mathbf{r}}) \right] + \hat{\mathbf{u}}_\vartheta \left[\sum_L \frac{1}{r}\rho_L(r)\frac{\partial}{\partial\vartheta}Y_L(\hat{\mathbf{r}}) \right]$$

$$+\hat{\mathbf{u}}_\varphi \left[\sum_L \frac{1}{r}\rho_L(r)\frac{1}{\sin\vartheta}\frac{\partial}{\partial\varphi}Y_L(\hat{\mathbf{r}}) \right]$$

$$=: \hat{\mathbf{u}}_r P_r + \hat{\mathbf{u}}_\vartheta P_\vartheta + \hat{\mathbf{u}}_\varphi P_\varphi . \tag{A.7.4}$$

In addition, we calculate the quantities

$$|\nabla\rho(\mathbf{r})| = \sqrt{\nabla\rho(\mathbf{r})\cdot\nabla\rho(\mathbf{r})} = \left[P_r^2 + P_\vartheta^2 + P_\varphi^2 \right]^{1/2} \tag{A.7.5}$$

and

$$\nabla|\nabla\rho(\mathbf{r})| = \frac{1}{2}\frac{1}{|\nabla\rho(\mathbf{r})|}\nabla\left(\nabla\rho(\mathbf{r})\cdot\nabla\rho(\mathbf{r})\right)$$

$$= \frac{1}{2}\frac{1}{|\nabla\rho(\mathbf{r})|}\nabla\left[P_r^2 + P_\vartheta^2 + P_\varphi^2 \right]$$

$$= \frac{1}{|\nabla\rho(\mathbf{r})|}\left[\hat{\mathbf{u}}_r \left(P_r\frac{\partial}{\partial r}P_r + P_\vartheta\frac{\partial}{\partial r}P_\vartheta + P_\varphi\frac{\partial}{\partial r}P_\varphi \right) \right.$$

$$+\hat{\mathbf{u}}_\vartheta \left(P_r\frac{1}{r}\frac{\partial}{\partial\vartheta}P_r + P_\vartheta\frac{1}{r}\frac{\partial}{\partial\vartheta}P_\vartheta + P_\varphi\frac{1}{r}\frac{\partial}{\partial\vartheta}P_\varphi \right)$$

$$\left. +\hat{\mathbf{u}}_\varphi \left(P_r\frac{1}{r}\frac{1}{\sin\vartheta}\frac{\partial}{\partial\varphi}P_r + P_\vartheta\frac{1}{r}\frac{1}{\sin\vartheta}\frac{\partial}{\partial\varphi}P_\vartheta + P_\varphi\frac{1}{r}\frac{1}{\sin\vartheta}\frac{\partial}{\partial\varphi}P_\varphi \right) \right]$$

$$= \frac{1}{|\nabla\rho(\mathbf{r})|}\left[\hat{\mathbf{u}}_r \left(P_r P_{rr} + P_\vartheta P_{\vartheta r} + P_\varphi P_{\varphi r} \right) + \hat{\mathbf{u}}_\vartheta \left(P_r P_{r\vartheta} + P_\vartheta P_{\vartheta\vartheta} + P_\varphi P_{\varphi\vartheta} \right) \right.$$

$$\left. +\hat{\mathbf{u}}_\varphi \left(P_r P_{r\varphi} + P_\vartheta P_{\vartheta\varphi} + P_\varphi P_{\varphi\varphi} \right) \right] . \tag{A.7.6}$$

In (A.7.4) to (A.7.6) we have used the first and second derivatives with respect to r, ϑ, and φ, i.e.

$$P_r = \sum_L \frac{\partial}{\partial r}\rho_L(r)Y_L(\hat{\mathbf{r}}) , \tag{A.7.7}$$

$$P_\vartheta = \sum_L \frac{1}{r}\rho_L(r)\frac{\partial}{\partial\vartheta}Y_L(\hat{\mathbf{r}}) , \tag{A.7.8}$$

$$P_\varphi = \sum_L \frac{1}{r} \rho_L(r) \frac{1}{\sin\vartheta} \frac{\partial}{\partial\varphi} Y_L(\hat{\mathbf{r}}) \,, \tag{A.7.9}$$

and

$$P_{rr} = \frac{\partial}{\partial r} P_r = \sum_L \frac{\partial^2}{\partial r^2} \rho_L(r) Y_L(\hat{\mathbf{r}}) \,, \tag{A.7.10}$$

$$P_{r\vartheta} = \frac{1}{r} \frac{\partial}{\partial\vartheta} P_r = \sum_L \frac{1}{r} \frac{\partial}{\partial r} \rho_L(r) \frac{\partial}{\partial\vartheta} Y_L(\hat{\mathbf{r}}) \,, \tag{A.7.11}$$

$$P_{r\varphi} = \frac{1}{r\sin\vartheta} \frac{\partial}{\partial\varphi} P_r = \sum_L \frac{1}{r} \frac{\partial}{\partial r} \rho_L(r) \frac{1}{\sin\vartheta} \frac{\partial}{\partial\varphi} Y_L(\hat{\mathbf{r}}) \,, \tag{A.7.12}$$

$$P_{\vartheta r} = \frac{\partial}{\partial r} P_\vartheta = \sum_L \frac{\partial}{\partial r} \left(\frac{1}{r} \rho_L(r) \right) \frac{\partial}{\partial\vartheta} Y_L(\hat{\mathbf{r}})$$
$$= \sum_L \left(\frac{1}{r} \frac{\partial}{\partial r} \rho_L(r) - \frac{1}{r^2} \rho_L(r) \right) \frac{\partial}{\partial\vartheta} Y_L(\hat{\mathbf{r}}) \,, \tag{A.7.13}$$

$$P_{\vartheta\vartheta} = \frac{1}{r} \frac{\partial}{\partial\vartheta} P_\vartheta = \sum_L \frac{1}{r^2} \rho_L(r) \frac{\partial^2}{\partial\vartheta^2} Y_L(\hat{\mathbf{r}}) \,, \tag{A.7.14}$$

$$P_{\vartheta\varphi} = \frac{1}{r\sin\vartheta} \frac{\partial}{\partial\varphi} P_\vartheta = \sum_L \frac{1}{r^2} \rho_L(r) \frac{1}{\sin\vartheta} \frac{\partial}{\partial\varphi} \frac{\partial}{\partial\vartheta} Y_L(\hat{\mathbf{r}}) \,, \tag{A.7.15}$$

$$P_{\varphi r} = \frac{\partial}{\partial r} P_\varphi = \sum_L \frac{\partial}{\partial r} \left(\frac{1}{r} \rho_L(r) \right) \frac{1}{\sin\vartheta} \frac{\partial}{\partial\varphi} Y_L(\hat{\mathbf{r}})$$
$$= \sum_L \left(\frac{1}{r} \frac{\partial}{\partial r} \rho_L(r) - \frac{1}{r^2} \rho_L(r) \right) \frac{1}{\sin\vartheta} \frac{\partial}{\partial\varphi} Y_L(\hat{\mathbf{r}}) \,, \tag{A.7.16}$$

$$P_{\varphi\vartheta} = \frac{1}{r} \frac{\partial}{\partial\vartheta} P_\varphi = \sum_L \frac{1}{r^2} \rho_L(r) \frac{\partial}{\partial\vartheta} \left(\frac{1}{\sin\vartheta} \frac{\partial}{\partial\varphi} \right) Y_L(\hat{\mathbf{r}})$$
$$= \sum_L \frac{1}{r^2} \rho_L(r) \left(-\frac{\cot\vartheta}{\sin\vartheta} \frac{\partial}{\partial\varphi} Y_L(\hat{\mathbf{r}}) + \frac{1}{\sin\vartheta} \frac{\partial}{\partial\vartheta} \frac{\partial}{\partial\varphi} Y_L(\hat{\mathbf{r}}) \right) \,, \tag{A.7.17}$$

$$P_{\varphi\varphi} = \frac{1}{r\sin\vartheta} \frac{\partial}{\partial\varphi} P_\varphi = \sum_L \frac{1}{r^2} \rho_L(r) \frac{1}{\sin^2\vartheta} \frac{\partial^2}{\partial\varphi^2} Y_L(\hat{\mathbf{r}}) \,. \tag{A.7.18}$$

Note the order of the indices as well as of the partial derivatives.

Finally, we will have to evaluate the Laplacian applied to the density (A.7.1). Since, in spherical coordinates,

$$\Delta = \frac{1}{r^2} \frac{\partial}{\partial r} \left(r^2 \frac{\partial}{\partial r} \right) - \frac{\mathbf{L}^2}{r^2}$$
$$= \frac{2}{r} \frac{\partial}{\partial r} + \frac{\partial^2}{\partial r^2} - \frac{\mathbf{L}^2}{r^2} \,, \tag{A.7.19}$$

we have

$$\Delta\rho(\mathbf{r}) = \sum_L \left(\frac{\partial^2}{\partial r^2}\rho_L(r) + \frac{2}{r}\frac{\partial}{\partial r}\rho_L(r) \right) Y_L(\hat{\mathbf{r}})$$

$$- \sum_L \frac{1}{r^2}\rho_L(r)\, l(l+1)Y_L(\hat{\mathbf{r}}) \,. \tag{A.7.20}$$

To sum up, we will need the following radial functions and derivatives

$$\rho_L(r), \qquad \frac{1}{r}\rho_L(r), \qquad \frac{1}{r^2}\rho_L(r) \,,$$

$$\frac{\partial}{\partial r}\rho_L(r), \qquad \frac{\partial^2}{\partial r^2}\rho_L(r), \qquad \frac{1}{r}\frac{\partial}{\partial r}\rho_L(r) \,,$$

as well as the following angular functions and derivatives

$$Y_L(\hat{\mathbf{r}}), \qquad \frac{\partial}{\partial\vartheta}Y_L(\hat{\mathbf{r}}), \qquad \frac{1}{\sin\vartheta}\frac{\partial}{\partial\varphi}Y_L(\hat{\mathbf{r}}),$$

$$\frac{\partial^2}{\partial\vartheta^2}Y_L(\hat{\mathbf{r}}), \qquad \frac{1}{\sin\vartheta}\frac{\partial}{\partial\varphi}\frac{\partial}{\partial\vartheta}Y_L(\hat{\mathbf{r}}), \qquad \frac{1}{\sin^2\vartheta}\frac{\partial^2}{\partial\varphi^2}Y_L(\hat{\mathbf{r}}),$$

$$\frac{\partial}{\partial\vartheta}\left(\frac{1}{\sin\vartheta}\frac{\partial}{\partial\varphi}Y_L(\hat{\mathbf{r}}) \right) \,.$$

References

1. N. W. Ashcroft and N. D. Mermin, *Solid State Physics* (Holt-Saunders, Philadelphia 1976)
2. T. Loucks, *Augmented Plane Wave Method* (Benjamin, New York 1967)

B

Details of the Envelope Functions

B.1 Calculation of Spherical Bessel Functions

The first section of this part of the appendix is devoted to the detailed calculation of the spherical Bessel functions. Here, we refer especially to Sect. 3.1, where the notations for these functions have been first noted and where a lot of identities for them were given. As a matter of fact, several notes on the evaluation of spherical Bessel functions exist in the literature, to which the reader is referred for further information [6, 11, 17].

In order to remove any leading dimension in the spherical Bessel functions we define, besides the barred functions given by (3.1.12) to (3.1.14), the following set

$$\hat{n}_l(\rho) := -\rho^{l+1} n_l(\rho) , \tag{B.1.1}$$

$$\hat{\jmath}_l(\rho) := \rho^{-l} \jmath_l(\rho) , \tag{B.1.2}$$

$$\hat{h}_l^{(1)}(\rho) := i\rho^{l+1} h_l^{(1)}(\rho) . \tag{B.1.3}$$

Just in the same manner as the barred functions they originate from the pure spherical Bessel functions by multiplication with a power of ρ or κ, respectively. Note that these function are all real for any argument ρ except for the spherical Hankel function which is real for negative ρ^2 only.

The most straightforward way to evaluate all spherical Bessel functions for a given argument and for all values of l up to a certain l_{max} consists of calculating two of them and derive the others by use of the well known recursion relations. For the pure functions these identities have been given by (3.1.53), from which the corresponding formulas for the above defined functions result as

$$(2l + 1)\hat{n}_l(\rho r) = \rho^2 \hat{n}_{l-1}(\rho) + \hat{n}_{l+1}(\rho) , \tag{B.1.4}$$

$$(2l + 1)\hat{\jmath}_l(\rho) = \hat{\jmath}_{l-1}(\rho) + \rho^2 \hat{\jmath}_{l+1}(\rho) , \tag{B.1.5}$$

$$(2l + 1)\hat{h}_l^{(1)}(\rho) = \rho^2 \hat{h}_{l-1}^{(1)}(\rho) + \hat{h}_{l+1}^{(1)}(\rho) . \tag{B.1.6}$$

V. Eyert: *Details of the Envelope Functions*, Lect. Notes Phys. **719**, 199–246 (2007)
DOI 10.1007/978-3-540-71007-3_6 © Springer-Verlag Berlin Heidelberg 2007

Since both the spherical Neumann function as well as the spherical Hankel function can be derived directly from the spherical Bessel function by invoking the identities (3.1.49) and (3.1.6), in the present context,

$$\hat{n}_l(\rho) = (-)^l \hat{j}_{-l-1}(\rho), \qquad \text{for } l = 0, \pm 1, \pm 2, \ldots , \tag{B.1.7}$$

and

$$\hat{h}_l^{(1)}(\rho) = \hat{n}_l(\rho) + i\rho^{2l+1}\hat{j}_l(\rho) , \tag{B.1.8}$$

we will first concentrate on the spherical Bessel function.

For the spherical Bessel functions many authors recommend to calculate them by downward recursion, hence, to use (B.1.5) in the form

$$\hat{j}_{l-1}(\rho) = (2l + 1)\hat{j}_l(\rho) - \rho^2 \hat{j}_{l+1}(\rho) . \tag{B.1.9}$$

This is a good choice in many cases since it avoids division by ρ^2 and thus numerical instabilities for small ρ.

To be concrete, we start calculating those two spherical Bessel functions with the highest l. For this purpose we make use of the series expansion [1, (10.1.2)]

$$\hat{j}_l(\rho) = \frac{\rho^l}{(2l + 1)!!} \sum_{\alpha=0}^{\infty} \frac{(-\frac{1}{2}\rho^2)^\alpha (2l + 1)!!}{\alpha!(2l + 1 + 2\alpha)!!} . \tag{B.1.10}$$

For the function defined in (B.1.2) this reads as

$$\hat{j}_l(\rho) = \sum_{\alpha=0}^{\infty} t_\alpha^l(\rho) , \tag{B.1.11}$$

with

$$t_\alpha^l(\rho) := \frac{(-\rho^2)^\alpha}{(2\alpha)!!(2l + 1 + 2\alpha)!!} . \tag{B.1.12}$$

The terms of the series fulfill the identities

$$t_0^l(\rho) = \frac{1}{(2l + 1)!!} , \tag{B.1.13}$$

$$t_\alpha^l(\rho) = \frac{1}{(2l + 1 + 2\alpha)} t_\alpha^{l-1}(\rho) , \tag{B.1.14}$$

and

$$t_\alpha^{l-1}(\rho) = \frac{-\rho^2}{2\alpha} t_{\alpha-1}^l(\rho) \qquad \text{for } \alpha \neq 0 . \tag{B.1.15}$$

Note that the series for $l = l_{max}$ and $l = l_{max} - 1$ may be calculated simultaneously.

Finally, these two functions are used to generate all $\hat{j}_l(\rho)$ for $l = l_{max} - 2, \ldots, -l_{max} - 1$ by downward recursion and to calculate the spherical Neumann functions with the help of (B.1.7).

The procedure just outlined is both accurate and stable for negative and small positive arguments ρ^2. However, for positive arguments ρ^2 the terms in the series (B.1.10) alternate in sign and for large arguments convergence will be very bad. For this reason, we prefer a different procedure for arguments $\rho^2 > l^2$. In this case, we adopt the explicit formulas for $l = 0$ and $l = 1$

$$\hat{j}_0(\rho) = \frac{\sin \rho}{\rho}, \qquad (B.1.16)$$

$$\hat{j}_1(\rho) = \frac{\sin \rho}{\rho^3} - \frac{\cos \rho}{\rho^2}. \qquad (B.1.17)$$

Here, we used the identities (3.1.20) and (3.1.21). Bessel functions for higher l are then calculated by upward recursion. For the Neumann functions the procedure is the same as for the negative energy case.

Next we turn to the situation where we want to calculate the Neumann functions without having the Bessel functions at hand. In this case upward recursion is still well suited to determine the high-l Neumann functions. As a starting point, we use the Neumann functions for $l = 0$ and $l = 1$, for which we get from the explicit formulas (3.1.18) and (3.1.19)

$$\hat{n}_0(\rho) = \cos \rho, \qquad (B.1.18)$$
$$\hat{n}_1(\rho) = \cos \rho + \rho \sin \rho. \qquad (B.1.19)$$

Finally we may calculate the spherical Hankel functions with the help of (B.1.8). Nevertheless, when we are interested in spherical Bessel and Hankel functions alone, but not in the spherical Neumann functions it is desirable to evaluate the Hankel functions directly. In this case we use the explicit formulas (3.1.22) and (3.1.23) to obtain

$$\hat{h}_0^{(1)}(\rho) = e^{i\rho}, \qquad (B.1.20)$$
$$\hat{h}_1^{(1)}(\rho) = e^{i\rho}(1 - i\rho). \qquad (B.1.21)$$

Functions for higher l are then calculated using the upward recursion (B.1.6).

B.2 Properties and Calculation of Cubic Harmonics

In this section we discuss some properties of the cubic harmonics, which are the real linear combinations of the complex spherical harmonics defined in many textbooks [12, 14]. Furthermore, we present several methods for calculating these functions as well as their harmonic polynomials.

The spherical harmonics

$$Y_{lm}(\vartheta, \varphi) = \begin{cases} (-)^m & \text{for } m \geq 0 \\ 1 & \text{for } m < 0 \end{cases} \left[\frac{2l + 1}{4\pi} \frac{(l - |m|)!}{(l + |m|)!} \right]^{1/2} P_l^{|m|}(\cos \vartheta) e^{im\varphi},$$

$$(B.2.1)$$

are complex eigenfunctions of the angular momentum operators L^2 and L_z with the eigenvalues $l(l+1)$ and m for integer l; $P_l^{|m|}(\cos\vartheta)$ denotes the associated Legendre functions [14, App. B].

For many purposes it is useful not to work with the complex spherical harmonics but with their linear combinations

$$
\begin{aligned}
\mathcal{Y}_{lm}^+(\vartheta,\varphi) &:= (-)^m\sqrt{2}\,\Re Y_{lm}(\vartheta,\varphi) \\
&= (-)^m\frac{1}{\sqrt{2}}\,[Y_{lm}(\vartheta,\varphi)+Y_{lm}^*(\vartheta,\varphi)] \\
&= \left\{\begin{array}{ll} 1 & \text{for } m>0 \\ (-)^m & \text{for } m<0 \end{array}\right\}\left[\frac{2l+1}{2\pi}\frac{(l-|m|)!}{(l+|m|)!}\right]^{1/2}P_l^{|m|}(\cos\vartheta)\cos m\varphi\,,
\end{aligned}
$$

$$(B.2.2)$$

and

$$
\begin{aligned}
\mathcal{Y}_{lm}^-(\vartheta,\varphi) &:= -\sqrt{2}\,\Im Y_{lm}(\vartheta,\varphi) \\
&= \frac{-1}{i\sqrt{2}}\,[Y_{lm}(\vartheta,\varphi)-Y_{lm}^*(\vartheta,\varphi)] \\
&= \left\{\begin{array}{ll} (-)^m & \text{for } m>0 \\ 1 & \text{for } m<0 \end{array}\right\}\left[\frac{2l+1}{2\pi}\frac{(l-|m|)!}{(l+|m|)!}\right]^{1/2}P_l^{|m|}(\cos\vartheta)(-\sin m\varphi)\,.
\end{aligned}
$$

$$(B.2.3)$$

They are usually referred to as the cubic harmonics. For the time being we are not interested in the function for $m=0$, which is a real function anyway. Using the identity

$$Y_{lm}^*(\vartheta,\varphi) = (-)^m Y_{l-m}(\vartheta,\varphi)\,, \tag{B.2.4}$$

we thus obtain

$$\mathcal{Y}_{lm}^+(\vartheta,\varphi) = \frac{1}{\sqrt{2}}\,[(-)^m Y_{lm}(\vartheta,\varphi)+Y_{l-m}(\vartheta,\varphi)]\,, \tag{B.2.5}$$

$$\mathcal{Y}_{lm}^-(\vartheta,\varphi) = \frac{1}{i\sqrt{2}}\,[-Y_{lm}(\vartheta,\varphi)+(-)^m Y_{l-m}(\vartheta,\varphi)]\,. \tag{B.2.6}$$

Hence, the real functions simply result as superpositions of the eigenstates for l,m and $l,-m$. As a consequence, they themselves will only be eigenstates of the operators L^2 and L_z^2. Indeed, straightforward calculation yields

$$L_z\mathcal{Y}_{lm}^+(\vartheta,\varphi) = i(-)^{m+1}m\mathcal{Y}_{lm}^-(\vartheta,\varphi)\,, \tag{B.2.7}$$

$$L_z\mathcal{Y}_{lm}^-(\vartheta,\varphi) = \frac{1}{i}(-)^{m+1}m\mathcal{Y}_{lm}^+(\vartheta,\varphi)\,. \tag{B.2.8}$$

Thus, the real functions are fully characterized by the quantum numbers l and $|m|$. In contrast, any distinction between m and $-m$ has no physical meaning. The same holds for the forefactors in the curly brackets of (B.2.2) and (B.2.3), which, at most, could be chosen in one or the other way for reasons of convenience.

Nevertheless, it might be useful to keep the sign of m in order to formally mark the functions $\mathcal{Y}_{lm}^+(\vartheta, \varphi)$ and $\mathcal{Y}_{lm}^-(\vartheta, \varphi)$. Using, in addition, the function for $m = 0$ in its original form as given by (B.2.1) we arrive at

$$\mathcal{Y}_{lm}(\vartheta, \varphi) = \beta_{l|m|} P_l^{|m|}(\cos \vartheta) \begin{cases} \cos |m|\varphi & \text{for } m \geq 0 \\ \sin |m|\varphi & \text{for } m < 0 \end{cases}, \qquad (B.2.9)$$

with the forefactors

$$\beta_{l|m|} = \left[\frac{2l+1}{2\pi} \frac{(l-|m|)!}{(l+|m|)!} \right]^{1/2} \qquad \text{for } m \neq 0, \qquad (B.2.10)$$

and

$$\beta_{l0} = \left[\frac{2l+1}{4\pi} \right]^{1/2} \qquad \text{for } m = 0. \qquad (B.2.11)$$

Still, we point out that the definition of the cubic harmonics contained in (B.2.5) and (B.2.6) represents a unitary transformation in the two dimensional space spanned by the complex spherical harmonics Y_{lm} and Y_{l-m}. It induces a rotation from these eigenstates of the operators L^2 and L_z to the states $\mathcal{Y}_{lm}$ and $\mathcal{Y}_{l-m}$. Hence, the orthonormality of the former set of functions is transferred to the latter one. From the conservation of all scalar products under such a unitary transformation

$$\begin{pmatrix} Y_{lm} \ (\vartheta, \varphi) \\ Y_{l-m} \ (\vartheta, \varphi) \end{pmatrix}^* \begin{pmatrix} Y_{lm} \ (\vartheta', \varphi') \\ Y_{l-m} \ (\vartheta', \varphi') \end{pmatrix} = \begin{pmatrix} \mathcal{Y}_{lm} \ (\vartheta, \varphi) \\ \mathcal{Y}_{l-m} \ (\vartheta, \varphi) \end{pmatrix} \begin{pmatrix} \mathcal{Y}_{lm} \ (\vartheta', \varphi') \\ \mathcal{Y}_{l-m} \ (\vartheta', \varphi') \end{pmatrix}, \qquad (B.2.12)$$

we furthermore get the important relation

$$\sum_{m=-l}^{+l} Y_{lm}^*(\vartheta, \varphi) Y_{lm}(\vartheta', \varphi') = \sum_{m=-l}^{+l} \mathcal{Y}_{lm}(\vartheta, \varphi) \mathcal{Y}_{lm}(\vartheta', \varphi'). \qquad (B.2.13)$$

Obviously, the expressions on both sides are real quantities. Note that (B.2.13) allows to write (3.6.1) and (3.6.8) as well as all other equations in Sect. 3.6 identically with the complex spherical harmonics (B.2.1) and with their real counterparts (B.2.2) and (B.2.3).

After these more fundamental considerations we will now present several methods for the numerical calculation of the cubic harmonics and their harmonic polynomials. The latter are defined by [20]

$$\tilde{\mathcal{Y}}_{lm}(\mathbf{r}) = r^l \mathcal{Y}_{lm}(\vartheta, \varphi)$$

$$= \beta_{l|m|} T_l^{|m|}(z, r) \begin{cases} C_{|m|}(x, y) & \text{for } m \geq 0 \\ S_{|m|}(x, y) & \text{for } m < 0 \end{cases}, \qquad (B.2.14)$$

with $\beta_{l|m|}$ given by (B.2.10) and (B.2.11) as well as

$$T_l^{|m|}(z,r) = \frac{r^l}{(r\sin\vartheta)^{|m|}} P_l^{|m|}(\cos\vartheta) , \tag{B.2.15}$$

$$C_{|m|}(x,y) = (r\sin\vartheta)^{|m|}\cos|m|\varphi , \tag{B.2.16}$$

$$S_{|m|}(x,y) = (r\sin\vartheta)^{|m|}\sin|m|\varphi . \tag{B.2.17}$$

This choice was motivated by the simple recursion relations of the functions C and S [20]

$$C_{|m|+1}(x,y) = xC_{|m|}(x,y) - yS_{|m|}(x,y) , \tag{B.2.18}$$
$$S_{|m|+1}(x,y) = xS_{|m|}(x,y) + yC_{|m|}(x,y) , \tag{B.2.19}$$

which follow directly from the addition theorems for the trigonometric functions. Furthermore, the recursion relations of the associated Legendre functions $P_l^{|m|}(\cos\vartheta)$ induce analogous relations for the $T_l^{|m|}$ [14, 20]

$$T_1^0 = z , \tag{B.2.20}$$
$$(l+1)T_{l+1}^0 = (2l+1)zT_l^0 - lr^2T_{l-1}^0 , \tag{B.2.21}$$
$$T_l^l = (2l-1)!! , \tag{B.2.22}$$
$$(x^2+y^2)T_l^{|m|+1} = (l+|m|)r^2T_{l-1}^{|m|} - (l-|m|)zT_l^{|m|} . \tag{B.2.23}$$

The last equation must be solved for $T_l^{|m|+1}$ in the case $(x^2+y^2) > z^2$ and for $T_l^{|m|}$ otherwise in order to avoid numerical instabilities.

This distinction in the evaluation of the $T_l^{|m|}$ is not needed in the method to be described now. Its derivation starts from the definition of the associated Legendre functions in terms of the Legendre polynomials [14, App. B]

$$P_l^{|m|}(u) = (1-u^2)^{\frac{|m|}{2}} \frac{\partial^{|m|}}{\partial u^{|m|}} P_l(u) , \tag{B.2.24}$$

with $u = \cos\vartheta$. The polynomials themselves may be constructed by use of Rodrigues's formula

$$P_l(u) = P_l^0(u) = \frac{1}{2^l l!} \frac{\partial^l}{\partial u^l}(u^2-1)^l . \tag{B.2.25}$$

Combining it with (B.2.15) we obtain

$$T_l^{|m|}(z,r) = r^{l-|m|} \frac{1}{2^l l!} \frac{\partial^{l+|m|}}{\partial u^{l+|m|}}(u^2-1)^l , \tag{B.2.26}$$

with $u = \frac{z}{r}$. Using the identity

$$\frac{\partial^{l+|m|}}{\partial u^{l+|m|}} = \left(\frac{\partial z}{\partial u}\frac{\partial}{\partial z}\right)^{l+|m|} = \left(r\frac{\partial}{\partial z}\right)^{l+|m|} = r^{l+|m|}\frac{\partial^{l+|m|}}{\partial z^{l+|m|}} , \tag{B.2.27}$$

we further note

$$T_l^{|m|}(z,r) = \frac{1}{2^l l!} r^{l-|m|} r^{l+|m|} \frac{\partial^{l+|m|}}{\partial z^{l+|m|}} (z^2 - r^2)^l r^{-2l}$$

$$= \frac{1}{2^l l!} \frac{\partial^{l+|m|}}{\partial z^{l+|m|}} (z^2 - r^2)^l . \tag{B.2.28}$$

The final detailed evaluation is done with the help of Binomi's law in the form

$$(z^2 - r^2)^l = \sum_{k=0}^{l} (-)^k \binom{l}{k} z^{2l-2k} r^{2k} , \tag{B.2.29}$$

which in our case leads to

$$\frac{\partial^{l+|m|}}{\partial z^{l+|m|}} (z^2 - r^2)^l = \sum_{\substack{k=0 \\ 2k \le l-|m|}}^{l} (-)^k \binom{l}{k} \frac{(2l-2k)!}{(l-|m|-2k)!} z^{l-|m|-2k} r^{2k} . \tag{B.2.30}$$

Thus, we arrive at the result

$$T_l^{|m|}(z,r) = \frac{1}{2^l} \sum_{\substack{k=0 \\ 2k \le l-|m|}}^{l} (-)^k \frac{(2l-2k)!}{k!(l-k)!(l-|m|-2k)!} z^{l-|m|-2k} r^{2k} . \tag{B.2.31}$$

The coefficients in this equation may be calculated numerically. Hence, (B.2.31) allows for an explicit evaluation of the $T_l^{|m|}$ up to arbitrary angular momentum l.

Finally, we present a third method to calculate the $T_l^{|m|}$, which is also based on the recursion relations for the associated Legendre functions. For either of the above discussed methods has its disadvantages: To be specific, the explicit formulation (B.2.31) might become numerically unstable since it contains sums and differences of large numbers. This might involve numerical inaccuracies for high values of l. In contrast, the recursion relations (B.2.20) to (B.2.23) do not allow for fast calculations due to the distinction of the different cases mentioned above. Furthermore, they might likewise become unstable for small argument r. These disadvantages are circumvented by the following method [17, Chap. 6.8]. It is essentially based on the recursion relation

$$(l+1-|m|)T_{l+1}^{|m|} = z(2l+1)T_l^{|m|} - r^2(l+|m|)T_{l-1}^{|m|} , \tag{B.2.32}$$

which may be derived from an analogous identity for the associated Legendre functions [14, App. B] and which is identical to (B.2.21) for $m = 0$. Furthermore, we note explicitly the special case $l = |m|$

$$T_{l+1}^l = z(2l+1)T_l^l , \tag{B.2.33}$$

where we have set $T_{l-1}^l \overset{!}{=} 0$. Finally, we use (B.2.22), which can be likewise written as

$$T_{l+1}^{l+1} = (2l+1)T_l^l . \tag{B.2.34}$$

It serves as a starting point, which allows to calculate $T_{|m|}^{|m|}$ for every $|m|$, then $T_{|m|+1}^{|m|}$ using (B.2.33), and, finally, with the help of the recursion relation (B.2.32), all the $T_l^{|m|}$ for $l > |m| + 1$.

At the very end we explicitly note the cubic harmonics for angular momenta up to $l = 3$ as

$$\tilde{\mathcal{Y}}_{00} = \frac{1}{\sqrt{4\pi}} \, , \tag{B.2.35}$$

$$\tilde{\mathcal{Y}}_{1-1} = \sqrt{\frac{3}{4\pi}} y \, ,$$

$$\tilde{\mathcal{Y}}_{10} = \sqrt{\frac{3}{4\pi}} z \, , \tag{B.2.36}$$

$$\tilde{\mathcal{Y}}_{11} = \sqrt{\frac{3}{4\pi}} x \, ,$$

$$\tilde{\mathcal{Y}}_{2-2} = \sqrt{\frac{15}{4\pi}} xy \, ,$$

$$\tilde{\mathcal{Y}}_{2-1} = \sqrt{\frac{15}{4\pi}} yz \, ,$$

$$\tilde{\mathcal{Y}}_{20} = \sqrt{\frac{5}{16\pi}} (3z^2 - r^2) \, , \tag{B.2.37}$$

$$\tilde{\mathcal{Y}}_{21} = \sqrt{\frac{15}{4\pi}} xz \, ,$$

$$\tilde{\mathcal{Y}}_{22} = \sqrt{\frac{15}{16\pi}} (x^2 - y^2) \, ,$$

$$\tilde{\mathcal{Y}}_{3-3} = \sqrt{\frac{35}{32\pi}} (3x^2y - y^3) \, ,$$

$$\tilde{\mathcal{Y}}_{3-2} = \sqrt{\frac{105}{4\pi}} xyz \, ,$$

$$\tilde{\mathcal{Y}}_{3-1} = \sqrt{\frac{21}{32\pi}} (5yz^2 - yr^2) \, ,$$

$$\tilde{\mathcal{Y}}_{30} = \sqrt{\frac{7}{16\pi}} (5z^3 - 3zr^2) \, , \tag{B.2.38}$$

$$\tilde{\mathcal{Y}}_{31} = \sqrt{\frac{21}{32\pi}} (5xz^2 - xr^2) \, ,$$

$$\tilde{\mathcal{Y}}_{32} = \sqrt{\frac{105}{16\pi}} (x^2 - y^2)z \, ,$$

$$\tilde{\mathcal{Y}}_{33} = \sqrt{\frac{35}{32\pi}} (x^3 - 3xy^2) \, .$$

B.3 Angular Derivatives of Cubic Harmonics

Having the cubic harmonics at hand we turn in the present section to their derivatives with respect to the angles ϑ and φ. These derivatives are needed, e.g. in the
course of applying the generalized gradient approximation. At the end of the section
we will deal with matrix elements of the gradient, which enter the calculation of
oscillator strengths. We start from (B.2.9),

$$\mathcal{Y}_{lm}(\vartheta, \varphi) = \beta_{l|m|} P_l^{|m|}(\cos \vartheta) \begin{cases} \cos |m|\varphi & \text{for } m > 0 \\ 1 & \text{for } m = 0 \\ \sin |m|\varphi & \text{for } m < 0 \end{cases}, \qquad (B.3.1)$$

from which the derivatives with respect to φ are easily calculated as

$$\frac{\partial}{\partial \varphi} \mathcal{Y}_{lm}(\vartheta, \varphi) = \beta_{l|m|} P_l^{|m|}(\cos \vartheta) \begin{cases} -|m|\sin |m|\varphi & \text{for } m > 0 \\ 0 & \text{for } m = 0 \\ +|m|\cos |m|\varphi & \text{for } m < 0 \end{cases}$$

$$= -m\mathcal{Y}_{l-m}(\vartheta, \varphi), \qquad (B.3.2)$$

and

$$\frac{\partial^2}{\partial \varphi^2} \mathcal{Y}_{lm}(\vartheta, \varphi) = \beta_{l|m|} P_l^{|m|}(\cos \vartheta) \begin{cases} -|m|^2 \cos |m|\varphi & \text{for } m > 0 \\ 0 & \text{for } m = 0 \\ -|m|^2 \sin |m|\varphi & \text{for } m < 0 \end{cases}$$

$$= -m^2 \mathcal{Y}_{lm}(\vartheta, \varphi). \qquad (B.3.3)$$

In order to access the derivatives with respect to ϑ we use $u = \cos \vartheta$ as in Sect.
B.2 and note

$$\frac{\partial}{\partial \vartheta} = \frac{\partial \cos \vartheta}{\partial \vartheta} \frac{\partial}{\partial \cos \vartheta} = -\sqrt{1 - u^2} \frac{\partial}{\partial u}. \qquad (B.3.4)$$

Concentrating on the associated Legendre functions, which alone cover the ϑ-depen-
dency of the cubic harmonics, we obtain from (B.2.24)

$$\frac{\partial}{\partial u} P_l^{|m|}(u) = -|m|u(1 - u^2)^{\frac{|m|}{2} - 1} \frac{\partial^{|m|}}{\partial u^{|m|}} P_l(u) + (1 - u^2)^{\frac{|m|}{2}} \frac{\partial^{|m|+1}}{\partial u^{|m|+1}} P_l(u)$$

$$= -|m| \frac{u}{1 - u^2} P_l^{|m|}(u) + \frac{1}{\sqrt{1 - u^2}} P_l^{|m|+1}(u), \qquad (B.3.5)$$

which, together with (B.3.4), leads to

$$\frac{\partial}{\partial \vartheta} P_l^{|m|}(\cos \vartheta) = |m|u(1 - u^2)^{\frac{|m|-1}{2}} \frac{\partial^{|m|}}{\partial u^{|m|}} P_l(u) - (1 - u^2)^{\frac{|m|+1}{2}} \frac{\partial^{|m|+1}}{\partial u^{|m|+1}} P_l(u)$$

$$= |m| \frac{u}{\sqrt{1 - u^2}} P_l^{|m|}(\cos \vartheta) - P_l^{|m|+1}(\cos \vartheta)$$

$$= |m| \cot \vartheta P_l^{|m|}(\cos \vartheta) - P_l^{|m|+1}(\cos \vartheta). \qquad (B.3.6)$$

Alternative expressions can be derived from the relations [14, App. B]

$$\frac{\partial}{\partial u} P_l^{|m|}(u) = -l\frac{u}{1-u^2} P_l^{|m|}(u) + (l+|m|)\frac{1}{1-u^2} P_{l-1}^{|m|}(u)$$

$$= (l+1)\frac{u}{1-u^2} P_l^{|m|}(u) - (l+1-|m|)\frac{1}{1-u^2} P_{l+1}^{|m|}(u) ,$$

(B.3.7)

which are connected by the recursion relation [14, App. B]

$$(2l+1)u P_l^{|m|}(u) = (l+|m|) P_{l-1}^{|m|}(u) + (l+1-|m|) P_{l+1}^{|m|}(u) ,$$

(B.3.8)

and which can be also used for $l = 0$ if the convention $P_{-1} = 0$ is used. A yet different formula arises from multiplying the first and second line of (B.3.7) by $l+1$ and l, respectively, and adding the resulting expressions, this leading to

$$\frac{\partial}{\partial u} P_l^{|m|}(u) = \frac{1}{2l+1}\left[(l+1)(l+|m|)\frac{1}{1-u^2} P_{l-1}^{|m|}(u)\right.$$

$$\left. -l(l+1-|m|)\frac{1}{1-u^2} P_{l+1}^{|m|}(u)\right].$$

(B.3.9)

Using the previous identities together with (B.3.4) we write for the derivatives with respect to ϑ

$$\frac{\partial}{\partial\vartheta} P_l^{|m|}(\cos\vartheta)$$

$$= l\frac{u}{\sqrt{1-u^2}} P_l^{|m|}(\cos\vartheta) - (l+|m|)\frac{1}{\sqrt{1-u^2}} P_{l-1}^{|m|}(\cos\vartheta)$$

$$= l\cot\vartheta P_l^{|m|}(\cos\vartheta) - (l+|m|)\frac{1}{\sin\vartheta} P_{l-1}^{|m|}(\cos\vartheta) ,$$

(B.3.10)

$$\frac{\partial}{\partial\vartheta} P_l^{|m|}(\cos\vartheta)$$

$$= -(l+1)\frac{u}{\sqrt{1-u^2}} P_l^{|m|}(\cos\vartheta) + (l+1-|m|)\frac{1}{\sqrt{1-u^2}} P_{l+1}^{|m|}(\cos\vartheta)$$

$$= -(l+1)\cot\vartheta P_l^{|m|}(\cos\vartheta) + (l+1-|m|)\frac{1}{\sin\vartheta} P_{l+1}^{|m|}(\cos\vartheta) ,$$

(B.3.11)

and

$$\frac{\partial}{\partial\vartheta} P_l^{|m|}(\cos\vartheta) = \frac{1}{2l+1}\frac{1}{\sin\vartheta}\left[-(l+1)(l+|m|) P_{l-1}^{|m|}(\cos\vartheta)\right.$$

$$\left. +l(l+1-|m|) P_{l+1}^{|m|}(\cos\vartheta)\right].$$

(B.3.12)

Note that for $l = |m|$ the respective terms on the right-hand sides of the previous identities vanish by definition of the associated Legendre functions.

For completeness, we combine (B.3.10) to (B.3.12) with the definition (B.3.1) of the cubic harmonics and write

$$\frac{\partial}{\partial\vartheta}\mathcal{Y}_{lm}(\vartheta,\varphi) = l\cot\vartheta\,\mathcal{Y}_{lm}(\vartheta,\varphi)$$

$$-(l+|m|)\frac{1}{\sin\vartheta}\frac{\beta_{l|m|}}{\beta_{l-1|m|}}\mathcal{Y}_{l-1m}(\vartheta,\varphi)\,, \qquad (B.3.13)$$

$$\frac{\partial}{\partial\vartheta}\mathcal{Y}_{lm}(\vartheta,\varphi) = -(l+1)\cot\vartheta\,\mathcal{Y}_{lm}(\vartheta,\varphi)$$

$$+(l+1-|m|)\frac{1}{\sin\vartheta}\frac{\beta_{l|m|}}{\beta_{l+1|m|}}\mathcal{Y}_{l+1m}(\vartheta,\varphi)\,, \qquad (B.3.14)$$

and

$$\frac{\partial}{\partial\vartheta}\mathcal{Y}_{lm}(\vartheta,\varphi) = -\frac{l+1}{2l+1}(l+|m|)\frac{1}{\sin\vartheta}\frac{\beta_{l|m|}}{\beta_{l-1|m|}}\mathcal{Y}_{l-1m}(\vartheta,\varphi)$$

$$+\frac{l}{2l+1}(l+1-|m|)\frac{1}{\sin\vartheta}\frac{\beta_{l|m|}}{\beta_{l+1|m|}}\mathcal{Y}_{l+1m}(\vartheta,\varphi)\,. \qquad (B.3.15)$$

Turning to the second derivative with respect to ϑ we start from (B.3.4) and write

$$\frac{\partial^2}{\partial\vartheta^2} = \frac{\partial}{\partial\vartheta}\left[-\sin\vartheta\frac{\partial}{\partial\cos\vartheta}\right]$$

$$= -\cos\vartheta\frac{\partial}{\partial\cos\vartheta} - \sin\vartheta\frac{\partial\cos\vartheta}{\partial\vartheta}\frac{\partial^2}{\partial\cos\vartheta^2}$$

$$= -\cos\vartheta\frac{\partial}{\partial\cos\vartheta} + \sin^2\vartheta\frac{\partial^2}{\partial\cos\vartheta^2}$$

$$= (1-u^2)\frac{\partial^2}{\partial u^2} - u\frac{\partial}{\partial u}\,. \qquad (B.3.16)$$

Combining this with the differential equation for the associated Legendre functions,

$$\left[\frac{d}{du}(1-u^2)\frac{d}{du} + l(l+1) - \frac{m^2}{1-u^2}\right]P_l^m(u)$$

$$= \left[(1-u^2)\frac{d^2}{du^2} - 2u\frac{d}{du} + l(l+1) - \frac{m^2}{1-u^2}\right]P_l^m(u) = 0\,, \qquad (B.3.17)$$

we are able to note the result

$$\frac{\partial^2}{\partial\vartheta^2}P_l^{|m|}(\cos\vartheta)$$

$$= \left[(1 - u^2)\frac{\partial^2}{\partial u^2} - u\frac{\partial}{\partial u}\right] P_l^{|m|}(u)$$

$$= \left[u\frac{\partial}{\partial u} - l(l+1) + \frac{|m|^2}{1 - u^2}\right] P_l^{|m|}(u)$$

$$= \left[-\frac{u}{\sqrt{1 - u^2}}\frac{\partial}{\partial \vartheta} - l(l+1) + \frac{|m|^2}{1 - u^2}\right] P_l^{|m|}(\cos \vartheta)$$

$$= -\cot \vartheta \frac{\partial}{\partial \vartheta} P_l^{|m|}(\cos \vartheta) - \left(l(l+1) - \frac{m^2}{\sin^2 \vartheta}\right) P_l^{|m|}(\cos \vartheta)$$

$$= -|m|\cot^2 \vartheta P_l^{|m|}(\cos \vartheta) + \cot \vartheta P_l^{|m|+1}(\cos \vartheta)$$
$$\quad - \left(l(l+1) - \frac{m^2}{\sin^2 \vartheta}\right) P_l^{|m|}(\cos \vartheta)$$

$$= -l\cot^2 \vartheta P_l^{|m|}(\cos \vartheta) + (l + |m|)\frac{\cot \vartheta}{\sin \vartheta} P_{l-1}^{|m|}(\cos \vartheta)$$
$$\quad - \left(l(l+1) - \frac{m^2}{\sin^2 \vartheta}\right) P_l^{|m|}(\cos \vartheta)$$

$$= (l+1)\cot^2 \vartheta P_l^{|m|}(\cos \vartheta) - (l + 1 - |m|)\frac{\cot \vartheta}{\sin \vartheta} P_{l+1}^{|m|}(\cos \vartheta)$$
$$\quad - \left(l(l+1) - \frac{m^2}{\sin^2 \vartheta}\right) P_l^{|m|}(\cos \vartheta) , \tag{B.3.18}$$

where, in the last steps, we have used (B.3.6) (B.3.10), and (B.3.11) for the first derivatives.

Special attention deserves the case $u = \pm 1$, hence, $\sin \vartheta = 0$, due to the $1/\sin \vartheta$ terms on the right-hand sides of (B.3.6), (B.3.10), (B.3.11), and (B.3.18). However, as is obvious from the first line of (B.3.6), for the first derivative the problem occurs only for $|m| = 1$, since for all other values the $\sin \vartheta$ terms cancel. Thus, treating the cases $|m| = 0$, $|m| = 1$, and $|m| > 1$ separately and starting with $|m| = 0$ we obtain from (B.3.6)

$$\frac{\partial}{\partial \vartheta} P_l^0(\cos \vartheta) = -\sqrt{1 - u^2}\frac{\partial}{\partial u} P_l(u) = -P_l^1(u) . \tag{B.3.19}$$

Since [14, App. B]

$$P_l^{|m|}(1) = P_l^{|m|}(-1) = 0 \qquad \text{for } |m| > 0 , \tag{B.3.20}$$

the right-hand side of (B.3.19) vanishes for $u = \pm 1$. In contrast, for $|m| > 1$ there are no divergences at all in (B.3.6). We are thus able to summarize the previous results to

$$\frac{\partial}{\partial \vartheta} P_l^{|m|}(1) = \frac{\partial}{\partial \vartheta} P_l^{|m|}(-1) = 0 \qquad \text{for } |m| \geq 0, |m| \neq 1 . \tag{B.3.21}$$

Finally, turning to $|m| = 1$ we obtain from (B.3.6)

$$\frac{\partial}{\partial\vartheta}P_l^1(\cos\vartheta) = u\frac{\partial}{\partial u}P_l(u) - (1-u^2)\frac{\partial^2}{\partial u^2}P_l(u) . \tag{B.3.22}$$

This expression can be evaluated with the help of the differential equation specifying the Legendre polynomials

$$\left[\frac{d}{du}(1-u^2)\frac{d}{du} + l(l+1)\right]P_l(u)$$
$$= \left[(1-u^2)\frac{d^2}{du^2} - 2u\frac{d}{du} + l(l+1)\right]P_l(u) = 0 , \tag{B.3.23}$$

which follows from that of the associated Legendre functions, (B.3.17), for $|m| = 0$. In particular, for $u = \pm 1$ (B.3.23) reduces to

$$u\frac{d}{du}P_l(u) = \frac{1}{2}l(l+1)P_l(u) . \tag{B.3.24}$$

Hence, using the identities [14, App. B]

$$P_l(1) = 1, \qquad P_l(-1) = (-)^l , \tag{B.3.25}$$

we are eventually able to combine (B.3.21) and (B.3.22) to

$$\frac{\partial}{\partial\vartheta}P_l^{|m|}(\cos\vartheta) = \delta_{|m|1}\frac{1}{2}l(l+1)(\cos\vartheta)^l \qquad \text{for } \cos\vartheta = \pm 1 . \tag{B.3.26}$$

Finally, for the second derivative we start from (B.3.18), which, for $u = \pm 1$, reduces to

$$\frac{\partial^2}{\partial\vartheta^2}P_l^{|m|}(\cos\vartheta) = -u\frac{\partial}{\partial u}P_l^{|m|}(u) . \tag{B.3.27}$$

Combining this with (B.3.5) and using (B.3.24) and (B.3.25), we arrive at the result

$$\frac{\partial^2}{\partial\vartheta^2}P_l^{|m|}(\cos\vartheta)$$
$$= |m|u^2(1-u^2)^{\frac{|m|}{2}-1}\frac{\partial^{|m|}}{\partial u^{|m|}}P_l(u) - u(1-u^2)^{\frac{|m|}{2}}\frac{\partial^{|m|+1}}{\partial u^{|m|+1}}P_l(u)$$
$$= \begin{cases} -u\frac{\partial}{\partial u}P_l(u) & \text{for } |m| = 0 \\ \frac{u^2}{\sqrt{1-u^2}}\frac{\partial}{\partial u}P_l(u) & \text{for } |m| = 1 \\ 2u^2\frac{\partial^2}{\partial u^2}P_l(u) & \text{for } |m| = 2 \\ 0 & \text{for } |m| > 2 \end{cases}$$
$$= \begin{cases} -\frac{1}{2}l(l+1)(\cos\vartheta)^l & \text{for } |m| = 0 \\ \frac{u}{\sqrt{1-u^2}}\frac{1}{2}l(l+1)(\cos\vartheta)^l & \text{for } |m| = 1 \\ \frac{1}{2}l(l+1)\left(l(l+1)-2\right)(\cos\vartheta)^l & \text{for } |m| = 2 \\ 0 & \text{for } |m| > 2 \end{cases} , \tag{B.3.28}$$

which indeed diverges for $|m| = 1$.

Finally, we combine (B.3.2), (B.3.10), (B.3.13), and (B.3.14) to note for the mixed derivative

$$\frac{\partial}{\partial\varphi}\frac{\partial}{\partial\vartheta}\mathcal{Y}_{lm}(\vartheta,\varphi)$$

$$= -m\beta_{l|m|}\left[l\cot\vartheta P_l^{|m|}(\cos\vartheta) - (l+|m|)\frac{1}{\sin\vartheta}P_{l-1}^{|m|}(\cos\vartheta)\right]$$

$$\begin{cases} \sin|m|\varphi & \text{for } m > 0 \\ 0 & \text{for } m = 0 \\ \cos|m|\varphi & \text{for } m < 0 \end{cases}$$

$$= -ml\cot\vartheta\mathcal{Y}_{l-m}(\vartheta,\varphi)$$

$$+m(l+|m|)\frac{1}{\sin\vartheta}\frac{\beta_{l|m|}}{\beta_{l-1|m|}}\mathcal{Y}_{l-1-m}(\vartheta,\varphi)$$

$$= m(l+1)\cot\vartheta\mathcal{Y}_{l-m}(\vartheta,\varphi)$$

$$-m(l+1-|m|)\frac{1}{\sin\vartheta}\frac{\beta_{l|m|}}{\beta_{l+1|m|}}\mathcal{Y}_{l+1-m}(\vartheta,\varphi)\,, \tag{B.3.29}$$

which concludes the calculation of angular derivatives of the cubic harmonics.

For the rest of this section, we aim at calculating the gradient of the cubic harmonics. In spherical coordinates, the gradient is given by

$$\nabla = \nabla_r + \nabla_{\vartheta,\varphi} = \hat{\mathbf{u}}_r\frac{\partial}{\partial r} + \hat{\mathbf{u}}_\vartheta\frac{1}{r}\frac{\partial}{\partial\vartheta} + \hat{\mathbf{u}}_\varphi\frac{1}{r\sin\vartheta}\frac{\partial}{\partial\varphi}\,, \tag{B.3.30}$$

where

$$\hat{\mathbf{u}}_r = \begin{pmatrix} \sin\vartheta\cos\varphi \\ \sin\vartheta\sin\varphi \\ \cos\vartheta \end{pmatrix}\,, \quad \hat{\mathbf{u}}_\vartheta = \begin{pmatrix} \cos\vartheta\cos\varphi \\ \cos\vartheta\sin\varphi \\ -\sin\vartheta \end{pmatrix}\,, \quad \hat{\mathbf{u}}_\varphi = \begin{pmatrix} -\sin\varphi \\ \cos\varphi \\ 0 \end{pmatrix}\,, \tag{B.3.31}$$

are the three standard unit vectors. Of course, for the spherical and cubic harmonics the radial derivative vanishes and we are left with only the angular contribution to the gradient. Inserting into (B.3.30) the intermediate results (B.3.2) and (B.3.13) to (B.3.15) we obtain

$$r\nabla_{\vartheta,\varphi}\mathcal{Y}_{lm}(\vartheta,\varphi)$$

$$= \hat{\mathbf{u}}_\vartheta\left[l\cot\vartheta\mathcal{Y}_{lm}(\vartheta,\varphi) - (l+|m|)\frac{1}{\sin\vartheta}\frac{\beta_{l|m|}}{\beta_{l-1|m|}}\mathcal{Y}_{l-1m}(\vartheta,\varphi)\right]$$

$$-\hat{\mathbf{u}}_\varphi\frac{m}{\sin\vartheta}\mathcal{Y}_{l-m}(\vartheta,\varphi)\,, \tag{B.3.32}$$

$$r\nabla_{\vartheta,\varphi}\mathcal{Y}_{lm}(\vartheta,\varphi)$$

$$= \hat{\mathbf{u}}_\vartheta\left[-(l+1)\cot\vartheta\mathcal{Y}_{lm}(\vartheta,\varphi) + (l+1-|m|)\frac{1}{\sin\vartheta}\frac{\beta_{l|m|}}{\beta_{l+1|m|}}\mathcal{Y}_{l+1m}(\vartheta,\varphi)\right]$$

$$-\hat{\mathbf{u}}_\varphi\frac{m}{\sin\vartheta}\mathcal{Y}_{l-m}(\vartheta,\varphi)\,, \tag{B.3.33}$$

and

$$r\nabla_{\vartheta,\varphi}\mathcal{Y}_{lm}(\vartheta,\varphi)$$

$$= \hat{\mathbf{u}}_{\vartheta}\left[-\frac{l+1}{2l+1}(l+|m|)\frac{1}{\sin\vartheta}\frac{\beta_{l|m|}}{\beta_{l-1|m|}}\mathcal{Y}_{l-1m}(\vartheta,\varphi)\right.$$

$$\left.+\frac{l}{2l+1}(l+1-|m|)\frac{1}{\sin\vartheta}\frac{\beta_{l|m|}}{\beta_{l+1|m|}}\mathcal{Y}_{l+1m}(\vartheta,\varphi)\right]$$

$$-\hat{\mathbf{u}}_{\varphi}\frac{m}{\sin\vartheta}\mathcal{Y}_{l-m}(\vartheta,\varphi)\ . \tag{B.3.34}$$

In general, we are interested in representing the result as a spherical-harmonics expansion

$$r\nabla_{\vartheta,\varphi}\mathcal{Y}_{l'm'}(\vartheta,\varphi) = \sum_{l''m''}\mathbf{D}_{L''L'}\mathcal{Y}_{l''m''}(\vartheta,\varphi)\ , \tag{B.3.35}$$

where $L = (l,m)$ as usual and the expansion coefficients are given by

$$\mathbf{D}_{LL'} = \int d\Omega\,\mathcal{Y}_{lm}(\vartheta,\varphi)\,(r\nabla_{\vartheta,\varphi})\,\mathcal{Y}_{l'm'}(\vartheta,\varphi)\ . \tag{B.3.36}$$

Of course, these coefficients could be calculated by inserting into the definition (B.3.36) one of (B.3.32) to (B.3.34). However, this leads to a sequence of straightforward but rather laborious calculations. A much simpler derivation starts from the definition of the matrix

$$\mathbf{G}_{LL'} = \int d^2\hat{\mathbf{r}}\,\mathcal{Y}_{lm}(\vartheta,\varphi)\hat{\mathbf{u}}_r\mathcal{Y}_{l'm'}(\vartheta,\varphi)\ , \tag{B.3.37}$$

which formally arises from a spherical-harmonics expansion analogous to (B.3.35), namely,

$$\hat{\mathbf{u}}_r\mathcal{Y}_{l'm'}(\vartheta,\varphi) = \sum_{l''m''}\mathbf{G}_{L''L'}\mathcal{Y}_{l''m''}(\vartheta,\varphi)\ . \tag{B.3.38}$$

Since, according to (B.2.14) and (B.2.36), the vector $\hat{\mathbf{u}}_r$ can be expressed in terms of the cubic harmonics for $l = 1$ as

$$\hat{\mathbf{u}}_r = \hat{\mathbf{r}} = \frac{\mathbf{r}}{r} = \sqrt{\frac{4\pi}{3}}\begin{pmatrix}\mathcal{Y}_{11}(\vartheta,\varphi)\\\mathcal{Y}_{1-1}(\vartheta,\varphi)\\\mathcal{Y}_{10}(\vartheta,\varphi)\end{pmatrix}\ , \tag{B.3.39}$$

the matrix $\mathbf{G}_{LL'}$ reduces to Gaunt coefficients with non-vanishing contributions only for $l' = l \pm 1$ as well as $m' = m$ or $m' = m \pm 1$; the terms with $l' = l$ vanish since the sum of all three angular momenta must be even for parity reasons.

The important point to notice is that both $\mathbf{G}_{LL'}$ and $\mathbf{D}_{LL'}$ arise as matrix elements of vector operators built with the cubic harmonics. Due to the Wigner-Eckart theorem [15] these matrix elements must be proportional to each other with the scaling factor depending only on l and l' but not on m or m'. As a consequence, we may readily note the intermediate result

$$\mathbf{D}_{LL'} = [\delta_{l'l-1}a_{l-1} + \delta_{l'l+1}a_{l+1}]\,\mathbf{G}_{LL'}\,, \tag{B.3.40}$$

where we have used the above conditions for the Gaunt integrals. The coefficients $a_{l\pm1}$ are given by

$$a_{l\pm1} = \delta_{l'l\pm1}\frac{\int d\Omega\, \mathcal{Y}_{l'\mp1m}(\vartheta,\varphi)\,(r\nabla_{\vartheta,\varphi})_\alpha\,\mathcal{Y}_{l'm'}(\vartheta,\varphi)}{\int d\Omega\, \mathcal{Y}_{l'\mp1m}(\vartheta,\varphi)\,(\hat{\mathbf{u}}_r)_\alpha\,\mathcal{Y}_{l'm'}(\vartheta,\varphi)}\,, \tag{B.3.41}$$

where α may be any of the Cartesian components; according to the Wigner-Eckart theorem the result for $a_{l\pm1}$ must be the same for all three components of the vectors. Furthermore, again due to the Wigner-Eckart theorem, the coefficients are independent of the values for m and m' as long as the selection rules are obeyed. In order to keep things as simple as possible, we opt for the z-component of $\hat{\mathbf{u}}_r$ and choose $m = m' = 0$. As a consequence, the dependence on the angle φ cancels completely from both integrals. Using (B.3.30) and (B.3.31), the definition (B.3.1) as well as $u = \cos\vartheta$ and (B.3.4), we are thus able to note the intermediate result

$$\begin{aligned}
a_{l\pm1} &= -\delta_{l'l\pm1}\frac{\int d\Omega\, \mathcal{Y}_{l'\mp10}(\vartheta,\varphi)\sin\vartheta\frac{\partial}{\partial\vartheta}\mathcal{Y}_{l'0}(\vartheta,\varphi)}{\int d\Omega\, \mathcal{Y}_{l'\mp10}(\vartheta,\varphi)\cos\vartheta\mathcal{Y}_{l'0}(\vartheta,\varphi)}\\[4pt]
&= -\delta_{l'l\pm1}\frac{\int_0^\pi \sin\vartheta d\vartheta\, P_{l'\mp1}^0(\cos\vartheta)\sin\vartheta\frac{\partial}{\partial\vartheta}P_{l'}^0(\cos\vartheta)}{\int_0^\pi \sin\vartheta d\vartheta\, P_{l'\mp1}^0(\cos\vartheta)\cos\vartheta P_{l'}^0(\cos\vartheta)}\\[4pt]
&= \delta_{l'l\pm1}\frac{\int_{-1}^1 du\, P_{l'\mp1}^0(u)(1-u^2)\frac{\partial}{\partial u}P_{l'}^0(u)}{\int_{-1}^1 du\, P_{l'\mp1}^0(u)uP_{l'}^0(u)}\\[4pt]
&= \delta_{l'l\pm1}\frac{\int_{-1}^1 du\, P_{l'\mp1}^0(u)\left[(l'+1)l'P_{l'-1}^0(u) - l'(l'+1)P_{l'+1}^0(u)\right]}{\int_{-1}^1 du\, P_{l'\mp1}^0(u)\left[l'P_{l'-1}^0(u) + (l'+1)P_{l'+1}^0(u)\right]}\,,
\end{aligned} \tag{B.3.42}$$

where, in the last step, we inserted the identities (B.3.8) and (B.3.9). Finally, taking into account the orthonormality of the associated Legendre functions [14, App. B],

$$\int_{-1}^1 du\, P_l^{|m|}(u)P_{l'}^{|m|}(u) = \delta_{ll'}\frac{2}{2l+1}\frac{(l+|m|)!}{(l-|m|)!}\,, \tag{B.3.43}$$

we arrive at the result

$$a_{l+1} = \delta_{l'l+1}(l'+1), \qquad a_{l-1} = \delta_{l'l-1}(-l')\,, \tag{B.3.44}$$

hence,

$$\mathbf{D}_{LL'} = [\delta_{l'l-1}(-l') + \delta_{l'l+1}(l'+1)]\,\mathbf{G}_{LL'}\,, \tag{B.3.45}$$

which is very useful for the calculation of oscillator strengths.

B.4 Summations in Real and Reciprocal Lattice

In this section we will derive a set of identities, which interconnect summations in real and reciprocal lattice and thus represent special applications of the general Poisson sum formula [16, p. 466]. The results will be of use especially for the calculations of Bloch sums of the envelope functions with the help of the Ewald method in Sect. 3.5.

We start from the well-known fact that every lattice-periodic function may be expanded in plane waves [2]. In particular, we note

$$\sum_\mu \delta(\mathbf{r} - \mathbf{R}_\mu) = \sum_n f_n e^{i\mathbf{K}_n \mathbf{r}} , \tag{B.4.1}$$

where the summation over n encounters all vectors of the reciprocal lattice and the coefficients f_n are given by

$$f_n = \frac{1}{\Omega_c} \int_{\Omega_c} d^3\mathbf{r} \, e^{-i\mathbf{K}_n \mathbf{r}} \sum_\mu \delta(\mathbf{r} - \mathbf{R}_\mu) = \frac{1}{\Omega_c} . \tag{B.4.2}$$

Ω_c denotes the volume of the real-space unit cell. From (B.4.2) we obtain

$$\sum_\mu \delta(\mathbf{r} - \mathbf{R}_\mu) = \sum_n \frac{1}{\Omega_c} e^{i\mathbf{K}_n \mathbf{r}} . \tag{B.4.3}$$

With this in mind, we may now easily derive the general Poisson sum formula

$$\sum_\mu f(\boldsymbol{\tau} - \mathbf{R}_\mu) = \int d^3\mathbf{r} \sum_\mu \delta(\mathbf{r} - \mathbf{R}_\mu) f(\boldsymbol{\tau} - \mathbf{r})$$

$$= \frac{1}{\Omega_c} \sum_n \int d^3\mathbf{r} \, e^{i\mathbf{K}_n \mathbf{r}} f(\boldsymbol{\tau} - \mathbf{r})$$

$$= \frac{1}{\Omega_c} \sum_n e^{i\mathbf{K}_n \boldsymbol{\tau}} \int d^3\mathbf{r} \, e^{-i\mathbf{K}_n(\boldsymbol{\tau} - \mathbf{r})} f(\boldsymbol{\tau} - \mathbf{r})$$

$$= \frac{1}{\Omega_c} \sum_n e^{i\mathbf{K}_n \boldsymbol{\tau}} \tilde{f}(\mathbf{K}_n) , \tag{B.4.4}$$

where $\tilde{f}$ is the Fourier-transform of f and the integral has to performed over all space. Furthermore $\boldsymbol{\tau}$ denotes an arbitrary real space vector.

In the same manner we derive the Poisson sum formula for the following two special applications. For the first one we start from

$$\sum_\mu e^{-i\mathbf{k}(\boldsymbol{\tau} - \mathbf{R}_\mu)} |\boldsymbol{\tau} - \mathbf{R}_\mu|^l e^{-(\boldsymbol{\tau} - \mathbf{R}_\mu)^2 \xi^2} Y_L(\widehat{\boldsymbol{\tau} - \mathbf{R}_\mu})$$

$$= \int d^3\mathbf{r} \sum_\mu \delta(\mathbf{r} - \mathbf{R}_\mu) |\boldsymbol{\tau} - \mathbf{r}|^l Y_L(\widehat{\boldsymbol{\tau} - \mathbf{r}}) e^{[-(\boldsymbol{\tau} - \mathbf{r})^2 \xi^2 - i\mathbf{k}(\boldsymbol{\tau} - \mathbf{r})]}$$

$$= \frac{1}{\Omega_c} \sum_n \int d^3r \; e^{i\mathbf{K}_n \mathbf{r}} |\boldsymbol{\tau} - \mathbf{r}|^l Y_L(\widehat{\boldsymbol{\tau} - \mathbf{r}}) e^{[-(\boldsymbol{\tau}-\mathbf{r})^2 \xi^2 - i\mathbf{k}(\boldsymbol{\tau}-\mathbf{r})]}$$

$$= \frac{1}{\Omega_c} \sum_n e^{i\mathbf{K}_n \boldsymbol{\tau}} \int d^3r \; |\boldsymbol{\tau} - \mathbf{r}|^l Y_L(\widehat{\boldsymbol{\tau} - \mathbf{r}}) e^{[-(\boldsymbol{\tau}-\mathbf{r})^2 \xi^2 - i(\mathbf{K}_n+\mathbf{k})(\boldsymbol{\tau}-\mathbf{r})]}$$

$$= \frac{1}{\Omega_c} \sum_n e^{i\mathbf{K}_n \boldsymbol{\tau}} \int d^3r \; |\mathbf{r}|^l Y_L(\hat{\mathbf{r}}) e^{[-\mathbf{r}^2 \xi^2 - i(\mathbf{K}_n+\mathbf{k})\mathbf{r}]} \; . \tag{B.4.5}$$

Using the identity [14, App. B] or [12, Sect. 16.8]

$$e^{-i(\mathbf{K}_n+\mathbf{k})\mathbf{r}} = 4\pi \sum_L (-i)^l j_l(|\mathbf{K}_n+\mathbf{k}||\mathbf{r}|) Y_L^*(\hat{\mathbf{r}}) Y_L(\widehat{\mathbf{K}_n+\mathbf{k}}) \tag{B.4.6}$$

with $L = (l, m)$ and the spherical Bessel function as given by (3.1.4) we calculate the integral in the last line of (B.4.5). Taking into account the orthonormality of the spherical harmonics we obtain

$$\int d^3r \; |\mathbf{r}|^l Y_L(\hat{\mathbf{r}}) e^{[-\mathbf{r}^2 \xi^2 - i(\mathbf{K}_n+\mathbf{k})\mathbf{r}]}$$

$$= 4\pi \sum_{L'} (-i)^{l'} Y_{L'}(\widehat{\mathbf{K}_n+\mathbf{k}}) \int d^3r \; |\mathbf{r}|^l Y_L(\hat{\mathbf{r}}) Y_{L'}^*(\hat{\mathbf{r}}) j_{l'}(|\mathbf{K}_n+\mathbf{k}||\mathbf{r}|) e^{-\mathbf{r}^2 \xi^2}$$

$$= 4\pi (-i)^l Y_L(\widehat{\mathbf{K}_n+\mathbf{k}}) \int dr \; r^{l+2} j_l(|\mathbf{K}_n+\mathbf{k}|r) e^{-r^2 \xi^2}$$

$$= 4\pi (-i)^l \sqrt{\frac{\pi}{2|\mathbf{K}_n+\mathbf{k}|}} Y_L(\widehat{\mathbf{K}_n+\mathbf{k}}) \int dr \; r^{l+\frac{3}{2}} J_{l+\frac{1}{2}}(|\mathbf{K}_n+\mathbf{k}|r) e^{-r^2 \zeta^2}$$

$$= 4\pi (-i)^l \sqrt{\frac{\pi}{2|\mathbf{K}_n+\mathbf{k}|}} Y_L(\widehat{\mathbf{K}_n+\mathbf{k}}) \frac{|\mathbf{K}_n+\mathbf{k}|^{l+\frac{1}{2}}}{(2\xi^2)^{l+\frac{3}{2}}} e^{-\frac{(\mathbf{K}_n+\mathbf{k})^2}{4\xi^2}}$$

$$\text{for } |\arg \xi| < \frac{\pi}{4}$$

$$= \pi^{\frac{3}{2}} (-i)^l 2^{-l} \xi^{-(2l+3)} |\mathbf{K}_n+\mathbf{k}|^l Y_L(\widehat{\mathbf{K}_n+\mathbf{k}}) e^{-\frac{(\mathbf{K}_n+\mathbf{k})^2}{4\xi^2}} \; . \tag{B.4.7}$$

Here, $J_{l+\frac{1}{2}}(\rho)$ denotes an ordinary Bessel function and the integral has been evaluated using [10, (6.631)] and [9, p. 394, (4)]. Combining (B.4.5) and (B.4.7) we are able to note the result

$$\sum_\mu e^{-i\mathbf{k}(\boldsymbol{\tau}-\mathbf{R}_\mu)} |\boldsymbol{\tau} - \mathbf{R}_\mu|^l e^{-(\boldsymbol{\tau}-\mathbf{R}_\mu)^2 \xi^2} Y_L(\widehat{\boldsymbol{\tau} - \mathbf{R}_\mu})$$

$$= \frac{\pi^{\frac{3}{2}}}{\Omega_c} (-i)^l 2^{-l} \xi^{-(2l+3)} \sum_n e^{i\mathbf{K}_n \boldsymbol{\tau}} |\mathbf{K}_n+\mathbf{k}|^l Y_L(\widehat{\mathbf{K}_n+\mathbf{k}}) e^{-\frac{(\mathbf{K}_n+\mathbf{k})^2}{4\xi^2}} \; . \tag{B.4.8}$$

In particular, we write down the Poisson transform for $l = 0$

$$\sum_\mu e^{-[(\boldsymbol{\tau}-\mathbf{R}_\mu)^2 \xi^2 + i\mathbf{k}(\boldsymbol{\tau}-\mathbf{R}_\mu)]} = \frac{\pi^{\frac{3}{2}}}{\Omega_c \xi^3} \sum_n e^{-[\frac{(\mathbf{K}_n+\mathbf{k})^2}{4\xi^2} - i\mathbf{K}_n \boldsymbol{\tau}]} \; , \tag{B.4.9}$$

which is identical to the formula (18) given by Ewald [4] or to (A.1.4) of [5].

The second abovementioned application of the Poisson sum formula is connected to the Fourier transform of a spherical Bessel function. We start from

$$\sum_{\mu} e^{-i\mathbf{k}(\boldsymbol{\tau}-\mathbf{R}_{\mu})} j_l(\kappa|\boldsymbol{\tau}-\mathbf{R}_{\mu}|) Y_L(\widehat{\boldsymbol{\tau}-\mathbf{R}_{\mu}})$$

$$= \int d^3r \sum_{\mu} \delta(\mathbf{r}-\mathbf{R}_{\mu}) e^{-i\mathbf{k}(\boldsymbol{\tau}-\mathbf{r})} j_l(\kappa|\boldsymbol{\tau}-\mathbf{r}|) Y_L(\widehat{\boldsymbol{\tau}-\mathbf{r}})$$

$$= \frac{1}{\Omega_c} \sum_{n} \int d^3r\, e^{i\mathbf{K}_n \mathbf{r}} e^{-i\mathbf{k}(\boldsymbol{\tau}-\mathbf{r})} j_l(\kappa|\boldsymbol{\tau}-\mathbf{r}|) Y_L(\widehat{\boldsymbol{\tau}-\mathbf{r}})$$

$$= \frac{1}{\Omega_c} \sum_{n} e^{i\mathbf{K}_n \boldsymbol{\tau}} \int d^3r\, e^{-i(\mathbf{K}_n+\mathbf{k})(\boldsymbol{\tau}-\mathbf{r})} j_l(\kappa|\boldsymbol{\tau}-\mathbf{r}|) Y_L(\widehat{\boldsymbol{\tau}-\mathbf{r}})$$

$$= \frac{1}{\Omega_c} \sum_{n} e^{i\mathbf{K}_n \boldsymbol{\tau}} \int d^3r\, e^{-i(\mathbf{K}_n+\mathbf{k})\mathbf{r}} j_l(\kappa|\mathbf{r}|) Y_L(\hat{\mathbf{r}}) \,. \tag{B.4.10}$$

Again, inserting the identity (B.4.6) and using the orthonormality of the spherical harmonics we obtain

$$\int d^3r\, e^{-i(\mathbf{K}_n+\mathbf{k})\mathbf{r}} j_l(\kappa|\mathbf{r}|) Y_L(\hat{\mathbf{r}})$$

$$= 4\pi \sum_{L'} (-i)^{l'} Y_{L'}(\widehat{\mathbf{K}_n+\mathbf{k}}) \int d^3r\, j_l(\kappa|\mathbf{r}|) Y_L(\hat{\mathbf{r}}) Y_{L'}^*(\hat{\mathbf{r}}) j_{l'}(|\mathbf{K}_n+\mathbf{k}||\mathbf{r}|)$$

$$= 4\pi (-i)^l Y_L(\widehat{\mathbf{K}_n+\mathbf{k}}) \int dr\, r^2 j_l(\kappa r) j_l(|\mathbf{K}_n+\mathbf{k}|r)$$

$$= 4\pi (-i)^l Y_L(\widehat{\mathbf{K}_n+\mathbf{k}}) \frac{\pi}{2} \frac{1}{\kappa} \frac{1}{|\mathbf{K}_n+\mathbf{k}|} \left[\delta(\kappa-|\mathbf{K}_n+\mathbf{k}|) - (-)^l \delta(\kappa+|\mathbf{K}_n+\mathbf{k}|) \right]$$

$$= 4\pi (-i)^l Y_L(\widehat{\mathbf{K}_n+\mathbf{k}}) \frac{\pi}{2} \frac{1}{\kappa^2} \left[\delta(\kappa-|\mathbf{K}_n+\mathbf{k}|) + (-)^l \delta(\kappa+|\mathbf{K}_n+\mathbf{k}|) \right] \,. \tag{B.4.11}$$

Here, we applied (3.3.46) to evaluate the integral over the product of the two spherical Bessel functions. Combining (B.4.10) and (B.4.11) we read off the result

$$\sum_{\mu} e^{-i\mathbf{k}(\boldsymbol{\tau}-\mathbf{R}_{\mu})} j_l(\kappa|\boldsymbol{\tau}-\mathbf{R}_{\mu}|) Y_L(\widehat{\boldsymbol{\tau}-\mathbf{R}_{\mu}})$$

$$= \frac{2\pi^2}{\Omega_c} (-i)^l \sum_{n} e^{i\mathbf{K}_n \boldsymbol{\tau}} Y_L(\widehat{\mathbf{K}_n+\mathbf{k}}) \frac{1}{\kappa^2}$$

$$\left[\delta(\kappa-|\mathbf{K}_n+\mathbf{k}|) + (-)^l \delta(\kappa+|\mathbf{K}_n+\mathbf{k}|) \right] \,. \tag{B.4.12}$$

Finally, in close analogy to the previous application to the spherical Bessel function, we use the Poisson sum formula in order to evaluate the lattice sum of barred spherical Hankel functions as defined in (3.1.14). We thus note

$$\sum_{\mu} e^{-i\mathbf{k}(\boldsymbol{\tau}-\mathbf{R}_{\mu})}\bar{h}_l^{(1)}(\kappa|\boldsymbol{\tau}-\mathbf{R}_{\mu}|)Y_L(\widehat{\boldsymbol{\tau}-\mathbf{R}_{\mu}})$$

$$= \int d^3r \sum_{\mu} \delta(\mathbf{r}-\mathbf{R}_{\mu})e^{-i\mathbf{k}(\boldsymbol{\tau}-\mathbf{r})}\bar{h}_l^{(1)}(\kappa|\boldsymbol{\tau}-\mathbf{r}|)Y_L(\widehat{\boldsymbol{\tau}-\mathbf{r}})$$

$$= \frac{1}{\Omega_c}\sum_n \int d^3r\, e^{i\mathbf{K}_n\mathbf{r}}e^{-i\mathbf{k}(\boldsymbol{\tau}-\mathbf{r})}\bar{h}_l^{(1)}(\kappa|\boldsymbol{\tau}-\mathbf{r}|)Y_L(\widehat{\boldsymbol{\tau}-\mathbf{r}})$$

$$= \frac{1}{\Omega_c}\sum_n e^{i\mathbf{K}_n\boldsymbol{\tau}} \int d^3r\, e^{-i(\mathbf{K}_n+\mathbf{k})(\boldsymbol{\tau}-\mathbf{r})}\bar{h}_l^{(1)}(\kappa|\boldsymbol{\tau}-\mathbf{r}|)Y_L(\widehat{\boldsymbol{\tau}-\mathbf{r}})$$

$$= \frac{1}{\Omega_c}\sum_n e^{i\mathbf{K}_n\boldsymbol{\tau}} \int d^3r\, e^{-i(\mathbf{K}_n+\mathbf{k})\mathbf{r}}\bar{h}_l^{(1)}(\kappa|\mathbf{r}|)Y_L(\hat{\mathbf{r}})\,. \qquad (\text{B}.4.13)$$

Proceding as before and using the identity (B.4.6) as well as the orthonormality of the spherical harmonics we obtain

$$\int d^3r\, e^{-i(\mathbf{K}_n+\mathbf{k})\mathbf{r}}\bar{h}_l^{(1)}(\kappa|\mathbf{r}|)Y_L(\hat{\mathbf{r}})$$

$$= 4\pi \sum_{L'}(-i)^{l'} Y_{L'}(\widehat{\mathbf{K}_n+\mathbf{k}}) \int d^3r\, \bar{h}_l^{(1)}(\kappa|\mathbf{r}|)Y_L(\hat{\mathbf{r}})Y_{L'}^*(\hat{\mathbf{r}})j_{l'}(|\mathbf{K}_n+\mathbf{k}||\mathbf{r}|)$$

$$= 4\pi(-i)^l Y_L(\widehat{\mathbf{K}_n+\mathbf{k}}) \int dr\, r^2\bar{h}_l^{(1)}(\kappa r)j_l(|\mathbf{K}_n+\mathbf{k}|r)$$

$$= 4\pi(-i)^l |\mathbf{K}_n+\mathbf{k}|^l Y_L(\widehat{\mathbf{K}_n+\mathbf{k}}) \int dr\, r^2\bar{h}_l^{(1)}(\kappa r)\bar{j}_l(|\mathbf{K}_n+\mathbf{k}|r)\,. \qquad (\text{B}.4.14)$$

The latter integral is of the type (3.3.42). Abbreviating $\mathbf{q}=\mathbf{K}_n+\mathbf{k}$, we write

$$\int_0^{\infty} dr\, r^2\bar{h}_l^{(1)}(\kappa r)\bar{j}_l(qr)$$

$$= \frac{s^2}{\kappa^2-q^2}\left[\bar{h}_l^{(1)}(\kappa s)\bar{j}_{l-1}(qs)-\kappa^2\bar{h}_{l-1}^{(1)}(\kappa s)\bar{j}_l(qs)\right]-\frac{1}{\kappa^2-q^2}\Bigg|_{s\to\infty}$$

$$= \frac{s^2}{\kappa^2-q^2}\left[\frac{\kappa^l q^{-l}}{s^2}e^{i(\kappa s-l\frac{\pi}{2})}\sin\left(qs-(l-1)\frac{\pi}{2}\right)\right.$$

$$\left. -\frac{\kappa^{l+1}q^{-l-1}}{s^2}e^{i(\kappa s-(l-1)\frac{\pi}{2})}\sin\left(qs-l\frac{\pi}{2}\right)\right]-\frac{1}{\kappa^2-q^2}\Bigg|_{s\to\infty}$$

$$= \frac{1}{\kappa^2-q^2}\left[\frac{\kappa^l}{q^l}e^{i(\kappa s-l\frac{\pi}{2})}\cos(qs-l\frac{\pi}{2})-i\frac{\kappa^{l+1}}{q^{l+1}}e^{i(\kappa s-l\frac{\pi}{2})}\sin(qs-l\frac{\pi}{2})-1\right]_{s\to\infty}$$

$$= -\frac{1}{\kappa^2-q^2}\,. \qquad (\text{B}.4.15)$$

Here we have used the identities (3.1.47) and (3.1.48) for the barred functions for large arguments. Furthermore, we have assumed a negative value of κ^2, which, in addition to ensuring an exponential decay of the functions in square brackets at large arguments, prohibits the prefactor from diverging. Combining (B.4.13) to (B.4.15) we thus arrive at the result

$$
\sum_{\mu} e^{-i\mathbf{k}(\boldsymbol{\tau}-\mathbf{R}_{\mu})} H_{l\kappa}(\boldsymbol{\tau}-\mathbf{R}_{\mu})
$$

$$
= \sum_{\mu} e^{-i\mathbf{k}(\boldsymbol{\tau}-\mathbf{R}_{\mu})} \bar{h}_l^{(1)}(\kappa|\boldsymbol{\tau}-\mathbf{R}_{\mu}|) Y_L(\widehat{\boldsymbol{\tau}-\mathbf{R}_{\mu}})
$$

$$
= -\frac{4\pi}{\Omega_c}(-i)^l \sum_{n} e^{i\mathbf{K}_n\boldsymbol{\tau}} |\mathbf{K}_n+\mathbf{k}|^l Y_L(\widehat{\mathbf{K}_n+\mathbf{k}}) \frac{1}{\kappa^2-|\mathbf{K}_n+\mathbf{k}|^2} , \quad (B.4.16)
$$

where, in the first line, we have insterted the definition (3.1.17). Just for completeness, we write down (B.4.16) explicitly for the case $l=0$,

$$
\sum_{\mu} e^{i\mathbf{k}\mathbf{R}_{\mu}} \frac{e^{i\kappa|\boldsymbol{\tau}-\mathbf{R}_{\mu}|}}{|\boldsymbol{\tau}-\mathbf{R}_{\mu}|} = -\frac{4\pi}{\Omega_c} \sum_{n} \frac{e^{i(\mathbf{K}_n+\mathbf{k})\boldsymbol{\tau}}}{\kappa^2-|\mathbf{K}_n+\mathbf{k}|^2} . \quad (B.4.17)
$$

B.5 Calculation of the Ewald Integral

On the following pages we will discuss several methods to evaluate the Ewald integral. It is defined by

$$
I_l = \frac{2}{\sqrt{\pi}} \int_{\frac{1}{2}\eta^{1/2}}^{\infty} \xi^{2l} e^{[-R^2\xi^2+\frac{\kappa^2}{4\xi^2}]} d\xi , \quad (B.5.1)
$$

and enters as part of the real-space lattice summations coming with the Ewald method described in Sect. 3.5. Since these Bloch sums have to be evaluated quite often it is reasonable to consider aspects of computational efficiency here, too.

A first important step in this direction is the derivation of a recursion relation for the Ewald integral. It is based on the identities

$$
\frac{d}{d\xi} \xi^{2l-1} = (2l-1)\xi^{2l-2} , \quad (B.5.2)
$$

$$
\frac{d}{d\xi} e^{[-R^2\xi^2+\frac{\kappa^2}{4\xi^2}]} = -2[R^2\xi+\frac{\kappa^2}{4\xi^3}]e^{[-R^2\xi^2+\frac{\kappa^2}{4\xi^2}]} , \quad (B.5.3)
$$

which are used to calculate I_{l-1}. Integration by parts yields

$$
(2l-1)I_{l-1} = (2l-1)\frac{2}{\sqrt{\pi}} \int_{\frac{1}{2}\eta^{1/2}}^{\infty} \xi^{2l-2} e^{[-R^2\xi^2+\frac{\kappa^2}{4\xi^2}]} d\xi
$$

$$
= \frac{2}{\sqrt{\pi}} \left[\xi^{2l-1} e^{[-R^2\xi^2+\frac{\kappa^2}{4\xi^2}]} \right]_{\frac{1}{2}\eta^{1/2}}^{\infty}
$$

$$
+2\frac{2}{\sqrt{\pi}} \int_{\frac{1}{2}\eta^{1/2}}^{\infty} \xi^{2l-1} \left[R^2\xi+\frac{\kappa^2}{4\xi^3} \right] e^{[-R^2\xi^2+\frac{\kappa^2}{4\xi^2}]} d\xi
$$

$$
= -\frac{2}{\sqrt{\pi}} (\frac{1}{2}\eta^{1/2})^{2l-1} e^{[-\frac{1}{4}\eta R^2+\frac{\kappa^2}{\eta}]} + 2R^2 I_l + 2\frac{\kappa^2}{4} I_{l-2} . \quad (B.5.4)
$$

Hence, we note the intermediate result

$$R^2 I_l = \frac{2l-1}{2} I_{l-1} - \frac{\kappa^2}{4} I_{l-2} + \frac{1}{2}\frac{2}{\sqrt{\pi}}\left(\frac{1}{2}\eta^{1/2}\right)^{2l-1} e^{[-\frac{1}{4}\eta R^2 + \frac{\kappa^2}{\eta}]} . \tag{B.5.5}$$

Defining the dimensionless quantity

$$\bar{I}_l = R^{2l+1} I_l , \tag{B.5.6}$$

and using the abbreviations

$$a = \frac{1}{2}\eta^{1/2} R \quad \text{and} \quad b = \frac{1}{4}\kappa^2 R^2 \tag{B.5.7}$$

we arrive at the final expression

$$\begin{aligned}
\bar{I}_l &= \frac{2l-1}{2}\bar{I}_{l-1} - \frac{1}{4}\kappa^2 R^2 \bar{I}_{l-2} + \frac{1}{2}\left(\frac{1}{2}\eta^{1/2}R\right)^{2l-1} e^{[-\frac{1}{4}\eta R^2 + \frac{\kappa^2}{\eta}]} \\
&= \frac{2l-1}{2}\bar{I}_{l-1} - b\bar{I}_{l-2} + \frac{1}{2}a^{2l-1}e^{[-a^2 + \frac{b}{a^2}]} .
\end{aligned} \tag{B.5.8}$$

As a result, (B.5.5) and (B.5.8) allow to calculate all the integrals for higher l from those two with lowest l. For this reason, we will from now on concentrate on the calculation of the low-l Ewald integrals. The usual choice is $l = 0$ and $l = -1$. The latter integral is needed for the derivative of the integral I_l with respect to κ^2, which reads as

$$\frac{d}{d\kappa^2} I_l = \frac{1}{4}\int_{\frac{1}{2}\eta^{1/2}}^{\infty} \xi^{2l-2} e^{[-R^2\xi^2 + \frac{\kappa^2}{4\xi^2}]} d\xi = \frac{1}{4}I_{l-1} . \tag{B.5.9}$$

One of the most efficient ways to calculate these low-l Ewald integrals takes advantage of the fact that they can be expressed by the complementary error function as well as the complex scaled complementary error function. Nowadays, efficient computer codes exist for the calculation of both functions [7, 8, 18]. The complementary error function is defined in terms of the error function [1, (7.1.1)]

$$\mathrm{erf}(z) := \frac{2}{\sqrt{\pi}}\int_0^z e^{-t^2} dt , \tag{B.5.10}$$

as [1, (7.1.2)]

$$\mathrm{erfc}(z) := \frac{2}{\sqrt{\pi}}\int_z^{\infty} e^{-t^2} dt := 1 - \mathrm{erf}(z) . \tag{B.5.11}$$

Here we have used the identity (3.4.1). From this and the previous definitions we write down the limiting cases

$$\mathrm{erf}(\infty) = \mathrm{erfc}(0) = 1 , \tag{B.5.12}$$

$$\mathrm{erf}(0) = \mathrm{erfc}(\infty) = 0 . \tag{B.5.13}$$

Finally, the complex scaled complementary error function is defined by [1, (7.1.3)]

$$w(z) = e^{-z^2}\text{erfc}(-iz) \,,$$ (B.5.14)

which may be alternatively written as

$$\text{erfc}(z) = e^{-z^2} w(iz) \,.$$ (B.5.15)

For the latter function we note the identity [1, (7.1.12)]

$$w(z^*) = w^*(-z) \,.$$ (B.5.16)

In order to evaluate the low-l Ewald integrals we follow the same lines as in Sect. 3.4 and define a new integration variable for the integrals entering (B.5.10) and (B.5.11) by

$$t_\pm = R\xi \pm i\frac{\kappa}{2\xi}, \quad \text{i.e.} \quad \frac{dt_\pm}{d\xi} = R \mp i\frac{\kappa}{2\xi^2} \,.$$ (B.5.17)

Especially for $\xi = \frac{1}{2}\eta^{1/2}$ we note

$$t_{0\pm} = \frac{1}{2}R\eta^{1/2} \pm i\frac{\kappa}{\eta^{1/2}} = a \pm i\frac{\sqrt{b}}{a} \,,$$ (B.5.18)

where in the second step we have used the abbreviations (B.5.7). We may now rewrite the indefinite integral as

$$\int e^{-t_\pm^2} dt_\pm = \int d\xi \left(R \mp i\frac{\kappa}{2\xi^2} \right) e^{-R^2\xi^2 + \frac{\kappa^2}{4\xi^2} \mp i\kappa R} \,,$$ (B.5.19)

this leading to the identity

$$\int d\xi \left(1 \mp i\frac{\kappa}{2R\xi^2} \right) e^{-R^2\xi^2 + \frac{\kappa^2}{4\xi^2}} = \frac{e^{\pm i\kappa R}}{R} \int e^{-t_\pm^2} dt_\pm \,.$$ (B.5.20)

On adding the expressions for t_+ and t_- and inserting the limits of integration we thus obtain the result [1, (7.4.33) and (7.4.34)]

$$
\begin{aligned}
I_0 &= \int_{\frac{1}{2}\eta^{1/2}}^{\infty} e^{[-R^2\xi^2 + \frac{\kappa^2}{4\xi^2}]} d\xi \\
&= \frac{1}{2R} \left[e^{i\kappa R} \int_{t_{0+}}^{\infty} e^{-t_+^2} dt_+ + e^{-i\kappa R} \int_{t_{0-}}^{\infty} e^{-t_-^2} dt_- \right] \\
&= \frac{\sqrt{\pi}}{4R} \left[e^{i\kappa R}\text{erfc}(\frac{1}{2}\eta^{1/2}R + i\frac{\kappa}{\eta^{1/2}}) + e^{-i\kappa R}\text{erfc}(\frac{1}{2}\eta^{1/2}R - i\frac{\kappa}{\eta^{1/2}}) \right] \\
&= \frac{\sqrt{\pi}}{4R} \left[e^{i\kappa R}e^{-\frac{1}{4}\eta R^2 + \frac{\kappa^2}{\eta} - i\kappa R} w(\frac{1}{2}i\eta^{1/2}R - \frac{\kappa}{\eta^{1/2}}) \right. \\
&\qquad\qquad \left. + e^{-i\kappa R}e^{-\frac{1}{4}\eta R^2 + \frac{\kappa^2}{\eta} + i\kappa R} w(\frac{1}{2}i\eta^{1/2}R + \frac{\kappa}{\eta^{1/2}}) \right] \\
&= \frac{\sqrt{\pi}}{4R} e^{-\frac{1}{4}\eta R^2 + \frac{\kappa^2}{\eta}} \left[w(\frac{1}{2}i\eta^{1/2}R - \frac{\kappa}{\eta^{1/2}}) + w(\frac{1}{2}i\eta^{1/2}R + \frac{\kappa}{\eta^{1/2}}) \right] \,.
\end{aligned}
$$ (B.5.21)

Here we used the identity (B.5.15). Again, taking into account the abbreviations (B.5.7) we note

$$
I_0 = \frac{\sqrt{\pi}}{4R} \left[e^{2i\sqrt{b}} \operatorname{erfc}(a + \frac{i\sqrt{b}}{a}) + e^{-2i\sqrt{b}} \operatorname{erfc}(a - \frac{i\sqrt{b}}{a}) \right]
$$

$$
= \frac{\sqrt{\pi}}{4R} e^{-a^2 + \frac{b}{a^2}} \left[w(ia - \frac{\sqrt{b}}{a}) + w(ia + \frac{\sqrt{b}}{a}) \right] . \tag{B.5.22}
$$

In the previous expressions we appended alternative writings in terms of the complex scaled complementary error function, which offers particular advantage in the case of positive energies κ^2 when the arguments of the complementary error functions are complex. Whereas standard routines for the calculation of the complementary error function are available only for real arguments there exist efficient computer codes for the calculation of $w(z)$ for any argument [7, 8, 18].

Finally, replacing in the third line of (B.5.21) the complementary error function by the error function itself we write

$$
I_0 = \frac{\sqrt{\pi}}{2} \frac{\cos(\kappa R)}{R} - \frac{\sqrt{\pi}}{4R} \left[e^{i\kappa R} \operatorname{erf}(\frac{1}{2}\eta^{1/2}R + i\frac{\kappa}{\eta^{1/2}}) \right.
$$
$$
\left. + e^{-i\kappa R} \operatorname{erf}(\frac{1}{2}\eta^{1/2}R - i\frac{\kappa}{\eta^{1/2}}) \right] , \tag{B.5.23}
$$

which is identical to the result already given by Ewald [4].

Last not least, for positive energies κ^2, i.e. $b > 0$, we use the identity (B.5.16) and reduce the respective last lines of (B.5.21) and (B.5.22) to

$$
I_0 = \frac{\sqrt{\pi}}{2R} e^{-\frac{1}{4}\eta R^2 + \frac{\kappa^2}{\eta}} \Re w(\frac{1}{2}i\eta^{1/2}R - \frac{\kappa}{\eta^{1/2}})
$$
$$
= \frac{\sqrt{\pi}}{2R} e^{-a^2 + \frac{b}{a^2}} \Re w(ia - \frac{\sqrt{b}}{a}) \quad \text{for} \quad \kappa^2 > 0 . \tag{B.5.24}
$$

Next we turn to the corresponding integral for $l = -1$. Using the identities (3.2.22) and [1, (7.1.19)]

$$
\frac{d}{dz} \operatorname{erfc}(z) = -\frac{2}{\sqrt{\pi}} e^{-z^2} , \tag{B.5.25}
$$

we arrive at

$$
\frac{\partial}{\partial \kappa} \operatorname{erfc}(\frac{1}{2}\eta^{1/2}R \pm i\frac{\kappa}{\eta^{1/2}}) = \mp \frac{2}{\sqrt{\pi}} \frac{i}{\eta^{1/2}} e^{-\frac{1}{4}\eta R^2 + \frac{\kappa^2}{\eta} \mp i\kappa R} , \tag{B.5.26}
$$

and, combining this with (B.5.9) and (B.5.21), we obtain

$$I_{-1} = \frac{2}{\kappa}\frac{\partial}{\partial\kappa}I_0$$

$$= \frac{\sqrt{\pi}}{2}\frac{iR}{\kappa R}\left[e^{i\kappa R}\mathrm{erfc}(\frac{1}{2}\eta^{1/2}R + \frac{i\kappa}{\eta^{1/2}}) - e^{-i\kappa R}\mathrm{erfc}(\frac{1}{2}\eta^{1/2}R - \frac{i\kappa}{\eta^{1/2}})\right]$$

$$- \frac{i}{\kappa R\eta^{1/2}}\left[e^{i\kappa R}e^{-\frac{1}{4}\eta R^2 + \frac{\kappa^2}{\eta} - i\kappa R} - e^{-i\kappa R}e^{-\frac{1}{4}\eta R^2 + \frac{\kappa^2}{\eta} + i\kappa R}\right]$$

$$= -\frac{\sqrt{\pi}}{2}\frac{1}{i\kappa}\left[e^{i\kappa R}\mathrm{erfc}(\frac{1}{2}\eta^{1/2}R + \frac{i\kappa}{\eta^{1/2}}) - e^{-i\kappa R}\mathrm{erfc}(\frac{1}{2}\eta^{1/2}R - \frac{i\kappa}{\eta^{1/2}})\right]$$

$$= -\frac{\sqrt{\pi}}{2}\frac{1}{i\kappa}\left[e^{i\kappa R}e^{-\frac{1}{4}\eta R^2 + \frac{\kappa^2}{\eta} - i\kappa R}w(\frac{1}{2}i\eta^{1/2}R - \frac{\kappa}{\eta^{1/2}})\right.$$

$$\left. - e^{-i\kappa R}e^{-\frac{1}{4}\eta R^2 + \frac{\kappa^2}{\eta} + i\kappa R}w(\frac{1}{2}i\eta^{1/2}R + \frac{\kappa}{\eta^{1/2}})\right]$$

$$= -\frac{\sqrt{\pi}}{2}\frac{1}{i\kappa}e^{-\frac{1}{4}\eta R^2 + \frac{\kappa^2}{\eta}}\left[w(\frac{1}{2}i\eta^{1/2}R - \frac{\kappa}{\eta^{1/2}}) - w(\frac{1}{2}i\eta^{1/2}R + \frac{\kappa}{\eta^{1/2}})\right] .$$

$$\text{(B.5.27)}$$

An alternative way of deriving this result consists of starting from (B.5.20) as for the calculation of I_0 but now taking the difference of the expressions for t_+ and t_-, this leading to

$$-\frac{i\kappa}{R}\int_{\frac{1}{2}\eta^{1/2}}^{\infty}\xi^{-2}e^{[-R^2\xi^2 + \frac{\kappa^2}{4\xi^2}]}d\xi = \frac{1}{R}\left[e^{i\kappa R}\int_{t_{0+}}^{\infty}e^{-t_+^2}dt_+ - e^{-i\kappa R}\int_{t_{0-}}^{\infty}e^{-t_-^2}dt_-\right] .$$

$$\text{(B.5.28)}$$

From this we get immediately

$$I_{-1} = -\frac{1}{i\kappa}\left[e^{i\kappa R}\int_{t_{0+}}^{\infty}e^{-t_+^2}dt_+ - e^{-i\kappa R}\int_{t_{0-}}^{\infty}e^{-t_-^2}dt_-\right]$$

$$= -\frac{\sqrt{\pi}}{2}\frac{1}{i\kappa}\left[e^{i\kappa R}\mathrm{erfc}(\frac{1}{2}\eta^{1/2}R + \frac{i\kappa}{\eta^{1/2}}) - e^{-i\kappa R}\mathrm{erfc}(\frac{1}{2}\eta^{1/2}R - \frac{i\kappa}{\eta^{1/2}})\right] ,$$

$$\text{(B.5.29)}$$

which is identical to (B.5.27).

Finally, using in (B.5.27) the error function instead of the complementary error function itself we obtain

$$I_{-1} = -\frac{\sqrt{\pi}}{\kappa}\sin(\kappa R) - \frac{\sqrt{\pi}}{2}\frac{1}{i\kappa}\left[e^{i\kappa R}\mathrm{erf}(\frac{1}{2}\eta^{1/2}R + i\frac{\kappa}{\eta^{1/2}})\right.$$

$$\left. - e^{-i\kappa R}\mathrm{erf}(\frac{1}{2}\eta^{1/2}R - i\frac{\kappa}{\eta^{1/2}})\right] . \quad \text{(B.5.30)}$$

As before we combine the result (B.5.27) with the abbreviations (B.5.7) and write

$$I_{-1} = -\frac{\sqrt{\pi}}{4}\frac{R}{i\sqrt{b}}\left[e^{2i\sqrt{b}}\mathrm{erfc}(a+\frac{i\sqrt{b}}{a}) - e^{-2i\sqrt{b}}\mathrm{erfc}(a-\frac{i\sqrt{b}}{a})\right]$$

$$= -\frac{\sqrt{\pi}}{4}\frac{R}{i\sqrt{b}}e^{-a^2+\frac{b}{a^2}}\left[w(ia-\frac{\sqrt{b}}{a}) - w(ia+\frac{\sqrt{b}}{a})\right] . \qquad (B.5.31)$$

Again, for the abovementioned reasons, we added in the respective last lines of (B.5.27) and (B.5.31) the formulation based on the complex scaled complementary error function.

For positive energies κ^2 we employ the identity (B.5.16) and rewrite the last lines of (B.5.27) and (B.5.31) as

$$I_{-1} = -\sqrt{\pi}\frac{1}{i\kappa}e^{-\frac{1}{4}\eta R^2+\frac{\kappa^2}{\eta}}\,\Im w(\frac{1}{2}i\eta^{1/2}R - \frac{\kappa}{\eta^{1/2}})$$

$$= -\frac{\sqrt{\pi}}{2}\frac{R}{\sqrt{b}}e^{-a^2+\frac{b}{a^2}}\,\Im w(ia-\frac{\sqrt{b}}{a}) \qquad \text{for} \quad \kappa^2 > 0 . \qquad (B.5.32)$$

For $\kappa^2 = 0$ we use l'Hospitals rule to evaluate I_{-1} and with the help of (B.5.27) we obtain

$$I_{-1}\,|_{\kappa^2=0} = -\frac{\sqrt{\pi}}{2}\frac{1}{i}\left[2iR\,\mathrm{erfc}(\frac{1}{2}\eta^{1/2}R) - 2\frac{2}{\sqrt{\pi}}\frac{i}{\eta^{1/2}}e^{-\frac{1}{4}\eta R^2}\right]$$

$$= R\left[\frac{1}{\frac{1}{2}\eta^{1/2}R}e^{-\frac{1}{4}\eta R^2} - \sqrt{\pi}\mathrm{erfc}(\frac{1}{2}\eta^{1/2}R)\right]$$

$$= R\left[\frac{1}{a}e^{-a^2} - \sqrt{\pi}\mathrm{erfc}(a)\right] . \qquad (B.5.33)$$

The method just outlined has proven very accurate and fast for a wide range of parameters R, κ^2 and η. Hence, its use is highly recommended and there would be no reason to discuss other methods to calculate the Ewald integral. If we do so in the following it is just for reasons of completeness. However, due to their limited accuracy these alternative schemes are no longer in use.

The second method to be discussed is limited to small absolute values of κ^2. It starts out from an expansion of the Ewald integral with respect to κ^2, i.e.

$$I_l = \sum_{\alpha=0}^{\infty} c_\alpha \kappa^{2\alpha} . \qquad (B.5.34)$$

The coefficients c_α can be easily calculated by applying the identity (B.5.9) to the recursion relation (B.5.5). This results in

$$R^2 I_l = 2(2l-1)\frac{\partial}{\partial\kappa^2}I_l - 4\kappa^2\frac{\partial^2}{\partial(\kappa^2)^2}I_l + (\frac{1}{2}\eta^{1/2})^{2l-1}e^{[-\frac{1}{4}\eta R^2+\frac{\kappa^2}{\eta}]} . \qquad (B.5.35)$$

Combining this with the ansatz (B.5.34) and using the expansion

$$e^{\frac{\kappa^2}{\eta}} = \sum_{\alpha=0}^{\infty} \frac{1}{\alpha!} \left(\frac{\kappa^2}{\eta}\right)^{\alpha} , \tag{B.5.36}$$

we get

$$\sum_{\alpha=0}^{\infty} \left[R^2 c_{\alpha} - \left(\frac{1}{2}\eta^{1/2}\right)^{2l-1} e^{-\frac{1}{4}\eta R^2} \frac{1}{\alpha!} \frac{1}{\eta^{\alpha}} \right] \kappa^{2\alpha}$$

$$= \sum_{\alpha=0}^{\infty} [2(2l-1)\alpha c_{\alpha} - 4\alpha(\alpha-1)c_{\alpha}] \kappa^{2\alpha-2} . \tag{B.5.37}$$

This gives rise to the following recursion relation for the coefficients

$$(2(2l+1) - 4\alpha)\alpha c_{\alpha} = R^2 c_{\alpha-1} - \left(\frac{1}{2}\eta^{1/2}\right)^{2l-1} e^{-\frac{1}{4}\eta R^2} \frac{1}{(\alpha-1)!} \frac{1}{\eta^{\alpha-1}} . \tag{B.5.38}$$

In particular, for $l = 0$ we note

$$-2\alpha(2\alpha-1)c_{\alpha} = R^2 c_{\alpha-1} - \frac{e^{-\frac{1}{4}\eta R^2}}{\frac{1}{2}\eta^{1/2}} \frac{1}{(\alpha-1)!} \frac{1}{\eta^{\alpha-1}} , \tag{B.5.39}$$

or

$$c_{\alpha}\kappa^{2\alpha} = \frac{-\kappa^2}{2\alpha(2\alpha-1)} \left[R^2 c_{\alpha-1}\kappa^{2\alpha-2} - \frac{e^{-\frac{1}{4}\eta R^2}}{\frac{1}{2}\eta^{1/2}} \frac{1}{(\alpha-1)!} \left(\frac{\kappa^2}{\eta}\right)^{\alpha-1} \right]$$

$$= \frac{-2b}{\alpha(2\alpha-1)} \left[c_{\alpha-1}\kappa^{2\alpha-2} - \frac{e^{-a^2}}{aR} \frac{1}{(\alpha-1)!} \left(\frac{b}{a^2}\right)^{\alpha-1} \right] , \tag{B.5.40}$$

with

$$c_0 = I_0 |_{\kappa^2=0} = \frac{\sqrt{\pi}}{2R} \text{erfc}\left(\frac{1}{2}\eta^{1/2}R\right) = \frac{\sqrt{\pi}}{2R} \text{erfc}(a) . \tag{B.5.41}$$

For $l = -1$ we get

$$-2\alpha(2\alpha+1)c_{\alpha} = R^2 c_{\alpha-1} - \frac{e^{-\frac{1}{4}\eta R^2}}{(\frac{1}{2}\eta^{1/2})^3} \frac{1}{(\alpha-1)!} \frac{1}{\eta^{\alpha-1}}, \tag{B.5.42}$$

or

$$c_{\alpha}\kappa^{2\alpha} = \frac{-\kappa^2}{2\alpha(2\alpha+1)} \left[R^2 c_{\alpha-1}\kappa^{2\alpha-2} - \frac{e^{-\frac{1}{4}\eta R^2}}{(\frac{1}{2}\eta^{1/2})^3} \frac{1}{(\alpha-1)!} \left(\frac{\kappa^2}{\eta}\right)^{\alpha-1} \right]$$

$$= \frac{-2b}{\alpha(2\alpha+1)} \left[c_{\alpha-1}\kappa^{2\alpha-2} - R\frac{e^{-a^2}}{a^3} \frac{1}{(\alpha-1)!} \left(\frac{b}{a^2}\right)^{\alpha-1} \right] . \tag{B.5.43}$$

The coefficient c_0 results from

$$c_0 = I_{-1}|_{\kappa^2=0} = -\sqrt{\pi}\,\mathrm{Rerfc}(\tfrac{1}{2}\eta^{1/2}R) + \frac{2}{\eta^{1/2}}e^{-(\frac{1}{2}\eta^{1/2}R)}$$

$$= -\sqrt{\pi}\,\mathrm{Rerfc}(a) + \frac{R}{a}e^{-a^2} . \tag{B.5.44}$$

If we prefer to calculate

$$\bar{I}_l = \sum_{\alpha=0}^{\infty}\bar{c}_\alpha\kappa^{2\alpha} = \sum_{\alpha=0}^{\infty}c_\alpha R^{2l+1}\kappa^{2\alpha} , \tag{B.5.45}$$

these coefficients would fulfill for $l = 0$

$$\bar{c}_\alpha\kappa^{2\alpha} = \frac{-2b}{\alpha(2\alpha-1)}\left[\bar{c}_{\alpha-1}\kappa^{2\alpha-2} - \frac{e^{-a^2}}{a}\frac{1}{(\alpha-1)!}(\frac{b}{a^2})^{\alpha-1}\right] , \tag{B.5.46}$$

with

$$\bar{c}_0 = \frac{\sqrt{\pi}}{2}\mathrm{erfc}(a) , \tag{B.5.47}$$

and for $l = -1$

$$\bar{c}_\alpha\kappa^{2\alpha} = \frac{-2b}{\alpha(2\alpha+1)}\left[\bar{c}_{\alpha-1}\kappa^{2\alpha-2} - \frac{e^{-a^2}}{a^3}\frac{1}{(\alpha-1)!}(\frac{b}{a^2})^{\alpha-1}\right] \tag{B.5.48}$$

with

$$\bar{c}_0 = -\sqrt{\pi}\mathrm{erfc}(a) + \frac{e^{-a^2}}{a} . \tag{B.5.49}$$

For small values of $|\kappa^2|$ the previous series expansion is stable and needs only a factor 1.5 to 2 more computer time as the direct calculation via the error function used in the first method. However, when $|\kappa^2|$ increases, convergence of the series takes more time and so does the whole calculation.

Next we present a third method, which, however, is numerically stable only for negative κ^2. Another disadvantage is its computer time demand, which is roughly a factor four to six more than by use of the error function. Still, due to the arguments given, it may at least serve for a comparison with the results of the other methods.

Again, we start from the general expression (B.5.1) and substitute the integration variable ξ by

$$x = R^2\xi^2 - \frac{1}{4}\eta R^2 \quad \text{i.e.} \quad \xi = \frac{1}{R}\sqrt{x + \frac{1}{4}\eta R^2} . \tag{B.5.50}$$

Hence, we have

$$\frac{\partial\xi}{\partial x} = \frac{1}{2R}\frac{1}{\sqrt{x + \frac{1}{4}\eta R^2}} . \tag{B.5.51}$$

From this it follows

$$I_l = \frac{1}{2} \int_0^\infty \frac{1}{R^{2l+1}} \left(\sqrt{x + \frac{1}{4}\eta R^2}\right)^{2l-1} e^{[-x - \frac{1}{4}\eta R^2 + \frac{\kappa^2 R^2}{4}(x + \frac{1}{4}\eta R^2)^{-1}]} dx$$

$$= \frac{1}{R^{2l+1}} \frac{1}{2e^{\frac{1}{4}\eta R^2}} \int_0^\infty \left(\sqrt{x + \frac{1}{4}\eta R^2}\right)^{2l-1} e^{-x} e^{\frac{\kappa^2 R^2}{4}(x + \frac{1}{4}\eta R^2)^{-1}} dx$$

$$= \frac{1}{R^{2l+1}} \frac{1}{2e^{a^2}} \int_0^\infty \left(\sqrt{x + a^2}\right)^{2l-1} e^{-x} e^{\frac{b}{x+a^2}} dx . \tag{B.5.52}$$

The final evaluation is done numerically by a Gauss-Laguerre integration, which represents the integral as a finite sum, i.e. [1, (25.4.45)]

$$\int_0^\infty e^{-x} f(x) dx = \sum_{i=1}^n w_i f(x_i) + R_n . \tag{B.5.53}$$

Here, the x_i are the zeros of the Laguerre-polynomial $L_n(x)$ of order n and the weights arise from the formula

$$w_i = \frac{x_i}{(n+1)^2 [L_{n+1}(x_i)]^2} . \tag{B.5.54}$$

Finally, the error R_n may be estimated from

$$R_n = \frac{(n!)^2}{(2n)!} f^{(2n)}(\xi) \quad \text{where} \quad 0 < \xi < \infty . \tag{B.5.55}$$

For fixed n the abscissae x_i and the weights w_i are calculated once and stored for further use. Experience has shown that $n = 32$ is sufficient for our purpose, in which case we have

$x_1 = .44489365833267021 * 10^{-01}$, $w_1 = .10921834195238497 * 10^{+00}$,

$x_2 = .23452610951961853 * 10^{+00}$, $w_2 = .21044310793881321 * 10^{+00}$,

$x_3 = .57688462930188644 * 10^{+00}$, $w_3 = .23521322966984803 * 10^{+00}$,

$x_4 = .10724487538178176 * 10^{+01}$, $w_4 = .19590333597288107 * 10^{+00}$,

$x_5 = .17224087764446454 * 10^{+01}$, $w_5 = .12998378628607177 * 10^{+00}$,

$x_6 = .25283367064257951 * 10^{+01}$, $w_6 = .70578623865717421 * 10^{-01}$,

$x_7 = .34922132730219948 * 10^{+01}$, $w_7 = .31760912509175059 * 10^{-01}$,

$x_8 = .46164567697497674 * 10^{+01}$, $w_8 = .11918214834838558 * 10^{-01}$,

$x_9 = .59039585041742439 * 10^{+01}$, $w_9 = .37388162946115247 * 10^{-02}$,

$x_{10} = .73581267331862410 * 10^{+01}$, $w_{10} = .98080330661495536 * 10^{-03}$,

$x_{11} = .89829409242125955 * 10^{+01}$, $w_{11} = .21486491880136428 * 10^{-03}$,

$x_{12} = .10783018632539973 * 10^{+02}$, $w_{12} = .39203419679879449 * 10^{-04}$,

$x_{13} = .12763697986742725 * 10^{+02}$, $w_{13} = .59345416128686309 * 10^{-05}$,

$x_{14} = .14931139755522556 * 10^{+02}$, $w_{14} = .74164045786675604 * 10^{-06}$,

$x_{15} = .17292454336715316 * 10^{+02}$, $w_{15} = .76045678791207707 * 10^{-07}$,

$$x_{16} = .19855860940336054 * 10^{+02} \ , \ w_{16} = .63506022266258105 * 10^{-08},$$
$$x_{17} = .22630889013196775 * 10^{+02} \ , \ w_{17} = .42813829710409264 * 10^{-09},$$
$$x_{18} = .25628636022459247 * 10^{+02} \ , \ w_{18} = .23058994918913375 * 10^{-10},$$
$$x_{19} = .28862101816323474 * 10^{+02} \ , \ w_{19} = .97993792887271036 * 10^{-12},$$
$$x_{20} = .32346629153964734 * 10^{+02} \ , \ w_{20} = .32378016577292753 * 10^{-13},$$
$$x_{21} = .36100494805751978 * 10^{+02} \ , \ w_{21} = .81718234434206828 * 10^{-15},$$
$$x_{22} = .40145719771539440 * 10^{+02} \ , \ w_{22} = .15421338333938256 * 10^{-16},$$
$$x_{23} = .44509207995754934 * 10^{+02} \ , \ w_{23} = .21197922901636264 * 10^{-18},$$
$$x_{24} = .49224394987308635 * 10^{+02} \ , \ w_{24} = .20544296737880544 * 10^{-20}, \quad \text{(B.5.56)}$$
$$x_{25} = .54333721333396909 * 10^{+02} \ , \ w_{25} = .13469825866373929 * 10^{-22},$$
$$x_{26} = .59892509162134019 * 10^{+02} \ , \ w_{26} = .56612941303973577 * 10^{-25},$$
$$x_{27} = .65975377287935046 * 10^{+02} \ , \ w_{27} = .14185605454630460 * 10^{-27},$$
$$x_{28} = .72687628090662713 * 10^{+02} \ , \ w_{28} = .19133754944542158 * 10^{-30},$$
$$x_{29} = .80187446977913524 * 10^{+02} \ , \ w_{29} = .11922487600982213 * 10^{-33},$$
$$x_{30} = .88735340417892402 * 10^{+02} \ , \ w_{30} = .26715112192401271 * 10^{-37},$$
$$x_{31} = .98829542868283966 * 10^{+02} \ , \ w_{31} = .13386169421062655 * 10^{-41},$$
$$x_{32} = .11175139809793770 * 10^{+03} \ , \ w_{32} = .45105361938989741 * 10^{-47} \ .$$

B.6 Calculation of the Complementary Ewald Integral

The following pages are devoted to the calculation of the complementary Ewald integral, which is defined as

$$I_l^{(c)} = \int_0^{\frac{1}{2}\eta^{1/2}} \xi^{2l} e^{[-R^2\xi^2 + \frac{\kappa^2}{4\xi^2}]} d\xi \ . \tag{B.6.1}$$

Taken together with the Ewald integral as given by (B.5.1) both add up, apart from prefactors, to the integral representation (3.4.16) of the Hankel envelope function. While the Ewald integral contains the short range parts of the latter, the complementary Ewald integral represents the long range contributions.

As has been already discussed in Sect. 3.4, care has to be taken with the contour used for the integration (B.6.1) in order to circumvent divergence of the integrand at the lower boundary. We will come back to this point lateron.

As for the calculation of the Ewald integral in App. B.5, we start out setting up recursion relations. Using the identities (B.5.2) and (B.5.3) we get from integration by parts

$$(2l - 1)I_{l-1}^{(c)} = (2l - 1) \int_0^{\frac{1}{2}\eta^{1/2}} \xi^{2l-2} e^{[-R^2\xi^2 + \frac{\kappa^2}{4\xi^2}]} d\xi$$

$$= \left[\xi^{2l-1} e^{[-R^2\xi^2 + \frac{\kappa^2}{4\xi^2}]} \right]_0^{\frac{1}{2}\eta^{1/2}}$$

$$+2 \int_0^{\frac{1}{2}\eta^{1/2}} \xi^{2l-1} \left[R^2\xi + \frac{\kappa^2}{4\xi^3} \right] e^{[-R^2\xi^2 + \frac{\kappa^2}{4\xi^2}]} d\xi$$

$$= (\frac{1}{2}\eta^{1/2})^{2l-1} e^{[-\frac{1}{4}\eta R^2 + \frac{\kappa^2}{\eta}]} + 2R^2 I_l^{(c)} + 2\frac{\kappa^2}{4} I_{l-2}^{(c)} . \tag{B.6.2}$$

Hence, we get the intermediate result

$$R^2 I_l^{(c)} = \frac{2l-1}{2} I_{l-1}^{(c)} - \frac{\kappa^2}{4} I_{l-2}^{(c)} - \frac{1}{2} \left(\frac{1}{2}\eta^{1/2} \right)^{2l-1} e^{[-\frac{1}{4}\eta R^2 + \frac{\kappa^2}{\eta}]} . \tag{B.6.3}$$

Defining the dimensionless quantity

$$\bar{I}_l^{(c)} = R^{2l+1} I_l^{(c)} , \tag{B.6.4}$$

and using the abbreviations (B.5.7) we obtain

$$\bar{I}_l^{(c)} = \frac{2l-1}{2} \bar{I}_{l-1}^{(c)} - \frac{1}{4}\kappa^2 R^2 \bar{I}_{l-2}^{(c)} - \frac{1}{2}(\frac{1}{2}\eta^{1/2}R)^{2l-1} e^{[-\frac{1}{4}\eta R^2 + \frac{\kappa^2}{\eta}]}$$

$$= \frac{2l-1}{2} \bar{I}_{l-1}^{(c)} - b\bar{I}_{l-2}^{(c)} - \frac{1}{2}a^{2l-1} e^{[-a^2 + \frac{b}{a^2}]} . \tag{B.6.5}$$

So far, things are almost identical to the procedure adopted in the App. B.5 for the derivation of recursion relations for the Ewald integral. However, due to the different integration boundaries the first term in (B.6.2) as well as the last term in (B.6.3) and (B.6.5) carry a different sign as in the respective equations of the App. B.5.

With the above recursion relations at hand we have to calculate only the two integrals with the lowest l values explicitly. Although we do not need the derivative with respect to κ^2, which made calculation of the Ewald integral for $l = -1$ necessary, we still opt for evaluation of the complementary Ewald integrals for $l = 0$ and $l = -1$ as in App. B.5 for consistency reasons.

In evaluating the integrals we will make use of the integral representation of the Hankel function as well as of the results for the Ewald integral as gained in Sect. 3.4 and App. B.5. In particular, combining (B.5.1) and (B.6.1) we note

$$I_l^{(c)} = \int_0^\infty \xi^{2l} e^{[-R^2\xi^2 + \frac{\kappa^2}{4\xi^2}]} d\xi - I_l . \tag{B.6.6}$$

Using, in addition, (3.4.15) and the result (B.5.23) we write for $l = 0$

$$I_0^{(c)} = \frac{\sqrt{\pi}}{2} \frac{e^{i\kappa R}}{R} - I_0$$

$$= \frac{\sqrt{\pi}}{2} \frac{i \sin(\kappa R)}{R} + \frac{\sqrt{\pi}}{4R} \left[e^{i\kappa R} \mathrm{erf}(\frac{1}{2}\eta^{1/2}R + i\frac{\kappa}{\eta^{1/2}}) \right.$$

$$\left. + e^{-i\kappa R} \mathrm{erf}(\frac{1}{2}\eta^{1/2}R - i\frac{\kappa}{\eta^{1/2}}) \right] , \tag{B.6.7}$$

which is identical to the result given by Ewald [4]. With the help of the identity [1, (7.1.9)]

$$\operatorname{erf}(-z) = -\operatorname{erf}(z) , \tag{B.6.8}$$

the previous result can be cast into the form

$$
\begin{aligned}
I_0^{(c)} &= \frac{\sqrt{\pi}}{4R} \left[e^{i\kappa R} - e^{-i\kappa R} \right] \\
&\quad - \frac{\sqrt{\pi}}{4R} \left[e^{i\kappa R}\operatorname{erf}(-\frac{1}{2}\eta^{1/2}R - i\frac{\kappa}{\eta^{1/2}}) - e^{-i\kappa R}\operatorname{erf}(\frac{1}{2}\eta^{1/2}R - i\frac{\kappa}{\eta^{1/2}}) \right] \\
&= \frac{\sqrt{\pi}}{4R} \left[e^{i\kappa R}\operatorname{erfc}(-\frac{1}{2}\eta^{1/2}R - i\frac{\kappa}{\eta^{1/2}}) - e^{-i\kappa R}\operatorname{erfc}(\frac{1}{2}\eta^{1/2}R - i\frac{\kappa}{\eta^{1/2}}) \right] \\
&= \frac{\sqrt{\pi}}{4R} \left[e^{2i\sqrt{b}}\operatorname{erfc}(-a - i\frac{\sqrt{b}}{a}) - e^{-2i\sqrt{b}}\operatorname{erfc}(a - i\frac{\sqrt{b}}{a}) \right] .
\end{aligned}
\tag{B.6.9}
$$

For positive energies κ^2, where we use spherical Neumann rather than spherical Hankel functions, we calculate the quantity

$$
\begin{aligned}
I_0^{(c)} &- \frac{\sqrt{\pi}}{2}\frac{i\sin(\kappa R)}{R} \\
&= \frac{\sqrt{\pi}}{2}\frac{\cos(\kappa R)}{R} - I_0 \\
&= \frac{\sqrt{\pi}}{4R} \left[e^{i\kappa R}\operatorname{erf}(\frac{1}{2}\eta^{1/2}R + i\frac{\kappa}{\eta^{1/2}}) + e^{-i\kappa R}\operatorname{erf}(\frac{1}{2}\eta^{1/2}R - i\frac{\kappa}{\eta^{1/2}}) \right] .
\end{aligned}
\tag{B.6.10}
$$

Here we have used the (B.5.23) for the Ewald integral. However, for numerical calculations we employ the alternative expression (B.5.24) in terms of the complex scaled complementary error function and write the complementary Ewald integral as in (B.6.7), this leading to

$$
\begin{aligned}
I_0^{(c)} &= \frac{\sqrt{\pi}}{2R} \left[e^{i\kappa R} - e^{-\frac{1}{4}\eta R^2 + \frac{\kappa^2}{\eta}}\,\Re w(\frac{1}{2}i\eta^{1/2}R - \frac{\kappa}{\eta^{1/2}}) \right] \\
&= \frac{\sqrt{\pi}}{2R} \left[e^{2i\sqrt{b}} - e^{-a^2 + \frac{b}{a^2}}\,\Re w(ia - \frac{\sqrt{b}}{a}) \right] .
\end{aligned}
\tag{B.6.11}
$$

Again, we used in the last line the abbreviations (B.5.7). Finally, for $\kappa^2 = 0$ (B.6.10) reduces to

$$
I_0^{(c)} = \frac{\sqrt{\pi}}{2R}\operatorname{erf}(\frac{1}{2}\eta^{1/2}R) = \frac{\sqrt{\pi}}{2R}\operatorname{erf}(a) .
\tag{B.6.12}
$$

For $l = -1$ we recall (3.1.52) and (3.1.28) and write

$$
I_{-1}^{(c)} = \frac{\sqrt{\pi}}{2}2R\bar{h}_{-1}^{(1)}(\kappa r) - I_{-1}
$$

$$= -\frac{\sqrt{\pi}}{2} 2R \frac{1}{i\kappa} \bar{h}_0^{(1)}(\kappa r) - I_{-1}$$

$$= -\frac{\sqrt{\pi}}{i\kappa} e^{i\kappa R} - I_{-1}$$

$$= -\frac{\sqrt{\pi}}{i\kappa} \cos(\kappa R) - \frac{\sqrt{\pi}}{2i\kappa} \left[e^{i\kappa R} \mathrm{erf}(\frac{1}{2}\eta^{1/2}R + i\frac{\kappa}{\eta^{1/2}}) \right.$$

$$\left. - e^{-i\kappa R}\mathrm{erf}(\frac{1}{2}\eta^{1/2}R - i\frac{\kappa}{\eta^{1/2}}) \right] , \quad \text{(B.6.13)}$$

where in the last step we have used the result (B.5.31). Combining this with the identity (B.6.8) we obtain

$$I_{-1}^{(c)} = -\frac{\sqrt{\pi}}{2i\kappa} \left[e^{i\kappa R} + e^{-i\kappa R} \right]$$

$$+ \frac{\sqrt{\pi}}{2i\kappa} \left[e^{i\kappa R}\mathrm{erf}(-\frac{1}{2}\eta^{1/2}R - i\frac{\kappa}{\eta^{1/2}}) + e^{-i\kappa R}\mathrm{erf}(\frac{1}{2}\eta^{1/2}R - i\frac{\kappa}{\eta^{1/2}}) \right]$$

$$= -\frac{\sqrt{\pi}}{2i\kappa} \left[e^{i\kappa R}\mathrm{erfc}(-\frac{1}{2}\eta^{1/2}R - i\frac{\kappa}{\eta^{1/2}}) + e^{-i\kappa R}\mathrm{erfc}(\frac{1}{2}\eta^{1/2}R - i\frac{\kappa}{\eta^{1/2}}) \right]$$

$$= -\frac{\sqrt{\pi}}{2i\kappa} \left[e^{2i\sqrt{b}}\mathrm{erfc}(-a - i\frac{\sqrt{b}}{a}) + e^{-2i\sqrt{b}}\mathrm{erfc}(a - i\frac{\sqrt{b}}{a}) \right] . \quad \text{(B.6.14)}$$

For positive energies κ^2 we combine the third line of (B.6.13) with the result (B.5.32) for the Ewald integral and note

$$I_{-1}^{(c)} = -\frac{\sqrt{\pi}}{i\kappa} e^{i\kappa R} - I_{-1}$$

$$= -\frac{\sqrt{\pi}}{i\kappa} e^{i\kappa R} + \frac{\sqrt{\pi}}{i\kappa} e^{-\frac{1}{4}\eta R^2 + \frac{\kappa^2}{\eta}} \Im w(\frac{1}{2}i\eta^{1/2}R - \frac{\kappa}{\eta^{1/2}})$$

$$= -\frac{\sqrt{\pi}}{i\kappa} \left[e^{2i\sqrt{b}} - e^{-a^2 + \frac{b}{a^2}} \Im w(ia - \frac{\sqrt{b}}{a}) \right] \quad \text{for} \quad \kappa^2 > 0 . \quad \text{(B.6.15)}$$

Finally, using l'Hospitals rule and recalling the result (B.5.33) we note for $\kappa^2 = 0$

$$I_{-1}^{(c)} \big|_{\kappa^2=0} = -\sqrt{\pi}R - I_{-1} \big|_{\kappa^2=0}$$

$$= -\sqrt{\pi}R - R \left[\frac{1}{\frac{1}{2}\eta^{1/2}R} e^{-\frac{1}{4}\eta R^2} - \sqrt{\pi}\mathrm{erfc}(\frac{1}{2}\eta^{1/2}R) \right]$$

$$= -R \left[\frac{1}{\frac{1}{2}\eta^{1/2}R} e^{-\frac{1}{4}\eta R^2} + \sqrt{\pi}\mathrm{erf}(\frac{1}{2}\eta^{1/2}R) \right]$$

$$= -R \left[\frac{1}{a} e^{-a^2} + \sqrt{\pi}\mathrm{erf}(a) \right] . \quad \text{(B.6.16)}$$

For $l = 1$ we would start out from the recursion relation (B.5.5)

$$R^2 I_1 = \frac{1}{2} I_0 - \frac{\kappa^2}{4} I_{-1} + \frac{1}{2} \left(\frac{1}{2} \eta^{1/2} \right) e^{[-\frac{1}{4}\eta R^2 + \frac{\kappa^2}{\eta}]} . \tag{B.6.17}$$

Combining this with (B.6.1) and furthermore using (3.4.17) and (3.1.29) we write

$$
\begin{aligned}
I_1^{(c)} &= \frac{\sqrt{\pi}}{2} \frac{1}{2R} \left(-\frac{e^{i\kappa R}}{R} \left(i\kappa - \frac{1}{R} \right) \right) - I_1 \\
&= \frac{\sqrt{\pi}}{2} \frac{e^{i\kappa R}}{2R^2} \left(\frac{1}{R} - i\kappa \right) - \frac{1}{2R^2} I_0 + \frac{\kappa^2}{4R^2} I_{-1} - \frac{1}{2R^2} \left(\frac{1}{2} \eta^{1/2} \right) e^{[-\frac{1}{4}\eta R^2 + \frac{\kappa^2}{\eta}]} \\
&= \frac{1}{2R^2} I_0^{(c)} - \frac{\sqrt{\pi}}{2} i\kappa \frac{e^{i\kappa R}}{2R^2} + \frac{\kappa^2}{4R^2} I_{-1} - \frac{1}{2R^2} \left(\frac{1}{2} \eta^{1/2} \right) e^{[-\frac{1}{4}\eta R^2 + \frac{\kappa^2}{\eta}]} ,
\end{aligned}
\tag{B.6.18}
$$

where in the last step we have used the identity (B.6.7). Using, in addition, the result (B.5.30) we arrive at

$$
\begin{aligned}
I_1^{(c)} &= \frac{1}{2R^2} \frac{\sqrt{\pi}}{2} \frac{i \sin(\kappa R)}{R} \\
&\quad + \frac{\sqrt{\pi}}{8R^3} \left[e^{i\kappa R} \mathrm{erf}(\frac{1}{2}\eta^{1/2} R + i\frac{\kappa}{\eta^{1/2}}) + e^{-i\kappa R} \mathrm{erf}(\frac{1}{2}\eta^{1/2} R - i\frac{\kappa}{\eta^{1/2}}) \right] \\
&\quad - \frac{\sqrt{\pi}}{2} i\kappa \frac{e^{i\kappa R}}{2R^2} - \frac{\kappa^2}{4R^2} \frac{\sqrt{\pi}}{\kappa} \sin(\kappa R) \\
&\quad - \frac{\sqrt{\pi}}{2} \frac{i\kappa}{4R^2} \left[e^{i\kappa R} \mathrm{erf}(\frac{1}{2}\eta^{1/2} R + i\frac{\kappa}{\eta^{1/2}}) - e^{-i\kappa R} \mathrm{erf}(\frac{1}{2}\eta^{1/2} R - i\frac{\kappa}{\eta^{1/2}}) \right] \\
&\quad - \frac{1}{2R^2} \left(\frac{1}{2}\eta^{1/2} \right) e^{[-\frac{1}{4}\eta R^2 + \frac{\kappa^2}{\eta}]} \\
&= \frac{\sqrt{\pi}}{2} \frac{i\kappa}{2R} \left[\frac{\sin(\kappa R)}{\kappa R^2} - \frac{\cos(\kappa R)}{R} \right] \\
&\quad + \frac{\sqrt{\pi}}{8R^3} \left[e^{i\kappa R} \mathrm{erf}(\frac{1}{2}\eta^{1/2} R + i\frac{\kappa}{\eta^{1/2}}) + e^{-i\kappa R} \mathrm{erf}(\frac{1}{2}\eta^{1/2} R - i\frac{\kappa}{\eta^{1/2}}) \right] \\
&\quad - \frac{\sqrt{\pi}}{2} \frac{i\kappa}{4R^2} \left[e^{i\kappa R} \mathrm{erf}(\frac{1}{2}\eta^{1/2} R + i\frac{\kappa}{\eta^{1/2}}) - e^{-i\kappa R} \mathrm{erf}(\frac{1}{2}\eta^{1/2} R - i\frac{\kappa}{\eta^{1/2}}) \right] \\
&\quad - \frac{1}{2R^2} \left(\frac{1}{2}\eta^{1/2} \right) e^{[-\frac{1}{4}\eta R^2 + \frac{\kappa^2}{\eta}]} \\
&= \frac{\sqrt{\pi}}{2} \frac{i\kappa^3}{2R} \bar{j}_1(\kappa R) \\
&\quad + \frac{\sqrt{\pi}}{8R^3} \left[e^{i\kappa R} \mathrm{erf}(\frac{1}{2}\eta^{1/2} R + i\frac{\kappa}{\eta^{1/2}}) + e^{-i\kappa R} \mathrm{erf}(\frac{1}{2}\eta^{1/2} R - i\frac{\kappa}{\eta^{1/2}}) \right] \\
&\quad - \frac{\sqrt{\pi}}{2} \frac{i\kappa}{4R^2} \left[e^{i\kappa R} \mathrm{erf}(\frac{1}{2}\eta^{1/2} R + i\frac{\kappa}{\eta^{1/2}}) - e^{-i\kappa R} \mathrm{erf}(\frac{1}{2}\eta^{1/2} R - i\frac{\kappa}{\eta^{1/2}}) \right] \\
&\quad - \frac{1}{2R^2} \left(\frac{1}{2}\eta^{1/2} \right) e^{[-\frac{1}{4}\eta R^2 + \frac{\kappa^2}{\eta}]} ,
\end{aligned}
\tag{B.6.19}
$$

which again could be rewritten in terms of the complex scaled complementary error function. Finally, with the identities (B.6.7) and (B.6.19) at hand we are able to evaluate all higher complementary Ewald integrals with the help of the recursion relation (B.6.3).

B.7 Calculation of $D_{L\kappa}^{(3)}$ and $K_{L\kappa}^{(3)}$

The quantity $D_{L\kappa}^{(3)}(\tau)$ as defined in Sect. 3.5 contains the long-range contributions of a single envelope Hankel function. According to (3.5.21) it is defined by

$$D_{L\kappa}^{(3)}(\tau) = -\frac{1}{\pi}\delta(\tau)\delta_{l0}I \; , \tag{B.7.1}$$

where

$$I = \int_{0e^{\frac{i}{2}(\arg\kappa - \pi + \omega)}}^{\frac{1}{2}\eta^{1/2}} e^{\frac{\kappa^2}{4\xi^2}}\, d\xi \; . \tag{B.7.2}$$

The calculation of this integral may be performed in a straightforward manner for negative energies κ^2 (see [5, App. A.2]). As already discussed in Sect. 3.4, the contour for the integral (B.7.2) may be chosen as to coincide completely with the real axis. However, for positive energies κ^2 care has to be taken in order to avoid the singularity at the origin. Hence, a suitable contour has to be chosen.

A different strategy consists of representing the integral as the difference of two others which both allow for a very simple evaluation. We write

$$
\begin{aligned}
I &= \lim_{r\to 0}\int_{0e^{\frac{i}{2}(\arg\kappa - \pi + \omega)}}^{\frac{1}{2}\eta^{1/2}} e^{[-r^2\xi^2 + \frac{\kappa^2}{4\xi^2}]}\, d\xi \\
&= \lim_{r\to 0}\int_{0e^{\frac{i}{2}(\arg\kappa - \pi + \omega)}}^{\infty e^{\frac{i}{2}(\arg\kappa - \omega)}} e^{[-r^2\xi^2 + \frac{\kappa^2}{4\xi^2}]}\, d\xi - \lim_{r\to 0}\int_{\frac{1}{2}\eta^{1/2}}^{\infty e^{\frac{i}{2}(\arg\kappa - \omega)}} e^{[-r^2\xi^2 + \frac{\kappa^2}{4\xi^2}]}\, d\xi \; .
\end{aligned}
\tag{B.7.3}
$$

Here we have, in addition, multiplied the integrand with the exponential $e^{-r^2\xi^2}$, the effect of which will be removed lateron by performing the limit $r \to 0$. As a consequence we are now able to identify the first integral on the right-hand side with the barred Hankel function for $l = 0$ as given by (3.4.5) and (3.4.6), respectively (except for a factor $\frac{2}{\sqrt{\pi}}$). In contrast, the second term is just the Ewald integral for $l = 0$, which can be evaluated as described in App. B.5. Hence, we have just replaced the long-range terms of a single Hankel envelope function with the difference of the Hankel function itself and its short range terms, which are represented by the Ewald integral.

Following now the recipes of App. B.5 to calculate the Ewald integral we first set the phase ω to $\arg\kappa$ in order to arrive at a contour which lies on the real axis for all $\xi \geq \frac{1}{2}\eta^{1/2}$. Then we may use the result (B.5.21) as well as (3.4.6) to get

$$I = \frac{\sqrt{\pi}}{2} \lim_{r \to 0} \frac{e^{i\kappa r}}{r}$$

$$- \frac{\sqrt{\pi}}{4} \lim_{r \to 0} \frac{1}{r} \left[e^{i\kappa r} \mathrm{erfc}(\frac{1}{2}\eta^{1/2}r + \frac{i\kappa}{\eta^{1/2}}) + e^{-i\kappa r} \mathrm{erfc}(\frac{1}{2}\eta^{1/2}r - \frac{i\kappa}{\eta^{1/2}}) \right]$$

$$= \frac{\sqrt{\pi}}{2} \lim_{r \to 0} \frac{1}{r} \left[e^{i\kappa r} - \frac{1}{2} \left(e^{i\kappa r} + e^{-i\kappa r} \right) + \frac{1}{2} e^{i\kappa r} \mathrm{erf}(\frac{1}{2}\eta^{1/2}r + \frac{i\kappa}{\eta^{1/2}}) \right.$$

$$\left. + \frac{1}{2} e^{-i\kappa r} \mathrm{erf}(\frac{1}{2}\eta^{1/2}r - \frac{i\kappa}{\eta^{1/2}}) \right] . \tag{B.7.4}$$

Here we have used the identity (B.5.11) to replace the complementary error function by the error function. Next we perform the limit $r \to 0$, in which case the square bracket on the right-hand side of (B.7.4) reduces to the expression

$$\left[\frac{1}{2} \mathrm{erf}(\frac{i\kappa}{\eta^{1/2}}) + \frac{1}{2} \mathrm{erf}(-\frac{i\kappa}{\eta^{1/2}}) \right] ,$$

which vanishes due to the identity [1, (7.1.9)]

$$\mathrm{erf}(-z) = -\mathrm{erf}(z) . \tag{B.7.5}$$

As a consequence, in the limit $r \to 0$ both the numerator and the denominator on the right-hand side of (B.7.4) vanish and we have to apply l'Hospitals rule to get the correct result. Using the identity (B.5.25) and, finally, performing the limit $r \to 0$ we arrive at

$$I = \frac{\sqrt{\pi}}{2} i\kappa - \frac{\sqrt{\pi}}{4} \left[i\kappa \mathrm{erfc}(\frac{i\kappa}{\eta^{1/2}}) - i\kappa \mathrm{erfc}(-\frac{i\kappa}{\eta^{1/2}}) \right] + \frac{1}{2} \eta^{1/2} e^{\kappa^2/\eta}$$

$$= \frac{\sqrt{\pi}}{2} i\kappa + \frac{\sqrt{\pi}}{2} i\kappa \mathrm{erf}(\frac{i\kappa}{\eta^{1/2}}) + \frac{1}{2} \eta^{1/2} e^{\kappa^2/\eta} , \tag{B.7.6}$$

where again we applied (B.5.11) and (B.7.5). As for the calculation of the Ewald integral we use, in addition, the formulation with the complex scaled complementary error function

$$I = \frac{\sqrt{\pi}}{2} i\kappa - \frac{\sqrt{\pi}}{4} i\kappa e^{\frac{\kappa^2}{\eta}} \left[w(\frac{-\kappa}{\eta^{1/2}}) - w(\frac{\kappa}{\eta^{1/2}}) \right] + \frac{1}{2} \eta^{1/2} e^{\kappa^2/\eta} . \tag{B.7.7}$$

As already discussed in Sect. B.5, it enables for an efficient calculation especially for positive energies κ^2, in which case no standard routines for the complementary error function are available. Using the identity (B.5.16) for this case we may rewrite (B.7.7) as

$$I = \frac{\sqrt{\pi}}{2} i\kappa + \frac{\sqrt{\pi}}{2} \kappa e^{\frac{\kappa^2}{\eta}} \Im w(\frac{-\kappa}{\eta^{1/2}}) + \frac{1}{2} \eta^{1/2} e^{\kappa^2/\eta} \qquad \text{for} \ \ \kappa^2 > 0 . \tag{B.7.8}$$

Finally, we combine (B.7.1), (B.7.6) and (B.7.8) to note the result

$$D_{L\kappa}^{(3)}(\boldsymbol{\tau}) = \delta(\boldsymbol{\tau})\delta_{l0}\left[\frac{i\kappa}{4\sqrt{\pi}}\left(\mathrm{erfc}(\frac{i\kappa}{\eta^{1/2}}) - \mathrm{erfc}(-\frac{i\kappa}{\eta^{1/2}})\right) - \frac{\eta^{1/2}}{2\pi}e^{\kappa^2/\eta} - \frac{i\kappa}{2\sqrt{\pi}}\right]$$

$$= -\delta(\boldsymbol{\tau})\delta_{l0}\left[\frac{i\kappa}{2\sqrt{\pi}}\mathrm{erf}(\frac{i\kappa}{\eta^{1/2}}) + \frac{\eta^{1/2}}{2\pi}e^{\kappa^2/\eta} + \frac{i\kappa}{2\sqrt{\pi}}\right]$$

$$= -\delta(\boldsymbol{\tau})\delta_{l0}\left[\frac{\kappa}{2\sqrt{\pi}}e^{\frac{\kappa^2}{\eta}}\Im w(\frac{-\kappa}{\eta^{1/2}}) + \frac{\eta^{1/2}}{2\pi}e^{\kappa^2/\eta} + \frac{i\kappa}{2\sqrt{\pi}}\right] \quad \text{for} \quad \kappa^2 > 0 \,.$$

$$(B.7.9)$$

There still exist alternative expressions for $D_{L\kappa}^{(3)}(\boldsymbol{\tau})$, which are based on series expansions for the error function and the exponential. These series expansions have been so far used in numerical calculations. However, as regards accuracy and speed we prefer the closed form (B.7.9).

We start from the series expansion of the error function [1, (7.1.5)]

$$\mathrm{erf}(z) = \frac{2}{\sqrt{\pi}}\sum_{\alpha=0}^{\infty}\frac{(-)^\alpha z^{2\alpha+1}}{\alpha!(2\alpha+1)} \,, \qquad (B.7.10)$$

from which we get

$$\frac{i\kappa}{2\sqrt{\pi}}\mathrm{erf}(\frac{i\kappa}{\eta^{1/2}}) = \frac{1}{\pi}\eta^{1/2}\left(\frac{i\kappa}{\eta^{1/2}}\right)\sum_{\alpha=0}^{\infty}\frac{(-)^\alpha(\frac{i\kappa}{\eta^{1/2}})^{2\alpha+1}}{\alpha!(2\alpha+1)}$$

$$= -\frac{1}{\pi}\eta^{1/2}\left(\frac{\kappa^2}{\eta}\right)\sum_{\alpha=0}^{\infty}\frac{(\frac{\kappa^2}{\eta})^\alpha}{\alpha!(2\alpha+1)} \,. \qquad (B.7.11)$$

Inserting this into (B.7.9) we arrive at the alternative expression

$$D_{L\kappa}^{(3)}(\boldsymbol{\tau}) = -\delta(\boldsymbol{\tau})\delta_{l0}\frac{\eta^{1/2}}{2\pi}\left[e^{\kappa^2/\eta} - 2\left(\frac{\kappa^2}{\eta}\right)\sum_{\alpha=0}^{\infty}\frac{(\frac{\kappa^2}{\eta})^\alpha}{\alpha!(2\alpha+1)}\right] - \delta(\boldsymbol{\tau})\delta_{l0}\frac{i\kappa}{2\sqrt{\pi}} \,,$$

$$(B.7.12)$$

which, except for a factor of -4π, is identical to that given by Williams, Kübler and Gelatt for use with Hankel functions [21], i.e. for negative energies κ^2. Nevertheless, we point out that our derivation as presented above holds for any value of the energy κ^2. However, it should be noted that $D_{L\kappa}^{(3)}(\boldsymbol{\tau})$ is a real quantity for negative energies only. This is not the case for positive energies due to the last term in (B.7.9) and (B.7.12).

Next we use, in addition, the series expansion for the exponential entering (B.7.9) and (B.7.12). Combining it directly with the series expansion of the error function we get for the terms in square brackets in (B.7.12)

$$e^{\kappa^2/\eta} - 2\left(\frac{\kappa^2}{\eta}\right)\sum_{\alpha=0}^{\infty}\frac{(\frac{\kappa^2}{\eta})^\alpha}{\alpha!(2\alpha+1)} = \sum_{\alpha=0}^{\infty}\frac{(\frac{\kappa^2}{\eta})^\alpha}{\alpha!} - 2\sum_{\alpha=0}^{\infty}\frac{(\frac{\kappa^2}{\eta})^{\alpha+1}}{\alpha!(2\alpha+1)}$$

$$= \sum_{\alpha=0}^{\infty}\frac{(\frac{\kappa^2}{\eta})^\alpha}{\alpha!} - 2\sum_{\alpha=0}^{\infty}\frac{\alpha(\frac{\kappa^2}{\eta})^\alpha}{\alpha!(2\alpha-1)}$$

$$= -\sum_{\alpha=0}^{\infty}\frac{(\frac{\kappa^2}{\eta})^\alpha}{\alpha!(2\alpha-1)} . \tag{B.7.13}$$

Inserting this into (B.7.12) we arrive at the third alternative expression

$$D_{L\kappa}^{(3)}(\boldsymbol{\tau}) = \delta(\boldsymbol{\tau})\delta_{l0}\frac{\eta^{1/2}}{2\pi}\sum_{\alpha=0}^{\infty}\frac{(\frac{\kappa^2}{\eta})^\alpha}{\alpha!(2\alpha-1)} - \delta(\boldsymbol{\tau})\delta_{l0}\frac{i\kappa}{2\sqrt{\pi}} . \tag{B.7.14}$$

As already discussed in Sect. 3.5, the quantity $D_{L\kappa}^{(3)}(\boldsymbol{\tau})$ is just the third contribution to the Bloch sum of Hankel envelope functions. However, there exist cases where we prefer to work with the Neumann envelope function instead. For instance, Neumann functions usually are chosen for positive energies κ^2. As also already mentioned in Sect. 3.5 the corresponding Bloch sum of Neumann envelope functions arises from the Bloch sum of Hankel envelope functions by just replacing the third contribution $D_{L\kappa}^{(3)}(\boldsymbol{\tau})$ by the quantity $K_{L\kappa}^{(3)}(\boldsymbol{\tau})$. These two are connected via the identity (3.5.31), which we repeat here for completeness

$$K_{L\kappa}^{(3)}(\boldsymbol{\tau}) = D_{L\kappa}^{(3)}(\boldsymbol{\tau}) + \delta(\boldsymbol{\tau})\delta_{l0}\frac{i\kappa}{2\sqrt{\pi}} . \tag{B.7.15}$$

Combining this with the previously derived results (B.7.9), (B.7.12) and (B.7.14) we arrive at the expression

$$K_{L\kappa}^{(3)}(\boldsymbol{\tau}) = \delta(\boldsymbol{\tau})\delta_{l0}\left[\frac{i\kappa}{4\sqrt{\pi}}\left(\mathrm{erfc}(\frac{i\kappa}{\eta^{1/2}}) - \mathrm{erfc}(-\frac{i\kappa}{\eta^{1/2}})\right) - \frac{\eta^{1/2}}{2\pi}e^{\kappa^2/\eta}\right]$$

$$= -\delta(\boldsymbol{\tau})\delta_{l0}\left[\frac{i\kappa}{2\sqrt{\pi}}\mathrm{erf}(\frac{i\kappa}{\eta^{1/2}}) + \frac{\eta^{1/2}}{2\pi}e^{\kappa^2/\eta}\right]$$

$$= -\delta(\boldsymbol{\tau})\delta_{l0}\left[\frac{\kappa}{2\sqrt{\pi}}e^{\frac{\kappa^2}{\eta}}\Im w(\frac{-\kappa}{\eta^{1/2}}) + \frac{\eta^{1/2}}{2\pi}e^{\kappa^2/\eta}\right] \quad \text{for} \quad \kappa^2 > 0$$

$$= -\delta(\boldsymbol{\tau})\delta_{l0}\frac{\eta^{1/2}}{2\pi}\left[e^{\kappa^2/\eta} - 2\left(\frac{\kappa^2}{\eta}\right)\sum_{\alpha=0}^{\infty}\frac{(\frac{\kappa^2}{\eta})^\alpha}{\alpha!(2\alpha+1)}\right]$$

$$= \delta(\boldsymbol{\tau})\delta_{l0}\frac{\eta^{1/2}}{2\pi}\sum_{\alpha=0}^{\infty}\frac{(\frac{\kappa^2}{\eta})^\alpha}{\alpha!(2\alpha-1)} . \tag{B.7.16}$$

The last line is identical to the result given by Williams, Janak, and Moruzzi [20], again except for a factor -4π.

Note that all expressions given in this section hold for any value of the energy κ^2. As already discussed in Sect. 2.1, working with Hankel or Neumann envelope functions is rather a matter of choice. However, we prefer Neumann envelope functions for positive energies κ^2. Then $D_{L\kappa}^{(3)}(\tau)$ is a complex quantity due to the last term in (B.7.9), (B.7.12) and (B.7.14), which cancels out in $K_{L\kappa}^{(3)}(\tau)$.

In closing this section, we calculate the derivatives of both $D_{L\kappa}^{(3)}(\tau)$ and $K_{L\kappa}^{(3)}(\tau)$ with respect to the energy κ^2. Using the identities (3.2.22), (B.5.11), (B.5.24), and [1, (7.1.20)]

$$\frac{d}{dz}w(z) = -2zw(z) + \frac{2i}{\sqrt{\pi}} , \tag{B.7.17}$$

we get from the alternative expressions (B.7.9), (B.7.12), and (B.7.14) for the derivatives

$$\dot{D}_{L\kappa}^{(3)}(\tau) = \frac{\partial}{\partial \kappa^2} D_{L\kappa}^{(3)}(\tau)$$

$$= \delta(\tau)\delta_{l0}\left[\frac{i}{8\sqrt{\pi}\kappa}\left(\operatorname{erfc}(\frac{i\kappa}{\eta^{1/2}}) - \operatorname{erfc}(-\frac{i\kappa}{\eta^{1/2}})\right) - \frac{i}{4\sqrt{\pi}\kappa}\right]$$

$$= -\delta(\tau)\delta_{l0}\left[\frac{i}{4\sqrt{\pi}\kappa}\operatorname{erf}(\frac{i\kappa}{\eta^{1/2}}) + \frac{i}{4\sqrt{\pi}\kappa}\right]$$

$$= -\delta(\tau)\delta_{l0}\left[\frac{1}{4\sqrt{\pi}\kappa}e^{\frac{\kappa^2}{\eta}}\Im w(\frac{-\kappa}{\eta^{1/2}}) + \frac{i}{4\sqrt{\pi}\kappa}\right] \qquad \text{for} \quad \kappa^2 > 0$$

$$= -\delta(\tau)\delta_{l0}\frac{1}{2\pi\eta^{1/2}}\left[e^{\kappa^2/\eta} - 2\sum_{\alpha=0}^{\infty}\frac{(\alpha+1)(\frac{\kappa^2}{\eta})^\alpha}{\alpha!(2\alpha+1)}\right] - \delta(\tau)\delta_{l0}\frac{i}{4\sqrt{\pi}\kappa}$$

$$= \delta(\tau)\delta_{l0}\frac{1}{2\pi\eta^{1/2}}\sum_{\alpha=0}^{\infty}\frac{(\frac{\kappa^2}{\eta})^\alpha}{\alpha!(2\alpha+1)} - \delta(\tau)\delta_{l0}\frac{i}{4\sqrt{\pi}\kappa} . \tag{B.7.18}$$

The energy derivative of $K_{L\kappa}^{(3)}(\tau)$ can be calculated directly from (B.7.15) as

$$\dot{K}_{L\kappa}^{(3)}(\tau) = \frac{\partial}{\partial \kappa^2}K_{L\kappa}^{(3)}(\tau) = \frac{\partial}{\partial \kappa^2}D_{L\kappa}^{(3)}(\tau) + \delta(\tau)\delta_{l0}\frac{i}{4\sqrt{\pi}\kappa} , \tag{B.7.19}$$

which, when combined with (B.7.18), leads to the explicit expressions

$$\dot{K}_{L\kappa}^{(3)}(\tau) = \delta(\tau)\delta_{l0}\left[\frac{i}{8\sqrt{\pi}\kappa}\left(\operatorname{erfc}(\frac{i\kappa}{\eta^{1/2}}) - \operatorname{erfc}(-\frac{i\kappa}{\eta^{1/2}})\right)\right]$$

$$= -\delta(\tau)\delta_{l0}\left[\frac{i}{4\sqrt{\pi}\kappa}\operatorname{erf}(\frac{i\kappa}{\eta^{1/2}})\right]$$

$$= -\delta(\tau)\delta_{l0}\left[\frac{1}{4\sqrt{\pi}\kappa}e^{\frac{\kappa^2}{\eta}}\Im w(\frac{-\kappa}{\eta^{1/2}})\right] \qquad \text{for} \quad \kappa^2 > 0$$

$$= -\delta(\tau)\delta_{l0}\frac{1}{2\pi\eta^{1/2}}\left[e^{\kappa^2/\eta} - 2\sum_{\alpha=0}^{\infty}\frac{(\alpha+1)(\frac{\kappa^2}{\eta})^\alpha}{\alpha!(2\alpha+1)}\right]$$

$$= \delta(\boldsymbol{\tau})\delta_{l0}\frac{1}{2\pi\eta^{1/2}}\sum_{\alpha=0}^{\infty}\frac{(\frac{\kappa^2}{\eta})^{\alpha}}{\alpha!(2\alpha+1)} \ . \tag{B.7.20}$$

Note especially the singularity of $D_{L\kappa}^{(3)}(\boldsymbol{\tau})$ for $\kappa^2 = 0$, which can be avoided by passing over to Neumann envelope functions, hence, to $K_{L\kappa}^{(3)}(\boldsymbol{\tau})$.

Finally, for $\kappa^2 = 0$ we apply l'Hospitals rule to the second line of (B.7.20), and, using the identities (B.7.5), (B.5.11) and (B.5.25), we obtain

$$\dot{K}_{L\kappa}^{(3)}(\boldsymbol{\tau}) \, |_{\kappa^2=0} = \delta(\boldsymbol{\tau})\delta_{l0}\frac{1}{2\pi\eta^{1/2}} \ . \tag{B.7.21}$$

B.8 Cutoff Radii for the Ewald Method

As outlined in Sect. 3.5, the Ewald method for calculating the Bloch sum of a function is based essentially on two ideas. First, by virtue of the Poisson transform, the real space lattice sum can be alternatively written as a reciprocal lattice sum. Second, if the function to be Bloch-summed can be split into two parts containing the short and long range contributions, these may be summed separately in real and reciprocal space, respectively. Whereas the original Bloch sum may suffer from very slow convergence, the separate summations usualy converge very quickly needing only of the order of 100 lattice vectors (in a three-dimensional lattice).

In practice it is useful to know the lattice vectors needed for the summation in advance. This allows to prepare for the corresponding arrays and to calculate auxiliary quantities on beforehand. However, it requires evaluation of the radii of those spheres, which contain all the necessary real and reciprocal lattice vectors. It is the purpose of the present section to determine these cutoff radii for the Bloch sum of the Hankel envelope functions as given by (3.5.18) to (3.5.21). According to these equations, the values of the cutoff radii depend on the Ewald parameter η, the required tolerance Δ for the convergence of the lattice sums, the interstitial energy κ^2, and the angular momentum l of the Hankel envelope function.

Turning first to the reciprocal space lattice sum as given by (3.5.19) we identify the function determining the convergence with the single terms in the sum

$$f_l(q) = q^l \frac{e^{\frac{\kappa^2-q^2}{\eta}}}{\kappa^2 - q^2} \ . \tag{B.8.1}$$

Here we have abbreviated $q = |\mathbf{K}_n + \mathbf{k}|$. In addition, we have ignored the exponential and the spherical harmonics entering the summation in (3.5.19). In setting up a criterion for successful convergence we might compare each term in the sum with the maximum term occuring so far and stop the summation once the ratio of the minimum and the maximum terms is below the tolerance Δ. However, in order to make things easier for the determination of the cutoff radius q_{cut} we use the ratio of the minimum and the value of the function at $q = 1$ for that purpose this being an even stronger criterion. To be specific, the reciprocal space cutoff radius is determined from the condition

$$f_l(q_{cut}) = \Delta f_l(q = 1) . \tag{B.8.2}$$

In order to evaluate q_{cut} we insert (B.8.1) into (B.8.2) and transform the latter in the following way

$$q_{cut}^l \left(1 - \kappa^2\right) e^{\frac{\kappa^2 - q_{cut}^2}{\eta}} = \Delta \left(q_{cut}^2 - \kappa^2\right) e^{\frac{\kappa^2 - 1}{\eta}} . \tag{B.8.3}$$

This can be reduced to

$$q_{cut}^l \left(1 - \kappa^2\right) e^{-\frac{q_{cut}^2}{\eta}} = \Delta \left(q_{cut}^2 - \kappa^2\right) e^{-\frac{1}{\eta}} . \tag{B.8.4}$$

Taking the logarithm we arrive at the condition

$$\frac{-q_{cut}^2}{\eta} + l \ln q_{cut} + \ln \left(1 - \kappa^2\right) = \frac{-1}{\eta} + \ln \Delta + \ln \left(q_{cut}^2 - \kappa^2\right) . \tag{B.8.5}$$

As a consequence, the cutoff radius q_{cut} arises as the zero of the function

$$F_l(q) = -q^2 + \eta \left[l \ln q + \ln \left(1 - \kappa^2\right) - \ln \left(q^2 - \kappa^2\right) - \ln \Delta\right] + 1 . \tag{B.8.6}$$

This function is shown in Fig. B.1 for a standard parameter setting. In particular, for $q = 1$ we obtain

$$F_l(q = 1) = -\eta \ln \Delta > 0 \quad \text{for } 0 < \Delta < 1 . \tag{B.8.7}$$

This value is easily identified in Fig. B.1 as that point, where all functions coincide. In contrast, for large values of q the functions $F_l(q)$ are dominated by the term $-q^2$ and eventually become negative. According to (B.8.6) and Fig. B.1 the values of the functions for $q > 1$ increase with increasing l and so do the zeroes of the functions, q_{cut}. We are thus on the safe side if we use the cutoff radius for the function with highest l for all functions.

In practice, we will start from $q = 1$ and proceed in steps of size $\eta^{1/2}$ to larger values of q until a sign change of the function $F_l(q)$ has been detected. Then the exact position of the zero can be easily identified by using bisection in the last interval.

The real-space cutoff radius r_{cut} is calculated along very similar lines. Here the function monitoring the convergence of the lattice sum is given by (3.5.20) as

$$f_l(r) = \frac{2}{\sqrt{\pi}} r^l \int_{\frac{1}{2}\eta^{1/2}}^{\infty} \xi^{2l} e^{-r^2 \xi^2 + \frac{\kappa^2}{4\xi^2}} d\xi , \tag{B.8.8}$$

where we have included the factor $2/\sqrt{\pi}$ for convenience. Again, we have ignored exponentials and spherical remaining prefactors. As before for the reciprocal space lattice sum, we define successful convergence in terms of the ratio of the function

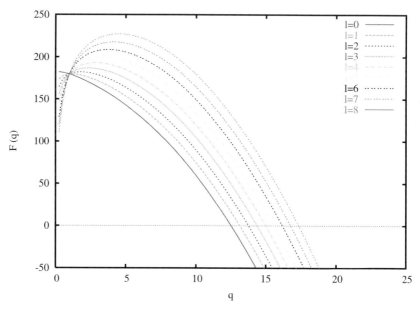

Fig. B.1. Function $F_l(q)$ as defined in (B.8.6) for $\kappa^2 = -25.0$, $\eta = 6.5$, $\Delta = 10^{-10}$, and different values of l. All parameters given in units of the Wigner-Seitz radius

$f_l(r)$ and its value at $r = 1$. Hence, the real-space cutoff radius arises from the condition

$$f_l(r_{cut}) = \Delta f_l(r = 1) . \tag{B.8.9}$$

Again, switching to the logarithm and defining the function

$$F_l(r) = \ln f_l(r) - \ln f_l(r = 1) - \ln \Delta , \tag{B.8.10}$$

we obtain the cutoff radius r_{cut} as the zero of this function. The function $F_l(r)$ is shown in Fig. B.2 for a standard parameter setting. For $r = 1$ it reduces to

$$F_l(r = 1) = -\ln \Delta > 0 \quad \text{for } 0 < \Delta < 1 . \tag{B.8.11}$$

For large values of r, the functions $F_l(r)$ are dominated by the r^2 contribution to the exponential in the integral entering (B.8.8) and, hence, eventually become negative. We may thus proceed in a similar way as for the reciprocal space cutoff radius and, starting from $r = 1$, increase r in steps of size $2/\eta^{1/2}$ until $F_l(r)$ changes sign. Finally, the exact position of the zero can be found by bisection. Note that, according to (B.8.10) and Fig. B.2, the values of the functions $F_l(r)$ increase with increasing l and so do the zeroes. For this reason, we may again stay with the maximum l value und use the resulting cutoff radius for all l.

In the course of the calculations we fall back on the evaluation of the Ewald integral as sketched in Sect. B.5. Using the definition (B.5.1) we may write the function $F_l(r)$ as

$$F_l(r) = \ln \frac{2}{\sqrt{\pi}} r^l \int_{\frac{1}{2}\eta^{1/2}}^{\infty} \xi^{2l} e^{-r^2\xi^2 + \frac{\kappa^2}{4\xi^2}} \, d\xi$$

$$- \ln \frac{2}{\sqrt{\pi}} \int_{\frac{1}{2}\eta^{1/2}}^{\infty} \xi^{2l} e^{-\xi^2 + \frac{\kappa^2}{4\xi^2}} \, d\xi - \ln \Delta$$

$$= l \ln r + \ln \frac{2}{\sqrt{\pi}} \int_{\frac{1}{2}\eta^{1/2}}^{\infty} \xi^{2l} e^{-r^2\xi^2 + \frac{\kappa^2}{4\xi^2}} \, d\xi$$

$$- \ln \frac{2}{\sqrt{\pi}} \int_{\frac{1}{2}\eta^{1/2}}^{\infty} \xi^{2l} e^{-\xi^2 + \frac{\kappa^2}{4\xi^2}} \, d\xi - \ln \Delta$$

$$= l \ln r + \ln I_l(r) - \ln I_l(1) - \ln \Delta , \tag{B.8.12}$$

where we have appended the argument r to the integrals. In practice, we will rather use the quantity

$$\bar{I}_l(r) = r^{2l+1} I_l(r) , \tag{B.8.13}$$

as defined in (B.5.6), and write

$$F_l(r) = l \ln r + \ln r^{-2l-1} \bar{I}_l(r) - \ln \bar{I}_l(1) - \ln \Delta$$
$$= -(l+1) \ln r + \ln \bar{I}_l(r) - \ln \bar{I}_l(1) - \ln \Delta . \tag{B.8.14}$$

We may thus use the machinery developed for the calculation of structure constants for the evaluation of the real-space cutoff radius.

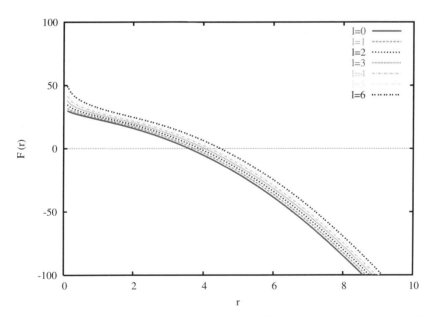

Fig. B.2. Function $F_l(r)$ as defined in (B.8.10) for $\kappa^2 = -25.0$, $\eta = 6.5$, $\Delta = 10^{-10}$, and different values of l. All parameters given in units of the Wigner-Seitz radius

Finally, the number of lattice vectors to be included in the summations are calculated from

$$N_{r,max} = \frac{\frac{4\pi}{3} r_{cut}^3}{\Omega_c} = \left(\frac{r_{cut}}{R_{WS}}\right)^3 , \tag{B.8.15}$$

and

$$N_{q,max} = \frac{\frac{4\pi}{3} q_{cut}^3}{(2\pi)^3} \Omega_c = \frac{2}{9\pi} (q_{cut} R_{WS})^3 , \tag{B.8.16}$$

respectively, where R_{WS} is the Wigner-Seitz radius and Ω_c denotes the volume of the unit cell.

B.9 Integrals over Augmented Pseudo Functions

As an alternative to the Hankel envelope functions we constructed the Hankel pseudo functions in Sect. 3.9, which do not diverge at their origin but show a smooth behaviour in all space. This was achieved by replacing the Hankel envelope function by the augmented pseudo function within the respective on-center sphere. The augmented pseudo function is defined by (3.9.2) and fulfills the radial Schrödinger equation (3.9.3) subject to the boundary condition (3.9.4) and the regularity at the origin. As already outlined in Sect. 3.9, the pseudo envelope functions are not augmented within the off-center spheres since there the envelope function itself already is smooth enough. Hence, within this latter type of spheres the one-center expansion of the envelope function results in the Bessel envelope function.

In the present section we define the radial integrals over the augmented pseudo functions and set up interrelations between them. In doing so, we benefit from the fact already mentioned in Sect. 3.11 that the construction of the augmented pseudo functions $\tilde{H}_{L\kappa}^0(\mathbf{r}_i)$ was done in complete analogy with the construction of the augmented Hankel functions $\tilde{H}_{L\kappa\sigma}(\mathbf{r}_i)$ in Sect. 2.1. Since the corresponding integrals with the augmented functions were derived in detail in the App. A.1 we merely have to rewrite those results in terms of the augmented pseudo functions $\tilde{H}_{L\kappa}^0(\mathbf{r}_i)$ and the Bessel envelope functions $J_{L\kappa}(\mathbf{r}_i)$.

To begin with, we use (3.9.2) to define the pseudo Hankel integral

$$
\begin{aligned}
S_{l\kappa i}^{(0)} &:= \langle \tilde{H}_{L\kappa}^0 | \tilde{H}_{L'\kappa}^0 \rangle_i \delta_{LL'} \\
&= \int_{\Omega_i} d^3 \mathbf{r}_i \, \tilde{H}_{L\kappa}^{0*}(\mathbf{r}_i) \tilde{H}_{L\kappa}^0(\mathbf{r}_i) \\
&= \int_0^{S_i} dr_i \, r_i^2 \tilde{h}_{l\kappa}^0(r_i) \tilde{h}_{l\kappa}^0(r_i) .
\end{aligned}
\tag{B.9.1}
$$

In addition, we have to calculate the mixed integrals which are made of two different functions. Again using the previously mentioned analogy with the evaluations given in the App. A.1 we just rewrite (A.1.11) and (A.1.8) and get

$$\langle \tilde{H}^0_{L\kappa_1} | \tilde{H}^0_{L'\kappa_2} \rangle_i \delta_{LL'}$$

$$= \langle \tilde{H}^0_{L'\kappa_2} | \tilde{H}^0_{L\kappa_1} \rangle_i \delta_{LL'}$$

$$= \int_0^{S_i} dr_i \, r_i^2 \tilde{h}^0_{l\kappa_1}(r_i) \tilde{h}^0_{l\kappa_2}(r_i)$$

$$= \frac{1}{E^{(0)}_{l\kappa_1 i} - E^{(0)}_{l\kappa_2 i}} W\{r_i \bar{h}_l(\kappa_1 r_i), r_i \bar{h}_l(\kappa_2 r_i); r_i\}|_{r_i = S_i} \,, \tag{B.9.2}$$

and

$$\langle \tilde{H}^0_{L\kappa_1} | J_{L'\kappa_2} \rangle_i \delta_{LL'}$$

$$= \langle J_{L'\kappa_2} | \tilde{H}^0_{L\kappa_1} \rangle_i \delta_{LL'}$$

$$= \int_0^{S_i} dr_i \, r_i^2 \tilde{h}^0_{l\kappa_1}(r_i) \bar{j}_{l\kappa_2}(r_i)$$

$$= \frac{1}{E^{(0)}_{l\kappa_1 i} - \kappa_2^2} W\{r_i \bar{h}_l(\kappa_1 r_i), r_i \bar{j}_l(\kappa_2 r_i); r_i\}|_{r_i = S_i} \,, \tag{B.9.3}$$

where the energy $E^{(0)}_{l\kappa i}$ results from solving the radial Schrödinger equations (3.9.3). Again the case $\kappa_1^2 = \kappa_2^2 = \kappa^2$ is of particular importance since then according to (3.2.40) the Wronskian entering (B.9.3) reduces to unity.

As for the augmented function in the App. A.1 we have by now reduced the number of quantities characterizing each pseudo function to two, namely the energy $E^{(0)}_{l\kappa i}$ and the integral $S^{(0)}_{l\kappa i}$.

For different values of the energy parameter it might still happen that the pseudo energy $E^{(0)}_{l\kappa_1 i}$ and the interstitial energy κ_2^2 of the Bessel envelope function are identical. In this case the augmented pseudo function $\tilde{h}^0_{l\kappa_1}$ and the Bessel envelope function $\bar{j}_{l\kappa_2}$ by construction would also be identical and we were able to note

$$\int_0^{S_i} dr_i \, r_i^2 \tilde{h}^0_{l\kappa_1}(r_i) \bar{j}_{l\kappa_2}(r_i)$$

$$= \int_0^{S_i} dr_i \, r_i^2 \tilde{h}^0_{l\kappa_1}(r_i) \tilde{h}^0_{l\kappa_1}(r_i) = S^{(0)}_{l\kappa_1 i}$$

$$= \int_0^{S_i} dr_i \, r_i^2 \bar{j}_{l\kappa_2}(r_i) \bar{j}_{l\kappa_2}(r_i) \,, \tag{B.9.4}$$

for $E^{(0)}_{l\kappa_1 i} = \kappa_2^2$.

B.10 Matrix Representation of Rotations

In this section we want to set up the matrix describing a real-space rotation about a rotation axis $\omega\mathbf{g}$, where $\mathbf{g}$ denotes the unit vector pointing along the axis and ω

is the rotation angle. In order to arrive at the rotation matrix $\mathcal{R}$ we suppose that it transforms an arbitrary vector $\mathbf{r}$ to a new vector $\mathbf{r}'$, i.e.

$$\mathbf{r}' = \mathcal{R} \cdot \mathbf{r} . \tag{B.10.1}$$

Next we define a coordinate system by the components of $\mathbf{r}$ parallel and perpendicular to $\mathbf{g}$ as well as the vector perpendicular to these two,

$$\mathbf{r}_{\parallel} := \mathbf{g} \, (\mathbf{g} \cdot \mathbf{r}) , \tag{B.10.2}$$

$$\mathbf{r}_{\perp} := \mathbf{r} - \mathbf{r}_{\parallel} = \mathbf{r} - \mathbf{g} \, (\mathbf{g} \cdot \mathbf{r}) , \tag{B.10.3}$$

$$\mathbf{g} \times \mathbf{r}_{\perp} := \mathbf{g} \times \mathbf{r} . \tag{B.10.4}$$

It is then trivial to state for the transformed vector the result

$$\mathbf{r}'_{\parallel} = \mathbf{r}_{\parallel} \tag{B.10.5}$$

$$\mathbf{r}'_{\perp} = \cos \omega \, \mathbf{r}_{\perp} + \sin \omega \, \mathbf{g} \times \mathbf{r} , \tag{B.10.6}$$

which when combined with (B.10.2) and (B.10.2) leads to [13]

$$\begin{aligned} \mathbf{r}' &= \mathbf{r}'_{\parallel} + \mathbf{r}'_{\perp} \\ &= \cos \omega \, \mathbf{r} + (1 - \cos \omega) \, \mathbf{g} \, (\mathbf{g} \cdot \mathbf{r}) + \sin \omega \, \mathbf{g} \times \mathbf{r} . \end{aligned} \tag{B.10.7}$$

With this intermediate result at hand we are in a position to derive the representation of the rotation matrix $\mathcal{R}$ in terms of a Cartesian coordinate system spanned by the three mutually orthogonal unit vectors $\mathbf{u}_x$, $\mathbf{u}_y$, and $\mathbf{u}_z$. To this end we rewrite (B.10.7) as

$$\begin{aligned} \mathbf{r}' = \Big[&\cos \omega \, \mathcal{I} + (1 - \cos \omega) \, \mathbf{g} \mathbf{g}^T \\ &+ \sin \omega \, \left((\mathbf{g} \times \mathbf{u}_x) \, \mathbf{u}_x^T + (\mathbf{g} \times \mathbf{u}_y) \, \mathbf{u}_y^T + (\mathbf{g} \times \mathbf{u}_z) \, \mathbf{u}_z^T \right) \Big] \mathbf{r} . \end{aligned} \tag{B.10.8}$$

and identify the terms in the square bracket as the desired rotation matrix $\mathcal{R}$. Evaluating

$$\mathbf{g} \times \mathbf{u}_x = g_z \, \mathbf{u}_y - g_y \, \mathbf{u}_z , \tag{B.10.9}$$

$$\mathbf{g} \times \mathbf{u}_y = -g_z \, \mathbf{u}_x + g_x \, \mathbf{u}_z , \tag{B.10.10}$$

$$\mathbf{g} \times \mathbf{u}_z = g_y \, \mathbf{u}_x - g_x \, \mathbf{u}_y , \tag{B.10.11}$$

we arrive at

$$\begin{aligned} (\mathbf{g} &\times \mathbf{u}_x) \, \mathbf{u}_x^T + (\mathbf{g} \times \mathbf{u}_y) \, \mathbf{u}_y^T + (\mathbf{g} \times \mathbf{u}_z) \, \mathbf{u}_z^T \\ &= g_z \, \mathbf{u}_y \, \mathbf{u}_x^T - g_y \, \mathbf{u}_z \, \mathbf{u}_x^T - g_z \, \mathbf{u}_x \, \mathbf{u}_y^T + g_x \, \mathbf{u}_z \, \mathbf{u}_y^T + g_y \, \mathbf{u}_x \, \mathbf{u}_z^T - g_x \, \mathbf{u}_y \, \mathbf{u}_z^T \\ &= \begin{pmatrix} 0 & -g_z & g_y \\ g_z & 0 & -g_x \\ -g_y & g_x & 0 \end{pmatrix} . \end{aligned} \tag{B.10.12}$$

Hence, we arrive at the final result [13]

$$\mathcal{R} = \cos\omega\,\mathcal{I} + (1 - \cos\omega)\begin{pmatrix} g_x g_x & g_x g_y & g_x g_z \\ g_y g_x & g_y g_y & g_y g_z \\ g_z g_x & g_z g_y & g_z g_z \end{pmatrix}$$

$$+ \sin\omega\begin{pmatrix} 0 & -g_z & g_y \\ g_z & 0 & -g_x \\ -g_y & g_x & 0 \end{pmatrix}. \tag{B.10.13}$$

For the special case $\omega = \pi$ this reduces to

$$\mathcal{M} = -\mathcal{I} + 2\begin{pmatrix} g_x g_x & g_x g_y & g_x g_z \\ g_y g_x & g_y g_y & g_y g_z \\ g_z g_x & g_z g_y & g_z g_z \end{pmatrix}. \tag{B.10.14}$$

Next we turn to the inverse problem, namely, the calculation of the rotation axis $\mathbf{g}$ and the angle ω from a given rotation matrix $\mathcal{R}$. The rotation angle is easily deduced from the trace of the matrix, which according to (B.10.13) reads as

$$tr\mathcal{R} = R_{xx} + R_{yy} + R_{zz} = 1 + 2\cos\omega. \tag{B.10.15}$$

Here, we have used the normalization of the vector $\mathbf{g}$. In addition, we can tell from the determinant of $\mathcal{R}$ whether the rotation is a proper or improper one, i.e. whether it contains the inversion or not.

Finally, the rotation axis $\mathbf{g}$ is deduced either from the antisymmetric part of $\mathcal{R}$ via

$$R_{zy} - R_{yz} = \sin\omega\,2g_x, \tag{B.10.16}$$
$$R_{xz} - R_{zx} = \sin\omega\,2g_y, \tag{B.10.17}$$
$$R_{yx} - R_{xy} = \sin\omega\,2g_z, \tag{B.10.18}$$

or else from the diagonal elements

$$R_{xx} = \cos\omega + (1 - \cos\omega)\,g_x^2, \tag{B.10.19}$$
$$R_{yy} = \cos\omega + (1 - \cos\omega)\,g_y^2, \tag{B.10.20}$$
$$R_{zz} = \cos\omega + (1 - \cos\omega)\,g_z^2. \tag{B.10.21}$$

The second alternative applies especially to $\omega = \pi$, in which case the sign of the components of $\mathbf{g}$ is irrelevant and we obtain

$$g_x^2 = \frac{1}{2}(R_{xx} + 1), \tag{B.10.22}$$

$$g_y^2 = \frac{1}{2}(R_{yy} + 1), \tag{B.10.23}$$

$$g_z^2 = \frac{1}{2}(R_{zz} + 1). \tag{B.10.24}$$

References

1. M. Abramowitz and I. A. Stegun, *Handbook of Mathematical Functions* (Dover, New York 1972)
2. N. W. Ashcroft and N. D. Mermin, *Solid State Physics* (Holt-Saunders, Philadelphia 1976)
3. P. E. Blöchl, Gesamtenergien, Kräfte und Metall-Halbleiter Grenzflächen. PhD thesis, Universität Stuttgart (1989)
4. P. P. Ewald, Ann. Phys. **64**, 253 (1921)
5. V. Eyert, Entwicklung und Implementation eines Full-Potential-ASW-Verfahrens. PhD thesis, Technische Hochschule Darmstadt (1991)
6. W. Gautschi, SIAM Rev. **9**, 24 (1967)
7. W. Gautschi, Commun. ACM **12**, 635 (1969)
8. W. Gautschi, SIAM J. Numer. Anal. **7**, 187 (1970)
9. C. D. Gelatt, Jr., H. Ehrenreich, and R. E. Watson, Phys. Rev. B **15**, 1613 (1977)
10. I. S. Gradshteyn and I. M. Ryzhik, *Table of Integrals, Series and Products* (Academic Press, New York 1980)
11. A. Jablonski, J. Comput. Phys. **111**, 256 (1994)
12. J. D. Jackson, *Classical Electrodynamics* (Wiley, New York 1975)
13. W. Ludwig, *Festkörperphysik* (Akad. Verlagsgesellschaft, Heidelberg 1970)
14. A. Messiah, *Quantum Mechanics*, vol 1 (North Holland, Amsterdam 1976)
15. A. Messiah, *Quantum Mechanics*, vol 2 (North Holland, Amsterdam 1978)
16. P. M. Morse and H. Feshbach, *Methods of Theoretical Physics* (McGraw-Hill, New York 1953)
17. W. H. Press, B. P. Flannery, S. A. Teukolsky, and W. T. Vetterling, *Numerical Recipes – The Art of Scientific Computing* (Cambridge University Press, Cambridge 1989)
18. G. P. M. Poppe and C. M. J. Wijers, ACM Trans. Math. Software **16**, 38 (1990)
19. G. N. Watson, *Theory of Bessel Functions* (University Press, Cambridge 1966)
20. A. R. Williams, J. F. Janak, and V. L. Moruzzi, Phys. Rev. B **6**, 4509 (1972)
21. A. R. Williams, J. Kübler, and C. D. Gelatt, Jr., Phys. Rev. B **19**, 6094 (1979)

C

Details of the Plane-Wave Based Full-Potential ASW Method

C.1 Practical Aspects of the Fourier Transform

In practice, the Fourier expansion of the pseudo function as well as of the pseudo charge density and potential will be performed by using standard Fast Fourier Transform (FFT) routines (see e.g. [4,7]). In order to make use of such routines we must rewrite the exponentials appearing in these expansions. For the pseudo function we start out from (4.2.36),

$$D_{L\kappa}^0(\mathbf{r}_i, \mathbf{k})e^{i\mathbf{k}\boldsymbol{\tau}_i} = \sum_n D_{L\kappa i}^0(\mathbf{K}_n + \mathbf{k})e^{i\mathbf{K}_n \mathbf{r}_i}e^{i\mathbf{k}\mathbf{r}} , \qquad (C.1.1)$$

where we have already extracted the phase factors as outlined at the end of Sect. 4.2. Note that the pseudo functions enter only as part of products and, hence, the second exponential on the right-hand side of (C.1.1) cancels out.

Of course, the finite set of reciprocal lattice vectors, which are included in the expansion (C.1.1), must be symmetric about the origin. In addition, most FFT routines require that the number of expansion coefficients, i.e. of reciprocal lattice vectors per dimension must be a power of two or, at least, must be an even number. Although both conditions can be obeyed only for an infinite lattice they should be fulfilled as best as possible. Defining a reciprocal offset vector by

$$\mathbf{K}_0 = -\sum_{i=1}^{d} \frac{N_i}{2}\mathbf{b}_i , \qquad (C.1.2)$$

where the vectors $\mathbf{b}_i$ are the reciprocal space primitive translations and the N_i denote the numbers of Fourier coefficients per dimension, we write the general reciprocal lattice vector of the finite set as

$$\mathbf{K}_n = \mathbf{K}_0 + \sum_{i=1}^{d} m_{ni}\mathbf{b}_i \qquad \text{with} m_{ni} \in [0 : N_i - 1] . \qquad (C.1.3)$$

Writing the vectors of the real-space mesh as

V. Eyert: *Details of the Plane-Wave Based Full-Potential ASW Method*, Lect. Notes Phys. **719**, 247–256 (2007)
DOI 10.1007/978-3-540-71007-3_7

$$\mathbf{r} = \sum_{i=1}^{d} \frac{n_{\mathbf{r}i}}{N_i} \mathbf{a}_i \qquad \text{with} n_{\mathbf{r}i} \in [0 : N_i - 1] \ , \tag{C.1.4}$$

where the vectors $\mathbf{a}_i$ are the real-space primitive translations, and, using the identity

$$\mathbf{a}_i \cdot \mathbf{b}_j = 2\pi \cdot \delta_{ij} \ , \tag{C.1.5}$$

we arrive at the intermediate result

$$\mathbf{K}_0 \mathbf{r} = -\pi \sum_{i=1}^{d} n_{\mathbf{r}i} \ . \tag{C.1.6}$$

The corresponding exponential thus trivially reduces to

$$e^{i\mathbf{K}_0\mathbf{r}} = \begin{cases} -1 & \text{for } \sum_{i=1}^{d} n_{\mathbf{r}i} \text{ odd} \\ +1 & \text{for } \sum_{i=1}^{d} n_{\mathbf{r}i} \text{ even} \end{cases} \ . \tag{C.1.7}$$

Combining the previous identities we write the first exponential on the right hand side of (C.1.1) as

$$\begin{aligned} e^{i\mathbf{K}_n\mathbf{r}_i} &= e^{-i\mathbf{K}_n\boldsymbol{\tau}_i} e^{i\mathbf{K}_0\mathbf{r}} e^{i\sum_{i=1}^{d} m_{ni}\mathbf{b}_i\mathbf{r}} \\ &= e^{-i\mathbf{K}_n\boldsymbol{\tau}_i} e^{i\mathbf{K}_0\mathbf{r}} e^{2\pi i \sum_{i=1}^{d} m_{ni}n_{\mathbf{r}i}/N_i} \ , \end{aligned} \tag{C.1.8}$$

where we have again used the identity (C.1.5). Note again that the pseudo functions appear only in products and thus the second exponential in (C.1.8) can also be omitted in the calculation of the pseudo function.

Finally, we can easily transfer the previous results to the pseudo part of the charge density or the potential. For the former we write according to (4.2.6)

$$\rho_{val,\sigma}^0(\mathbf{r}) = \sum_t \rho_{val,\sigma}^0(\mathbf{K}_t) e^{i\mathbf{K}_t\mathbf{r}} \ , \tag{C.1.9}$$

and apply (C.1.8) with $\boldsymbol{\tau}_i = 0$ to the exponential.

C.2 Calculation of the Auxiliary Density

The prefactors g_{Ki} entering the definition (4.4.17) are determined by the condition (4.4.18), which contains the integral

$$I = \int_0^S r^{2k+2} \left(1 - \left(\frac{r}{S} \right)^2 \right)^3 dr. \tag{C.2.1}$$

Substituting $x = \frac{r}{S}$ we get by repeated partial integration the result

$$I = S^{2k+3} \int_0^1 x^{2k+2}(1 - x^2)^3 dx$$

$$= \frac{S^{2k+3}}{2k+3} x^{2k+3}(1-x^2)^3 \Big|_0^1 + 6\frac{S^{2k+3}}{2k+3} \int_0^1 x^{2k+4}(1-x^2)^2 dx$$

$$= 6\frac{S^{2k+3}}{(2k+3)(2k+5)} x^{2k+5}(1-x^2)^2 \Big|_0^1$$

$$+ 24\frac{S^{2k+3}}{(2k+3)(2k+5)} \int_0^1 x^{2k+6}(1-x^2) dx$$

$$= 24\frac{S^{2k+3}}{(2k+3)(2k+5)(2k+7)} x^{2k+7}(1-x^2) \Big|_0^1$$

$$+ 48\frac{S^{2k+3}}{(2k+3)(2k+5)(2k+7)} \int_0^1 x^{2k+8} dx$$

$$= 48\frac{S^{2k+3}}{(2k+3)(2k+5)(2k+7)(2k+9)} \quad . \tag{C.2.2}$$

C.3 Fourier Transform of the Auxiliary Density

For the Fourier transform of the auxiliary density as given by (4.4.20) we need the integral

$$I = \int_0^S dr r^2 \rho_{aux,L}(r) j_l(qr) \; . \tag{C.3.1}$$

Using the representation (4.4.7) of the auxiliary density and substituting

$$x = qr, x_0 = qS \; ,$$

we get for the above integral

$$I = g_{Lj} \int_0^S dr r^{l+2} j_l(qr) \left(1 - \left(\frac{r}{S}\right)^2 \right)^3$$

$$= g_{Lj} \frac{1}{q^{l+3}} \int_0^{x_0} dx x^{l+2} j_l(x) \left(1 - \left(\frac{x}{x_0}\right)^2 \right)^3$$

$$=: g_{Lj} \frac{1}{q^{l+3}} I' \; . \tag{C.3.2}$$

In order to evaluate this integral we fall back on the identity (3.3.5) and, using multiple partial integration, we arrive at

$$I' = x^{l+2} j_{l+1}(x) \left(1 - \left(\frac{x}{x_0}\right)^2 \right)^3 \Big|_0^{x_0} + \frac{6}{x_0^2} \int_0^{x_0} dx x^{l+3} j_{l+1}(x) \left(1 - \left(\frac{x}{x_0}\right)^2 \right)^2$$

$$= \frac{6}{x_0^2} x^{l+3} j_{l+2}(x) \left(1 - \left(\frac{x}{x_0} \right)^2 \right)^2 \Big|_0^{x_0} + \frac{24}{x_0^4} \int_0^{x_0} dx x^{l+4} j_{l+2}(x) \left(1 - \left(\frac{x}{x_0} \right)^2 \right)$$

$$= \frac{24}{x_0^4} x^{l+4} j_{l+3}(x) \left(1 - \left(\frac{x}{x_0} \right)^2 \right) \Big|_0^{x_0} + \frac{48}{x_0^6} \int_0^{x_0} dx x^{l+5} j_{l+3}(x)$$

$$= \frac{48}{x_0^6} x^{l+5} j_{l+4}(x) \Big|_0^{x_0}$$

$$= 48 x_0^{l-1} j_{l+4}(x) \ . \tag{C.3.3}$$

From this we get the result

$$I = 48 g_{Lj} \frac{1}{q^{l+3}} (qS)^{l-1} j_{l+4}(qS) = 48 g_{Lj} S^{l-1} \frac{1}{q^4} j_{l+4}(qS) \ . \tag{C.3.4}$$

C.4 Alternative Auxiliary Density

The prefactors g_{Ki} entering the definition (4.4.16) are determined by the condition (4.4.17), which contains the integral

$$I_k(S) = \frac{2}{\sqrt{\pi}} 2^k \left(\frac{1}{2} \eta^{1/2} \right)^{2k} \int_0^S r^{2k+2} e^{-\frac{1}{4} \eta r^2} \, dr \ . \tag{C.4.1}$$

The calculation of this integral is performed along similar lines as that of the Ewald integral in App. B.5. Hence, we first note the identities

$$\frac{d}{dr} r^{2k+1} = (2k+1) r^{2k} \ . \tag{C.4.2}$$

$$\frac{d}{dr} e^{-\frac{1}{4} \eta r^2} = -\frac{\eta}{2} r e^{-\frac{1}{4} \eta r^2} \ . \tag{C.4.3}$$

Integrating (C.4.1) by parts we get

$$-\frac{\eta}{2} I_k(S) = \frac{2}{\sqrt{\pi}} 2^k \left(\frac{1}{2} \eta^{1/2} \right)^{2k} \int_0^S r^{2k+1} \left(-\frac{\eta}{2} r \right) e^{-\frac{1}{4} \eta r^2} \, dr$$

$$= \frac{2}{\sqrt{\pi}} 2^k \left(\frac{1}{2} \eta^{1/2} \right)^{2k} \left[r^{2k+1} e^{-\frac{1}{4} \eta r^2} \right]_0^S$$

$$- \frac{2}{\sqrt{\pi}} 2^k \left(\frac{1}{2} \eta^{1/2} \right)^{2k} (2k+1) \int_0^S r^{2k} e^{-\frac{1}{4} \eta r^2} \, dr$$

$$= \frac{2}{\sqrt{\pi}} 2^k \left(\frac{2}{\eta^{1/2}} \right) \left(\frac{1}{2} \eta^{1/2} S \right)^{2k+1} e^{-\frac{1}{4} \eta S^2}$$

$$- (2k+1) 2 \left(\frac{1}{2} \eta^{1/2} \right)^2 I_{k-1}(S) \ . \tag{C.4.4}$$

Using the abbreviation

$$a = \frac{1}{2}\eta^{1/2}S , \qquad (C.4.5)$$

the previous recursion relation reads as

$$I_k(S) = (2k+1)I_{k-1}(S) - \frac{2}{\sqrt{\pi}}2^k\frac{4}{\eta^{3/2}}a^{2k+1}e^{-a^2} . \qquad (C.4.6)$$

By now we are left with the integral

$$I_{-1}(S) = \frac{2}{\sqrt{\pi}}\frac{1}{2}\left(\frac{2}{\eta^{1/2}}\right)^2\int_0^S e^{-\frac{1}{4}\eta r^2} dr . \qquad (C.4.7)$$

Substituting

$$t = \frac{1}{2}\eta^{1/2}r \quad \text{and} \quad dt = \frac{1}{2}\eta^{1/2}dr, \qquad (C.4.8)$$

we write

$$\begin{aligned}
I_{-1}(S) &= \frac{2}{\sqrt{\pi}}\left(\frac{2}{\eta^{1/2}}\right)^2\frac{1}{\eta^{1/2}}\int_0^{\frac{1}{2}\eta^{1/2}S} e^{-t^2} dt \\
&= \frac{4}{\eta^{3/2}}\text{erf}(\frac{1}{2}\eta^{1/2}S) \\
&= \frac{4}{\eta^{3/2}}\text{erf}(a) , \qquad (C.4.9)
\end{aligned}$$

where we have used the definition (B.5.10) of the error function as well as the abbreviation (C.4.5).

For the special case $S \to \infty$ the recursion relation (C.4.6) and the result (C.4.9) reduce to

$$I_k(\infty) = (2k+1)I_{k-1}(\infty) , \qquad (C.4.10)$$

and

$$I_{-1}(\infty) = \frac{4}{\eta^{3/2}} , \qquad (C.4.11)$$

where, in the last line, we have taken the identity (B.5.12) into account. Finally, combining (C.4.10) and (C.4.11) we arrive at the familiar result

$$I_k(\infty) = (2k+1)!!\frac{4}{\eta^{3/2}} . \qquad (C.4.12)$$

C.5 The Initial Electron Density

In order to enter the self-consistency cycle of a full-potential calculation and to achieve fast convergence we need a good initial guess for the electron density or the potential. As a matter of fact, both functions are well approximated by the overlapping densities or potentials of free atoms [6]. As a consequence, we can use them to evaluate the muffin-tin radii, which have to be known before the actual calculations start [2,3]. Once the radii are given we are able to set up the electron density and potential within each atomic sphere. However, in order to do so we need a representation of an arbitrary function given relative to the atomic site $\boldsymbol{\tau}_i$ in terms of functions given relative to a different site $\boldsymbol{\tau}_j$. This problem was first formulated by Löwdin and the solution is referred to as Löwdin's α-function method [5]. It allows to calculate a function, which is written as the product of a radial function and a spherical harmonics and centered at a given site, as a spherical harmonics expansion about another site. To be more specific, we represent the function

$$\Phi_L(\mathbf{r}_i) = \varphi_l(r_i)Y_L(\hat{\mathbf{r}}_i) , \tag{C.5.1}$$

given relative to site $\boldsymbol{\tau}_i$, as

$$\Phi_L(\mathbf{r}_j) = \sum_K \alpha_{KL}(r_j; a)Y_K(\hat{\mathbf{r}}_j) , \tag{C.5.2}$$

where $L = (l, m)$ and $K = (k, m')$. Of course the site $\boldsymbol{\tau}_j$ must be different from $\boldsymbol{\tau}_i$ and, hence, the distance $a := |\boldsymbol{\tau}_j - \boldsymbol{\tau}_i|$ be greater than zero.

A computationally efficient as well as stable form of the α-functions entering (C.5.2) was given by Duff [1]. For simplicity Duff concentrated on the special situation, where separation of the two sites $\boldsymbol{\tau}_i$ and $\boldsymbol{\tau}_j$ is along the z axis. In this case $m' = m$ and we note

$$
\begin{aligned}
&\alpha_{KL}(r_j; a) \\
&= \frac{F_{KL}}{a^2} \sum_{p=0}^{\frac{l-m-\delta(l,m)}{2}} \sum_{q=0}^{2p+\delta(l,m)} \beta(l,m;p,q) \left(1 - \frac{r_j^2}{a^2}\right)^{2p+\delta(l,m)-q} \\
&\quad \sum_{p'=0}^{\frac{k+m-\delta(k,m)}{2}} \left(\frac{r_j}{a}\right)^{m-1-2p'-\delta(k,m)} \sum_{q'=0}^{2p'+\delta(k,m)} \beta(k,-m;p',q') \\
&\quad \left(1 + \frac{r_j^2}{a^2}\right)^{2p'+\delta(k,m)-q'} \\
&\quad (-1)^{q'} \int_{|a-r_j|}^{a+r_j} dr'' \, r'' \, \varphi_l(r'') \left(\frac{r''}{a}\right)^{2(q+q'-p)-m-\delta(l,m)} ,
\end{aligned}
\tag{C.5.3}
$$

where

$$\delta(i, j) = \begin{cases} 1 & \text{for } i+j \text{ odd} \\ 0 & \text{for } i+j \text{ even} \end{cases} , \tag{C.5.4}$$

$$\beta(l, m; p, q)$$

$$= \frac{(-1)^p \; (l + m + \delta(l, m) + 2p)!}{2^{2p+\delta(l,m)} \left(\frac{l+m+\delta(l,m)+2p}{2}\right)! \left(\frac{l-m-\delta(l,m)-2p}{2}\right)! (2p + \delta(l, m) - q)! q!} ,$$

$$\text{(C.5.5)}$$

and

$$F_{KL} = (-1)^{[l+m+\delta(l,m)+k+m-\delta(k,m)]/2} 2^{-(l+k+1)}$$

$$\sqrt{\frac{(2l + 1)(2k + 1)(l - m)!(k + m)!}{(k - m)!(l + m)!}} . \qquad \text{(C.5.6)}$$

Since in the present context we aim at a good initial guess for the density and potential it is sufficient to take only the spherical symmetric part of all functions into consideration. Hence, we assume $l = m = 0$ in (C.5.1) and we restrict the expansion (C.5.2) to only the first term, i.e. the one with $k = m = 0$. In this case, the functions δ and β as given by (C.5.4) and (C.5.5) reduce to 0 and 1, respectively. Furthermore, we have

$$F_{(0,0)(0,0)} = \frac{1}{2} . \qquad \text{(C.5.7)}$$

Finally, of all the summations in (C.5.3) only the first term survives and we thus have the result

$$\alpha_{(0,0)(0,0)}(r_j; a) = \frac{1}{2r_j a} \int_{|a-r_j|}^{a+r_j} dr'' \; r'' \; \varphi_0(r'') . \qquad \text{(C.5.8)}$$

As a check we consider the special case of a constant function, $\varphi_0(r'') = c$, for which we note

$$\alpha_{(0,0)(0,0)}(r_j; a) = \frac{c}{2r_j a} \left[\frac{(a + r_j)^2}{2} - \frac{(a - r_j)^2}{2}\right] = c . \qquad \text{(C.5.9)}$$

In particular, we use this result for small values of r_j, where the integration interval is likewise small and, hence, the function can be approximated by a constant, $\varphi_0(r'') \approx \varphi_0(a)$. As a consequence, the α-function does not diverge for $r_j \to 0$ but we have

$$\alpha_{(0,0)(0,0)}(r_j = 0; a) = \varphi_0(a) , \qquad \text{(C.5.10)}$$

for any function $\varphi_0(r'')$. This result can be likewise derived from l'Hospitals rule. A different example is the function $\varphi_0(r'') = 1/r''$, for which we obtain

$$\alpha_{(0,0)(0,0)}(r_j; a) = \frac{2r_j}{2r_j a} = \frac{1}{a} . \qquad \text{(C.5.11)}$$

In practice, we use the α-function expansion to represent the tails coming from the electron densities and potentials of all neighbouring spheres inside a particular sphere. In other words, the α-function expansion enters as part of a sum over all other atomic sites. It is thus useful to define, with the help of (C.5.8), the quantity

$$A(r_j) := \sum_{\substack{i \\ i \neq j}} r_j \alpha_{(0,0)(0,0)}(r_j; |\boldsymbol{\tau}_j - \boldsymbol{\tau}_i|)$$

$$= \frac{1}{2} \sum_{\substack{i \\ i \neq j}} \frac{1}{|\boldsymbol{\tau}_j - \boldsymbol{\tau}_i|} \int_{|\boldsymbol{\tau}_j - \boldsymbol{\tau}_i| - r_j}^{|\boldsymbol{\tau}_j - \boldsymbol{\tau}_i| + r_j} dr'' \, r'' \, \varphi_0(r'')$$

$$= \frac{1}{2} \int_{-r_j}^{r_j} dr'' \sum_{\substack{i \\ i \neq j}} \frac{|\boldsymbol{\tau}_j - \boldsymbol{\tau}_i| + r''}{|\boldsymbol{\tau}_j - \boldsymbol{\tau}_i|} \varphi_0(|\boldsymbol{\tau}_j - \boldsymbol{\tau}_i| + r'')$$

$$=: \frac{1}{2} \int_{-r_j}^{r_j} dr'' \, \chi_0(r'') \,. \tag{C.5.12}$$

Here we have in the second line used the fact that the radii of all spheres are smaller than the distance between two sphere centers. Hence, the quantity $|\boldsymbol{\tau}_j - \boldsymbol{\tau}_i| - r_j$ is always positive. Furthermore, we have in the last step combined the sum over all neighbouring atoms into a single function $\chi_0(r'')$, which depends only on the radial component inside the central sphere. Once we have performed the sum over i we are left with only a single integral over this radial function. In practice, we will evaluate this integral numerically using the intervals given by the intraatomic radial mesh (A.5.1). In particular, we will use the extended Simpson rule [7, (4.1.13)], which results from successively applying the standard Simpson rule [7, (4.1.4)] to non-overlapping pairs of intervals. As a consequence, we are able to calculate the integral entering (C.5.12) for the n'th point of the radial mesh, $r_{j,n}$, recursively from the corresponding integral for $r_{j,n-2}$, i.e.

$$A(r_{j,n}) = \frac{1}{2} \int_{-r_{j,n}}^{r_{j,n}} dr'' \, \chi_0(r'')$$

$$= A(r_{j,n-2}) + \frac{1}{2} \int_{-r_{j,n}}^{-r_{j,n-2}} dr'' \, \chi_0(r'') + \frac{1}{2} \int_{r_{j,n-2}}^{r_{j,n}} dr'' \, \chi_0(r'') \,. \tag{C.5.13}$$

Next, evaluating the two integrals with the help of the standard Simpson rule and taking into account the nonlinearity of the intraatomic radial mesh (A.5.1) in a similar way as in (A.5.6) we obtain the recursion relation

$$A(r_{j,n})$$
$$= A(r_{j,n-2})$$
$$+ \frac{1}{6} \left[\chi_0(-r_{j,n}) \frac{dr}{dx} \Big|_{x=n} + 4\chi_0(-r_{j,n-1}) \frac{dr}{dx} \Big|_{x=n-1} + \chi_0(-r_{j,n-2}) \frac{dr}{dx} \Big|_{x=n-2} \right]$$
$$+ \frac{1}{6} \left[\chi_0(r_{j,n-2}) \frac{dr}{dx} \Big|_{x=n-2} + 4\chi_0(r_{j,n-1}) \frac{dr}{dx} \Big|_{x=n-1} + \chi_0(r_{j,n}) \frac{dr}{dx} \Big|_{x=n} \right]$$
$$= A(r_{j,n-2}) + \frac{1}{6} \left[\left(\chi_0(r_{j,n-2}) + \chi_0(-r_{j,n-2}) \right) \frac{dr}{dx} \Big|_{x=n-2} \right]$$

$$+4\Big(\chi_0(r_{j,n-1}) + \chi_0(-r_{j,n-1})\Big)\frac{dr}{dx}\Big|_{x=n-1}$$

$$+\Big(\chi_0(r_{j,n}) + \chi_0(-r_{j,n})\Big)\frac{dr}{dx}\Big|_{x=n}\Big]. \qquad (C.5.14)$$

Since, by the definition (A.5.1), the first mesh point is at the origin of the atomic sphere, i.e. $r_{j,n=1} = 0$, we are able to calculate the α-functions for all odd values of n by using (C.5.14) with the initial value

$$A(r_{j,n=1}) = \sum_{\substack{i\\i\neq j}} r_{j,n=1}\alpha_{(0,0)(0,0)}(r_{j,n=1}; |\boldsymbol{\tau}_j - \boldsymbol{\tau}_i|)$$

$$= \sum_{\substack{i\\i\neq j}} r_{j,n=1}\varphi_0(|\boldsymbol{\tau}_j - \boldsymbol{\tau}_i|)$$

$$= 0, \qquad (C.5.15)$$

as resulting from (C.5.10) and (C.5.12). In particular, combining (C.5.14) and (C.5.15) we obtain

$$A(r_{j,n=3})$$

$$= A(r_{j,n=1}) + \frac{1}{6}\Big[\Big(\chi_0(r_{j,n=1}) + \chi_0(-r_{j,n=1})\Big)\frac{dr}{dx}\Big|_{x=1}$$

$$+4\Big(\chi_0(r_{j,n=2}) + \chi_0(-r_{j,n=2})\Big)\frac{dr}{dx}\Big|_{x=2}$$

$$+\Big(\chi_0(r_{j,n=3}) + \chi_0(-r_{j,n=3})\Big)\frac{dr}{dx}\Big|_{x=3}\Big]$$

$$= \frac{1}{6}\Big[4\Big(\chi_0(r_{j,n=2}) + \chi_0(-r_{j,n=2})\Big)\frac{dr}{dx}\Big|_{x=2}$$

$$+\Big(\chi_0(r_{j,n=3}) + \chi_0(-r_{j,n=3})\Big)\frac{dr}{dx}\Big|_{x=3}\Big]. \qquad (C.5.16)$$

Here we have used the fact that the first mesh point is at the origin of the atomic sphere, i.e. $r_{j,n=1} = 0$. For even values of n we use instead of (C.5.15) the initial value

$$A(r_{j,n=2})$$

$$= \frac{1}{2}\int_{-r_{j,2}}^{r_{j,2}} dr'' \,\chi_0(r'')$$

$$= \frac{1}{6}\Big[\chi_0(-r_{j,n=2})\frac{dr}{dx}\Big|_{x=2} + 4\chi_0(r_{j,n=1})\frac{dr}{dx}\Big|_{x=1} + \chi_0(r_{j,n=2})\frac{dr}{dx}\Big|_{x=2}\Big]$$

$$= \frac{1}{6}\left[4\chi_0(r_{j,n=1})\frac{dr}{dx}\Big|_{x=1} + \left(\chi_0(r_{j,n=2}) + \chi_0(-r_{j,n=2})\right)\frac{dr}{dx}\Big|_{x=2}\right] . \quad \text{(C.5.17)}$$

Finally, we have to divide each $A(r_{j,n})$ by $r_{j,n}$ in order to arrive at the summed α-function itself, i.e.

$$\frac{1}{r_j}A(r_j) = \sum_{\substack{i \\ i\neq j}} \alpha_{(0,0)(0,0)}(r_j; |\boldsymbol{\tau}_j - \boldsymbol{\tau}_i|) . \quad \text{(C.5.18)}$$

While this has to be done for all $r_{j,n} > 0$, we obtain the result for $r_{j,n=1} = 0$ directly from (C.5.10) as

$$\sum_{\substack{i \\ i\neq j}} \alpha_{(0,0)(0,0)}(r_{j,n=1} = 0; |\boldsymbol{\tau}_j - \boldsymbol{\tau}_i|) = \sum_{\substack{i \\ i\neq j}} \varphi_0(|\boldsymbol{\tau}_j - \boldsymbol{\tau}_i|) = \chi_0(r_{j,n=1} = 0) . \quad \text{(C.5.19)}$$

To conclude, we have calculated the sum over α-functions for all pairs (i,j) of atoms. In order to calculate the electron density or the potential within a particular muffin-tin we first truncate the free atom electronic density and potential at the muffin-tin radius. In a second step, we add to this the sum over α-function expansions from the neighbouring spheres. This gives a unique and reliable description of both the density and the potential within all the muffin-tin spheres. Finally, in order to get a good guess for the interstitial region, we integrate all the muffin-tin densities and distribute the remaining electronic density homogeneously over the interstitial region. For the potential, the Hartree-contribution is set to zero while the exchange-correlation potential is that of the constant electronic density and thus likewise a constant.

References

1. K. J. Duff, Intern. J. Quant. Chem. **5**, 111 (1971)
2. V. Eyert and K.-H. Höck, Phys. Rev. B **57**, 12727 (1998)
3. O. Jepsen and O. K. Andersen, Z. Phys. B **97**, 35 (1995)
4. D. A. Langs, J. Appl. Cryst. **29**, 481 (1996)
5. P. O. Löwdin, Adv. Phys. **5**, 1 (1956)
6. L. F. Mattheiss, Phys. Rev. **133**, A1399 (1964)
7. W. H. Press, B. P. Flannery, S. A. Teukolsky, and W. T. Vetterling, *Numerical Recipes – The Art of Scientific Computing* (Cambridge University Press, Cambridge 1989)

D

Brillouin-Zone Integration

In the present chapter we will discuss several methods for Brillouin-zone integration. After defining the basic issues we will outline, in particular, the definition of special points and turn to the histogram method, the high-precision sampling method, and the linear tetrahedron method. In doing so, we will largely follow the work of Lehmann and Taut, Jepsen and Andersen, Monkhorst and Pack, Lambin and Vigneron, Methfessel and Paxton, as well as Blöchl [4, 5, 8, 11–15]. Nevertheless, a lot of workers have dealt with the issue and for this reason the previous list is by far from complete.

D.1 Basic Notions

The objective of the present chapter is to calculate integrals of the general form

$$G_A(E) = \frac{1}{\Omega_{BZ}} \sum_n \int_{\Omega_{BZ}} d^3\mathbf{k} \, \frac{A_n(\mathbf{k})}{E - \varepsilon_n(\mathbf{k}) + i0^+} \,, \tag{D.1.1}$$

where the $\mathbf{k}$-space integration extends over the first Brillouin zone with volume Ω_{BZ}; n is the band index labelling different single-particle wave functions $\psi_n(\mathbf{k})$ and the corresponding band energies $\varepsilon_n(\mathbf{k})$. In addition, the $A_n(\mathbf{k})$ are expectation values of a single-particle operator A as calculated with the wave functions,

$$A_n(\mathbf{k}) = \langle \psi_n(\mathbf{k})|A|\psi_n(\mathbf{k}) \rangle \,. \tag{D.1.2}$$

In particular, we assume that the wave functions $\psi_n(\mathbf{k})$ are (orthonormalized) eigenfunctions of a $\mathbf{k}$-dependent Hamiltonian H and $\varepsilon_n(\mathbf{k})$ the corresponding eigenvalues. Obviously, contributions from different bands are independent of each other. The matrix elements $A_n(\mathbf{k})$ may be regarded as generalized spectral densities and $G_A(E)$ as the corresponding (retarded) Green's function of the operator A; here 0^+ is a positive infinitesimal. Note that in the present context we have suppressed

V. Eyert: *Brillouin-Zone Integration*, Lect. Notes Phys. **719**, 257–305 (2007)
DOI 10.1007/978-3-540-71007-3_8

the spin dependence of the operator A, the band energies $\varepsilon_n(\mathbf{k})$ and the matrix elements $A_n(\mathbf{k})$, which can be easily reintroduced.

As well known, by employing the Dirac identity

$$\frac{1}{E - \varepsilon_n(\mathbf{k}) \pm i0^+} = \mathcal{P}\frac{1}{E - \varepsilon_n(\mathbf{k})} \mp i\pi\delta(E - \varepsilon_n(\mathbf{k})) , \tag{D.1.3}$$

where $\mathcal{P}$ denotes the Cauchy principal value, we may readily split Green's function into two parts

$$G_A(E) = R_A(E) - i\pi I_A(E) , \tag{D.1.4}$$

with

$$R_A(E) = \frac{1}{\Omega_{BZ}} \sum_n \mathcal{P} \int_{\Omega_{BZ}} d^3\mathbf{k} \, \frac{A_n(\mathbf{k})}{E - \varepsilon_n(\mathbf{k})} , \tag{D.1.5}$$

and

$$-\pi I_A(E) = -\frac{\pi}{\Omega_{BZ}} \sum_n \int_{\Omega_{BZ}} d^3\mathbf{k} \, A_n(\mathbf{k})\delta(E - \varepsilon_n(\mathbf{k})) . \tag{D.1.6}$$

In particular, if the generalized spectral densities $A_n(\mathbf{k})$ are real quantities, the contributions (D.1.5) and (D.1.6) are just the real and imaginary part of the Green's function. While all derivations presented in the rest of this chapter will hold for the general case of complex spectral densities we will nevertheless speak of the real and imaginary part in order to keep the wording simple. As is also well known, the two parts of the Green's function are connected via the Kramers-Kronig relation, which results from combining (D.1.5) and (D.1.6) to write

$$R_A(E) = \frac{1}{\Omega_{BZ}} \sum_n \mathcal{P} \int dE' \frac{1}{E - E'} \int_{\Omega_{BZ}} d^3\mathbf{k} \, A_n(\mathbf{k})\delta(E' - \varepsilon_n(\mathbf{k}))$$

$$= \mathcal{P} \int dE' \frac{1}{E - E'} I_A(E') . \tag{D.1.7}$$

Of course, we may use the Kramers-Kronig relation to avoid the direct calculation of the real part. However, a numerical evaluation suffers from the disadvantage that it needs an energy mesh, which extends from $-\infty$ to $+\infty$ and at the same time has to be rather dense in order to account for the fine structures of the imaginary part. For this reason, we opt for a direct calculation of both the real and imaginary part.

Most prominent examples for the operator A are $A = \mathcal{I}$ and $A = H$. The first one leads directly to the density of states

$$\rho(E) = I_{\mathcal{I}}(E) = \frac{1}{\Omega_{BZ}} \sum_n \int_{\Omega_{BZ}} d^3\mathbf{k} \, \delta(E - \varepsilon_n(\mathbf{k})) , \tag{D.1.8}$$

and the total charge

$$Q = \int dE \, \rho(E)f(E) , \tag{D.1.9}$$

where $f(E)$ denotes the Fermi function

$$f(E) = \frac{1}{e^{\beta(E-\mu)} + 1} , \tag{D.1.10}$$

with $\beta = \frac{1}{k_B T}$ and μ being the chemical potential. In contrast, the second choice gives rise to the energy weighted density of states

$$\rho(E)E = I_H(E) = \frac{1}{\Omega_{BZ}} \sum_n \int_{\Omega_{BZ}} d^3k \, \varepsilon_n(\mathbf{k})\delta(E - \varepsilon_n(\mathbf{k})) , \tag{D.1.11}$$

and the total band energy

$$E_B = \int dE \, \rho(E)E f(E) . \tag{D.1.12}$$

As a matter of fact, for many operators of physical interest (D.1.2) leads to matrix elements, which depend on the $\mathbf{k}$-vector and on the band index n only through the band energy $\varepsilon_n(\mathbf{k})$. Both of the operators $A = \mathcal{I}$ and $A = H$ fall into this category. For operators of this type we could then write

$$A_n(\mathbf{k}) = \tilde{A}(\varepsilon_n(\mathbf{k})) . \tag{D.1.13}$$

As a consequence, we will be able to separate the $\mathbf{k}$-space integration and the energy dependence in (D.1.1) and write

$$G_A(E) = \frac{1}{\Omega_{BZ}} \sum_n \int_{\Omega_{BZ}} d^3k \, \frac{\tilde{A}(\varepsilon_n(\mathbf{k}))}{E - \varepsilon_n(\mathbf{k}) + i0^+} . \tag{D.1.14}$$

In particular, we get for the imaginary part from (D.1.6) and (D.1.8)

$$I_A(E) = \frac{1}{\Omega_{BZ}} \sum_n \int_{\Omega_{BZ}} d^3k \, \tilde{A}(\varepsilon_n(\mathbf{k}))\delta(E - \varepsilon_n(\mathbf{k}))$$
$$= \tilde{A}(E)\rho(E) . \tag{D.1.15}$$

Thus, given the function $\tilde{A}(E)$ we need only Green's function $G_{\mathcal{I}}(E)$ to evaluate the Brillouin-zone integral.

Still, there exists an alternative formulation of the Green's function $G_A(E)$, which will be quite useful in combination with interpolation schemes for the band structure as, e.g. the linear tetrahedron method. In order to derive this formulation we consider the constant energy surface for each band, $S(\varepsilon_n)$ and define the unit vector perpendicular to the local surface element as

$$\mathbf{u}_\perp = \frac{\nabla_\mathbf{k}\varepsilon_n(\mathbf{k})}{|\nabla_\mathbf{k}\varepsilon_n(\mathbf{k})|} . \tag{D.1.16}$$

This allows to write the $\mathbf{k}$-space volume element as

$$d^3k = dS \, \mathbf{u}_\perp \cdot d\mathbf{k}_\perp = dS \, \frac{\nabla_\mathbf{k}\varepsilon_n(\mathbf{k})}{|\nabla_\mathbf{k}\varepsilon_n(\mathbf{k})|} \cdot d\mathbf{k}_\perp = d\varepsilon_n(\mathbf{k}) \, dS \, \frac{1}{|\nabla_\mathbf{k}\varepsilon_n(\mathbf{k})|} , \tag{D.1.17}$$

where $\mathbf{k}_\perp$ denotes the component of the $\mathbf{k}$-vector perpendicular to the constant energy surface, hence, parallel to the plane normal $\mathbf{u}_\perp$. We thus obtain for Green's function (D.1.1) the alternative expression

$$G_A(E) = \frac{1}{\Omega_{BZ}} \sum_n \int d\varepsilon_n(\mathbf{k}) \int_{S(\varepsilon_n)} dS \, \frac{1}{|\nabla_\mathbf{k}\varepsilon_n(\mathbf{k})|} \frac{A_n(\mathbf{k})}{E - \varepsilon_n(\mathbf{k}) + i0^+} \,, \quad (D.1.18)$$

which, for the real and imaginary parts (D.1.5) and (D.1.6), reduces to

$$R_A(E) = \frac{1}{\Omega_{BZ}} \sum_n \mathcal{P} \int d\varepsilon_n(\mathbf{k}) \int_{S(\varepsilon_n)} dS \, \frac{A_n(\mathbf{k})}{|\nabla_\mathbf{k}\varepsilon_n(\mathbf{k})|} \frac{1}{E - \varepsilon_n(\mathbf{k})} \,, \quad (D.1.19)$$

and

$$\begin{aligned}
I_A(E) &= \frac{1}{\Omega_{BZ}} \sum_n \int d\varepsilon_n(\mathbf{k}) \int_{S(\varepsilon_n)} dS \, \frac{A_n(\mathbf{k})}{|\nabla_\mathbf{k}\varepsilon_n(\mathbf{k})|} \delta(E - \varepsilon_n(\mathbf{k})) \\
&= \frac{1}{\Omega_{BZ}} \sum_n \int_{S(\varepsilon_n)} dS \, \frac{A_n(\mathbf{k})}{|\nabla_\mathbf{k}\varepsilon_n(\mathbf{k})|} \delta(E - \varepsilon_n(\mathbf{k})) \,. \quad (D.1.20)
\end{aligned}$$

Hence, the imaginary part of Green's function arises just as an integral over the constant energy surface. Note that the results (D.1.18) to (D.1.20) are fully equivalent to (D.1.1), (D.1.5), and (D.1.6); they differ only by the writing of the infinitesimal volume element. Since the electronic bands $\varepsilon_n(\mathbf{k})$ are periodic functions in $\mathbf{k}$-space, bounded from both below and above, there must be points in each unit cell, where the gradient $\nabla_\mathbf{k}\varepsilon_n(\mathbf{k})$ vanishes. Although the integrands in (D.1.18) to (D.1.20) diverge at these points it can be shown — at least for $A = \mathcal{I}$ — that these so-called van Hove singularities are integrable in three dimensions and give rise to kinks in the density of states [2].

The actual calculation of Brillouin-zone integrals of the type (D.1.1) is governed by two requirements: First, we will be able to evaluate the wave functions $\psi_n(\mathbf{k})$, the band energies $\varepsilon_n(\mathbf{k})$, and the matrix elements $A_n(\mathbf{k})$ entering the integrals only at a finite number N_{tot} of $\mathbf{k}$-points. Second, we want to express the integral (D.1.1) or, equivalently, its real and imaginary parts, (D.1.5) and (D.1.6), as a weighted sum of the matrix elements (D.1.2) as calculated at the finite set of $\mathbf{k}$-points, i.e. we would like to write

$$G_A(E) = \sum_{ni} A_n(\mathbf{k}_i) w_{ni}^G(E) \,, \quad (D.1.21)$$

where the weights $w_{ni}^G(E)$ in general are complex. Using (D.1.4) we readily obtain

$$R_A(E) = \sum_{ni} A_n(\mathbf{k}_i) w_{ni}^R(E) \,, \quad (D.1.22)$$

$$I_A(E) = \sum_{ni} A_n(\mathbf{k}_i) w_{ni}^I(E) \,, \quad (D.1.23)$$

with

$$w_{ni}^G(E) = w_{ni}^R(E) - i\pi w_{ni}^I(E) \ . \tag{D.1.24}$$

As they stand, (D.1.21) to (D.1.23) are not very helpful since the weights are as yet unspecified. However, it is important to realize that integration is a linear operation and, hence, the weights are independent of a particular set of matrix elements $A_n(\mathbf{k})$ [5]. In contrast, they are exclusively determined by the band structure $\varepsilon_n(\mathbf{k})$. As a consequence, the issue of performing accurate Brillouin-zone integrations has been reduced to the appropriate calculation of weights $w_{ni}^G(E)$.

In order to attack this problem we start considering the simplest possible choice of operator $A = \mathcal{I}$. Combining (D.1.5), (D.1.6), (D.1.19), (D.1.20), (D.1.22), and (D.1.23) we obtain for the real and imaginary parts

$$\begin{aligned}
R_{\mathcal{I}}(E) &= \frac{1}{\Omega_{BZ}} \sum_n \mathcal{P} \int_{\Omega_{BZ}} d^3\mathbf{k} \, \frac{1}{E - \varepsilon_n(\mathbf{k})} \\
&= \frac{1}{\Omega_{BZ}} \sum_n \mathcal{P} \int d\varepsilon_n(\mathbf{k}) \int_{S(\varepsilon_n)} dS \, \frac{1}{|\nabla_{\mathbf{k}}\varepsilon_n(\mathbf{k})|} \frac{1}{E - \varepsilon_n(\mathbf{k})} \\
&\overset{!}{=} \sum_{ni} w_{ni}^R(E) \ ,
\end{aligned} \tag{D.1.25}$$

and

$$\begin{aligned}
I_{\mathcal{I}}(E) &= \frac{1}{\Omega_{BZ}} \sum_n \int_{\Omega_{BZ}} d^3\mathbf{k} \, \delta(E - \varepsilon_n(\mathbf{k})) \\
&= \frac{1}{\Omega_{BZ}} \sum_n \int_{S(\varepsilon_n)} dS \, \frac{1}{|\nabla_{\mathbf{k}}\varepsilon_n(\mathbf{k})|} \delta(E - \varepsilon_n(\mathbf{k})) \\
&\overset{!}{=} \sum_{ni} w_{ni}^I(E) \ ,
\end{aligned} \tag{D.1.26}$$

and from this the identities

$$\begin{aligned}
\sum_i^{N_{tot}} w_{ni}^R(E) &= \frac{1}{\Omega_{BZ}} \mathcal{P} \int_{\Omega_{BZ}} d^3\mathbf{k} \, \frac{1}{E - \varepsilon_n(\mathbf{k})} \\
&= \frac{1}{\Omega_{BZ}} \mathcal{P} \int d\varepsilon_n(\mathbf{k}) \int_{S(\varepsilon_n)} dS \, \frac{1}{|\nabla_{\mathbf{k}}\varepsilon_n(\mathbf{k})|} \frac{1}{E - \varepsilon_n(\mathbf{k})} \ ,
\end{aligned} \tag{D.1.27}$$

and

$$\begin{aligned}
\sum_i^{N_{tot}} w_{ni}^I(E) &= \frac{1}{\Omega_{BZ}} \int_{\Omega_{BZ}} d^3\mathbf{k} \, \delta(E - \varepsilon_n(\mathbf{k})) \\
&= \frac{1}{\Omega_{BZ}} \int_{S(\varepsilon_n)} dS \, \frac{1}{|\nabla_{\mathbf{k}}\varepsilon_n(\mathbf{k})|} \delta(E - \varepsilon_n(\mathbf{k})) \ . \tag{D.1.28}
\end{aligned}$$

Here we have assumed that the weights for different bands are independent of each other. This is justified by the fact that the contributions from different bands to the Green's function are also independent; different bands may be simply regarded as different functions in $\mathbf{k}$-space.

Obviously, (D.1.27) and (D.1.28) establish sum rules for the weights $w_{ni}^R(E)$ and $w_{ni}^I(E)$. However, the calculation of the weights themselves is an open issue. In order to resolve it we make again use of the fact that the weights are independent of the actual choice of the matrix elements $A_n(\mathbf{k})$. This freedom allows to define a set of filter functions $w_j(\mathbf{k})$, which could be tailored such that the calculation of single weights becomes feasible [4, 5]. To be specific, we combine (D.1.5), (D.1.6), (D.1.19), (D.1.20), (D.1.22), and (D.1.23) with the matrix elements $A_n(\mathbf{k})$ replaced by filter functions and note

$$
\sum_i^{N_{tot}} w_j(\mathbf{k}_i) w_{ni}^R(E) = \frac{1}{\Omega_{BZ}} \mathcal{P} \int_{\Omega_{BZ}} d^3\mathbf{k}\; w_j(\mathbf{k}) \frac{1}{E - \varepsilon_n(\mathbf{k})}
$$
$$
= \frac{1}{\Omega_{BZ}} \mathcal{P} \int d\varepsilon_n(\mathbf{k}) \int_{S(\varepsilon_n)} dS\; \frac{w_j(\mathbf{k})}{|\nabla_{\mathbf{k}}\varepsilon_n(\mathbf{k})|} \frac{1}{E - \varepsilon_n(\mathbf{k})} \qquad \forall_{j,n} ,
$$

(D.1.29)

and

$$
\sum_i^{N_{tot}} w_j(\mathbf{k}_i) w_{ni}^I(E) = \frac{1}{\Omega_{BZ}} \int_{\Omega_{BZ}} d^3\mathbf{k}\; w_j(\mathbf{k}) \delta(E - \varepsilon_n(\mathbf{k}))
$$
$$
= \frac{1}{\Omega_{BZ}} \int_{S(\varepsilon_n)} dS\; \frac{w_j(\mathbf{k})}{|\nabla_{\mathbf{k}}\varepsilon_n(\mathbf{k})|} \delta(E - \varepsilon_n(\mathbf{k})) \qquad \forall_{j,n} .
$$

(D.1.30)

Of course, in order to allow for a unique solution of this linear equation system there should be as many filter functions $w_j(\mathbf{k})$ as there are $\mathbf{k}$-points in the sum. Furthermore, to guarantee for a homogeneous sampling of all portions of the Brillouin zone the set of filter functions should at best fulfill the (local) sum rule

$$
\sum_j^{N_{tot}} w_j(\mathbf{k}) = 1 \qquad \forall_{\mathbf{k}} .
$$

(D.1.31)

Obviously, the linear equation systems (D.1.29) and (D.1.30) are trivially solved by the simple choice

$$
w_j(\mathbf{k}_i) = \delta_{ij} ,
$$

(D.1.32)

in which case we obtain

$$
w_{nj}^R(E) = \frac{1}{\Omega_{BZ}} \mathcal{P} \int_{\Omega_{BZ}} d^3\mathbf{k}\; w_j(\mathbf{k}) \frac{1}{E - \varepsilon_n(\mathbf{k})}
$$
$$
= \frac{1}{\Omega_{BZ}} \mathcal{P} \int d\varepsilon_n(\mathbf{k}) \int_{S(\varepsilon_n)} dS\; \frac{w_j(\mathbf{k})}{|\nabla_{\mathbf{k}}\varepsilon_n(\mathbf{k})|} \frac{1}{E - \varepsilon_n(\mathbf{k})} \qquad \forall_n ,
$$

(D.1.33)

and

$$w_{nj}^I(E) = \frac{1}{\Omega_{BZ}} \int_{\Omega_{BZ}} d^3\mathbf{k}\, w_j(\mathbf{k})\delta(E - \varepsilon_n(\mathbf{k}))$$

$$= \frac{1}{\Omega_{BZ}} \int_{S(\varepsilon_n)} dS\, \frac{w_j(\mathbf{k})}{|\nabla_\mathbf{k}\varepsilon_n(\mathbf{k})|}\delta(E - \varepsilon_n(\mathbf{k})) \qquad \forall_n . \qquad (D.1.34)$$

As before, the latter two are connected via the Kramers-Kronig relation, hence, we write

$$w_{nj}^R(E) = \mathcal{P} \int dE'\, \frac{1}{E - E'} w_{nj}^I(E') . \qquad (D.1.35)$$

While being equivalent to (D.1.7) this form bears an important advantage, which traces back to the fact that the weights $w_{nj}^R(E)$ and $w_{nj}^I(E)$ have a much simpler form than the real and imaginary part of the Green's function and may be even known analytically. As a consequence, there is a good chance that the energy integration coming with (D.1.35) may be also performed analytically and, hence, the real parts of the weights be determined exactly.

It is important to note that the choice (D.1.32) fixes only the values of the filter functions at the discrete of points $\mathbf{k}_i$ while leaving their shape in between unspecified. In general, as we will see in the subsequent sections, different methods for Brillouin-zone integration are distinguished by their different choice of the filter functions as well as their interpolation of the band structure $\varepsilon_n(\mathbf{k})$ between the points $\mathbf{k}_i$. Before presenting the most prominent methods we will discuss how optimal sets of these discrete $\mathbf{k}$-points can be selected.

D.2 Special Points

The selection of a set of discrete $\mathbf{k}$-points for Brillouin-zone sampling is governed by three requirements. First, there should be as few points as possible in order to reduce the computational cost to a minimum. Second, the points should be evenly spread over the first Brillouin zone. Third, the $\mathbf{k}$-points may fulfill additional requirements as, e.g. orthogonality of a set of plane waves when calculated at these points.

First attempts towards selection of optimal $\mathbf{k}$-points are due to Baldereschi who proposed to use only one point, which he called the mean-value point [3]. Chadi and Cohen proposed a set of only few so-called special points to be used especially in calculations for semiconductors [6].

Lateron, Monkhorst and Pack proposed a set of points, which included the set given by Chadi and Cohen but allowed for a simple construction scheme common to all space groups [15]. This new set fulfills all the above criteria and has become widely used. In addition, it has been combined with the linear tetrahedron method and thereby increased the applicability of the latter a lot [5].

The actual setup of the special points by Monkhorst and Pack is straightforward. Given the divisions N_1, N_2, and N_3 of the reciprocal space primitive translations $\mathbf{b}_i$ the special points are defined by

$$
\mathbf{k}_{I_1,I_2,I_3} = \sum_{i=1}^{3} \frac{2I_i - N_i - 2 + I_{off,i}}{2N_i} \mathbf{b}_i
$$

$$
= \sum_{i=1}^{3} (2I_i - N_i - 2 + I_{off,i}) \mathbf{d}_i \qquad \text{for } 1 \leq I_i \leq N_i, \qquad \text{(D.2.1)}
$$

where, in the second step, we have used the scaled vectors

$$
\mathbf{d}_i = \frac{1}{2N_i} \mathbf{b}_i . \qquad \text{(D.2.2)}
$$

Monkhorst and Pack proposed to use even divisions N_i and the offset $I_{off,i} = 1$, in which case the numerators run from $1 - N_i$ to $N_i - 1$ and high-symmetry points are avoided. In contrast, for $I_{off,i} = N_i$ the numerators are in the range $0, \ldots, 2N_i - 2$ and the high-symmetry points are included [7].

The discussion about the possible inclusion of the high-symmetry points traces back to the early days of the tetrahedron method when the tetrahedra were constructed in the irreducible wedge of the first Brillouin zone only. This led to the question of how to correctly weight the surface points and in some case errors showed up [10]. However, in more recent implementations the tetrahedra are constructed using the special points by Monkhorst and Pack before taking into account crystal symmetry [9]. This way weighting errors are avoided and the issue of including high-symmetry points is without meaning. Yet, their inclusion is recommended since it leads to a smaller number of points in the irreducible wedge without loss of accuracy.

D.3 Simple-Sampling Method

Since the sampling methods to be described in this and the subsequent section were always used in combination with the special points proposed by Chadi and Cohen or Monkhorst and Pack they are often called special point methods. However, this is somewhat misleading for sampling methods could in principle be used with any set of $\mathbf{k}$-points. Moreover, as already mentioned, interpolation methods are also used with the special point sets. For this reason, we prefer the name sampling or histogram method.

As outlined at the end of Sect. D.1, methods for Brillouin-zone integration are characterized by their particular choice of filter function $w_j(\mathbf{k})$ and the way they approximate the band dispersion $\varepsilon_n(\mathbf{k})$ between the discrete points $\mathbf{k}_i$.

In the simple-sampling method we use

$$
w_j(\mathbf{k}) = \Theta_j(\mathbf{k}) , \qquad \text{(D.3.1)}
$$

where $\Theta_j(\mathbf{k})$ is a step function defined as

$$\Theta_j(\mathbf{k}) = \begin{cases} 1 & \text{for } |\mathbf{k} - \mathbf{k}_j| \le |\mathbf{k} - \mathbf{k}_i| \\ 0 & \text{elsewhere} \end{cases} . \tag{D.3.2}$$

Hence, the step function $\Theta_j(\mathbf{k})$ is unity within the Voronoi polyhedron centered at $\mathbf{k}_j$ and vanishes outside. Note that all Voronoi polyhedra are mutually exclusive, have volume Ω_{BZ}/N_{tot}, and, taken together, fill space completely. The condition (D.1.31) is thus fulfilled.

For the band structure $\varepsilon_n(\mathbf{k})$ a similar definition as for the filter functions is used, namely,

$$\varepsilon_n(\mathbf{k}) = \sum_i^{N_{tot}} \varepsilon_n(\mathbf{k}_i)\Theta_i(\mathbf{k}) \qquad \forall_n . \tag{D.3.3}$$

Thus the dispersion is assumed to be constant within each Voronoi polyhedron.

Next, inserting (D.3.1) and (D.3.3) into (D.1.33) and (D.1.34) we arrive at the following expression for the weights

$$\begin{aligned} w_{nj}^R(E) &= \frac{1}{\Omega_{BZ}} \frac{\Omega_{BZ}}{N_{tot}} \frac{1}{E - \varepsilon_n(\mathbf{k}_j)} \\ &= \frac{1}{N_{tot}} \frac{1}{E - \varepsilon_n(\mathbf{k}_j)} , \end{aligned} \tag{D.3.4}$$

and

$$\begin{aligned} w_{nj}^I(E) &= \frac{1}{\Omega_{BZ}} \frac{\Omega_{BZ}}{N_{tot}} \delta(E - \varepsilon_n(\mathbf{k}_j)) \\ &= \frac{1}{N_{tot}} \delta(E - \varepsilon_n(\mathbf{k}_j)) . \end{aligned} \tag{D.3.5}$$

As mentioned at the end of Sect. D.1 we may alternatively use the Kramers-Kronig relation (D.1.35). Doing so we trivially obtain

$$\begin{aligned} w_{nj}^R(E) &= \mathcal{P} \int dE' \frac{1}{E - E'} w_{nj}^I(E') \\ &= \frac{1}{N_{tot}} \mathcal{P} \int dE' \frac{1}{E - E'} \delta(E' - \varepsilon_n(\mathbf{k}_j)) \\ &= \frac{1}{N_{tot}} \frac{1}{E - \varepsilon_n(\mathbf{k}_j)} , \end{aligned} \tag{D.3.6}$$

which is, as expected, identical to (D.3.4).

Finally, combining (D.3.5) with (D.1.26) we arrive at the total density of states

$$\rho(E) = I_{\mathcal{I}}(E) = \frac{1}{N_{tot}} \sum_{ni} \delta(E - \varepsilon_n(\mathbf{k}_i)) . \tag{D.3.7}$$

It is thus given as a scaled sum of δ-peaks located at the eigenvalues, which have been calculated at the discrete set of special points. For all energies between the peaks the density of states vanishes. This is a consequence of the above assumption of a constant band dispersion within each Voronoi polyhedron. Although it allows to accumulate the density of states for each **k**-point independently the approximation of a constant dispersion is too crude and leads to a rather statistical determination of the density of states and related quantities. As an example, we show in Fig. D.1 the density of states of Na as calculated using the simple-sampling method with $30 \times 30 \times 30$ **k**-points. While the square-root behaviour typical for free-electron like metals is still visible the result clearly reflects the statistical nature of the simple-sampling scheme. The noise produced by this procedure could be suppressed only with a very huge number of **k**-points. As a consequence, the simple-sampling method is rather inefficient. Nevertheless, note that for the calculation of integrated densities of states the statistical errors average out to a large extent this leading to reasonable results.

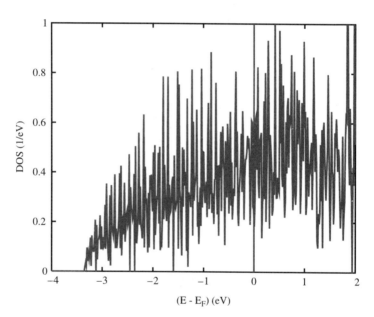

Fig. D.1. Density of states of Na as calculated with the simple-sampling method with $30 \times 30 \times 30$ **k**-points

D.4 High-Precision Sampling Method

Even if the simple-sampling method yields results, which are less satisfactory than those obtained by the more sophisticated schemes it still offers the important advantage that it does not require to store all the band energies at all **k**-points. This is due to the simple form of the integration weights as well as the fact that the

Brillouin integrals are calculated by accumulation. It is thus well justified to keep the simple-sampling methods as a basis for further developments.

Most attempts to improve the simple-sampling schemes concentrated on the δ-peak structure of the weights (D.3.5), which transfers to the density of states as given by (D.3.7) and which causes the bad signal-to-noise ratio of the simple-sampling method as revealed by Fig. D.1. Of course, the most straightforward approach to overcome this weakness consists of replacing each δ-peak by a Gaussian of finite width, which leads to much smoother curves.

A different and very systematic approach was proposed by Methfessel and Paxton under the name high-precision sampling [14]. It is based on an expansion of the δ-distribution in a complete set of functions. Methfessel and Paxton opted for Hermite polynomials, hence

$$\delta(x) = \sum_{n=0}^{\infty} \alpha_n H_{2n}(x) e^{-x^2}, \tag{D.4.1}$$

where the odd terms were omitted since the δ-distribution is even.

As well known [1, Chap. 22], the Hermite polynomials form a complete set of functions, which are orthonormalized to Gaussian weights, i.e.

$$\int_{-\infty}^{\infty} dx\, e^{-x^2} H_n(x) H_m(x) = n! 2^n \delta_{nm}. \tag{D.4.2}$$

They may be explicitly calculated using Rodrigues's formula

$$H_n(x) = (-)^n e^{x^2} \frac{d^n}{dx^n} e^{-x^2}, \qquad n = 0, \dots, \infty, \tag{D.4.3}$$

and fulfill the recurrence relation

$$H_{n+1}(x) = 2x H_n(x) - 2n H_{n-1}(x), \tag{D.4.4}$$

with

$$H_0(x) = 1 \qquad \text{and} \qquad H_1(x) = 2x. \tag{D.4.5}$$

From this we obtain the explicit values

$$H_n(0) = \begin{cases} 0 & \text{for } n \text{ odd} \\ (-)^{\frac{n}{2}} \frac{n!}{\frac{n}{2}!} & \text{for } n \text{ even} \end{cases}. \tag{D.4.6}$$

Finally, we note the identity

$$\frac{d}{dx} H_n(x) = 2n H_{n-1}(x), \tag{D.4.7}$$

which together with the recurrence relation leads to

$$\frac{d}{dx} H_n(x) e^{-x^2} = 2n H_{n-1}(x) e^{-x^2} - 2x H_n(x) e^{-x^2}$$

$$= -e^{-x^2} H_{n+1}(x). \tag{D.4.8}$$

With the orthonormalization (D.4.2) at hand we are ready to calculate the coefficients entering (D.4.1) as

$$
\begin{aligned}
\alpha_n &= \int_{-\infty}^{\infty} dx\, \delta(x) H_{2n}(x) \frac{1}{n! 2^n \sqrt{\pi}} \\
&= \frac{H_{2n}(0)}{(2n)! 4^n \sqrt{\pi}} \\
&= \frac{(-)^n}{n! 4^n \sqrt{\pi}} \ .
\end{aligned}
\tag{D.4.9}
$$

Having represented the δ-distribution as a series expansion we may approximate it by keeping only a finite number of terms, i.e.

$$
\delta(x) \approx D_N(x) = \sum_{n=0}^{N} \alpha_n H_{2n}(x) e^{-x^2} \ .
\tag{D.4.10}
$$

Since the Hermite polynomials are rather smooth functions truncation of the series results in an effective and systematic smoothing of the function $D_N(x)$.

For energy-integrated functions we turn from the δ-distribution to the step function

$$
\Theta(x) = \begin{cases} 0 & \text{for } x < 0 \\ 1 & \text{for } x > 0 \end{cases} ,
\tag{D.4.11}
$$

which is related to the δ-distribution by

$$
\Theta(x) = \int_{-\infty}^{x} dx'\, \delta(x') \ .
\tag{D.4.12}
$$

In order to approximate the step function in consistence with the above approximation for the δ-distribution we insert (D.4.10) into (D.4.12) and write

$$
\begin{aligned}
\Theta(x) \approx S_N(x) &= \int_{-\infty}^{x} dx'\, D_N(x') \\
&= \sum_{n=0}^{N} \alpha_n \int_{-\infty}^{x} dx'\, H_{2n}(x') e^{-x'^2} \ .
\end{aligned}
\tag{D.4.13}
$$

Using the identity

$$
\int_{-\infty}^{x} dx'\, e^{-x'^2} = \int_{-\infty}^{0} dx'\, e^{-x'^2} + \int_{0}^{x} dx'\, e^{-x'^2} = \frac{\sqrt{\pi}}{2}(1 + \mathrm{erf}(x)) ,
\tag{D.4.14}
$$

for the first term of the series as well as (D.4.8) we arrive at the result

$$
S_0(x) = \frac{1}{2}(1 + \mathrm{erf}(x)) ,
\tag{D.4.15}
$$

$$
S_N(x) = S_0(x) - \sum_{n=1}^{N} \alpha_n H_{2n-1}(x) e^{-x^2} \ .
\tag{D.4.16}
$$

Note that these formulas deviate from the corresponding ones given by Methfessel and Paxton since they used $1 - \Theta(x)$ instead of $\Theta(x)$. The functions $D_N(x)$ and $S_N(x)$ are displayed in Fig. D.2 for $N = 0, \ldots, 5$. The trend towards the δ-distribution and the step function, respectively, for increasing N is clearly visible.

After these preparations we obtain for the integration weights

$$w_{nj}^I(E) = \frac{1}{N_{tot}} D_N(E - \varepsilon_n(\mathbf{k}_j)) \,, \tag{D.4.17}$$

and may thus write down the density of states as

$$\rho(E) = \frac{1}{N_{tot}} \sum_{ni} D_N(E - \varepsilon_n(\mathbf{k}_i)) \,. \tag{D.4.18}$$

To conclude, by approximating the δ-distribution in the way outlined above we have converted the density of states from a sum of δ-peaks to a smooth function and thereby suppressed the statistical noise. This becomes obvious from Fig. D.3, which displays the density of states of Na as calculated using the high-precision sampling method with the same number of **k**-points as used for the generation of Fig. D.1. Obviously, the statistical nature has been completely abandoned. However, there are still some oscillations in the curve especially at higher energies. In addition, the smoothing coming with the truncation of the series causes distinct deviations from the square-root behaviour at the lower band edge.

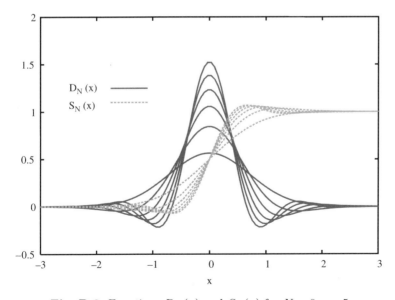

Fig. D.2. Functions $D_N(x)$ and $S_N(x)$ for $N = 0, \ldots, 5$

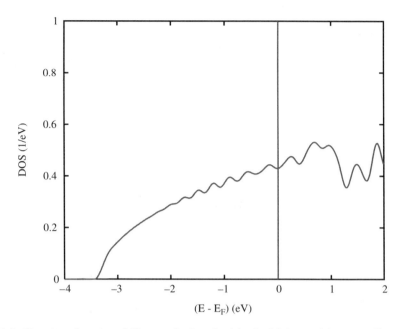

Fig. D.3. Density of states of Na as calculated with the high-precision sampling method with $30 \times 30 \times 30$ **k**-points

D.5 Linear Tetrahedron Method

In this section we turn to the interpolation schemes, which by construction are superiour to the sampling methods described in the previous sections in that they give much more accurate results for the same number of **k**-points. In particular, we will deal with the linear tetrahedron method. In general, the application of interpolation methods proceeds in three steps. First, the Brillouin zone is divided into non-overlapping and space-filling microcells, which usually are parallelepipeds or tetrahedra. The wave functions $\psi_n(\mathbf{k})$, band energies $\varepsilon_n(\mathbf{k})$, and matrix elements $A_n(\mathbf{k})$ are calculated at the vertices of these microcells. Second, the variation of the band energies within each microcell is approximated by an analytical expression, which in most cases is just the linear interpolation between the vertices. However, also quadratic schemes have been implemented. This second step corresponds to determining the filter functions and the approximation to the band dispersion as outlined at the end of Sect. D.1. Finally, the Brillouin-zone integrals are calculated as a sum of the analytical results for the integrals over all microcells.

The linear tetrahedron method to be discussed in this section was first proposed by Lehmann et al., Jepsen and Andersen as well as by Lehmann and Taut [8, 12, 13]. Since then it has witnessed several extensions and improvements [4, 5]. In the original scheme only the irreducible wedge of the first Brillouin zone was filled with tetrahedra. In some situations this led to weighting problems due to surface effects [10]. In addition, the setup of the tetrahedra was different for each Bravais lattice and, hence, made the method difficult to apply in practice.

With the introduction of the special points by Monkhorst and Pack a different strategy became standard. Now the whole reciprocal space unit cell is divided into the microcells in a way common to all lattices by first creating parallelepipeds between the special points and then uniquely dividing each parallelepiped into six tetrahedra [9]. Symmetry considerations are applied only for evaluating the band energies at the vertices of the microcells. In contrast, the integration is always taken over the full Brillouin zone. The new scheme is visualized in Fig. D.4, which shows the division of a parallelepiped into tetrahedra. Since all parallelepipeds are divided in the same way, the calculations become much simpler. In order to avoid roundoff errors the shortest diagonal of the parallelepiped is chosen as the edge common to all tetrahedra (between the vertices 3 and 6 in Fig. D.4). Due to the use of the same division for all parallelepipeds the weighting error mentioned before is avoided.

Next we turn to the calculation of Brillouin-zone integrals as a sum of integrals over all microcells. In doing so, we will specifically opt for tetrahedra. To be concrete, we aim at calculating integrals of the type (D.1.5) or (D.1.19) and (D.1.6) or (D.1.20), respectively, i.e.

$$
\begin{aligned}
R_a(E) &= \frac{1}{\Omega_{BZ}} \mathcal{P} \int_{\Omega_{BZ}} d^3\mathbf{k} \, \frac{a(\mathbf{k})}{E - \varepsilon(\mathbf{k})} \\
&= \frac{1}{\Omega_{BZ}} \mathcal{P} \int d\varepsilon(\mathbf{k}) \int_{S(\varepsilon)} dS \, \frac{a(\mathbf{k})}{|\nabla_\mathbf{k}\varepsilon(\mathbf{k})|} \frac{1}{E - \varepsilon(\mathbf{k})} \\
&= \sum_T r_{a,T}(E) \,,
\end{aligned}
\tag{D.5.1}
$$

and

$$
\begin{aligned}
I_a(E) &= \frac{1}{\Omega_{BZ}} \int_{\Omega_{BZ}} d^3\mathbf{k} \, a(\mathbf{k})\delta(E - \varepsilon(\mathbf{k})) \\
&= \frac{1}{\Omega_{BZ}} \int_{S(\varepsilon)} dS \, \frac{a(\mathbf{k})}{|\nabla_\mathbf{k}\varepsilon(\mathbf{k})|} \delta(E - \varepsilon(\mathbf{k})) \\
&= \sum_T i_{a,T}(E) \,,
\end{aligned}
\tag{D.5.2}
$$

where $a(\mathbf{k})$, may be any function in $\mathbf{k}$-space evaluated at the points of the discrete mesh. In particular, it may be one of the matrix elements $A_n(\mathbf{k})$ or any of the filter functions $w_j(\mathbf{k})$. Note that we have suppressed the band index in order to keep things simple. As already indicated in (D.5.1) and (D.5.2), within the tetrahedron method the above integrals arise as sums of contributions from all tetrahedra with

$$
\begin{aligned}
r_{a,T}(E) &= \frac{1}{\Omega_{BZ}} \mathcal{P} \int_{\Omega_T} d^3\mathbf{k} \, \frac{a(\mathbf{k})}{E - \varepsilon(\mathbf{k})} \\
&= \frac{1}{\Omega_{BZ}} \mathcal{P} \int d\varepsilon(\mathbf{k}) \int_{S(\varepsilon)\in T} dS \, \frac{a(\mathbf{k})}{|\nabla_\mathbf{k}\varepsilon(\mathbf{k})|} \frac{1}{E - \varepsilon(\mathbf{k})} \,,
\end{aligned}
\tag{D.5.3}
$$

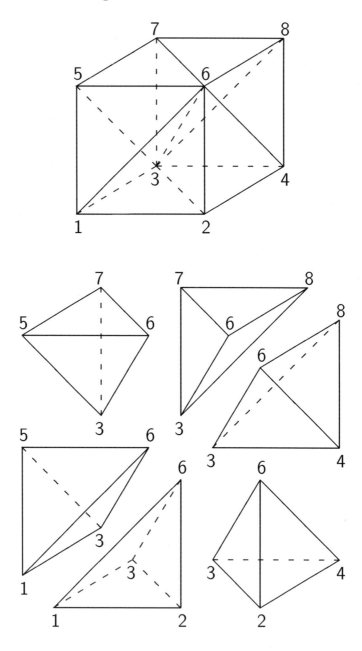

Fig. D.4. Division of a parallelepiped into tetrahedra (after [5])

and

$$i_{a,T}(E) = \frac{1}{\Omega_{BZ}} \int_{\Omega_T} d^3\mathbf{k}\, a(\mathbf{k})\delta(E - \varepsilon(\mathbf{k}))$$

$$= \frac{1}{\Omega_{BZ}} \int_{S(\varepsilon)\in T} dS \frac{a(\mathbf{k})}{|\nabla_{\mathbf{k}}\varepsilon(\mathbf{k})|}\delta(E - \varepsilon(\mathbf{k})) , \tag{D.5.4}$$

where Ω_T denotes the region of the tetrahedron and the surface integral extends over that portion of the constant energy surface, which is completely inside the tetrahedron. Again, the latter two quantities are connected by the Kramers-Kronig relation

$$r_{a,T}(E) = \mathcal{P} \int dE' \frac{1}{E - E'} i_{a,T}(E') , \tag{D.5.5}$$

which holds separately for each tetrahedron. In the following, we will start evaluating the imaginary part, i.e. the integrals of the type (D.5.4). This is motivated by the fact that the imaginary part enters both the density of states and number of states, whereas the real part is needed for the weights only. In deriving the formulas, we will follow the detailed description given by Lehmann and Taut [13] but will also include the recent developments by Blöchl et al. [4,5]. Furthermore, we will benefit from the work by Lambin and Vigneron, who presented complete results for both the real and imaginary part [11].

Each tetrahedron is defined by its vertices $\mathbf{k_0}$, $\mathbf{k_1}$, $\mathbf{k_2}$, and $\mathbf{k_3}$. Labelling of the vertices is such that the band energies at these points,

$$\varepsilon_i = \varepsilon(\mathbf{k}_i) \qquad i = 0, 1, 2, 3 , \tag{D.5.6}$$

are ordered, i.e. $\varepsilon_0 \leq \varepsilon_1 \leq \varepsilon_2 \leq \varepsilon_3$. In order to fully specify the geometry of the tetrahedron it is useful to define reduced $\mathbf{k}$-vectors

$$\mathbf{k}'_i = \mathbf{k}_i - \mathbf{k}_0 \qquad \text{for } i = 0, 1, 2, 3, \qquad \mathbf{k}'_0 \equiv \mathbf{0} , \tag{D.5.7}$$

and their contragredient vectors

$$\mathbf{r}_i = \frac{1}{6\Omega_T} \left(\mathbf{k}'_j \times \mathbf{k}'_k \right) \varepsilon_{ijk} , \tag{D.5.8}$$

where

$$\Omega_T = \frac{1}{6}\mathbf{k}'_1 \cdot \left(\mathbf{k}'_2 \times \mathbf{k}'_3 \right) , \tag{D.5.9}$$

is the volume of a single tetrahedron. Here we have used the fact that the Brillouin zone is divided into parallelepipeds of equal shape, which themselves are divided into tetrahedra of equal size. From the previous definitions we have

$$\mathbf{r}_i \cdot \mathbf{k}'_j = \delta_{ij} . \tag{D.5.10}$$

Next, we write the linear interpolation of the band dispersion inside the tetrahedron from the values at the vertices as

$$\bar{\varepsilon}(\mathbf{k}) = \varepsilon_0 + \mathbf{b}_\varepsilon \cdot \mathbf{k}' \,, \tag{D.5.11}$$

where

$$\mathbf{k}' = \mathbf{k} - \mathbf{k}_0 \,, \tag{D.5.12}$$

and

$$\mathbf{b}_\varepsilon = \sum_{i=1}^{3} (\varepsilon_i - \varepsilon_0) \, \mathbf{r}_i =: \sum_{i=1}^{3} \varepsilon_{i0} \mathbf{r}_i \,. \tag{D.5.13}$$

Note that the so constructed linear approximation is continuous at the surfaces of the tetrahedron. From (D.5.11) we have

$$\nabla_{\mathbf{k}} \bar{\varepsilon}(\mathbf{k}) = \nabla_{\mathbf{k}'} \bar{\varepsilon}(\mathbf{k}') = \mathbf{b}_\varepsilon \,. \tag{D.5.14}$$

This result is related to the fact that due to the linearization of the band structure within the tetrahedron the constant energy surface is a plane. Furthermore, since the gradient as given by (D.5.13) is completely determined by the band energies at the corners it is constant throughout the tetrahedron.

In the same manner as for the band dispersion we use the linear interpolation of the function $a(\mathbf{k})$,

$$\bar{a}(\mathbf{k}) = a_0 + \mathbf{b}_a \cdot \mathbf{k}' \,, \tag{D.5.15}$$

where

$$\mathbf{b}_a = \sum_{i=1}^{3} (a_i - a_0) \, \mathbf{r}_i =: \sum_{i=1}^{3} a_{i0} \mathbf{r}_i \,, \tag{D.5.16}$$

and

$$a_i = a(\mathbf{k}_i) \qquad i = 0, 1, 2, 3 \,. \tag{D.5.17}$$

After these preparations we are able to write the integrals (D.5.3) and (D.5.4) as

$$
\begin{aligned}
r_{a,T}(E) &= \frac{1}{\Omega_{BZ}} \mathcal{P} \int_{\varepsilon_0}^{\varepsilon_3} d\bar{\varepsilon}(\mathbf{k}) \int_{S(\bar{\varepsilon}) \in T} dS \, \frac{\bar{a}(\mathbf{k})}{|\nabla_{\mathbf{k}} \bar{\varepsilon}(\mathbf{k})|} \frac{1}{E - \bar{\varepsilon}(\mathbf{k})} \\
&= \frac{1}{\Omega_{BZ}} \frac{1}{|\mathbf{b}_\varepsilon|} \mathcal{P} \int_{\varepsilon_0}^{\varepsilon_3} d\bar{\varepsilon}(\mathbf{k}) \, \frac{1}{E - \bar{\varepsilon}(\mathbf{k})} \int_{S(\bar{\varepsilon}) \in T} dS \, (a_0 + \mathbf{b}_a \cdot \mathbf{k}') \\
&= \frac{1}{\Omega_{BZ}} \frac{1}{|\mathbf{b}_\varepsilon|} \mathcal{P} \int_{\varepsilon_0}^{\varepsilon_3} d\bar{\varepsilon}(\mathbf{k}) \, \frac{1}{E - \bar{\varepsilon}(\mathbf{k})} \left[a_0 \int_{S(\bar{\varepsilon}) \in T} dS + \mathbf{b}_a \int_{S(\bar{\varepsilon}) \in T} dS \, \mathbf{k}' \right] \,,
\end{aligned}
$$

$$\tag{D.5.18}$$

and

$$i_{a,T}(E) = \frac{1}{\Omega_{BZ}} \int_{S(\bar{\varepsilon})\in T} dS \, \frac{\bar{a}(\mathbf{k})}{|\nabla_{\mathbf{k}}\varepsilon(\mathbf{k})|} \delta(E - \bar{\varepsilon}(\mathbf{k}))$$

$$= \frac{1}{\Omega_{BZ}} \frac{1}{|\mathbf{b}_\varepsilon|} \int_{S(\bar{\varepsilon})\in T} dS \, (a_0 + \mathbf{b}_a \cdot \mathbf{k}') \, \delta(E - \bar{\varepsilon}(\mathbf{k}))$$

$$= \frac{1}{\Omega_{BZ}} \frac{1}{|\mathbf{b}_\varepsilon|} \left[a_0 \int_{S(E)\in T} dS + \mathbf{b}_a \int_{S(E)\in T} dS \, \mathbf{k}' \right] . \qquad (D.5.19)$$

Here we have replaced both the band dispersion $\varepsilon(\mathbf{k})$ and the function $a(\mathbf{k})$ by their linear interpolations. In addition, we have splitted the surface integrals into two contributions. While the integral in the first term in square brackets is just the area of the portion of the constant energy surface inside the tetrahedron,

$$f := \int_{S(E)\in T} dS , \qquad (D.5.20)$$

the integral entering the second term is this area times the vector of the center of gravity of this surface,

$$f \, \mathbf{k}_{cg} := \int_{S(E)\in T} dS \, \mathbf{k}' . \qquad (D.5.21)$$

Inserting these expressions into (D.5.18) and (D.5.19) we arrive at the intermediate results

$$r_{a,T}(E) = \frac{1}{\Omega_{BZ}} \frac{1}{|\mathbf{b}_\varepsilon|} \mathcal{P} \int_{\varepsilon_0}^{\varepsilon_3} d\bar{\varepsilon}(\mathbf{k}) \, \frac{f}{E - \bar{\varepsilon}(\mathbf{k})} \, [a_0 + \mathbf{b}_a \cdot \mathbf{k}_{cg}] , \qquad (D.5.22)$$

and

$$i_{a,T}(E) = \frac{1}{\Omega_{BZ}} \frac{f}{|\mathbf{b}_\varepsilon|} \, [a_0 + \mathbf{b}_a \cdot \mathbf{k}_{cg}] . \qquad (D.5.23)$$

In order to explicitly calculate the area and the center of gravity we distinguish several cases depending on the value of the energy E relative to the energies ε_i at the vertices. The situation is sketched in Fig. D.5, which shows a single tetrahedron spanned by the points $\mathbf{k_0}$, $\mathbf{k_1}$, $\mathbf{k_2}$, and $\mathbf{k_3}$ and the constant energy surfaces S_0, S_1, and S_3, which are shaded in Fig. D.5 and defined by their corners

$$\mathbf{k}_{ij}^E = \mathbf{k}'_j + \frac{E_j}{\varepsilon_{ij}} \left(\mathbf{k}'_i - \mathbf{k}'_j \right) = \frac{E_j \mathbf{k}'_i - E_i \mathbf{k}'_j}{\varepsilon_{ij}} = \mathbf{k}_{ji}^E$$
$$\text{for } i, j = 0, 1, 2, 3, \; i \neq j . \qquad (D.5.24)$$

Note that in order to enhance the clarity of the figure we have ignored the fact that the gradient is constant within a tetrahedron and, hence, all constant energy surfaces within a tetrahedron should be parallel. The areas of the shaded triangles are easily calculated as

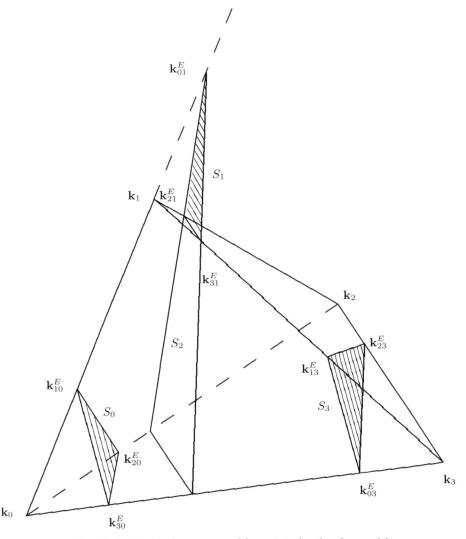

Fig. D.5. Tetrahedron spanned by points $\mathbf{k_0}$, $\mathbf{k_1}$, $\mathbf{k_2}$, and $\mathbf{k_3}$

$$
\begin{aligned}
f_0 = \int_{S_0} dS &= \frac{1}{2} \left| \left(\mathbf{k}_{20}^E - \mathbf{k}_{10}^E \right) \times \left(\mathbf{k}_{30}^E - \mathbf{k}_{10}^E \right) \right| \\
&= \frac{1}{2} E_0^2 \left| \frac{\mathbf{k}_2' \times \mathbf{k}_3'}{\varepsilon_{20}\varepsilon_{30}} + \frac{\mathbf{k}_3' \times \mathbf{k}_1'}{\varepsilon_{30}\varepsilon_{10}} + \frac{\mathbf{k}_1' \times \mathbf{k}_2'}{\varepsilon_{10}\varepsilon_{20}} \right| \\
&= \frac{1}{2} E_0^2 6 \Omega_T \left| \frac{\mathbf{r}_1}{\varepsilon_{20}\varepsilon_{30}} + \frac{\mathbf{r}_2}{\varepsilon_{10}\varepsilon_{30}} + \frac{\mathbf{r}_3}{\varepsilon_{10}\varepsilon_{20}} \right| \\
&= 3 \Omega_T \frac{E_0^2}{\varepsilon_{10}\varepsilon_{20}\varepsilon_{30}} \left| \mathbf{b}_\varepsilon \right| ,
\end{aligned}
\tag{D.5.25}
$$

$$f_1 = \int_{S_1} dS = \frac{1}{2} \left| \left(\mathbf{k}_{21}^E - \mathbf{k}_{01}^E \right) \times \left(\mathbf{k}_{31}^E - \mathbf{k}_{01}^E \right) \right|$$

$$= \frac{1}{2} E_1^2 \left| \frac{\mathbf{k}_2' \times \mathbf{k}_3' + \mathbf{k}_3' \times \mathbf{k}_1' + \mathbf{k}_1' \times \mathbf{k}_2'}{\varepsilon_{21} \varepsilon_{31}} - \frac{\mathbf{k}_3' \times \mathbf{k}_1'}{\varepsilon_{31} \varepsilon_{01}} - \frac{\mathbf{k}_1' \times \mathbf{k}_2'}{\varepsilon_{21} \varepsilon_{01}} \right|$$

$$= \frac{1}{2} E_1^2 6 \Omega_T \left| \frac{\mathbf{r}_1 + \mathbf{r}_2 + \mathbf{r}_3}{\varepsilon_{21} \varepsilon_{31}} + \frac{\mathbf{r}_2}{\varepsilon_{31} \varepsilon_{10}} + \frac{\mathbf{r}_3}{\varepsilon_{21} \varepsilon_{10}} \right|$$

$$= 3 \Omega_T \frac{E_1^2}{\varepsilon_{10} \varepsilon_{21} \varepsilon_{31}} \left| \varepsilon_{10} \mathbf{r}_1 + (\varepsilon_{10} + \varepsilon_{21}) \mathbf{r}_2 + (\varepsilon_{10} + \varepsilon_{31}) \mathbf{r}_3 \right|$$

$$= 3 \Omega_T \frac{E_1^2}{\varepsilon_{10} \varepsilon_{21} \varepsilon_{31}} \left| \mathbf{b}_\varepsilon \right| , \tag{D.5.26}$$

$$f_3 = \int_{S_3} dS = \frac{1}{2} \left| \left(\mathbf{k}_{13}^E - \mathbf{k}_{03}^E \right) \times \left(\mathbf{k}_{23}^E - \mathbf{k}_{03}^E \right) \right|$$

$$= \frac{1}{2} E_3^2 \left| \frac{\mathbf{k}_1' \times \mathbf{k}_2' + \mathbf{k}_3' \times \mathbf{k}_1' + \mathbf{k}_2' \times \mathbf{k}_3'}{\varepsilon_{13} \varepsilon_{23}} - \frac{\mathbf{k}_2' \times \mathbf{k}_3'}{\varepsilon_{03} \varepsilon_{23}} - \frac{\mathbf{k}_3' \times \mathbf{k}_1'}{\varepsilon_{03} \varepsilon_{13}} \right|$$

$$= \frac{1}{2} E_3^2 6 \Omega_T \left| \frac{\mathbf{r}_1 + \mathbf{r}_2 + \mathbf{r}_3}{\varepsilon_{31} \varepsilon_{32}} - \frac{\mathbf{r}_1}{\varepsilon_{30} \varepsilon_{32}} - \frac{\mathbf{r}_2}{\varepsilon_{30} \varepsilon_{31}} \right|$$

$$= 3 \Omega_T \frac{E_3^2}{\varepsilon_{30} \varepsilon_{31} \varepsilon_{32}} \left| (\varepsilon_{30} - \varepsilon_{31}) \mathbf{r}_1 + (\varepsilon_{30} - \varepsilon_{32}) \mathbf{r}_2 + \varepsilon_{30} \mathbf{r}_3 \right|$$

$$= 3 \Omega_T \frac{E_3^2}{\varepsilon_{30} \varepsilon_{31} \varepsilon_{32}} \left| \mathbf{b}_\varepsilon \right| . \tag{D.5.27}$$

In addition, we obtain for the centers of gravity the general expression

$$\mathbf{k}_{cg,j} = \frac{1}{3} \sum_{\substack{i=0 \\ i \neq j}}^3 \mathbf{k}_{ij}^E = \mathbf{k}_j' + \frac{E_j}{3} \sum_{\substack{i=0 \\ i \neq j}}^3 \frac{\mathbf{k}_i' - \mathbf{k}_j'}{\varepsilon_{ij}}$$

$$= \frac{1}{3} \sum_{\substack{i=0 \\ i \neq j}}^3 \frac{E_j \mathbf{k}_i' - E_i \mathbf{k}_j'}{\varepsilon_{ij}} \qquad \text{for } j = 0, 1, 3 . \tag{D.5.28}$$

Combining it with (D.5.16) and taking into account the identity (D.5.10) we arrive at

$$\mathbf{b}_a \cdot \mathbf{k}_{cg,j} = \sum_{l=1}^3 a_{l0} \mathbf{r}_l \cdot \mathbf{k}_j' + \frac{E_j}{3} \sum_{l=1}^3 \sum_{\substack{i=0 \\ i \neq j}}^3 \frac{a_{l0}}{\varepsilon_{ij}} \mathbf{r}_l \cdot \left(\mathbf{k}_i' - \mathbf{k}_j' \right)$$

$$= a_{j0} + \frac{E_j}{3} \sum_{\substack{i=0 \\ i \neq j}}^3 \frac{a_{i0} - a_{j0}}{\varepsilon_{ij}}$$

$$= a_{j0} + \frac{E_j}{3} \sum_{\substack{i=0 \\ i \neq j}}^3 \frac{a_{ij}}{\varepsilon_{ij}}$$

$$= \frac{1}{3}\sum_{l=1}^{3} a_{l0} \sum_{\substack{i=0\\i\neq j}}^{3} \frac{E_j \mathbf{r}_l \cdot \mathbf{k}_i' - E_i \mathbf{r}_l \cdot \mathbf{k}_j'}{\varepsilon_{ij}}$$

$$= \frac{1}{3}\sum_{\substack{i=0\\i\neq j}}^{3} \frac{E_j a_{i0} - E_i a_{j0}}{\varepsilon_{ij}} \qquad \text{for } j = 0,1,3\,, \qquad (D.5.29)$$

hence,

$$a_0 + \mathbf{b}_a \cdot \mathbf{k}_{cg,j} = a_j + \frac{E_j}{3}\sum_{\substack{i=0\\i\neq j}}^{3} \frac{a_{ij}}{\varepsilon_{ij}}$$

$$= \frac{1}{3}\sum_{\substack{i=0\\i\neq j}}^{3} \frac{E_j a_i - E_i a_j}{\varepsilon_{ij}} \qquad \text{for } j = 0,1,3\,. \qquad (D.5.30)$$

The latter expression reduces exactly to one if all the a_i are unity. In addition to the triangles S_0 and S_3 we will need the area and center of gravity of the open quadrangle S_2, which is shown in the center of Fig. D.5. As is obvious from the figure, this quadrangle appears as the difference of the triangles S_0 and S_1. Hence, the area of the surface S_2 arises as

$$f_2 = f_0 - f_1\,, \qquad (D.5.31)$$

and for the center of gravity we note

$$f_2\,\mathbf{k}_{cg,2} = \int_{S_2} dS\,\mathbf{k}' = \int_{S_0} dS\,\mathbf{k}' - \int_{S_1} dS\,\mathbf{k}' = f_0\,\mathbf{k}_{cg,0} - f_1\,\mathbf{k}_{cg,1}\,. \qquad (D.5.32)$$

However, as we will see lateron representing the quadrangle S_2 as the *difference* of two triangles bears the danger of potential numerical instabilities. This observation led Lambin and Vigneron to propose an alternative approach expressing the area of the quadrangle as the *sum* of triangles. In particular, they presented a symmetric formula including those four triangles, which are built from three of the corners of the quadrangle. With the four corners given by $\mathbf{k}_{20}^E$, $\mathbf{k}_{21}^E$, $\mathbf{k}_{30}^E$, and $\mathbf{k}_{31}^E$ we thus note

$$f_{2,a} = \int_{S_{2,a}} dS$$

$$= \frac{1}{2}\left|\left(\mathbf{k}_{21}^E - \mathbf{k}_{20}^E\right) \times \left(\mathbf{k}_{30}^E - \mathbf{k}_{20}^E\right)\right|$$

$$= \frac{1}{2}\left|\frac{E_1 E_0 \mathbf{k}_2' \times \mathbf{k}_3' - E_2 E_0 \mathbf{k}_1' \times \mathbf{k}_3'}{\varepsilon_{21}\varepsilon_{30}} + \frac{E_2 E_0 \mathbf{k}_1' \times \mathbf{k}_2'}{\varepsilon_{21}\varepsilon_{20}} + \frac{E_0^2 \mathbf{k}_3' \times \mathbf{k}_2'}{\varepsilon_{20}\varepsilon_{30}}\right|$$

$$= 3E_0\Omega_T\left|\frac{E_1\mathbf{r}_1 + E_2\mathbf{r}_2}{\varepsilon_{21}\varepsilon_{30}} + \frac{E_2\mathbf{r}_3}{\varepsilon_{21}\varepsilon_{20}} - \frac{E_0\mathbf{r}_1}{\varepsilon_{20}\varepsilon_{30}}\right|$$

$$= 3E_0\Omega_T\frac{\left|E_1\varepsilon_{20}\mathbf{r}_1 + E_2\varepsilon_{20}\mathbf{r}_2 + E_2\varepsilon_{30}\mathbf{r}_3 - E_0\varepsilon_{21}\mathbf{r}_1\right|}{\varepsilon_{20}\varepsilon_{21}\varepsilon_{30}}$$

$$= -3E_0 E_2\Omega_T\frac{|\mathbf{b}_\varepsilon|}{\varepsilon_{20}\varepsilon_{21}\varepsilon_{30}}\,, \qquad (D.5.33)$$

and

$$
\begin{aligned}
f_{2,b} &= \int_{S_{2,b}} dS \\
&= \frac{1}{2}\left|\left(\mathbf{k}_{21}^{E} - \mathbf{k}_{31}^{E}\right) \times \left(\mathbf{k}_{30}^{E} - \mathbf{k}_{31}^{E}\right)\right| \\
&= \frac{1}{2}\left| \frac{E_1 E_0 \mathbf{k}_2' \times \mathbf{k}_3' - E_2 E_0 \mathbf{k}_1' \times \mathbf{k}_3'}{\varepsilon_{21}\varepsilon_{30}} - \frac{E_0 E_3 \mathbf{k}_3' \times \mathbf{k}_1'}{\varepsilon_{30}\varepsilon_{31}} \right. \\
&\qquad\qquad \left. - \frac{E_1^2 \mathbf{k}_2' \times \mathbf{k}_3' - E_1 E_3 \mathbf{k}_2' \times \mathbf{k}_1' - E_1 E_2 \mathbf{k}_1' \times \mathbf{k}_3'}{\varepsilon_{21}\varepsilon_{31}} \right| \\
&= 3\Omega_T \left| E_0 \frac{E_1 \mathbf{r}_1 + E_2 \mathbf{r}_2}{\varepsilon_{21}\varepsilon_{30}} - E_0 \frac{E_3 \mathbf{r}_2}{\varepsilon_{30}\varepsilon_{31}} - E_1 \frac{E_1 \mathbf{r}_1 + E_3 \mathbf{r}_3 + E_2 \mathbf{r}_2}{\varepsilon_{21}\varepsilon_{31}} \right| \\
&= 3\Omega_T \left| E_0 \frac{E_1 \varepsilon_{31}\mathbf{r}_1 + E_2 \varepsilon_{31}\mathbf{r}_2 - E_3 \varepsilon_{21}\mathbf{r}_2}{\varepsilon_{21}\varepsilon_{30}\varepsilon_{31}} \right. \\
&\qquad\qquad \left. - E_0 \frac{E_1 \varepsilon_{30}\mathbf{r}_1 + E_1 \varepsilon_{30}\mathbf{r}_3 + E_1 \varepsilon_{30}\mathbf{r}_2}{\varepsilon_{21}\varepsilon_{30}\varepsilon_{31}} + E_1 \frac{\mathbf{b}_\varepsilon}{\varepsilon_{21}\varepsilon_{31}} \right| \\
&= -3 E_1 E_3 \Omega_T \frac{|\mathbf{b}_\varepsilon|}{\varepsilon_{21}\varepsilon_{30}\varepsilon_{31}} \ , \tag{D.5.34}
\end{aligned}
$$

as well as

$$
\begin{aligned}
f_{2,a'} &= \int_{S_{2,a'}} dS \\
&= \frac{1}{2}\left|\left(\mathbf{k}_{20}^{E} - \mathbf{k}_{21}^{E}\right) \times \left(\mathbf{k}_{31}^{E} - \mathbf{k}_{21}^{E}\right)\right| \\
&= \frac{1}{2}\left| \frac{E_0 E_1 \mathbf{k}_2' \times \mathbf{k}_3' - E_0 E_3 \mathbf{k}_2' \times \mathbf{k}_1'}{\varepsilon_{20}\varepsilon_{31}} + \frac{E_0 E_2 \mathbf{k}_2' \times \mathbf{k}_1'}{\varepsilon_{20}\varepsilon_{21}} \right. \\
&\qquad\qquad \left. + \frac{E_1^2 \mathbf{k}_3' \times \mathbf{k}_2' - E_1 E_2 \mathbf{k}_3' \times \mathbf{k}_1' - E_3 E_1 \mathbf{k}_1' \times \mathbf{k}_2'}{\varepsilon_{21}\varepsilon_{31}} \right| \\
&= 3\Omega_T \left| E_0 \frac{E_1 \mathbf{r}_1 + E_3 \mathbf{r}_3}{\varepsilon_{20}\varepsilon_{31}} - E_0 \frac{E_2 \mathbf{r}_3}{\varepsilon_{20}\varepsilon_{21}} - E_1 \frac{E_1 \mathbf{r}_1 + E_2 \mathbf{r}_2 + E_3 \mathbf{r}_3}{\varepsilon_{21}\varepsilon_{31}} \right| \\
&= 3\Omega_T \left| E_0 \frac{E_1 \varepsilon_{21}\mathbf{r}_1 + E_3 \varepsilon_{21}\mathbf{r}_3 - E_2 \varepsilon_{31}\mathbf{r}_3}{\varepsilon_{21}\varepsilon_{30}\varepsilon_{31}} \right. \\
&\qquad\qquad \left. - E_0 \frac{E_1 \varepsilon_{20}\mathbf{r}_1 + E_1 \varepsilon_{20}\mathbf{r}_3 + E_1 \varepsilon_{20}\mathbf{r}_2}{\varepsilon_{20}\varepsilon_{21}\varepsilon_{31}} + E_1 \frac{\mathbf{b}_\varepsilon}{\varepsilon_{21}\varepsilon_{31}} \right| \\
&= -3 E_1 E_2 \Omega_T \frac{|\mathbf{b}_\varepsilon|}{\varepsilon_{20}\varepsilon_{21}\varepsilon_{31}} \ , \tag{D.5.35}
\end{aligned}
$$

and

$$
\begin{aligned}
f_{2,b'} &= \int_{S_{2,b'}} dS \\
&= \frac{1}{2}\left|\left(\mathbf{k}_{20}^{E} - \mathbf{k}_{30}^{E}\right) \times \left(\mathbf{k}_{31}^{E} - \mathbf{k}_{30}^{E}\right)\right|
\end{aligned}
$$

$$= \frac{1}{2} \left| \frac{E_0 E_1 \mathbf{k}_2' \times \mathbf{k}_3' - E_0 E_3 \mathbf{k}_2' \times \mathbf{k}_1'}{\varepsilon_{20}\varepsilon_{31}} - \frac{E_0^2 \mathbf{k}_2' \times \mathbf{k}_3'}{\varepsilon_{20}\varepsilon_{30}} - \frac{E_3 E_0 \mathbf{k}_1' \times \mathbf{k}_3'}{\varepsilon_{30}\varepsilon_{31}} \right|$$

$$= 3 E_0 \Omega_T \left| \frac{E_1 \mathbf{r}_1 + E_3 \mathbf{r}_3}{\varepsilon_{20}\varepsilon_{31}} - \frac{E_0 \mathbf{r}_1}{\varepsilon_{20}\varepsilon_{30}} + \frac{E_3 \mathbf{r}_2}{\varepsilon_{30}\varepsilon_{31}} \right|$$

$$= 3 E_0 \Omega_T \frac{|E_1 \varepsilon_{30} \mathbf{r}_1 + E_3 \varepsilon_{30} \mathbf{r}_3 - E_0 \varepsilon_{31} \mathbf{r}_1 + E_3 \varepsilon_{20} \mathbf{r}_2|}{\varepsilon_{20}\varepsilon_{30}\varepsilon_{31}}$$

$$= -3 E_0 E_3 \Omega_T \frac{|\mathbf{b}_\varepsilon|}{\varepsilon_{20}\varepsilon_{30}\varepsilon_{31}} \ . \tag{D.5.36}$$

The minus signs in the last lines take care of the fact that for this constant energy surface the energy E is above the corner energies ε_0 and ε_1 but below the energies ε_2 and ε_3. Note that for the limiting cases $E = \varepsilon_1$ and $E = \varepsilon_2$ two of the four triangles vanish. Finally, using (D.5.24) we write for the centers of gravity

$$\mathbf{k}_{cg,2,a} = \frac{1}{3} \left[\mathbf{k}_{20}^E + \mathbf{k}_{21}^E + \mathbf{k}_{30}^E \right]$$

$$= \frac{1}{3} \left[\mathbf{k}_1' + \frac{E_0}{\varepsilon_{20}} \mathbf{k}_2' + \frac{E_1}{\varepsilon_{21}} (\mathbf{k}_2' - \mathbf{k}_1') + \frac{E_0}{\varepsilon_{30}} \mathbf{k}_3' \right]$$

$$= \frac{1}{3} \left[\frac{E_0}{\varepsilon_{20}} \mathbf{k}_2' + \frac{E_1}{\varepsilon_{21}} \mathbf{k}_2' - \frac{E_2}{\varepsilon_{21}} \mathbf{k}_1' + \frac{E_0}{\varepsilon_{30}} \mathbf{k}_3' \right] \ , \tag{D.5.37}$$

$$\mathbf{k}_{cg,2,b} = \frac{1}{3} \left[\mathbf{k}_{21}^E + \mathbf{k}_{30}^E + \mathbf{k}_{31}^E \right]$$

$$= \frac{1}{3} \left[2\mathbf{k}_1' + \frac{E_1}{\varepsilon_{21}} (\mathbf{k}_2' - \mathbf{k}_1') + \frac{E_0}{\varepsilon_{30}} \mathbf{k}_3' + \frac{E_1}{\varepsilon_{31}} (\mathbf{k}_3' - \mathbf{k}_1') \right]$$

$$= \frac{1}{3} \left[\frac{E_1}{\varepsilon_{21}} \mathbf{k}_2' - \frac{E_2}{\varepsilon_{21}} \mathbf{k}_1' + \frac{E_0}{\varepsilon_{30}} \mathbf{k}_3' + \frac{E_1}{\varepsilon_{31}} \mathbf{k}_3' - \frac{E_3}{\varepsilon_{31}} \mathbf{k}_1' \right] \ , \tag{D.5.38}$$

$$\mathbf{k}_{cg,2,a'} = \frac{1}{3} \left[\mathbf{k}_{20}^E + \mathbf{k}_{21}^E + \mathbf{k}_{31}^E \right]$$

$$= \frac{1}{3} \left[2\mathbf{k}_1' + \frac{E_0}{\varepsilon_{20}} \mathbf{k}_2' + \frac{E_1}{\varepsilon_{21}} (\mathbf{k}_2' - \mathbf{k}_1') + \frac{E_1}{\varepsilon_{31}} (\mathbf{k}_3' - \mathbf{k}_1') \right]$$

$$= \frac{1}{3} \left[\frac{E_0}{\varepsilon_{20}} \mathbf{k}_2' + \frac{E_1}{\varepsilon_{21}} \mathbf{k}_2' - \frac{E_2}{\varepsilon_{21}} \mathbf{k}_1' + \frac{E_1}{\varepsilon_{31}} \mathbf{k}_3' - \frac{E_3}{\varepsilon_{31}} \mathbf{k}_1' \right] \ , \tag{D.5.39}$$

$$\mathbf{k}_{cg,2,b'} = \frac{1}{3} \left[\mathbf{k}_{20}^E + \mathbf{k}_{30}^E + \mathbf{k}_{31}^E \right]$$

$$= \frac{1}{3} \left[\mathbf{k}_1' + \frac{E_0}{\varepsilon_{20}} \mathbf{k}_2' + \frac{E_0}{\varepsilon_{30}} \mathbf{k}_3' + \frac{E_1}{\varepsilon_{31}} (\mathbf{k}_3' - \mathbf{k}_1') \right]$$

$$= \frac{1}{3} \left[\frac{E_0}{\varepsilon_{20}} \mathbf{k}_2' + \frac{E_0}{\varepsilon_{30}} \mathbf{k}_3' + \frac{E_1}{\varepsilon_{31}} \mathbf{k}_3' - \frac{E_3}{\varepsilon_{31}} \mathbf{k}_1' \right] \ . \tag{D.5.40}$$

The areas of each two triangles, which share a diagonal, i.e. of the triangles $S_{2,a}$ and $S_{2,b}$ as well as $S_{2,a'}$ and $S_{2,b'}$, add to the total area. Of course, both representations are equivalent and may be combined or else used separately. Here, we prefer the representation in terms of the triangles $S_{2,a}$ and $S_{2,b}$.

Combining (D.5.37) and (D.5.38) with (D.5.16) and taking into account the identity (D.5.10) we arrive at

$$
\begin{aligned}
\mathbf{b}_a \cdot \mathbf{k}_{cg,2,a} &= \frac{1}{3} \sum_{l=1}^{3} a_{l0} \mathbf{r}_l \cdot \left[\mathbf{k}_1' + \frac{E_0}{\varepsilon_{20}} \mathbf{k}_2' + \frac{E_1}{\varepsilon_{21}} (\mathbf{k}_2' - \mathbf{k}_1') + \frac{E_0}{\varepsilon_{30}} \mathbf{k}_3' \right] \\
&= \frac{1}{3} \left[a_{10} + \frac{E_0}{\varepsilon_{20}} a_{20} + \frac{E_1}{\varepsilon_{21}} a_{21} + \frac{E_0}{\varepsilon_{30}} a_{30} \right] \\
&= \frac{1}{3} \left[-\frac{E_2}{\varepsilon_{21}} a_{10} + \left(\frac{E_0}{\varepsilon_{20}} + \frac{E_1}{\varepsilon_{21}} \right) a_{20} + \frac{E_0}{\varepsilon_{30}} a_{30} \right] ,
\end{aligned} \tag{D.5.41}
$$

$$
\begin{aligned}
\mathbf{b}_a \cdot \mathbf{k}_{cg,2,b} &= \frac{1}{3} \sum_{l=1}^{3} a_{l0} \mathbf{r}_l \cdot \left[2\mathbf{k}_1' + \frac{E_1}{\varepsilon_{21}} (\mathbf{k}_2' - \mathbf{k}_1') + \frac{E_0}{\varepsilon_{30}} \mathbf{k}_3' + \frac{E_1}{\varepsilon_{31}} (\mathbf{k}_3' - \mathbf{k}_1') \right] \\
&= \frac{1}{3} \left[2a_{10} + \frac{E_1}{\varepsilon_{21}} a_{21} + \frac{E_0}{\varepsilon_{30}} a_{30} + \frac{E_1}{\varepsilon_{31}} a_{31} \right] \\
&= \frac{1}{3} \left[-\left(\frac{E_2}{\varepsilon_{21}} + \frac{E_3}{\varepsilon_{31}} \right) a_{10} + \frac{E_1}{\varepsilon_{21}} a_{20} + \left(\frac{E_0}{\varepsilon_{30}} + \frac{E_1}{\varepsilon_{31}} \right) a_{30} \right] ,
\end{aligned}
$$

$$
\tag{D.5.42}
$$

hence,

$$
\begin{aligned}
a_0 + \mathbf{b}_a \cdot \mathbf{k}_{cg,2,a} &= \frac{1}{3} \left[2a_0 + a_1 + E_0 \frac{a_{20}}{\varepsilon_{20}} + E_1 \frac{a_{21}}{\varepsilon_{21}} + E_0 \frac{a_{30}}{\varepsilon_{30}} \right] \\
&= a_0 + \frac{1}{3} \left[a_{10} - \varepsilon_{10} \frac{a_{21}}{\varepsilon_{21}} + E_0 \left(\frac{a_{20}}{\varepsilon_{20}} + \frac{a_{21}}{\varepsilon_{21}} + \frac{a_{30}}{\varepsilon_{30}} \right) \right] ,
\end{aligned} \tag{D.5.43}
$$

$$
\begin{aligned}
a_0 + \mathbf{b}_a \cdot \mathbf{k}_{cg,2,b} &= \frac{1}{3} \left[2a_1 + a_0 + E_1 \frac{a_{21}}{\varepsilon_{21}} + E_0 \frac{a_{30}}{\varepsilon_{30}} + E_1 \frac{a_{31}}{\varepsilon_{31}} \right] \\
&= a_1 + \frac{1}{3} \left[a_{01} + \varepsilon_{10} \frac{a_{30}}{\varepsilon_{30}} + E_1 \left(\frac{a_{21}}{\varepsilon_{21}} + \frac{a_{30}}{\varepsilon_{30}} + \frac{a_{31}}{\varepsilon_{31}} \right) \right] .
\end{aligned}
$$

$$
\tag{D.5.44}
$$

Obviously, if all the a_i are unity the square brackets in the respective second lines vanish and the right-hand sides reduce to unity.

With the previous intermediate results at hand we are able to treat the following cases separately. As mentioned above we will for the time being concentrate on the imaginary part and, hence, insert one of (D.5.25) to (D.5.27) as well as (D.5.30) into (D.5.23). In addition, we will use (D.5.33) and (D.5.34) as well as (D.5.43) and (D.5.44).

1. $E \leq \varepsilon_0$

In this case the constant energy surface is completely outside the tetrahedron. Both f and $f \, \mathbf{k}_{cg}$ are zero and we note

$$
i_{a,T}(E) = 0 , \tag{D.5.45}
$$

as well as

$$\int_{\varepsilon_0}^{E} dE'\, i_{a,T}(E') = 0 . \tag{D.5.46}$$

2. $\varepsilon_0 \leq E \leq \varepsilon_1$

The constant energy surface is the leftmost triangle S_0 displayed in Fig. D.5. Inserting (D.5.25) and (D.5.30) into (D.5.23) we obtain for the integral the result[2pt]

$$
\begin{aligned}
i_{a,T}(E) &= \frac{1}{\Omega_{BZ}} \frac{f_0}{|\mathbf{b}_\varepsilon|} \left(a_0 + \mathbf{b}_a \cdot \mathbf{k}_{cg,0} \right) \\
&= \frac{\Omega_T}{\Omega_{BZ}} \frac{E_0^2}{\varepsilon_{10}\varepsilon_{20}\varepsilon_{30}} \left(3a_0 + E_0 \sum_{i=1}^{3} \frac{a_{i0}}{\varepsilon_{i0}} \right) .[2pt]
\end{aligned}
\tag{D.5.47}
$$

For the energy integrated integral we readily write[2pt]

$$\int_{\varepsilon_0}^{E} dE'\, i_{a,T}(E') = \frac{\Omega_T}{\Omega_{BZ}} \frac{E_0^3}{\varepsilon_{10}\varepsilon_{20}\varepsilon_{30}} \left(a_0 + \frac{E_0}{4} \sum_{i=1}^{3} \frac{a_{i0}}{\varepsilon_{i0}} \right) , [2pt] \tag{D.5.48}$$

which is zero for $E = \varepsilon_0$. In contrast, for $E = \varepsilon_1$ we note

$$\int_{\varepsilon_0}^{\varepsilon_1} dE'\, i_{a,T}(E') = \frac{\Omega_T}{\Omega_{BZ}} \frac{\varepsilon_{10}^2}{\varepsilon_{20}\varepsilon_{30}} \left(a_0 + \frac{\varepsilon_{10}}{4} \sum_{i=1}^{3} \frac{a_{i0}}{\varepsilon_{i0}} \right) . \tag{D.5.49}$$

3. $\varepsilon_1 \leq E \leq \varepsilon_2$

The constant energy surface is the open quadrangle S_2 shown in the center of Fig. D.5 with the area and the center of gravity given by (D.5.31) and (D.5.32). We thus write the integral as

$$
\begin{aligned}
i_{a,T}(E) &= \frac{1}{\Omega_{BZ}} \frac{f_2}{|\mathbf{b}_\varepsilon|} \left(a_0 + \mathbf{b}_a \cdot \mathbf{k}_{cg,2} \right) \\
&= \frac{1}{\Omega_{BZ}} \frac{f_0}{|\mathbf{b}_\varepsilon|} \left(a_0 + \mathbf{b}_a \cdot \mathbf{k}_{cg,0} \right) - \frac{1}{\Omega_{BZ}} \frac{f_1}{|\mathbf{b}_\varepsilon|} \left(a_0 + \mathbf{b}_a \cdot \mathbf{k}_{cg,1} \right) \\
&= \frac{\Omega_T}{\Omega_{BZ}} \frac{E_0^2}{\varepsilon_{10}\varepsilon_{20}\varepsilon_{30}} \left(3a_0 + E_0 \sum_{i=1}^{3} \frac{a_{i0}}{\varepsilon_{i0}} \right) \\
&\quad - \frac{\Omega_T}{\Omega_{BZ}} \frac{E_1^2}{\varepsilon_{10}\varepsilon_{21}\varepsilon_{31}} \left(3a_1 + E_1 \sum_{\substack{i=0 \\ i \neq 1}}^{3} \frac{a_{i1}}{\varepsilon_{i1}} \right) .
\end{aligned}
\tag{D.5.50}
$$

On energy integration this leads to

$$\int_{\varepsilon_0}^{E} dE'\, i_{a,T}(E') = \frac{\Omega_T}{\Omega_{BZ}} \frac{E_0^3}{\varepsilon_{10}\varepsilon_{20}\varepsilon_{30}} \left(a_0 + \frac{E_0}{4} \sum_{i=1}^{3} \frac{a_{i0}}{\varepsilon_{i0}} \right)$$

$$-\frac{\Omega_T}{\Omega_{BZ}}\frac{E_1^3}{\varepsilon_{10}\varepsilon_{21}\varepsilon_{31}}\left(a_1+\frac{E_1}{4}\sum_{\substack{i=0\\i\neq1}}^{3}\frac{a_{i1}}{\varepsilon_{i1}}\right)+C\ .\quad\text{(D.5.51)}$$

The constant C must be zero since the integral reduces to the correct expression (D.5.49) for $E=\varepsilon_1$. For $E=\varepsilon_2$ we note

$$\int_{\varepsilon_0}^{\varepsilon_2}dE'\,i_{a,T}(E')=\frac{\Omega_T}{\Omega_{BZ}}\frac{\varepsilon_{20}^2}{\varepsilon_{10}\varepsilon_{30}}\left(a_0+\frac{\varepsilon_{20}}{4}\sum_{i=1}^{3}\frac{a_{i0}}{\varepsilon_{i0}}\right)$$
$$-\frac{\Omega_T}{\Omega_{BZ}}\frac{\varepsilon_{21}^2}{\varepsilon_{10}\varepsilon_{31}}\left(a_1+\frac{\varepsilon_{21}}{4}\sum_{\substack{i=0\\i\neq1}}^{3}\frac{a_{i1}}{\varepsilon_{i1}}\right)\ .\quad\text{(D.5.52)}$$

Complications may arise from the previous expressions in the limit $\varepsilon_{10}\to0$. In this case, we make use of the alternative approach based on (D.5.33), (D.5.34), (D.5.43), and (D.5.44). Inserting these into (D.5.23) we note

$$i_{a,T}(E)$$
$$=\frac{1}{\Omega_{BZ}}\frac{f_2}{|\mathbf{b}_\varepsilon|}\left(a_0+\mathbf{b}_a\cdot\mathbf{k}_{cg,2}\right)$$
$$=\frac{1}{\Omega_{BZ}}\frac{f_{2,a}}{|\mathbf{b}_\varepsilon|}\left(a_0+\mathbf{b}_a\cdot\mathbf{k}_{cg,2,a}\right)+\frac{1}{\Omega_{BZ}}\frac{f_{2,b}}{|\mathbf{b}_\varepsilon|}\left(a_0+\mathbf{b}_a\cdot\mathbf{k}_{cg,2,b}\right)$$
$$=\frac{\Omega_T}{\Omega_{BZ}}\frac{E_0\varepsilon_{20}-E_0^2}{\varepsilon_{20}\varepsilon_{21}\varepsilon_{30}}\left[3a_0+a_{10}-\varepsilon_{10}\frac{a_{21}}{\varepsilon_{21}}+E_0\left(\frac{a_{20}}{\varepsilon_{20}}+\frac{a_{21}}{\varepsilon_{21}}+\frac{a_{30}}{\varepsilon_{30}}\right)\right]$$
$$+\frac{\Omega_T}{\Omega_{BZ}}\frac{E_1\varepsilon_{31}-E_1^2}{\varepsilon_{21}\varepsilon_{30}\varepsilon_{31}}\left[3a_1-a_{10}+\varepsilon_{10}\frac{a_{30}}{\varepsilon_{30}}+E_1\left(\frac{a_{21}}{\varepsilon_{21}}+\frac{a_{30}}{\varepsilon_{30}}+\frac{a_{31}}{\varepsilon_{31}}\right)\right]\ .$$
$$\text{(D.5.53)}$$

In particular, for $E=\varepsilon_1$ this result becomes identical to (D.5.47). Integrating with respect to the energy we obtain

$$\int_{\varepsilon_0}^{E}dE'\,i_{a,T}(E')$$
$$=\frac{\Omega_T}{\Omega_{BZ}}\frac{\frac{1}{2}E_0^2\varepsilon_{20}-\frac{1}{3}E_0^3}{\varepsilon_{20}\varepsilon_{21}\varepsilon_{30}}\left(3a_0+a_{10}-\varepsilon_{10}\frac{a_{21}}{\varepsilon_{21}}\right)$$
$$+\frac{\Omega_T}{\Omega_{BZ}}\frac{\frac{1}{3}E_0^3\varepsilon_{20}-\frac{1}{4}E_0^4}{\varepsilon_{20}\varepsilon_{21}\varepsilon_{30}}\left(\frac{a_{20}}{\varepsilon_{20}}+\frac{a_{21}}{\varepsilon_{21}}+\frac{a_{30}}{\varepsilon_{30}}\right)$$
$$+\frac{\Omega_T}{\Omega_{BZ}}\frac{\frac{1}{2}E_1^2\varepsilon_{31}-\frac{1}{3}E_1^3}{\varepsilon_{21}\varepsilon_{30}\varepsilon_{31}}\left(3a_1-a_{10}+\varepsilon_{10}\frac{a_{30}}{\varepsilon_{30}}\right)$$
$$+\frac{\Omega_T}{\Omega_{BZ}}\frac{\frac{1}{3}E_1^3\varepsilon_{31}-\frac{1}{4}E_1^4}{\varepsilon_{21}\varepsilon_{30}\varepsilon_{31}}\left(\frac{a_{21}}{\varepsilon_{21}}+\frac{a_{30}}{\varepsilon_{30}}+\frac{a_{31}}{\varepsilon_{31}}\right)+C\ ,\quad\text{(D.5.54)}$$

which for $E = \varepsilon_1$ reduces to

$$
\int_{\varepsilon_0}^{\varepsilon_1} dE' \, i_{a,T}(E')
$$

$$
= \frac{\Omega_T}{\Omega_{BZ}} \frac{\varepsilon_{10}^2}{\varepsilon_{20}\varepsilon_{21}\varepsilon_{30}} \left[\left(\frac{1}{2}\varepsilon_{20} - \frac{1}{3}\varepsilon_{10} \right) \left(3a_0 + a_{10} - \varepsilon_{10}\frac{a_{21}}{\varepsilon_{21}} \right) \right.
$$

$$
\left. + \left(\frac{1}{3}\varepsilon_{10}\varepsilon_{20} - \frac{1}{4}\varepsilon_{10}^2 \right) \left(\frac{a_{20}}{\varepsilon_{20}} + \frac{a_{21}}{\varepsilon_{21}} + \frac{a_{30}}{\varepsilon_{30}} \right) \right] + C
$$

$$
= \frac{\Omega_T}{\Omega_{BZ}} \frac{\varepsilon_{10}^2}{\varepsilon_{20}\varepsilon_{21}\varepsilon_{30}} \left[\left(\frac{1}{3}\varepsilon_{21} + \frac{1}{6}\varepsilon_{20} \right) \left(3a_0 + a_{10} - \varepsilon_{10}\frac{a_{21}}{\varepsilon_{21}} \right) \right.
$$

$$
\left. + \left(\frac{1}{4}\varepsilon_{10}\varepsilon_{21} + \frac{1}{12}\varepsilon_{10}\varepsilon_{20} \right) \left(\sum_{i=1}^{3} \frac{a_{i0}}{\varepsilon_{i0}} - \frac{a_{10}}{\varepsilon_{10}} + \frac{a_{21}}{\varepsilon_{21}} \right) \right] + C
$$

$$
= \frac{\Omega_T}{\Omega_{BZ}} \frac{\varepsilon_{10}^2}{\varepsilon_{20}\varepsilon_{21}\varepsilon_{30}} \left[\varepsilon_{21}a_0 + \frac{1}{2}\varepsilon_{20}a_0 + \varepsilon_{21}\frac{\varepsilon_{10}}{4} \sum_{i=1}^{3} \frac{a_{i0}}{\varepsilon_{i0}} + \frac{1}{12}\varepsilon_{10}\varepsilon_{20} \sum_{i=1}^{3} \frac{a_{i0}}{\varepsilon_{i0}} \right.
$$

$$
\left. + \frac{1}{12}(\varepsilon_{21} + \varepsilon_{20}) \left(a_{10} - \varepsilon_{10}\frac{a_{21}}{\varepsilon_{21}} \right) \right] + C
$$

$$
= \frac{\Omega_T}{\Omega_{BZ}} \frac{\varepsilon_{10}^2}{\varepsilon_{20}\varepsilon_{30}} \left[a_0 + \frac{\varepsilon_{10}}{4} \sum_{i=1}^{3} \frac{a_{i0}}{\varepsilon_{i0}} + \frac{1}{12} \frac{\varepsilon_{21}a_{10} - \varepsilon_{10}a_{21}}{\varepsilon_{21}} \right]
$$

$$
+ \frac{\Omega_T}{\Omega_{BZ}} \frac{\varepsilon_{10}^2}{\varepsilon_{21}\varepsilon_{30}} \left[\frac{1}{2}a_0 + \frac{\varepsilon_{10}}{12} \sum_{i=1}^{3} \frac{a_{i0}}{\varepsilon_{i0}} + \frac{1}{12} \frac{\varepsilon_{21}a_{10} - \varepsilon_{10}a_{21}}{\varepsilon_{21}} \right] + C \, .
$$

$$
\text{(D.5.55)}
$$

Comparing this to (D.5.49) we arrive at

$$
C = -\frac{\Omega_T}{\Omega_{BZ}} \frac{\varepsilon_{10}^2}{\varepsilon_{21}\varepsilon_{30}} \left[\frac{1}{2}a_0 + \frac{\varepsilon_{10}}{12} \sum_{i=1}^{3} \frac{a_{i0}}{\varepsilon_{i0}} + \frac{1}{12} \frac{\varepsilon_{21}a_{10} - \varepsilon_{10}a_{21}}{\varepsilon_{21}} \right]
$$

$$
- \frac{\Omega_T}{\Omega_{BZ}} \frac{\varepsilon_{10}^2}{\varepsilon_{20}\varepsilon_{30}} \frac{1}{12} \frac{\varepsilon_{21}a_{10} - \varepsilon_{10}a_{21}}{\varepsilon_{21}} \, . \qquad \text{(D.5.56)}
$$

4. $\varepsilon_2 \leq E \leq \varepsilon_3$

The constant energy surface is the rightmost triangle S_3 displayed in Fig. D.5. Inserting (D.5.27) and (D.5.30) into (D.5.23) we obtain for the integral the result

$$
i_{a,T}(E) = \frac{1}{\Omega_{BZ}} \frac{f_3}{|\mathbf{b}_\varepsilon|} (a_0 + \mathbf{b}_a \cdot \mathbf{k}_{cg,3})
$$

$$
= \frac{\Omega_T}{\Omega_{BZ}} \frac{E_3^2}{\varepsilon_{30}\varepsilon_{31}\varepsilon_{32}} \left(3a_3 + E_3 \sum_{i=0}^{2} \frac{a_{i3}}{\varepsilon_{i3}} \right). \qquad \text{(D.5.57)}
$$

For the energy integrated integral we readily note

$$\int_{\varepsilon_0}^{E} dE' \, i_{a,T}(E') = C + \frac{\Omega_T}{\Omega_{BZ}} \frac{E_3^3}{\varepsilon_{30}\varepsilon_{31}\varepsilon_{32}} \left(a_3 + \frac{E_3}{4} \sum_{i=0}^{2} \frac{a_{i3}}{\varepsilon_{i3}} \right) , \qquad \text{(D.5.58)}$$

where the second term vanishes for $E = \varepsilon_3$. However, in this case the integral on the left-hand side can be easily calculated by starting from the first line of the integral (D.5.2), inserting into this the linear interpolation (D.5.15) of $a(\mathbf{k})$ and integrating over all energies. Hence, we write

$$\begin{aligned} C &= \int_{\varepsilon_0}^{\varepsilon_3} dE' \, i_{a,T}(E') \\ &= \frac{1}{\Omega_{BZ}} \int_{\varepsilon_0}^{\varepsilon_3} dE' \int_{\Omega_T} d^3k \, \bar{a}(\mathbf{k}) \delta(E - \varepsilon(\mathbf{k})) \\ &= \frac{1}{\Omega_{BZ}} \int_{\Omega_T} d^3k' \, (a_0 + \mathbf{b}_a \cdot \mathbf{k}') \\ &= \frac{\Omega_T}{\Omega_{BZ}} \left(a_0 + \frac{1}{4} \sum_{i=0}^{3} \mathbf{b}_a \cdot \mathbf{k}'_i \right) \\ &= \frac{\Omega_T}{\Omega_{BZ}} \left(a_0 + \frac{1}{4} \sum_{i=0}^{3} a_{i0} \right) \\ &= \frac{\Omega_T}{4\Omega_{BZ}} \sum_{i=0}^{3} a_i , \end{aligned} \qquad \text{(D.5.59)}$$

where we have used the center of gravity of the whole tetrahedron and, in the second but last step, (D.5.16). Note that we obtain the same result from (D.5.49) for the special case $\varepsilon_0 < \varepsilon_1 = \varepsilon_2 = \varepsilon_3$.

5. $\varepsilon_3 \le E$

In this case the constant energy surface is completely outside the tetrahedron. Both f and $f \, \mathbf{k}_{cg}$ are zero and we note

$$i_{a,T}(E) = 0 . \qquad \text{(D.5.60)}$$

However, on integrating over the energy we include the whole tetrahedron and the result is identical to (D.5.58) for $E = \varepsilon_3$, i.e. to the result (D.5.59),

$$\begin{aligned} \int_{\varepsilon_0}^{E} dE' \, i_{a,T}(E') &= \frac{\Omega_T}{\Omega_{BZ}} \left(a_0 + \frac{1}{4} \sum_{i=0}^{3} a_{i0} \right) \\ &= \frac{\Omega_T}{4\Omega_{BZ}} \sum_{i=0}^{3} a_i . \end{aligned} \qquad \text{(D.5.61)}$$

It is important to note that neither of the previous results depends on the particular shape of the tetrahedra. In contrast, the integrals $i_{a,T}(E)$ and their energy integrals solely depend on the values of the energy dispersion $\varepsilon(\mathbf{k})$ and the functions $a(\mathbf{k})$ at the corners of the tetrahedron. This makes the formulas very easy to implement.

Having arrived at these general results we note especially the contribution of a single tetrahedron to the density of states

$$\rho_T(E) = i_{1,T}(E) \,, \tag{D.5.62}$$

and its energy integrated counterpart, namely, the number of states

$$n_T(E) = \int_{\varepsilon_0}^{E} dE' \, \rho_T(E') = \int_{\varepsilon_0}^{E} dE' \, i_{1,T}(E') \,. \tag{D.5.63}$$

They arise from the previous results by setting the values of the function $a(\mathbf{k})$ at the vertices to unity. Again we distinguish the five cases corresponding to different values of the energy.

1. $E \leq \varepsilon_0$

$$\rho_T(E) = 0 \,, \tag{D.5.64}$$
$$n_T(E) = 0 \,. \tag{D.5.65}$$

2. $\varepsilon_0 \leq E \leq \varepsilon_1$

$$\rho_T(E) = 3\frac{\Omega_T}{\Omega_{BZ}} \frac{E_0^2}{\varepsilon_{10}\varepsilon_{20}\varepsilon_{30}} \,, \tag{D.5.66}$$

$$n_T(E) = \frac{\Omega_T}{\Omega_{BZ}} \frac{E_0^3}{\varepsilon_{10}\varepsilon_{20}\varepsilon_{30}} \,. \tag{D.5.67}$$

3. $\varepsilon_1 \leq E \leq \varepsilon_2$

$$\rho_T(E) = 3\frac{\Omega_T}{\Omega_{BZ}} \left[\frac{E_0^2}{\varepsilon_{10}\varepsilon_{20}\varepsilon_{30}} - \frac{E_1^2}{\varepsilon_{10}\varepsilon_{21}\varepsilon_{31}} \right] \,, \tag{D.5.68}$$

$$n_T(E) = \frac{\Omega_T}{\Omega_{BZ}} \left[\frac{E_0^3}{\varepsilon_{10}\varepsilon_{20}\varepsilon_{30}} - \frac{E_1^3}{\varepsilon_{10}\varepsilon_{21}\varepsilon_{31}} \right] \,. \tag{D.5.69}$$

In order to avoid the singularity at $\varepsilon_{10} = 0$ we use $E_0 = E_1 + \varepsilon_{10}$ as well as the identity

$$\varepsilon_{21}\varepsilon_{31} - \varepsilon_{20}\varepsilon_{30} = \varepsilon_{20}\varepsilon_{31} - \varepsilon_{10}\varepsilon_{31} - \varepsilon_{20}\varepsilon_{31} - \varepsilon_{20}\varepsilon_{10}$$
$$= -\varepsilon_{10}\left(\varepsilon_{20} + \varepsilon_{31}\right)$$
$$= -\varepsilon_{10}\left(\varepsilon_{21} + \varepsilon_{30}\right) \,, \tag{D.5.70}$$

and rewrite (D.5.68) and (D.5.69) as

$$\rho_T(E) = 3\frac{\Omega_T}{\Omega_{BZ}} \left[\frac{\varepsilon_{10}^2 + 2\varepsilon_{10}E_1 + E_1^2}{\varepsilon_{10}\varepsilon_{20}\varepsilon_{30}} - \frac{E_1^2}{\varepsilon_{10}\varepsilon_{21}\varepsilon_{31}} \right]$$
$$= 3\frac{\Omega_T}{\Omega_{BZ}} \frac{1}{\varepsilon_{20}\varepsilon_{30}} \left[\varepsilon_{10} + 2E_1 + \frac{\varepsilon_{21}\varepsilon_{31} - \varepsilon_{20}\varepsilon_{30}}{\varepsilon_{10}\varepsilon_{21}\varepsilon_{31}} E_1^2 \right]$$
$$= 3\frac{\Omega_T}{\Omega_{BZ}} \frac{1}{\varepsilon_{20}\varepsilon_{30}} \left[\varepsilon_{10} + 2E_1 - \frac{\left(\varepsilon_{20} + \varepsilon_{31}\right)E_1^2}{\varepsilon_{21}\varepsilon_{31}} \right] \,, \tag{D.5.71}$$

and

$$
\begin{aligned}
n_T(E) &= \frac{\Omega_T}{\Omega_{BZ}} \left[\frac{\varepsilon_{10}^3 + 3\varepsilon_{10}^2 E_1 + 3\varepsilon_{10}E_1^2 + E_1^3}{\varepsilon_{10}\varepsilon_{20}\varepsilon_{30}} - \frac{E_1^3}{\varepsilon_{10}\varepsilon_{21}\varepsilon_{31}} \right] \\
&= \frac{\Omega_T}{\Omega_{BZ}} \frac{1}{\varepsilon_{20}\varepsilon_{30}} \left[\varepsilon_{10}^2 + 3\varepsilon_{10}E_1 + 3E_1^2 + \frac{\varepsilon_{21}\varepsilon_{31} - \varepsilon_{20}\varepsilon_{30}}{\varepsilon_{10}\varepsilon_{21}\varepsilon_{31}} E_1^3 \right] \\
&= \frac{\Omega_T}{\Omega_{BZ}} \frac{1}{\varepsilon_{20}\varepsilon_{30}} \left[\varepsilon_{10}^2 + 3\varepsilon_{10}E_1 + 3E_1^2 - \frac{(\varepsilon_{20} + \varepsilon_{31}) E_1^3}{\varepsilon_{21}\varepsilon_{31}} \right] .
\end{aligned}
$$

$$(D.5.72)$$

In contrast, starting from (D.5.53), (D.5.54), and (D.5.56), which avoid the singularity from the outset, we obtain the results

$$
\begin{aligned}
\rho_T(E) &= 3\frac{\Omega_T}{\Omega_{BZ}} \frac{E_0\varepsilon_{20} - E_0^2}{\varepsilon_{20}\varepsilon_{21}\varepsilon_{30}} + 3\frac{\Omega_T}{\Omega_{BZ}} \frac{E_1\varepsilon_{31} - E_1^2}{\varepsilon_{21}\varepsilon_{30}\varepsilon_{31}} \\
&= 3\frac{\Omega_T}{\Omega_{BZ}} \frac{1}{\varepsilon_{21}\varepsilon_{30}} \left[E_0 + E_1 - \frac{E_0^2}{\varepsilon_{20}} - \frac{E_1^2}{\varepsilon_{31}} \right] \\
&= -3\frac{\Omega_T}{\Omega_{BZ}} \frac{1}{\varepsilon_{21}\varepsilon_{30}} \left[\frac{E_0 E_2}{\varepsilon_{20}} + \frac{E_1 E_3}{\varepsilon_{31}} \right] ,
\end{aligned}
$$

$$(D.5.73)$$

and

$$
\begin{aligned}
n_T(E) &= \frac{\Omega_T}{\Omega_{BZ}} \frac{\frac{3}{2}E_0^2\varepsilon_{20} - E_0^3}{\varepsilon_{20}\varepsilon_{21}\varepsilon_{30}} + \frac{\Omega_T}{\Omega_{BZ}} \frac{\frac{3}{2}E_1^2\varepsilon_{31} - E_1^3}{\varepsilon_{21}\varepsilon_{30}\varepsilon_{31}} - \frac{1}{2}\frac{\Omega_T}{\Omega_{BZ}} \frac{\varepsilon_{10}^2}{\varepsilon_{21}\varepsilon_{30}} \\
&= \frac{\Omega_T}{\Omega_{BZ}} \frac{1}{\varepsilon_{21}\varepsilon_{30}} \left[\frac{3}{2}E_0^2 + \frac{3}{2}E_1^2 - \frac{E_0^3}{\varepsilon_{20}} - \frac{E_1^3}{\varepsilon_{31}} - \frac{1}{2}\varepsilon_{10}^2 \right] ,
\end{aligned}
$$

$$(D.5.74)$$

which can be easily shown to be identical to (D.5.71) and (D.5.72).

4. $\varepsilon_2 \leq E \leq \varepsilon_3$

$$
\rho_T(E) = 3\frac{\Omega_T}{\Omega_{BZ}} \frac{E_3^2}{\varepsilon_{30}\varepsilon_{31}\varepsilon_{32}} ,
$$

$$(D.5.75)$$

$$
n_T(E) = \frac{\Omega_T}{\Omega_{BZ}} \left[1 + \frac{E_3^3}{\varepsilon_{30}\varepsilon_{31}\varepsilon_{32}} \right] .
$$

$$(D.5.76)$$

5. $\varepsilon_3 \leq E$

$$
\rho_T(E) = 0 ,
$$

$$(D.5.77)$$

$$
n_T(E) = \frac{\Omega_T}{\Omega_{BZ}} .
$$

$$(D.5.78)$$

In passing we mention that, apart from the factor Ω_T/Ω_{BZ}, (D.5.66) and (D.5.75) are identical to (B6) of the paper by Lambin and Vigneron [11]. In addition, (D.5.73) is equivalent to their result, which is based on the triangles $S_{2,a'}$ and $S_{2,b'}$ rather than on the unprimed ones.

In closing this section we turn to the calculation of the weights $w_{nj}^R(E)$ and $w_{nj}^I(E)$, which, as outlined at the end of Sect. D.1, allow to turn the Brillouin-zone integrals into sums over **k**-points. In order to calculate these weights we introduced a set of filter functions $w_j(\mathbf{k})$, as many as there are **k**-points in the discrete set. With the simple choice $w_j(\mathbf{k}_i) = \delta_{ij}$, specified by (D.1.32), we were able to reduce calculation of the weights to (D.1.33) and (D.1.34). Yet, the shape of the filter functions between the discrete **k**-points was still open this hindering the final determination of the weights. Within the tetrahedron method we are in a position to make up for this issue and use a linear interpolation for the filter functions. As already mentioned after (D.5.2), we may use all the previous results with the general function $a(\mathbf{k})$ replaced by a filter function $w_j(\mathbf{k})$. Specific to the latter are the values at the vertices of the tetrahedron, which, however, are given by (D.1.32). We thus note

$$w_{j,i} = w_j(\mathbf{k}_i) = \delta_{ij} \qquad i,j = 0,1,2,3 \,. \tag{D.5.79}$$

With this choice the linear interpolation reads as

$$\bar{w}_j(\mathbf{k}) = w_{j,0} + \mathbf{b}_{w_j} \cdot \mathbf{k}' = \delta_{j0} + \mathbf{b}_{w_j} \cdot \mathbf{k}' \,, \tag{D.5.80}$$

where

$$\mathbf{b}_{w_j} = \sum_{i=1}^{3} (w_{j,i} - w_{j,0})\, \mathbf{r}_i =: \sum_{i=1}^{3} w_{j,i0}\mathbf{r}_i = \sum_{i=1}^{3} (\delta_{ji} - \delta_{j0})\, \mathbf{r}_i \,. \tag{D.5.81}$$

The linear interpolation is shown schematically in Fig. D.6 for the two-dimensional case.

Inserting the choice (D.5.79) into the results (D.5.45) to (D.5.61) we are able to derive the following results for the energy dependent weights as well as their energy-integrated counterparts. In doing so we suppress the band index and again distinguish the five different cases listed above.

1. $E \leq \varepsilon_0$
   (D.5.45) and (D.5.46) lead to the general formulas

$$w_{j,T}^I(E) = 0 \qquad \text{for } j = 0,1,2,3 \,, \tag{D.5.82}$$

$$\int_{\varepsilon_0}^{E} dE' \, w_{j,T}^I(E') = 0 \qquad \text{for } j = 0,1,2,3 \,. \tag{D.5.83}$$

2. $\varepsilon_0 \leq E \leq \varepsilon_1$
   From (D.5.47) and (D.5.48) we have the general results

$$w_{j,T}^I(E) = \frac{\Omega_T}{\Omega_{BZ}} \frac{E_0^2}{\varepsilon_{10}\varepsilon_{20}\varepsilon_{30}} \left(3w_{j,0} + E_0 \sum_{i=1}^{3} \frac{w_{j,i0}}{\varepsilon_{i0}} \right) \,, \tag{D.5.84}$$

and

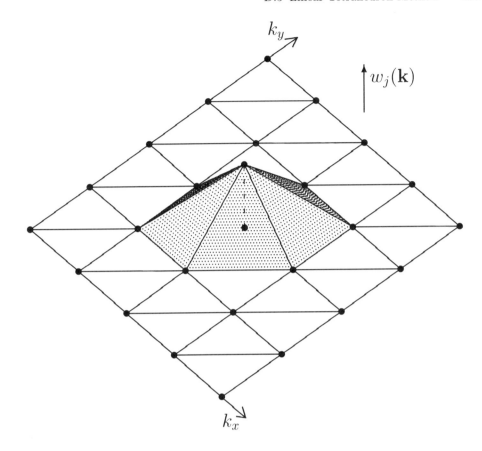

Fig. D.6. Two-dimensional, schematic illustration of the filter function $w_j(\mathbf{k})$ arising from linear interpolating between values $w_j(\mathbf{k}_i) = \delta_{ij}$ (after [5])

$$\int_{\varepsilon_0}^{E} dE'\, w_{j,T}^I(E') = \frac{\Omega_T}{\Omega_{BZ}} \frac{E_0^3}{\varepsilon_{10}\varepsilon_{20}\varepsilon_{30}} \left(w_{j,0} + \frac{E_0}{4} \sum_{i=1}^{3} \frac{w_{j,i0}}{\varepsilon_{i0}} \right) . \qquad (D.5.85)$$

From these we obtain the specific formulas

$$w_{0,T}^I(E) = \frac{\Omega_T}{\Omega_{BZ}} \frac{E_0^2}{\varepsilon_{10}\varepsilon_{20}\varepsilon_{30}} \left(3 - E_0 \sum_{i=1}^{3} \frac{1}{\varepsilon_{i0}} \right) , \qquad (D.5.86)$$

$$w_{j,T}^I(E) = \frac{\Omega_T}{\Omega_{BZ}} \frac{E_0^3}{\varepsilon_{j0}\varepsilon_{10}\varepsilon_{20}\varepsilon_{30}} \qquad \text{for } j = 1, 2, 3 . \qquad (D.5.87)$$

and

$$\int_{\varepsilon_0}^{E} dE'\, w_{0,T}^I(E') = \frac{\Omega_T}{4\Omega_{BZ}} \frac{E_0^3}{\varepsilon_{10}\varepsilon_{20}\varepsilon_{30}} \left(4 - E_0 \sum_{i=1}^{3} \frac{1}{\varepsilon_{i0}} \right) , \qquad (D.5.88)$$

$$\int_{\varepsilon_0}^{E} dE' \, w_{j,T}^I(E') = \frac{\Omega_T}{4\Omega_{BZ}} \frac{E_0^4}{\varepsilon_{j0}\varepsilon_{10}\varepsilon_{20}\varepsilon_{30}} \qquad \text{for } j = 1, 2, 3.$$

(D.5.89)

3. $\varepsilon_1 \leq E \leq \varepsilon_2$

The general results obtained from (D.5.50) and (D.5.51) read

$$w_{j,T}^I(E) = \frac{\Omega_T}{\Omega_{BZ}} \frac{E_0^2}{\varepsilon_{10}\varepsilon_{20}\varepsilon_{30}} \left(3w_{j,0} + E_0 \sum_{i=1}^{3} \frac{w_{j,i0}}{\varepsilon_{i0}} \right)$$

$$- \frac{\Omega_T}{\Omega_{BZ}} \frac{E_1^2}{\varepsilon_{10}\varepsilon_{21}\varepsilon_{31}} \left(3w_{j,1} + E_1 \sum_{\substack{i=0 \\ i \neq 1}}^{3} \frac{w_{j,i1}}{\varepsilon_{i1}} \right), \qquad \text{(D.5.90)}$$

and

$$\int_{\varepsilon_0}^{E} dE' \, w_{j,T}^I(E') = \frac{\Omega_T}{\Omega_{BZ}} \frac{E_0^3}{\varepsilon_{10}\varepsilon_{20}\varepsilon_{30}} \left(w_{j,0} + \frac{E_0}{4} \sum_{i=1}^{3} \frac{w_{j,i0}}{\varepsilon_{i0}} \right)$$

$$- \frac{\Omega_T}{\Omega_{BZ}} \frac{E_1^3}{\varepsilon_{10}\varepsilon_{21}\varepsilon_{31}} \left(w_{j,1} + \frac{E_1}{4} \sum_{\substack{i=0 \\ i \neq 1}}^{3} \frac{w_{j,i1}}{\varepsilon_{i1}} \right).$$

(D.5.91)

Specifically,

$$w_{0,T}^I(E) = \frac{\Omega_T}{\Omega_{BZ}} \left[\frac{E_0^2}{\varepsilon_{10}\varepsilon_{20}\varepsilon_{30}} \left(3 - E_0 \sum_{i=1}^{3} \frac{1}{\varepsilon_{i0}} \right) + \frac{E_1^3}{\varepsilon_{10}^2\varepsilon_{21}\varepsilon_{31}} \right],$$

(D.5.92)

$$w_{1,T}^I(E) = \frac{\Omega_T}{\Omega_{BZ}} \left[\frac{E_0^3}{\varepsilon_{10}^2\varepsilon_{20}\varepsilon_{30}} - \frac{E_1^2}{\varepsilon_{10}\varepsilon_{21}\varepsilon_{31}} \left(3 - E_1 \sum_{\substack{i=0 \\ i \neq 1}}^{3} \frac{1}{\varepsilon_{i1}} \right) \right],$$

(D.5.93)

$$w_{j,T}^I(E) = \frac{\Omega_T}{\Omega_{BZ}} \left[\frac{E_0^3}{\varepsilon_{j0}\varepsilon_{10}\varepsilon_{20}\varepsilon_{30}} - \frac{E_1^3}{\varepsilon_{j1}\varepsilon_{10}\varepsilon_{21}\varepsilon_{31}} \right] \qquad \text{for } j = 2, 3,$$

(D.5.94)

and

$$\int_{\varepsilon_0}^{E} dE' \, w_{0,T}^I(E') = \frac{\Omega_T}{4\Omega_{BZ}} \left[\frac{E_0^3}{\varepsilon_{10}\varepsilon_{20}\varepsilon_{30}} \left(4 - E_0 \sum_{i=1}^{3} \frac{1}{\varepsilon_{i0}} \right) + \frac{E_1^4}{\varepsilon_{10}^2\varepsilon_{21}\varepsilon_{31}} \right].$$

(D.5.95)

$$
\int_{\varepsilon_0}^{E} dE'\, w_{1,T}^{I}(E') = \frac{\Omega_T}{4\Omega_{BZ}} \left[\frac{E_0^4}{\varepsilon_{10}^2 \varepsilon_{20} \varepsilon_{30}} - \frac{E_1^3}{\varepsilon_{10} \varepsilon_{21} \varepsilon_{31}} \left(4 - E_1 \sum_{\substack{i=0 \\ i\neq 1}}^{3} \frac{1}{\varepsilon_{i1}} \right) \right],
$$

$$(\text{D.5.96})$$

$$
\int_{\varepsilon_0}^{E} dE'\, w_{j,T}^{I}(E') = \frac{\Omega_T}{4\Omega_{BZ}} \left[\frac{E_0^4}{\varepsilon_{j0}\varepsilon_{10}\varepsilon_{20}\varepsilon_{30}} - \frac{E_1^4}{\varepsilon_{j1}\varepsilon_{10}\varepsilon_{21}\varepsilon_{31}} \right] \qquad \text{for } j = 2,3 \,.
$$

$$(\text{D.5.97})$$

As before, we want to avoid the singularity arising from $\varepsilon_{10} = 0$. Again replacing E_0 by $E_1 + \varepsilon_{10}$ as well as E_1 by $E_0 - \varepsilon_{10}$ and using the identity (D.5.70) we rewrite (D.5.92) to (D.5.97) as

$$
\begin{aligned}
w_{0,T}^{I}(E) &= \frac{\Omega_T}{\Omega_{BZ}} \left[\frac{E_0^2}{\varepsilon_{10}\varepsilon_{20}\varepsilon_{30}} \left(3 - E_0 \sum_{i=1}^{3} \frac{1}{\varepsilon_{i0}} \right) \right. \\
&\qquad \left. - \frac{\varepsilon_{10}^3 - 3\varepsilon_{10}^2 E_0 + 3\varepsilon_{10} E_0^2 - E_0^3}{\varepsilon_{10}^2 \varepsilon_{21}\varepsilon_{31}} \right] \\
&= \frac{\Omega_T}{\Omega_{BZ}} \frac{1}{\varepsilon_{21}\varepsilon_{31}} \left[-\varepsilon_{10} + 3E_0 + 3E_0^2 \frac{\varepsilon_{21}\varepsilon_{31} - \varepsilon_{20}\varepsilon_{30}}{\varepsilon_{10}\varepsilon_{20}\varepsilon_{30}} \right. \\
&\qquad \left. + E_0^3 \frac{\varepsilon_{20}\varepsilon_{30} - \varepsilon_{21}\varepsilon_{31}}{\varepsilon_{10}^2 \varepsilon_{20}\varepsilon_{30}} - E_0^3 \frac{\varepsilon_{21}\varepsilon_{31}\left(\varepsilon_{20} + \varepsilon_{30}\right)}{\varepsilon_{10}\varepsilon_{20}^2 \varepsilon_{30}^2} \right] \\
&= \frac{\Omega_T}{\Omega_{BZ}} \frac{1}{\varepsilon_{21}\varepsilon_{31}} \left[-\varepsilon_{10} + 3E_0 - 3E_0^2 \frac{\varepsilon_{20} + \varepsilon_{31}}{\varepsilon_{20}\varepsilon_{30}} \right. \\
&\qquad \left. + E_0^3 \frac{\varepsilon_{20}\varepsilon_{30}\left(\varepsilon_{20} + \varepsilon_{31}\right) - \varepsilon_{21}\varepsilon_{31}\left(\varepsilon_{20} + \varepsilon_{31} + \varepsilon_{10}\right)}{\varepsilon_{10}\varepsilon_{20}^2 \varepsilon_{30}^2} \right] \\
&= \frac{\Omega_T}{\Omega_{BZ}} \frac{1}{\varepsilon_{21}\varepsilon_{31}} \left[-\varepsilon_{10} + 3E_0 - 3E_0^2 \frac{\varepsilon_{20} + \varepsilon_{31}}{\varepsilon_{20}\varepsilon_{30}} \right. \\
&\qquad \left. + E_0^3 \frac{\left(\varepsilon_{20} + \varepsilon_{31}\right)^2 - \varepsilon_{21}\varepsilon_{31}}{\varepsilon_{20}^2 \varepsilon_{30}^2} \right], \qquad (\text{D.5.98})
\end{aligned}
$$

$$
\begin{aligned}
w_{1,T}^{I}(E) &= \frac{\Omega_T}{\Omega_{BZ}} \left[\frac{\varepsilon_{10}^3 + 3\varepsilon_{10}^2 E_1 + 3\varepsilon_{10} E_1^2 + E_1^3}{\varepsilon_{10}^2 \varepsilon_{20}\varepsilon_{30}} \right. \\
&\qquad \left. - \frac{E_1^2}{\varepsilon_{10}\varepsilon_{21}\varepsilon_{31}} \left(3 - E_1 \sum_{\substack{i=0 \\ i\neq 1}}^{3} \frac{1}{\varepsilon_{i1}} \right) \right] \\
&= \frac{\Omega_T}{\Omega_{BZ}} \frac{1}{\varepsilon_{20}\varepsilon_{30}} \left[\varepsilon_{10} + 3E_1 + 3E_1^2 \frac{\varepsilon_{21}\varepsilon_{31} - \varepsilon_{20}\varepsilon_{30}}{\varepsilon_{10}\varepsilon_{21}\varepsilon_{31}} \right.
\end{aligned}
$$

$$+E_1^3 \frac{\varepsilon_{21}\varepsilon_{31} - \varepsilon_{20}\varepsilon_{30}}{\varepsilon_{10}^2\varepsilon_{21}\varepsilon_{31}} + E_1^3 \frac{\varepsilon_{20}\varepsilon_{30}\left(\varepsilon_{21} + \varepsilon_{31}\right)}{\varepsilon_{10}\varepsilon_{21}^2\varepsilon_{31}^2}\Bigg]$$

$$= \frac{\Omega_T}{\Omega_{BZ}} \frac{1}{\varepsilon_{20}\varepsilon_{30}}\Bigg[\varepsilon_{10} + 3E_1 - 3E_1^2\frac{\varepsilon_{20} + \varepsilon_{31}}{\varepsilon_{21}\varepsilon_{31}}$$

$$-E_1^3\frac{\varepsilon_{21}\varepsilon_{31}\left(\varepsilon_{20} + \varepsilon_{31}\right) - \varepsilon_{20}\varepsilon_{30}\left(\varepsilon_{20} + \varepsilon_{31} - \varepsilon_{10}\right)}{\varepsilon_{10}\varepsilon_{21}^2\varepsilon_{31}^2}\Bigg]$$

$$= \frac{\Omega_T}{\Omega_{BZ}} \frac{1}{\varepsilon_{20}\varepsilon_{30}}\Bigg[\varepsilon_{10} + 3E_1 - 3E_1^2\frac{\varepsilon_{20} + \varepsilon_{31}}{\varepsilon_{21}\varepsilon_{31}}$$

$$+E_1^3\frac{\left(\varepsilon_{20} + \varepsilon_{31}\right)^2 - \varepsilon_{20}\varepsilon_{30}}{\varepsilon_{21}^2\varepsilon_{31}^2}\Bigg], \tag{D.5.99}$$

$$w_{j,T}^I(E) = \frac{\Omega_T}{\Omega_{BZ}}\Bigg[\frac{\varepsilon_{10}^3 + 3\varepsilon_{10}^2 E_1 + 3\varepsilon_{10}E_1^2 + E_1^3}{\varepsilon_{j0}\varepsilon_{10}\varepsilon_{20}\varepsilon_{30}} - \frac{E_1^3}{\varepsilon_{j1}\varepsilon_{10}\varepsilon_{21}\varepsilon_{31}}\Bigg]$$

$$= \frac{\Omega_T}{\Omega_{BZ}} \frac{1}{\varepsilon_{j0}\varepsilon_{20}\varepsilon_{30}}\Bigg[\varepsilon_{10}^2 + 3\varepsilon_{10}E_1 + 3E_1^2$$

$$+E_1^3\frac{\varepsilon_{j1}\varepsilon_{21}\varepsilon_{31} - \left(\varepsilon_{j1} + \varepsilon_{10}\right)\varepsilon_{20}\varepsilon_{30}}{\varepsilon_{j1}\varepsilon_{10}\varepsilon_{21}\varepsilon_{31}}\Bigg]$$

$$= \frac{\Omega_T}{\Omega_{BZ}} \frac{1}{\varepsilon_{j0}\varepsilon_{20}\varepsilon_{30}}\Bigg[\varepsilon_{10}^2 + 3\varepsilon_{10}E_1 + 3E_1^2$$

$$-E_1^3\frac{\varepsilon_{20}\varepsilon_{30} + \varepsilon_{j1}\left(\varepsilon_{20} + \varepsilon_{31}\right)}{\varepsilon_{j1}\varepsilon_{21}\varepsilon_{31}}\Bigg]$$

$$\text{for } j = 2, 3, \tag{D.5.100}$$

and

$$\int_{\varepsilon_0}^{E} dE'\, w_{0,T}^I(E')$$

$$= \frac{\Omega_T}{4\Omega_{BZ}}\Bigg[\frac{E_0^3}{\varepsilon_{10}\varepsilon_{20}\varepsilon_{30}}\left(4 - E_0 \sum_{i=1}^{3}\frac{1}{\varepsilon_{i0}}\right)$$

$$+\frac{\varepsilon_{10}^4 - 4\varepsilon_{10}^3 E_0 + 6\varepsilon_{10}^2 E_0^2 - 4\varepsilon_{10}E_0^3 + E_0^4}{\varepsilon_{10}^2\varepsilon_{21}\varepsilon_{31}}\Bigg]$$

$$= \frac{\Omega_T}{4\Omega_{BZ}} \frac{1}{\varepsilon_{21}\varepsilon_{31}}\Bigg[\varepsilon_{10}^2 - 4\varepsilon_{10}E_0 + 6E_0^2 + 4E_0^3\frac{\varepsilon_{21}\varepsilon_{31} - \varepsilon_{20}\varepsilon_{30}}{\varepsilon_{10}\varepsilon_{20}\varepsilon_{30}}$$

$$+E_0^4\frac{\varepsilon_{20}\varepsilon_{30} - \varepsilon_{21}\varepsilon_{31}}{\varepsilon_{10}^2\varepsilon_{20}\varepsilon_{30}} - E_0^4\frac{\varepsilon_{21}\varepsilon_{31}\left(\varepsilon_{20} + \varepsilon_{30}\right)}{\varepsilon_{10}\varepsilon_{20}^2\varepsilon_{30}^2}\Bigg]$$

$$
= \frac{\Omega_T}{4\Omega_{BZ}} \frac{1}{\varepsilon_{21}\varepsilon_{31}} \left[\varepsilon_{10}^2 - 4\varepsilon_{10}E_0 + 6E_0^2 - 4E_0^3 \frac{\varepsilon_{20} + \varepsilon_{31}}{\varepsilon_{20}\varepsilon_{30}} \right.
$$

$$
\left. + E_0^4 \frac{\varepsilon_{20}\varepsilon_{30}\left(\varepsilon_{20} + \varepsilon_{31}\right) - \varepsilon_{21}\varepsilon_{31}\left(\varepsilon_{20} + \varepsilon_{31} + \varepsilon_{10}\right)}{\varepsilon_{10}\varepsilon_{20}^2\varepsilon_{30}^2} \right]
$$

$$
= \frac{\Omega_T}{4\Omega_{BZ}} \frac{1}{\varepsilon_{21}\varepsilon_{31}} \left[\varepsilon_{10}^2 - 4\varepsilon_{10}E_0 + 6E_0^2 - 4E_0^3 \frac{\varepsilon_{20} + \varepsilon_{31}}{\varepsilon_{20}\varepsilon_{30}} \right.
$$

$$
\left. + E_0^4 \frac{\left(\varepsilon_{20} + \varepsilon_{31}\right)^2 - \varepsilon_{21}\varepsilon_{31}}{\varepsilon_{20}^2\varepsilon_{30}^2} \right], \tag{D.5.101}
$$

$$
\int_{\varepsilon_0}^{E} dE' \, w_{1,T}^I(E')
$$

$$
= \frac{\Omega_T}{4\Omega_{BZ}} \left[\frac{\varepsilon_{10}^4 + 4\varepsilon_{10}^3 E_1 + 6\varepsilon_{10}^2 E_1^2 + 4\varepsilon_{10}E_1^3 + E_1^4}{\varepsilon_{10}^2\varepsilon_{20}\varepsilon_{30}} \right.
$$

$$
\left. - \frac{E_1^3}{\varepsilon_{10}\varepsilon_{21}\varepsilon_{31}} \left(4 - E_1 \sum_{\substack{i=0 \\ i \neq 1}}^{3} \frac{1}{\varepsilon_{i1}} \right) \right]
$$

$$
= \frac{\Omega_T}{4\Omega_{BZ}} \frac{1}{\varepsilon_{20}\varepsilon_{30}} \left[\varepsilon_{10}^2 + 4\varepsilon_{10}E_1 + 6E_1^2 + 4E_1^3 \frac{\varepsilon_{21}\varepsilon_{31} - \varepsilon_{20}\varepsilon_{30}}{\varepsilon_{10}\varepsilon_{21}\varepsilon_{31}} \right.
$$

$$
\left. + E_1^4 \frac{\varepsilon_{21}\varepsilon_{31} - \varepsilon_{20}\varepsilon_{30}}{\varepsilon_{10}^2\varepsilon_{21}\varepsilon_{31}} + E_1^4 \frac{\varepsilon_{20}\varepsilon_{30}\left(\varepsilon_{21} + \varepsilon_{31}\right)}{\varepsilon_{10}\varepsilon_{21}^2\varepsilon_{31}^2} \right]
$$

$$
= \frac{\Omega_T}{4\Omega_{BZ}} \frac{1}{\varepsilon_{20}\varepsilon_{30}} \left[\varepsilon_{10}^2 + 4\varepsilon_{10}E_1 + 6E_1^2 - 4E_1^3 \frac{\varepsilon_{20} + \varepsilon_{31}}{\varepsilon_{21}\varepsilon_{31}} \right.
$$

$$
\left. - E_1^4 \frac{\varepsilon_{21}\varepsilon_{31}\left(\varepsilon_{20} + \varepsilon_{31}\right) - \varepsilon_{20}\varepsilon_{30}\left(\varepsilon_{20} + \varepsilon_{31} - \varepsilon_{10}\right)}{\varepsilon_{10}\varepsilon_{21}^2\varepsilon_{31}^2} \right]
$$

$$
= \frac{\Omega_T}{4\Omega_{BZ}} \frac{1}{\varepsilon_{20}\varepsilon_{30}} \left[\varepsilon_{10}^2 + 4\varepsilon_{10}E_1 + 6E_1^2 - 4E_1^3 \frac{\varepsilon_{20} + \varepsilon_{31}}{\varepsilon_{21}\varepsilon_{31}} \right.
$$

$$
\left. + E_1^4 \frac{\left(\varepsilon_{20} + \varepsilon_{31}\right)^2 - \varepsilon_{20}\varepsilon_{30}}{\varepsilon_{21}^2\varepsilon_{31}^2} \right], \tag{D.5.102}
$$

$$
\int_{\varepsilon_0}^{E} dE' \, w_{j,T}^I(E')
$$

$$
= \frac{\Omega_T}{4\Omega_{BZ}} \left[\frac{\varepsilon_{10}^4 + 4\varepsilon_{10}^3 E_1 + 6\varepsilon_{10}^2 E_1^2 + 4\varepsilon_{10}E_1^3 + E_1^4}{\varepsilon_{j0}\varepsilon_{10}\varepsilon_{20}\varepsilon_{30}} \right.
$$

$$
\left. - \frac{E_1^4}{\varepsilon_{j1}\varepsilon_{10}\varepsilon_{21}\varepsilon_{31}} \right]
$$

$$= \frac{\Omega_T}{4\Omega_{BZ}} \frac{1}{\varepsilon_{j0}\varepsilon_{20}\varepsilon_{30}} \left[\varepsilon_{10}^3 + 4\varepsilon_{10}^2 E_1 + 6\varepsilon_{10}E_1^2 + 4E_1^3 \right.$$

$$\left. + E_1^4 \frac{\varepsilon_{j1}\varepsilon_{21}\varepsilon_{31} - (\varepsilon_{j1} + \varepsilon_{10})\,\varepsilon_{20}\varepsilon_{30}}{\varepsilon_{j1}\varepsilon_{10}\varepsilon_{21}\varepsilon_{31}} \right]$$

$$= \frac{\Omega_T}{4\Omega_{BZ}} \frac{1}{\varepsilon_{j0}\varepsilon_{20}\varepsilon_{30}} \left[\varepsilon_{10}^3 + 4\varepsilon_{10}^2 E_1 + 6\varepsilon_{10}E_1^2 + 4E_1^3 \right.$$

$$\left. - E_1^4 \frac{\varepsilon_{20}\varepsilon_{30} + \varepsilon_{j1}\,(\varepsilon_{20} + \varepsilon_{31})}{\varepsilon_{j1}\varepsilon_{21}\varepsilon_{31}} \right]$$

$$\text{for } j = 2, 3. \tag{D.5.103}$$

Alternatively, we start from (D.5.53), (D.5.54), and (D.5.56), which avoid the singularity from the outset, and obtain

$$w_{j,T}^I(E) = \frac{\Omega_T}{\Omega_{BZ}} \frac{E_0\varepsilon_{20} - E_0^2}{\varepsilon_{20}\varepsilon_{21}\varepsilon_{30}} \left[3w_{j,0} + w_{j,10} - \varepsilon_{10}\frac{w_{j,21}}{\varepsilon_{21}} \right.$$

$$\left. + E_0 \left(\frac{w_{j,20}}{\varepsilon_{20}} + \frac{w_{j,21}}{\varepsilon_{21}} + \frac{w_{j,30}}{\varepsilon_{30}} \right) \right]$$

$$+ \frac{\Omega_T}{\Omega_{BZ}} \frac{E_1\varepsilon_{31} - E_1^2}{\varepsilon_{21}\varepsilon_{30}\varepsilon_{31}} \left[3w_{j,1} - w_{j,10} + \varepsilon_{10}\frac{w_{j,30}}{\varepsilon_{30}} \right.$$

$$\left. + E_1 \left(\frac{w_{j,21}}{\varepsilon_{21}} + \frac{w_{j,30}}{\varepsilon_{30}} + \frac{w_{j,31}}{\varepsilon_{31}} \right) \right], \tag{D.5.104}$$

and

$$\int_{\varepsilon_0}^E dE' \, w_{j,T}^I(E')$$

$$= \frac{\Omega_T}{\Omega_{BZ}} \frac{\frac{1}{2}E_0^2\varepsilon_{20} - \frac{1}{3}E_0^3}{\varepsilon_{20}\varepsilon_{21}\varepsilon_{30}} \left(3w_{j,0} + w_{j,10} - \varepsilon_{10}\frac{w_{j,21}}{\varepsilon_{21}} \right)$$

$$+ \frac{\Omega_T}{\Omega_{BZ}} \frac{\frac{1}{3}E_0^3\varepsilon_{20} - \frac{1}{4}E_0^4}{\varepsilon_{20}\varepsilon_{21}\varepsilon_{30}} \left(\frac{w_{j,20}}{\varepsilon_{20}} + \frac{w_{j,21}}{\varepsilon_{21}} + \frac{w_{j,30}}{\varepsilon_{30}} \right)$$

$$+ \frac{\Omega_T}{\Omega_{BZ}} \frac{\frac{1}{2}E_1^2\varepsilon_{31} - \frac{1}{3}E_1^3}{\varepsilon_{21}\varepsilon_{30}\varepsilon_{31}} \left(3w_{j,1} - w_{j,10} + \varepsilon_{10}\frac{w_{j,30}}{\varepsilon_{30}} \right)$$

$$+ \frac{\Omega_T}{\Omega_{BZ}} \frac{\frac{1}{3}E_1^3\varepsilon_{31} - \frac{1}{4}E_1^4}{\varepsilon_{21}\varepsilon_{30}\varepsilon_{31}} \left(\frac{w_{j,21}}{\varepsilon_{21}} + \frac{w_{j,30}}{\varepsilon_{30}} + \frac{w_{j,31}}{\varepsilon_{31}} \right)$$

$$- \frac{\Omega_T}{\Omega_{BZ}} \frac{\varepsilon_{10}^2}{\varepsilon_{21}\varepsilon_{30}} \left[\frac{1}{2}w_{j,0} + \frac{\varepsilon_{10}}{12}\sum_{i=1}^{3}\frac{w_{j,i0}}{\varepsilon_{i0}} + \frac{1}{12}\frac{\varepsilon_{21}w_{j,10} - \varepsilon_{10}w_{j,21}}{\varepsilon_{21}} \right]$$

$$- \frac{\Omega_T}{\Omega_{BZ}} \frac{\varepsilon_{10}^2}{\varepsilon_{20}\varepsilon_{30}} \frac{1}{12}\frac{\varepsilon_{21}w_{j,10} - \varepsilon_{10}w_{j,21}}{\varepsilon_{21}}. \tag{D.5.105}$$

These formulas lead to the following specific results

$$w_{0,T}^I(E) = \frac{\Omega_T}{\Omega_{BZ}} \frac{E_0\varepsilon_{20} - E_0^2}{\varepsilon_{20}\varepsilon_{21}\varepsilon_{30}} \left(2 - E_0\left(\frac{1}{\varepsilon_{20}} + \frac{1}{\varepsilon_{30}}\right)\right)$$

$$+ \frac{\Omega_T}{\Omega_{BZ}} \frac{E_1\varepsilon_{31} - E_1^2}{\varepsilon_{21}\varepsilon_{30}\varepsilon_{31}} \left(1 - E_0\frac{1}{\varepsilon_{30}}\right)$$

$$= \frac{\Omega_T}{\Omega_{BZ}} \frac{E_0 E_2}{\varepsilon_{20}\varepsilon_{21}\varepsilon_{30}} \left(\frac{E_2}{\varepsilon_{20}} + \frac{E_3}{\varepsilon_{30}}\right) + \frac{\Omega_T}{\Omega_{BZ}} \frac{E_1 E_3^2}{\varepsilon_{21}\varepsilon_{30}^2\varepsilon_{31}}, \quad \text{(D.5.106)}$$

$$w_{1,T}^I(E) = \frac{\Omega_T}{\Omega_{BZ}} \frac{E_0\varepsilon_{20} - E_0^2}{\varepsilon_{20}\varepsilon_{21}\varepsilon_{30}} \left(1 - E_1\frac{1}{\varepsilon_{21}}\right)$$

$$+ \frac{\Omega_T}{\Omega_{BZ}} \frac{E_1\varepsilon_{31} - E_1^2}{\varepsilon_{21}\varepsilon_{30}\varepsilon_{31}} \left(2 - E_1\left(\frac{1}{\varepsilon_{21}} + \frac{1}{\varepsilon_{31}}\right)\right)$$

$$= \frac{\Omega_T}{\Omega_{BZ}} \frac{E_0 E_2^2}{\varepsilon_{20}\varepsilon_{21}^2\varepsilon_{30}} + \frac{\Omega_T}{\Omega_{BZ}} \frac{E_1 E_3}{\varepsilon_{21}\varepsilon_{30}\varepsilon_{31}} \left(\frac{E_2}{\varepsilon_{21}} + \frac{E_3}{\varepsilon_{31}}\right), \quad \text{(D.5.107)}$$

$$w_{2,T}^I(E) = \frac{\Omega_T}{\Omega_{BZ}} \frac{E_0\varepsilon_{20} - E_0^2}{\varepsilon_{20}\varepsilon_{21}\varepsilon_{30}} \left(-\varepsilon_{10}\frac{1}{\varepsilon_{21}} + E_0\left(\frac{1}{\varepsilon_{20}} + \frac{1}{\varepsilon_{21}}\right)\right)$$

$$+ \frac{\Omega_T}{\Omega_{BZ}} \frac{E_1\varepsilon_{31} - E_1^2}{\varepsilon_{21}\varepsilon_{30}\varepsilon_{31}} \left(E_1\frac{1}{\varepsilon_{21}}\right)$$

$$= -\frac{\Omega_T}{\Omega_{BZ}} \frac{E_0 E_2}{\varepsilon_{20}\varepsilon_{21}\varepsilon_{30}} \left(\frac{E_0}{\varepsilon_{20}} + \frac{E_1}{\varepsilon_{21}}\right) - \frac{\Omega_T}{\Omega_{BZ}} \frac{E_1^2 E_3}{\varepsilon_{21}^2\varepsilon_{30}\varepsilon_{31}},$$

$$\text{(D.5.108)}$$

$$w_{3,T}^I(E) = \frac{\Omega_T}{\Omega_{BZ}} \frac{E_0\varepsilon_{20} - E_0^2}{\varepsilon_{20}\varepsilon_{21}\varepsilon_{30}} \left(E_0\frac{1}{\varepsilon_{30}}\right)$$

$$+ \frac{\Omega_T}{\Omega_{BZ}} \frac{E_1\varepsilon_{31} - E_1^2}{\varepsilon_{21}\varepsilon_{30}\varepsilon_{31}} \left(\varepsilon_{10}\frac{1}{\varepsilon_{30}} + E_1\left(\frac{1}{\varepsilon_{30}} + \frac{1}{\varepsilon_{31}}\right)\right)$$

$$= -\frac{\Omega_T}{\Omega_{BZ}} \frac{E_0^2 E_2}{\varepsilon_{20}\varepsilon_{21}\varepsilon_{30}^2} - \frac{\Omega_T}{\Omega_{BZ}} \frac{E_1 E_3}{\varepsilon_{21}\varepsilon_{30}\varepsilon_{31}} \left(\frac{E_0}{\varepsilon_{30}} + \frac{E_1}{\varepsilon_{31}}\right),$$

$$\text{(D.5.109)}$$

as well as

$$\int_{\varepsilon_0}^E dE'\, w_{0,T}^I(E')$$

$$= \frac{\Omega_T}{\Omega_{BZ}} \frac{E_0^2\varepsilon_{20} - \frac{2}{3}E_0^3}{\varepsilon_{20}\varepsilon_{21}\varepsilon_{30}} - \frac{\Omega_T}{\Omega_{BZ}} \frac{\frac{1}{3}E_0^3\varepsilon_{20} - \frac{1}{4}E_0^4}{\varepsilon_{20}\varepsilon_{21}\varepsilon_{30}} \left(\frac{1}{\varepsilon_{20}} + \frac{1}{\varepsilon_{30}}\right)$$

$$+ \frac{\Omega_T}{\Omega_{BZ}} \frac{\frac{1}{2}E_1^2\varepsilon_{31} - \frac{1}{3}E_1^3}{\varepsilon_{21}\varepsilon_{30}\varepsilon_{31}} \left(1 - \varepsilon_{10}\frac{1}{\varepsilon_{30}}\right) - \frac{\Omega_T}{\Omega_{BZ}} \frac{\frac{1}{3}E_1^3\varepsilon_{31} - \frac{1}{4}E_1^4}{\varepsilon_{21}\varepsilon_{30}\varepsilon_{31}} \left(\frac{1}{\varepsilon_{30}}\right)$$

$$- \frac{\Omega_T}{\Omega_{BZ}} \frac{\varepsilon_{10}^2}{\varepsilon_{21}\varepsilon_{30}} \left[\frac{1}{2} - \frac{\varepsilon_{10}}{12} \sum_{i=1}^3 \frac{1}{\varepsilon_{i0}} - \frac{1}{12}\right] + \frac{\Omega_T}{\Omega_{BZ}} \frac{\varepsilon_{10}^2}{\varepsilon_{20}\varepsilon_{30}} \frac{1}{12}$$

$$= \frac{\Omega_T}{\Omega_{BZ}} \frac{E_0^2 \varepsilon_{20} - \frac{2}{3} E_0^3}{\varepsilon_{20}\varepsilon_{21}\varepsilon_{30}} - \frac{\Omega_T}{\Omega_{BZ}} \frac{\frac{1}{3}E_0^3 \varepsilon_{20} - \frac{1}{4}E_0^4}{\varepsilon_{20}\varepsilon_{21}\varepsilon_{30}} \left(\frac{1}{\varepsilon_{20}} + \frac{1}{\varepsilon_{30}} \right)$$

$$+ \frac{\Omega_T}{\Omega_{BZ}} \frac{\frac{1}{2}E_1^2 \varepsilon_{31} - \frac{1}{3}E_1^3}{\varepsilon_{21}\varepsilon_{30}^2} - \frac{\Omega_T}{\Omega_{BZ}} \frac{\frac{1}{3}E_1^3 \varepsilon_{31} - \frac{1}{4}E_1^4}{\varepsilon_{21}\varepsilon_{30}^2\varepsilon_{31}}$$

$$- \frac{\Omega_T}{\Omega_{BZ}} \frac{\varepsilon_{10}^2}{\varepsilon_{21}\varepsilon_{30}} \left[\frac{1}{6} + \frac{1}{12} \frac{\varepsilon_{31}}{\varepsilon_{30}} \right], \tag{D.5.110}$$

$$\int_{\varepsilon_0}^{E} dE'\, w_{1,T}^I(E')$$

$$= \frac{\Omega_T}{\Omega_{BZ}} \frac{\frac{1}{2}E_0^2 \varepsilon_{20} - \frac{1}{3}E_0^3}{\varepsilon_{20}\varepsilon_{21}\varepsilon_{30}} \left(1 + \varepsilon_{10} \frac{1}{\varepsilon_{21}} \right) - \frac{\Omega_T}{\Omega_{BZ}} \frac{\frac{1}{3}E_0^3 \varepsilon_{20} - \frac{1}{4}E_0^4}{\varepsilon_{20}\varepsilon_{21}\varepsilon_{30}} \frac{1}{\varepsilon_{21}}$$

$$+ \frac{\Omega_T}{\Omega_{BZ}} \frac{E_1^2 \varepsilon_{31} - \frac{2}{3}E_1^3}{\varepsilon_{21}\varepsilon_{30}\varepsilon_{31}} - \frac{\Omega_T}{\Omega_{BZ}} \frac{\frac{1}{3}E_1^3 \varepsilon_{31} - \frac{1}{4}E_1^4}{\varepsilon_{21}\varepsilon_{30}\varepsilon_{31}} \left(\frac{1}{\varepsilon_{21}} + \frac{1}{\varepsilon_{31}} \right)$$

$$- \frac{\Omega_T}{\Omega_{BZ}} \frac{\varepsilon_{10}^2}{\varepsilon_{21}\varepsilon_{30}} \left[\frac{1}{12} + \frac{1}{12} \frac{\varepsilon_{20}}{\varepsilon_{21}} \right] - \frac{\Omega_T}{\Omega_{BZ}} \frac{\varepsilon_{10}^2}{\varepsilon_{20}\varepsilon_{30}} \frac{1}{12} \frac{\varepsilon_{20}}{\varepsilon_{21}}$$

$$= \frac{\Omega_T}{\Omega_{BZ}} \frac{\frac{1}{2}E_0^2 \varepsilon_{20} - \frac{1}{3}E_0^3}{\varepsilon_{21}^2\varepsilon_{30}} - \frac{\Omega_T}{\Omega_{BZ}} \frac{\frac{1}{3}E_0^3 \varepsilon_{20} - \frac{1}{4}E_0^4}{\varepsilon_{20}\varepsilon_{21}^2\varepsilon_{30}}$$

$$+ \frac{\Omega_T}{\Omega_{BZ}} \frac{E_1^2 \varepsilon_{31} - \frac{2}{3}E_1^3}{\varepsilon_{21}\varepsilon_{30}\varepsilon_{31}} - \frac{\Omega_T}{\Omega_{BZ}} \frac{\frac{1}{3}E_1^3 \varepsilon_{31} - \frac{1}{4}E_1^4}{\varepsilon_{21}\varepsilon_{30}\varepsilon_{31}} \left(\frac{1}{\varepsilon_{21}} + \frac{1}{\varepsilon_{31}} \right)$$

$$- \frac{\Omega_T}{\Omega_{BZ}} \frac{\varepsilon_{10}^2}{\varepsilon_{21}\varepsilon_{30}} \left[\frac{1}{6} + \frac{1}{12} \frac{\varepsilon_{20}}{\varepsilon_{21}} \right], \tag{D.5.111}$$

$$\int_{\varepsilon_0}^{E} dE'\, w_{2,T}^I(E')$$

$$= -\frac{\Omega_T}{\Omega_{BZ}} \frac{\frac{1}{2}E_0^2 \varepsilon_{20} - \frac{1}{3}E_0^3}{\varepsilon_{20}\varepsilon_{21}\varepsilon_{30}} \frac{\varepsilon_{10}}{\varepsilon_{21}} + \frac{\Omega_T}{\Omega_{BZ}} \frac{\frac{1}{3}E_0^3 \varepsilon_{20} - \frac{1}{4}E_0^4}{\varepsilon_{20}\varepsilon_{21}\varepsilon_{30}} \left(\frac{1}{\varepsilon_{20}} + \frac{1}{\varepsilon_{21}} \right)$$

$$+ \frac{\Omega_T}{\Omega_{BZ}} \frac{\frac{1}{3}E_1^3 \varepsilon_{31} - \frac{1}{4}E_1^4}{\varepsilon_{21}\varepsilon_{30}\varepsilon_{31}} \frac{1}{\varepsilon_{21}}$$

$$- \frac{\Omega_T}{\Omega_{BZ}} \frac{\varepsilon_{10}^2}{\varepsilon_{21}\varepsilon_{30}} \left[\frac{\varepsilon_{10}}{12} \frac{1}{\varepsilon_{20}} - \frac{1}{12} \frac{\varepsilon_{10}}{\varepsilon_{21}} \right] + \frac{\Omega_T}{\Omega_{BZ}} \frac{\varepsilon_{10}^2}{\varepsilon_{20}\varepsilon_{30}} \frac{1}{12} \frac{\varepsilon_{10}}{\varepsilon_{21}}$$

$$= -\frac{\Omega_T}{\Omega_{BZ}} \frac{\frac{1}{2}E_0^2 \varepsilon_{20} - \frac{1}{3}E_0^3}{\varepsilon_{20}\varepsilon_{21}\varepsilon_{30}} \frac{\varepsilon_{10}}{\varepsilon_{21}} + \frac{\Omega_T}{\Omega_{BZ}} \frac{\frac{1}{3}E_0^3 \varepsilon_{20} - \frac{1}{4}E_0^4}{\varepsilon_{20}\varepsilon_{21}\varepsilon_{30}} \left(\frac{1}{\varepsilon_{20}} + \frac{1}{\varepsilon_{21}} \right)$$

$$+ \frac{\Omega_T}{\Omega_{BZ}} \frac{\frac{1}{3}E_1^3 \varepsilon_{31} - \frac{1}{4}E_1^4}{\varepsilon_{21}^2\varepsilon_{30}\varepsilon_{31}} + \frac{\Omega_T}{\Omega_{BZ}} \frac{1}{12} \frac{\varepsilon_{10}^3}{\varepsilon_{21}^2\varepsilon_{30}}, \tag{D.5.112}$$

$$\int_{\varepsilon_0}^{E} dE'\, w_{3,T}^I(E')$$

$$= \frac{\Omega_T}{\Omega_{BZ}} \frac{\frac{1}{3}E_0^3 \varepsilon_{20} - \frac{1}{4}E_0^4}{\varepsilon_{20}\varepsilon_{21}\varepsilon_{30}} \frac{1}{\varepsilon_{30}} + \frac{\Omega_T}{\Omega_{BZ}} \frac{\frac{1}{2}E_1^2 \varepsilon_{31} - \frac{1}{3}E_1^3}{\varepsilon_{21}\varepsilon_{30}\varepsilon_{31}} \frac{\varepsilon_{10}}{\varepsilon_{30}}$$

$$+\frac{\Omega_T}{\Omega_{BZ}}\frac{\frac{1}{3}E_1^3\varepsilon_{31}-\frac{1}{4}E_1^4}{\varepsilon_{21}\varepsilon_{30}\varepsilon_{31}}\left(\frac{1}{\varepsilon_{30}}+\frac{1}{\varepsilon_{31}}\right)-\frac{\Omega_T}{\Omega_{BZ}}\frac{\varepsilon_{10}^2}{\varepsilon_{21}\varepsilon_{30}}\left[\frac{\varepsilon_{10}}{12}\frac{1}{\varepsilon_{30}}\right]$$

$$=\frac{\Omega_T}{\Omega_{BZ}}\frac{\frac{1}{3}E_0^3\varepsilon_{20}-\frac{1}{4}E_0^4}{\varepsilon_{20}\varepsilon_{21}\varepsilon_{30}^2}+\frac{\Omega_T}{\Omega_{BZ}}\frac{\frac{1}{2}E_1^2\varepsilon_{31}-\frac{1}{3}E_1^3}{\varepsilon_{21}\varepsilon_{30}\varepsilon_{31}}\frac{\varepsilon_{10}}{\varepsilon_{30}}$$

$$+\frac{\Omega_T}{\Omega_{BZ}}\frac{\frac{1}{3}E_1^3\varepsilon_{31}-\frac{1}{4}E_1^4}{\varepsilon_{21}\varepsilon_{30}\varepsilon_{31}}\left(\frac{1}{\varepsilon_{30}}+\frac{1}{\varepsilon_{31}}\right)-\frac{\Omega_T}{\Omega_{BZ}}\frac{1}{12}\frac{\varepsilon_{10}^3}{\varepsilon_{21}\varepsilon_{30}^2}. \tag{D.5.113}$$

4. $\varepsilon_2 \leq E \leq \varepsilon_3$

The general results obtained from (D.5.57) and (D.5.58) read

$$w_{j,T}^I(E) = \frac{\Omega_T}{\Omega_{BZ}}\frac{E_3^2}{\varepsilon_{30}\varepsilon_{31}\varepsilon_{32}}\left(3w_{j,3}+E_3\sum_{i=0}^{2}\frac{w_{j,i3}}{\varepsilon_{i3}}\right) \tag{D.5.114}$$

and

$$\int_{\varepsilon_0}^{E}dE'\,w_{j,T}^I(E') = \frac{\Omega_T}{\Omega_{BZ}}\left[\frac{1}{4}\sum_{i=0}^{3}w_{j,i}+\frac{E_3^3}{\varepsilon_{30}\varepsilon_{31}\varepsilon_{32}}\left(w_{j,3}+\frac{E_3}{4}\sum_{i=0}^{2}\frac{w_{j,i3}}{\varepsilon_{i3}}\right)\right]. \tag{D.5.115}$$

Specifically,

$$w_{j,T}^I(E) = -\frac{\Omega_T}{\Omega_{BZ}}\frac{E_3^3}{\varepsilon_{3j}\varepsilon_{30}\varepsilon_{31}\varepsilon_{32}}\qquad\text{for }j=0,1,2, \tag{D.5.116}$$

$$w_{3,T}^I(E) = \frac{\Omega_T}{\Omega_{BZ}}\frac{E_3^2}{\varepsilon_{30}\varepsilon_{31}\varepsilon_{32}}\left(3+E_3\sum_{i=0}^{2}\frac{1}{\varepsilon_{3i}}\right). \tag{D.5.117}$$

and

$$\int_{\varepsilon_0}^{E}dE'\,w_{j,T}^I(E') = \frac{\Omega_T}{4\Omega_{BZ}}\left[1-\frac{E_3^4}{\varepsilon_{3j}\varepsilon_{30}\varepsilon_{31}\varepsilon_{32}}\right]\qquad\text{for }j=0,1,2, \tag{D.5.118}$$

$$\int_{\varepsilon_0}^{E}dE'\,w_{3,T}^I(E') = \frac{\Omega_T}{4\Omega_{BZ}}\left[1+\frac{E_3^3}{\varepsilon_{30}\varepsilon_{31}\varepsilon_{32}}\left(4+E_3\sum_{i=0}^{2}\frac{1}{\varepsilon_{3i}}\right)\right]. \tag{D.5.119}$$

5. $\varepsilon_3 \leq E$

We have from (D.5.60) and (D.5.61)

$$w_{j,T}^I(E) = 0\qquad\text{for }j=0,1,2,3, \tag{D.5.120}$$

and

$$\int_{\varepsilon_0}^{E}dE'\,w_{j,T}^I(E') = \frac{\Omega_T}{4\Omega_{BZ}}\sum_{i=0}^{3}w_{j,i}$$

$$= \frac{\Omega_T}{4\Omega_{BZ}}\qquad\text{for }j=0,1,2,3. \tag{D.5.121}$$

As for the general formulas above, the results (D.5.86) and (D.5.87) as well as (D.5.116) and (D.5.117) are identical to those given in (B1) to (B4) of the paper by Lambin and Vigneron [11]. Moreover, (D.5.106) to (D.5.109) are equivalent to the respective results given in their paper.

Note that according to the sum rule (D.1.28) the sum of all four weights is identical to the contribution (D.5.62) of the respective tetrahedron to the density of states. An analogous relation holds for the integrated weights and the number of states. Both relations are easily checked for the five cases distinguished above, i.e. for the formulas (D.5.82) to (D.5.121) and the respective expressions (D.5.64) to (D.5.78) for the density of states and number of states. In computer codes, these identities can be used to check the correctness and numerical accuracy of the calculation of the weights and integrated weights.

In addition to the specific formulas, Lambin and Vigneron presented a general expression for the weights, namely,

$$
w_{j,T}^I(E) = \frac{\Omega_T}{\Omega_{BZ}} \frac{E_j^2}{\prod_{\substack{i=0 \\ i \neq j}}^3 \varepsilon_{ij}} \sum_{\substack{k=0 \\ k \neq j}}^3 \frac{E_k}{\varepsilon_{jk}} \Theta(E_j) + \frac{\Omega_T}{\Omega_{BZ}} \sum_{\substack{k=0 \\ k \neq j}}^3 \frac{E_k^3}{\prod_{\substack{i=0 \\ i \neq k}}^3 \varepsilon_{ik}} \frac{\Theta(E_k)}{\varepsilon_{jk}} , \quad (D.5.122)
$$

which, however, is of no use in numerical calculations. Nevertheless, it could be used as a starting point for the evaluation of the real part of the weights via the Kramers-Kronig relation. In the present context we will not deal with the evaluation of the real part but will just present the results in the form published by Lambin and Vigneron, i.e.

$$
w_{j,T}^R(E) = \frac{\Omega_T}{\Omega_{BZ}} \frac{E_j^2}{\prod_{\substack{i=0 \\ i \neq j}}^3 \varepsilon_{ij}} \left[1 + \sum_{\substack{k=0 \\ k \neq j}}^3 \frac{E_k}{\varepsilon_{jk}} \ln |E_j| \right] + \frac{\Omega_T}{\Omega_{BZ}} \sum_{\substack{k=0 \\ k \neq j}}^3 \frac{E_k^3}{\prod_{\substack{i=0 \\ i \neq k}}^3 \varepsilon_{ik}} \frac{\ln |E_k|}{\varepsilon_{jk}} .
$$
$$(D.5.123)$$

Specifically, this reads as

$$
w_{0,T}^R(E) = \frac{\Omega_T}{\Omega_{BZ}} \frac{E_0^2}{\varepsilon_{10}\varepsilon_{20}\varepsilon_{30}} \left[1 + \left(\frac{E_1}{\varepsilon_{01}} + \frac{E_2}{\varepsilon_{02}} + \frac{E_3}{\varepsilon_{03}} \right) \ln |E_0| \right]
$$
$$
+ \frac{\Omega_T}{\Omega_{BZ}} \left[\frac{E_1^3}{\varepsilon_{01}^2 \varepsilon_{21}\varepsilon_{31}} \ln |E_1| + \frac{E_2^3}{\varepsilon_{02}^2 \varepsilon_{12}\varepsilon_{32}} \ln |E_2| + \frac{E_3^3}{\varepsilon_{03}^2 \varepsilon_{13}\varepsilon_{23}} \ln |E_3| \right] ,
$$
$$(D.5.124)$$

$$
w_{1,T}^R(E) = \frac{\Omega_T}{\Omega_{BZ}} \frac{E_1^2}{\varepsilon_{01}\varepsilon_{21}\varepsilon_{31}} \left[1 + \left(\frac{E_0}{\varepsilon_{10}} + \frac{E_2}{\varepsilon_{12}} + \frac{E_3}{\varepsilon_{13}} \right) \ln |E_1| \right]
$$
$$
+ \frac{\Omega_T}{\Omega_{BZ}} \left[\frac{E_0^3}{\varepsilon_{10}^2 \varepsilon_{20}\varepsilon_{30}} \ln |E_0| + \frac{E_2^3}{\varepsilon_{02}\varepsilon_{12}^2 \varepsilon_{32}} \ln |E_2| + \frac{E_3^3}{\varepsilon_{03}\varepsilon_{13}^2 \varepsilon_{23}} \ln |E_3| \right] ,
$$
$$(D.5.125)$$

$$
w_{2,T}^R(E) = \frac{\Omega_T}{\Omega_{BZ}} \frac{E_2^2}{\varepsilon_{02}\varepsilon_{12}\varepsilon_{32}} \left[1 + \left(\frac{E_0}{\varepsilon_{20}} + \frac{E_1}{\varepsilon_{21}} + \frac{E_3}{\varepsilon_{23}} \right) \ln |E_2| \right]
$$

$$+\frac{\Omega_T}{\Omega_{BZ}}\left[\frac{E_0^3}{\varepsilon_{10}\varepsilon_{20}^2\varepsilon_{30}}\ln|E_0|+\frac{E_1^3}{\varepsilon_{01}\varepsilon_{21}^2\varepsilon_{31}}\ln|E_1|+\frac{E_3^3}{\varepsilon_{03}\varepsilon_{13}\varepsilon_{23}^2}\ln|E_3|\right],$$

$$\text{(D.5.126)}$$

$$w_{3,T}^R(E)=\frac{\Omega_T}{\Omega_{BZ}}\frac{E_3^2}{\varepsilon_{03}\varepsilon_{13}\varepsilon_{23}}\left[1+\left(\frac{E_0}{\varepsilon_{30}}+\frac{E_1}{\varepsilon_{31}}+\frac{E_2}{\varepsilon_{32}}\right)\ln|E_3|\right]$$

$$+\frac{\Omega_T}{\Omega_{BZ}}\left[\frac{E_0^3}{\varepsilon_{10}\varepsilon_{20}\varepsilon_{30}^2}\ln|E_0|+\frac{E_1^3}{\varepsilon_{01}\varepsilon_{21}\varepsilon_{31}^2}\ln|E_1|+\frac{E_2^3}{\varepsilon_{02}\varepsilon_{12}\varepsilon_{32}^2}\ln|E_2|\right].$$

$$\text{(D.5.127)}$$

Although generally valid these expressions are useful only when the four corner energies are all different. If this condition is not fulfilled, the following formulas should be used instead. Different cases are distinguished depending on how many corner energies are identical. Again, we recall the results published by Lambin and Vigneron without going into details. In doing so we still assume that the four corner energies are ordered according to $\varepsilon_0\le\varepsilon_1\le\varepsilon_2\le\varepsilon_3$.

1. $\varepsilon_k\ne\varepsilon_i=\varepsilon_j\ne\varepsilon_l,\ \varepsilon_k\ne\varepsilon_l$

   In this case, where i,j,k,l may refer to any of the four corners, we obtain

$$w_{i,T}^R(E)=w_{j,T}^R(E)$$

$$=\frac{\Omega_T}{\Omega_{BZ}}\sum_{m=k,l}\frac{E_m^3}{\varepsilon_{mn}\varepsilon_{mi}^3}\ln|E_m|$$

$$+\frac{\Omega_T}{\Omega_{BZ}}\frac{E_i}{\varepsilon_{ik}\varepsilon_{il}}\left\{\frac{1}{2}+\sum_{m=k,l}\frac{E_m}{\varepsilon_{im}}+\left[\sum_{m=k,l}\frac{E_m^2}{\varepsilon_{im}^2}+\frac{E_k}{\varepsilon_{ik}}\frac{E_l}{\varepsilon_{il}}\right]\ln|E_i|\right\}$$

$$\text{where } n\ne m, n\ne i, n\ne j\ .\qquad\text{(D.5.128)}$$

and

$$w_{m,T}^R(E)$$

$$=\frac{\Omega_T}{\Omega_{BZ}}\frac{E_m^2}{\varepsilon_{mi}^2\varepsilon_{nm}}\left\{1+\left[2\frac{E_i}{\varepsilon_{mi}}+\frac{E_n}{\varepsilon_{mn}}\right]\ln|E_m|\right\}$$

$$+\frac{\Omega_T}{\Omega_{BZ}}\frac{E_i^2}{\varepsilon_{im}^2\varepsilon_{ni}}\left\{1+\left[2\frac{E_m}{\varepsilon_{im}}+\frac{E_n}{\varepsilon_{in}}\right]\ln|E_i|\right\}+\frac{\Omega_T}{\Omega_{BZ}}\frac{E_n^3}{\varepsilon_{nm}^2\varepsilon_{ni}}\ln|E_n|$$

$$\text{for } m=k,l \text{ and } n\ne m, n\ne i, n\ne j\ .\qquad\text{(D.5.129)}$$

In particular, we distinguish the following specific situations.

a) $\varepsilon_0=\varepsilon_1<\varepsilon_2<\varepsilon_3$

$$w_{0,T}^R(E)=w_{1,T}^R(E)$$

$$=\frac{\Omega_T}{\Omega_{BZ}}\frac{E_2^3}{\varepsilon_{23}\varepsilon_{20}^3}\ln|E_2|+\frac{\Omega_T}{\Omega_{BZ}}\frac{E_3^3}{\varepsilon_{32}\varepsilon_{30}^3}\ln|E_3|$$

$$+\frac{\Omega_T}{\Omega_{BZ}}\frac{E_0}{\varepsilon_{02}\varepsilon_{03}}\left\{\frac{1}{2}+\frac{E_2}{\varepsilon_{02}}+\frac{E_3}{\varepsilon_{03}}+\left[\frac{E_2^2}{\varepsilon_{02}^2}+\frac{E_3^2}{\varepsilon_{03}^2}+\frac{E_2}{\varepsilon_{02}}\frac{E_3}{\varepsilon_{03}}\right]\ln|E_0|\right\},$$

$$\text{(D.5.130)}$$

$$w_{2,T}^R(E)$$
$$=\frac{\Omega_T}{\Omega_{BZ}}\frac{E_2^2}{\varepsilon_{20}^2\varepsilon_{32}}\left\{1+\left[2\frac{E_0}{\varepsilon_{20}}+\frac{E_3}{\varepsilon_{23}}\right]\ln|E_2|\right\}$$
$$+\frac{\Omega_T}{\Omega_{BZ}}\frac{E_0^2}{\varepsilon_{02}^2\varepsilon_{30}}\left\{1+\left[2\frac{E_2}{\varepsilon_{02}}+\frac{E_3}{\varepsilon_{03}}\right]\ln|E_0|\right\}+\frac{\Omega_T}{\Omega_{BZ}}\frac{E_3^3}{\varepsilon_{32}^2\varepsilon_{30}^2}\ln|E_3|,$$

$$\text{(D.5.131)}$$

$$w_{3,T}^R(E)$$
$$=\frac{\Omega_T}{\Omega_{BZ}}\frac{E_3^2}{\varepsilon_{30}^2\varepsilon_{23}}\left\{1+\left[2\frac{E_0}{\varepsilon_{30}}+\frac{E_2}{\varepsilon_{32}}\right]\ln|E_3|\right\}$$
$$+\frac{\Omega_T}{\Omega_{BZ}}\frac{E_0^2}{\varepsilon_{03}^2\varepsilon_{20}}\left\{1+\left[2\frac{E_3}{\varepsilon_{03}}+\frac{E_2}{\varepsilon_{02}}\right]\ln|E_0|\right\}+\frac{\Omega_T}{\Omega_{BZ}}\frac{E_2^3}{\varepsilon_{23}^2\varepsilon_{20}^2}\ln|E_2|.$$

$$\text{(D.5.132)}$$

b) $\varepsilon_0<\varepsilon_1=\varepsilon_2<\varepsilon_3$

$$w_{0,T}^R(E)$$
$$=\frac{\Omega_T}{\Omega_{BZ}}\frac{E_0^2}{\varepsilon_{01}^2\varepsilon_{30}}\left\{1+\left[2\frac{E_1}{\varepsilon_{01}}+\frac{E_3}{\varepsilon_{03}}\right]\ln|E_0|\right\}$$
$$+\frac{\Omega_T}{\Omega_{BZ}}\frac{E_1^2}{\varepsilon_{10}^2\varepsilon_{31}}\left\{1+\left[2\frac{E_0}{\varepsilon_{10}}+\frac{E_3}{\varepsilon_{13}}\right]\ln|E_1|\right\}+\frac{\Omega_T}{\Omega_{BZ}}\frac{E_3^3}{\varepsilon_{30}^2\varepsilon_{31}^2}\ln|E_3|,$$

$$\text{(D.5.133)}$$

$$w_{1,T}^R(E)=w_{2,T}^R(E)$$
$$=\frac{\Omega_T}{\Omega_{BZ}}\frac{E_0^3}{\varepsilon_{03}\varepsilon_{01}^3}\ln|E_0|+\frac{\Omega_T}{\Omega_{BZ}}\frac{E_3^3}{\varepsilon_{30}\varepsilon_{31}^3}\ln|E_3|$$
$$+\frac{\Omega_T}{\Omega_{BZ}}\frac{E_1}{\varepsilon_{10}\varepsilon_{13}}\left\{\frac{1}{2}+\frac{E_0}{\varepsilon_{10}}+\frac{E_3}{\varepsilon_{13}}+\left[\frac{E_0^2}{\varepsilon_{10}^2}+\frac{E_3^2}{\varepsilon_{13}^2}+\frac{E_0}{\varepsilon_{10}}\frac{E_3}{\varepsilon_{13}}\right]\ln|E_1|\right\},$$

$$\text{(D.5.134)}$$

$$w_{3,T}^R(E)$$
$$=\frac{\Omega_T}{\Omega_{BZ}}\frac{E_3^2}{\varepsilon_{31}^2\varepsilon_{03}}\left\{1+\left[2\frac{E_1}{\varepsilon_{31}}+\frac{E_0}{\varepsilon_{30}}\right]\ln|E_3|\right\}$$
$$+\frac{\Omega_T}{\Omega_{BZ}}\frac{E_1^2}{\varepsilon_{13}^2\varepsilon_{01}}\left\{1+\left[2\frac{E_3}{\varepsilon_{13}}+\frac{E_0}{\varepsilon_{10}}\right]\ln|E_1|\right\}+\frac{\Omega_T}{\Omega_{BZ}}\frac{E_0^3}{\varepsilon_{03}^2\varepsilon_{01}^2}\ln|E_0|.$$

$$\text{(D.5.135)}$$

c) $\varepsilon_0<\varepsilon_1<\varepsilon_2=\varepsilon_3$

$$w_{0,T}^R(E)$$

$$= \frac{\Omega_T}{\Omega_{BZ}} \frac{E_0^2}{\varepsilon_{02}^2 \varepsilon_{10}} \left\{ 1 + \left[2\frac{E_2}{\varepsilon_{02}} + \frac{E_1}{\varepsilon_{01}} \right] \ln|E_0| \right\}$$

$$+ \frac{\Omega_T}{\Omega_{BZ}} \frac{E_2^2}{\varepsilon_{20}^2 \varepsilon_{12}} \left\{ 1 + \left[2\frac{E_0}{\varepsilon_{20}} + \frac{E_1}{\varepsilon_{21}} \right] \ln|E_2| \right\} + \frac{\Omega_T}{\Omega_{BZ}} \frac{E_1^3}{\varepsilon_{10}^2 \varepsilon_{12}^2} \ln|E_1| \, ,$$

$$(D.5.136)$$

$$w_{1,T}^R(E)$$

$$= \frac{\Omega_T}{\Omega_{BZ}} \frac{E_1^2}{\varepsilon_{12}^2 \varepsilon_{01}} \left\{ 1 + \left[2\frac{E_2}{\varepsilon_{12}} + \frac{E_0}{\varepsilon_{10}} \right] \ln|E_1| \right\}$$

$$+ \frac{\Omega_T}{\Omega_{BZ}} \frac{E_2^2}{\varepsilon_{21}^2 \varepsilon_{02}} \left\{ 1 + \left[2\frac{E_1}{\varepsilon_{21}} + \frac{E_0}{\varepsilon_{20}} \right] \ln|E_2| \right\} + \frac{\Omega_T}{\Omega_{BZ}} \frac{E_0^3}{\varepsilon_{01}^2 \varepsilon_{02}^2} \ln|E_0| \, ,$$

$$(D.5.137)$$

$$w_{2,T}^R(E) = w_{3,T}^R(E)$$

$$= \frac{\Omega_T}{\Omega_{BZ}} \frac{E_0^3}{\varepsilon_{01} \varepsilon_{02}^3} \ln|E_0| + \frac{\Omega_T}{\Omega_{BZ}} \frac{E_1^3}{\varepsilon_{10} \varepsilon_{12}^3} \ln|E_1|$$

$$+ \frac{\Omega_T}{\Omega_{BZ}} \frac{E_2}{\varepsilon_{20} \varepsilon_{21}} \left\{ \frac{1}{2} + \frac{E_0}{\varepsilon_{20}} + \frac{E_1}{\varepsilon_{21}} + \left[\frac{E_0^2}{\varepsilon_{20}^2} + \frac{E_1^2}{\varepsilon_{21}^2} + \frac{E_0}{\varepsilon_{20}} \frac{E_1}{\varepsilon_{21}} \right] \ln|E_2| \right\} .$$

$$(D.5.138)$$

2. $\varepsilon_0 = \varepsilon_1 = \varepsilon_2 < \varepsilon_3$

   Again, we distinguish two different formulas,

$$w_{0,T}^R(E) = w_{1,T}^R(E) = w_{2,T}^R(E)$$

$$= \frac{\Omega_T}{\Omega_{BZ}} \frac{E_3^3}{\varepsilon_{30}^4} \ln\left|\frac{E_3}{E_0}\right| + \frac{\Omega_T}{\Omega_{BZ}} \frac{6E_3^2 - 3E_3\varepsilon_{30} + 2\varepsilon_{30}^2}{6\varepsilon_{30}^3} \, , \qquad (D.5.139)$$

$$w_{3,T}^R(E) = \frac{\Omega_T}{\Omega_{BZ}} 3\frac{E_0 E_3^2}{\varepsilon_{30}^4} \ln\left|\frac{E_0}{E_3}\right| - \frac{\Omega_T}{\Omega_{BZ}} \frac{3}{2} E_0 \frac{2E_3 - \varepsilon_{30}}{\varepsilon_{30}^3} - \frac{\Omega_T}{\Omega_{BZ}} \frac{1}{\varepsilon_{30}} \, .$$

$$(D.5.140)$$

3. $\varepsilon_0 = \varepsilon_1 < \varepsilon_2 = \varepsilon_3$

   For the symmetric case we have

$$w_{0,T}^R(E) = w_{1,T}^R(E)$$

$$= \frac{\Omega_T}{\Omega_{BZ}} 3\frac{E_0 E_2^2}{\varepsilon_{20}^4} \ln\left|\frac{E_0}{E_2}\right| - \frac{\Omega_T}{\Omega_{BZ}} \frac{3}{2} E_0 \frac{2E_2 - \varepsilon_{20}}{\varepsilon_{20}^3} - \frac{\Omega_T}{\Omega_{BZ}} \frac{1}{\varepsilon_{20}} \, ,$$

$$(D.5.141)$$

$$w_{2,T}^R(E) = w_{3,T}^R(E)$$

$$= \frac{\Omega_T}{\Omega_{BZ}} 3\frac{E_0^2 E_2}{\varepsilon_{20}^4} \ln\left|\frac{E_2}{E_0}\right| + \frac{\Omega_T}{\Omega_{BZ}} \frac{3}{2} E_2 \frac{2E_0 + \varepsilon_{20}}{\varepsilon_{20}^3} + \frac{\Omega_T}{\Omega_{BZ}} \frac{1}{\varepsilon_{20}} \, .$$

$$(D.5.142)$$

4. $\varepsilon_0 < \varepsilon_1 = \varepsilon_2 = \varepsilon_3$

For this case we note

$$w_{0,T}^{R}(E) = \frac{\Omega_T}{\Omega_{BZ}} \, 3 \frac{E_0^2 E_1}{\varepsilon_{10}^4} \ln\left|\frac{E_1}{E_0}\right| + \frac{\Omega_T}{\Omega_{BZ}} \frac{3}{2} E_1 \frac{2E_0 + \varepsilon_{10}}{\varepsilon_{10}^3} + \frac{\Omega_T}{\Omega_{BZ}} \frac{1}{\varepsilon_{10}},$$
(D.5.143)

$$w_{1,T}^{R}(E) = w_{2,T}^{R}(E) = w_{3,T}^{R}(E)$$
$$= \frac{\Omega_T}{\Omega_{BZ}} \frac{E_0^3}{\varepsilon_{10}^4} \ln\left|\frac{E_0}{E_1}\right| - \frac{\Omega_T}{\Omega_{BZ}} \frac{6E_0^2 + 3E_0\varepsilon_{10} + 2\varepsilon_{10}^2}{6\varepsilon_{10}^3}.$$
(D.5.144)

5. $\varepsilon_0 = \varepsilon_1 = \varepsilon_2 = \varepsilon_3$

Finally, we write

$$w_{0,T}^{R}(E) = w_{1,T}^{R}(E) = w_{2,T}^{R}(E) = w_{3,T}^{R}(E) = \frac{\Omega_T}{\Omega_{BZ}} \frac{1}{4E_0}.$$
(D.5.145)

Here, we have corrected for typos in (A4) and (A5) of the paper by Lambin and Vigneron and added plus and minus signs in front of the $\frac{3}{2}$-factor in (D.5.141) and (D.5.142)/(D.5.143), respectively. The previous formulas are identical to those given by Oppeneer and Lodder [16]. With (D.5.122) to (D.5.145) at hand we are in a position to calculate both the real and imaginary part of the Green's function without making use of the Kramers-Kronig relation.

Using the formulas derived in this section as well as the regular arrangement of the tetrahedra as described at the beginning one is able to reduce the char-

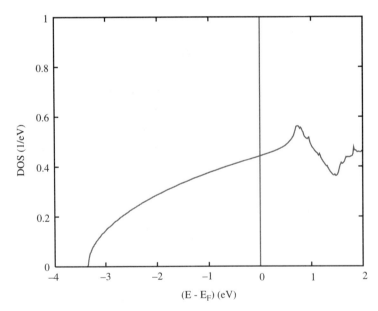

Fig. D.7. Density of states of Na as calculated with the linear tetrahedron method with $30 \times 30 \times 30$ **k**-points

acteristic error of the linear tetrahedron method from Δ^2 known from the classical arrangement of tetrahedra in the irreducible part of the Brillouin zone to $\exp(-\frac{1}{\Delta})$ [4, 5]. Here, Δ denotes the mean distance of the **k**-points in the discrete mesh. However, while this finding holds for semiconductors and insulators, for metals still the Δ^2-behaviour is observed. This situation is dealt with in the subsequent section.

Finally, the superiority of the linear tetrahedron method is clearly revealed by Fig. D.7, which displays the density of states of Na as calculated using the linear tetrahedron method with the same number of **k**-points as used in Figs. D.1 and D.3. Again, we observe considerable progress. While the high-precision sampling method was able to suppress the statistical noise of the simple sampling method, both the broadening and the oscillations observed in Fig. D.3 have now vanished.

D.6 Higher-Order Corrections

In an attempt to improve the linear tetrahedron method, Blöchl was able to assign the Δ^2-error resulting for metals to the linearization inside the tetrahedra and the finite filling of the bands [4, 5]. This is easily explained with the help of Fig. D.8, which shows a typical band dispersion. Obviously, the linear approximation places the interpolated band above the exact band in the low-energy region. In contrast, at higher energies, where the curvature of the band is opposite, the interpolation lies below the exact band. While for filled bands as seen in semiconductors and insulators both errors cancel to a high degree, the error in the lower portion of the band is left uncompensated in metals causing the typical Δ^2-behaviour. Note

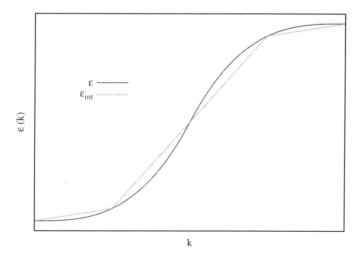

Fig. D.8. Schematic representation of the error introduced by linear interpolation. Exact and linearly interpolated bands are indicated by ε and ε_{int}, respectively (after [5])

that this is just the leading contribution to the error. While concentrating on this term, Blöchl was able to derive a rather simple correction formula, which removed the Δ^2-contribution and resulted in a much faster convergence of the method [4,5]. Without going into details we just give the correction formula as

$$\Delta w_{j,T}^I(E) = \frac{1}{40}\frac{d\rho_T(E)}{dE}\sum_{i=0}^{3}\varepsilon_{ij} , \qquad (D.6.1)$$

$$\int_{\varepsilon_0}^{E} dE' \, \Delta w_{j,T}^I(E') = \frac{1}{40}\rho_T(E)\sum_{i=0}^{3}\varepsilon_{ij} , \qquad (D.6.2)$$

where $j = 0, 1, 2, 3$. In this context it is important to note that the sum of all four corrections in each tetrahedron vanishes,

$$\sum_{j=0}^{3}\Delta w_{j,T}^I(E) = \frac{1}{40}\frac{d\rho_T(E)}{dE}\sum_{j=0}^{3}\sum_{i=0}^{3}\varepsilon_{ij} \overset{!}{=} 0 , \qquad (D.6.3)$$

$$\sum_{j=0}^{3}\int_{\varepsilon_0}^{E} dE' \, \Delta w_{j,T}^I(E') = \frac{1}{40}\rho_T(E)\sum_{j=0}^{3}\sum_{i=0}^{3}\varepsilon_{ij} \overset{!}{=} 0 . \qquad (D.6.4)$$

As a consequence, since, as pointed out at the end of the previous section, the sum of the weights and integrated weights leads to the density of states and number of states, respectively, the corrections (D.6.1) and (D.6.2), which usually go under the name Blöchl corrections, do not affect these latter quantities. In contrast, they introduce only a redistribution of the weights in each tetrahedron.

Of course, we may again distinguish the five cases corresponding to different positions of the energy relative to the corner energies. However, this merely consists of inserting the respective expressions for the density of states, (D.5.64), (D.5.66), (D.5.68), (D.5.71), (D.5.73), (D.5.75), and (D.5.77) as well as their energy derivatives into (D.6.1) and (D.6.2) and, hence, will not be explicited here.

References

1. M. Abramowitz and I. A. Stegun, *Handbook of Mathematical Functions* (Dover, New York 1972)
2. N. W. Ashcroft and N. D. Mermin, *Solid State Physics* (Holt-Saunders, Philadelphia 1976)
3. A. Baldereschi, Phys. Rev. B **7**, 5212 (1973)
4. P. E. Blöchl, Gesamtenergien, Kräfte und Metall-Halbleiter Grenzflächen. PhD thesis, Universität Stuttgart (1989)
5. P. E. Blöchl, O. Jepsen, and O. K. Andersen, Phys. Rev. B **49**, 16223 (1994)
6. D. J. Chadi and M. L. Cohen, Phys. Rev. B **8**, 5747 (1973)
7. J. Hama and M. Watanabe, J. Phys.: Cond. Matt. **4**, 4583 (1992)
8. O. Jepsen and O. K. Andersen, Solid State Commun. **9**, 1763 (1971)
9. O. Jepsen and O. K. Andersen, Phys. Rev. B **29**, 5965 (1984)

10. L. Kleinman, Phys. Rev. B **28**, 1139 (1983)
11. P. Lambin and J. P. Vigneron, Phys. Rev. B **29**, 3430 (1984)
12. G. Lehmann, P. Rennert, M. Taut, and H. Wonn, phys. stat. sol. **37**, K27 (1970)
13. G. Lehmann and M. Taut, phys. stat. sol. (b) **54**, 469 (1972)
14. M. S. Methfessel and A. T. Paxton, Phys. Rev. B **40**, 3616 (1989)
15. H. J. Monkhorst and J. D. Pack, Phys. Rev. B **13**, 5188 (1976)
16. P. M. Oppeneer and A. Lodder, J. Phys. F **17**, 1885 (1987)

Further Reading

D. E. Amos, ACM Trans. Math. Software **6**, 365 (1980)

D. E. Amos, ACM Trans. Math. Software **6**, 420 (1980)

D. E. Amos, ACM Trans. Math. Software **9**, 467 (1983)

D. E. Amos, ACM Trans. Math. Software **9**, 525 (1983)

O. K. Andersen and O. Jepsen, Phys. Rev. Lett. **53**, 2571 (1984)

O. K. Andersen, O. Jepsen, and G. Krier, Exact Muffin-Tin Orbital Theory. In: *Methods of Electronic Structure Calculations*, ed by V. Kumar, O. K. Andersen, and A. Mookerjee (World Scientific, Singapore 1994) pp 63–124

O. K. Andersen and R. V. Kasowski, Phys. Rev. B **4**, 1064 (1971)

O. K. Andersen and R. V. Kasowski, Solid State Commun. **11**, 799 (1972)

O. K. Andersen, Z. Pawlowska, and O. Jepsen, Phys. Rev. B **34**, 5253 (1986)

O. K. Andersen, H. L. Skriver, H. Nohl, and B. Johansson, Pure Appl. Chem. **52**, 93 (1979)

O. K. Andersen and R. G. Woolley, Mol. Phys. **26**, 905 (1973)

R. F. W. Bader and P. M. Beddall, J. Chem. Phys. **56**, 3320 (1972)

U. von Barth, Different Approximations Within Density-Functional Theory, Their Advantages and Limitations. In: *Methods of Electronic Structure Calculations*, ed by V. Kumar, O. K. Andersen, and A. Mookerjee (World Scientific, Singapore 1994) pp 21–62

U. von Barth and A. Pedroza, Phys. Scr. **32**, 353 (1985)

A. D. Becke, Phys. Rev. A **38**, 3098 (1988)

A. D. Becke, J. Chem. Phys. **98**, 5648 (1993)

A. D. Becke and K. E. Edgecombe, J. Chem. Phys. **92**, 5397 (1990)

C. L. Berman and L. Greengard, J. Math. Phys. **35**, 6036 (1994)

G. Bester and M. Fähnle, J. Phys.: Cond. Matt. **13**, 11541 (2001)

M. A. Blanco, M. Flórez, and M. Bermejo, J. Mol. Struct. (Theochem) **419**, 19 (1997)

F. Bloch, Z. Phys. **52**, 555 (1929)

P. E. Blöchl, Phys. Rev. B **50**, 17953 (1994)

M. Born and R. Oppenheimer, Ann. Phys. (Leipzig) **84**, 457 (1927)

N. Börnsen, G. Bester, B. Meyer, and M. Fähnle, J. Alloys Comp. **308**, 1 (2000)

N. Börnsen, B. Meyer, O. Grotheer, and M. Fähnle, J. Phys.: Cond. Matt. **11**, L287 (1999)

L. P. Bouckaert, R. Smoluchowski, and E. P. Wigner, Phys. Rev. **50**, 58 (1936)

C. J. Bradley and A. P. Cracknell, *The Mathematical Theory of Symmetry in Solids* (Clarendon Press, Oxford 1972)

R. P. Brent, *Algorithms for Minimization without Derivatives* (Prentice Hall, Englewood Cliffs 1973)

M. S. S. Brooks and P. J. Kelly, Phys. Rev. Lett. **51**, 1708 (1983)

K. Burke, J. P. Perdew, and M. Ernzerhof, J. Chem. Phys. **109**, 3760 (1998)

G. D. Byrne and C. A. Hall, *Numerical Solution of Systems of Nonlinear Algebraic Equations* (Academic Press, New York, London 1973)

J. Callaway and N. H. March, Density-Functional Methods: Theory and Applications. In: *Solid State Physics* vol 38, ed by F. Seitz, D. Turnbull, and H. Ehrenreich (Academic Press, Orlando 1983) pp 136–223

R. Car and M. Parrinello, Phys. Rev. Lett. **55**, 2471 (1985)

R. Car and M. Parrinello, Phys. Rev. Lett. **60**, 204 (1988)

F. Casula and F. Herman, J. Chem. Phys. **78**, 858 (1983)

D. M. Ceperley, Phys. Rev. B **18**, 3126 (1978)

D. M. Ceperley and B. J. Alder, Phys. Rev. Lett. **45**, 566 (1980)

D. J. Chadi, Phys. Rev. B **16**, 1746 (1977)

J. P. A. Charlesworth and W. Yeung, Comput. Phys. Commun. **88**, 186 (1995)

J. Chen, J. B. Krieger, Y. Li, and G. J. Iafrate, Phys. Rev. A **54**, 3939 (1996)

C. Chiccoli, S. Lorenzutta, and G. Maino, J. Comput. Phys. **78**, 278 (1988)

C. Chiccoli, S. Lorenzutta, and G. Maino, Computing **45**, 269 (1990)

C. H. Choi, J. Ivanic, M. S. Gordon, and K. Rüdenberg, J. Chem. Phys. **111**, 8825 (1999)

H. J. H. Clercx and P. P. J. M. Schram, J. Math. Phys. **34**, 5292 (1993)

M. H. Cohen and F. Reif, Quadrupole Effects in Nuclear Magnetic Resonance Studies of Solids. In: *Solid State Physics*, vol 5, ed by F. Seitz and D. Turnbull (Academic Press, Orlando 1970) pp 321–438

P. T. Coleridge, J. Molenaar, and A. Lodder, J. Phys. C **15**, 6943 (1982)

P. Cortona, S. Doniach, and C. Sommers, Phys. Rev. B **31**, 2842 (1985)

M. Danos and L. C. Maximon, J. Math. Phys. **6**, 766 (1965)

H.-Q. Ding, N. Karasawa, and W. A. Goddard, Chem. Phys. Lett. **196**, 6 (1992)

J. F. Dobson, J. Chem. Phys. **94**, 4328 (1991)

R. M. Dreizler and J. de Providência, *Density-Functional Methods in Physics* (Plenum Press, New York 1985)

R. Dronskowski and P. E. Blöchl, J. Phys. Chem. **97**, 8617 (1993)

N. Elyashar and D. D. Koelling, Phys. Rev. B **13**, 5362 (1976)

E. Engel and S. H. Vosko, Phys. Rev. B **47**, 2800 (1993)

E. Engel and S. H. Vosko, Phys. Rev. B **50**, 10498 (1994)

P. P. Ewald, Ann. Phys. **49**, 1 (1916)

V. Eyert, Entwicklung und Implementation eines Full-Potential-ASW-Verfahrens. PhD thesis, Technische Hochschule Darmstadt (1991)

V. Eyert, Electronic structure calculations for crystalline materials. In: *Density-Functional Methods: Applications in Chemistry and Materials Science*, ed by M. Springborg (Wiley, Chichester 1997) pp 233–304

V. Eyert, Octahedral Deformations and Metal-Insulator Transition in Transition Metal Chalcogenides. Habilitation thesis, University of Augsburg (1998)

R. Feder, F. Rosicky, and B. Ackermann, Z. Phys. **52**, 31 (1983)

B. U. Felderhof and R. B. Jones, J. Math. Phys. **28**, 836 (1987)

R. P. Feynman, Phys. Rev. **56**, 340 (1939)

G. E. Forsythe and C. B. Moler, *Computer Solution of Linear Algebraic Systems* (Prentice Hall, Englewood Cliffs 1967); *Computer-Verfahren für lineare algebraische Systeme* (Oldenbourg, München 1971)

D. L. Foulis, J. Math. Phys. **34**, 2004 (1993)

L. Fritsche and J. Yuan, Phys. Rev. A **57**, 3425 (1998)

S. Froyen, Phys. Rev. B **39**, 3168 (1989)

K. Fuchs, Proc. Roy. Soc. A **151**, 585 (1935)

R. Gaspár, Acta Phys. Hung. **3**, 263 (1954)

W. Gautschi, ACM Trans. Math. Software **5**, 466 (1979)

W. Gautschi, ACM Trans. Math. Software **5**, 482 (1979)

D. K. Ghosh, Phys. Rev. Lett. **27**, 1584 (1971)

G. Gilat and N. R. Bharatiya, Phys. Rev. B **12**, 3479 (1975)

S. Goedecker, Phys. Rev. B **48**, 17573 (1993)

D. E. Goldberg, *Genetic Algorithms in Search, Optimization, and Machine Learning* (Addison-Wesley, Reading 1989)

H. Gollisch and L. Fritsche, phys. stat. sol. (b) **86**, 145 (1978)

A. Gonis, Phys. Rev. B **33**, 5914 (1986)

A. Görling, Phys. Rev. A **47**, 2783 (1993)

A. Görling, Phys. Rev. B **53**, 7024 (1996)

A. Görling and M. Levy, Phys. Rev. A **50**, 196 (1994)

O. Gunnarsson, J. Phys. F **6**, 587 (1976)

O. Gunnarsson, J. Harris, and R. O. Jones, Phys. Rev. B **15**, 3027 (1977)

O. Gunnarsson, J. Harris, and R. O. Jones, J. Chem. Phys. **67**, 3970 (1977)

O. Gunnarsson and R. O. Jones, Phys. Rev. B **31**, 7588 (1985)

O. Gunnarsson and B. I. Lundqvist, Phys. Rev. B **13**, 4274 (1976)

O. Gunnarsson, B. I. Lundqvist, and J. W. Wilkins, Phys. Rev. B **10**, 1319 (1974)

D. Hackenbracht, Berechnete elektronische und thermomechanische Eigenschaften einiger La-In und Al-Ni - Verbindungen. Diploma thesis, Ruhr-Universität Bochum (1979)

J. Hama, M. Watanabe, and T. Kato, J. Phys.: Cond. Matt. **2**, 7445 (1990)

A. Haug, *Theoretical Solid State Physics* (Pergamon Press, New York 1972)

J. Harris, Density-Functional Calculations for Atomic Clusters. In: *The Electronic Structure of Complex Systems*, ed by P. Phariseau and W. Temmerman (Plenum Press, New York 1984) pp 141–182

J. Harris, Phys. Rev. B **31**, 1770 (1985)

J. Harris and G. S. Painter, Phys. Rev. B **22**, 2614 (1980)

W. A. Harrison, *Electronic Structure and the Properties of Solids* (Freeman, San Fransisco 1980)

L. Hedin, Phys. Rev. **139**, A796 (1965)

V. Heine, Electronic Structure from the Point of View of the Local Atomic Environment. In: *Solid State Physics*, vol 35, ed by H. Ehrenreich, F. Seitz, and D. Turnbull (Academic Press, New York 1980) pp 1–127

D. M. Heyes, Phys. Lett. A **187**, 273 (1994)

R. Hoffmann, *Solids and Surfaces: A Chemist's View of Bonding in Extended Structures* (VCH Verlagsgesellschaft, New York 1988)

G. Hummer, Chem. Phys. Lett. **235**, 297 (1995)

J. Ivanic and K. Rüdenberg, J. Phys. Chem. **100**, 6342 (1996)

D. A. H. Jacobs, *The State of the Art in Numerical Analysis* (Academic Press, London 1977)

J. F. Janak, Phys. Rev. B **9**, 3985 (1974)

O. Jepsen, J. Madsen, and O. K. Andersen, Phys. Rev. B **26**, 2790 (1982)

R. O. Jones and O. Gunnarsson, Rev. Mod. Phys. **61**, 689 (1989)

R. Jones and A. Sayyash, J. Phys. C **19**, L653 (1986)

W. Jones and N. H. March, *Theoretical Solid State Physics* (Wiley, London 1973)

S. Kaprzyk and P. E. Mijnarends, J. Phys. C **19**, 1283 (1986)

R. V. Kasowski, Phys. Rev. B **25**, 4189 (1982)

R. V. Kasowski, M.-H. Tsai, T. N. Rhodin, and D. D. Chambliss, Phys. Rev. B **34**, 2656 (1986)

J. Keller, J. Phys. C **4**, L85 (1971)

C. Kittel, *Introduction to Solid State Physics* (Wiley, New York 1986)

C. Kittel, *Quantum Theory of Solids* (Wiley, New York, 1987)

J. Köhler, Polarer, magneto-optischer Kerr-Effekt in kollinear und nicht-kollinear magnetischen, metallischen Verbindungen. PhD thesis, Technische Universität Darmstadt (1998)

D. D. Koelling, J. Comput. Phys. **67**, 253 (1986)

D. D. Koelling and G. O. Arbman, J. Phys. F **5**, 2041 (1975)

D. D. Koelling and B. N. Harmon, J. Phys. C **10**, 3107 (1977)

J. Korringa, Phys. Rep. **238**, 341 (1994)

G. F. Koster, Space Groups and Their Representations. In: *Solid State Physics*, vol 5, ed by F. Seitz and D. Turnbull (Academic Press, Orlando 1970) pp 173–256

T. Kotani, Phys. Rev. B **50**, 14816 (1994)

T. Kotani, Phys. Rev. Lett. **74**, 2989 (1995)

T. Kotani and H. Akai, Phys. Rev. B **52**, 17153 (1995)

T. Kotani and H. Akai, Phys. Rev. B **54**, 16502 (1996)

O. V. Kovalev, *Irreducible Representations of Space Groups* (Gordon and Breach, New York 1965)

A. Kratzer and W. Franz, *Transzendente Funktionen* (Akademische Verlagsgesellschaft, Leipzig 1960)

J. B. Krieger, Y. Li, and G. J. Iafrate, Phys. Lett. A **146**, 256 (1990)

J. B. Krieger, Y. Li, and G. J. Iafrate, Phys. Lett. A **148**, 470 (1990)

J. B. Krieger, Y. Li, and G. J. Iafrate, Phys. Rev. A **45**, 101 (1992)

J. B. Krieger, Y. Li, and G. J. Iafrate, Phys. Rev. A **46**, 5453 (1992)

B. C. H. Krutzen and F. Springelkamp, J. Phys.: Cond. Matt. **1**, 8369 (1989)

E. S. Kryachko, E. V. Ludeña, *Energy Density-Functional Theory of Many-Electron Systems* (Kluwer Academic Publishers, Dordrecht 1990)

R. Kutteh, E. Aprà, and J. Nichols, Chem. Phys. Lett. **238**, 173 (1995)

P. Lambin and P. Senet, Intern. J. Quant. Chem. **46**, 101 (1993)

W. R. L. Lambrecht and O. K. Andersen, Phys. Rev. B **34**, 2439 (1986)

G. Lehmann, phys. stat. sol. **38**, 151 (1970)

H. van Leuken, Electronic Structure of Metallic Multilayers. PhD thesis, University of Amsterdam (1991)

M. Levy, Proc. Natl. Acad. Sci. (USA) **76**, 6062 (1979)

M. Levy, Phys. Rev. A **26**, 1200 (1982)

M. Levy and J. P. Perdew, Int. J. Quant. Chem. **49**, 539 (1994)

Y. Li, J. B. Krieger, M. R. Norman, and G. J. Iafrate, Phys. Rev. B **44**, 10437 (1991)

E. H. Lieb, Intern. J. Quant. Chem. **24**, 243 (1983)

W. Ludwig and C. Falter, *Symmetries in Physics* (Springer, Berlin 1996)

S. Lundqvist and N. H. March, *Theory of the Inhomogeneous Electron Gas* (Plenum Press, New York 1983)

A. H. MacDonald, J. Phys. C **16**, 3869 (1983)

A. H. MacDonald, W. E. Pickett, and D. D. Koelling, J. Phys. C **13**, 2675 (1980)

A. H. MacDonald and S. H. Vosko, J. Phys. C **12**, 2977 (1979)

A. H. MacDonald, S. H. Vosko, and P. T. Coleridge, J. Phys. C **12**, 2991 (1979)

A. R. Mackintosh and O. K. Andersen, The electronic structure of transition metals In: *Electrons at the Fermi Surface*, ed by M. Springford (University Press, Cambridge 1980) pp 149–224

J. M. MacLaren, D. P. Clougherty, and R. C. Albers, Phys. Rev. B **42**, 3205 (1990)

W. Magnus, F. Oberhettinger, and R. P. Soni, *Formulas and Theorems for the Special Functions of Mathematical Physics* (Springer, Berlin 1966)

L. F. Mattheiss and D. R. Hamann, Phys. Rev. B **33**, 823 (1986)

M. S. Methfessel, Phys. Rev. B **38**, 1537 (1988)

M. S. Methfessel and J. Kübler, J. Phys. F **12**, 141 (1982)

M. S. Methfessel, C. O. Rodriguez, and O. K. Andersen, Phys. Rev. B **40**, 2009 (1989)

M. S. Methfessel and M. van Schilfgaarde, Phys. Rev. B **48**, 4937 (1993)

M. S. Methfessel, M. van Schilfgaarde, and R. A. Casali, A full-Potential LMTO Method Based on Smooth Hankel Functions. In: *Electronic Structure and Physical Properties of Solids. The Uses of the LMTO Method*, ed by H. Dreyssé (Springer, Berlin, Heidelberg, 2000) pp 114–147

M. S. Methfessel, M. van Schilfgaarde, and M. Scheffler, Phys. Rev. Lett. **70**, 29 (1993)

M. S. Milgram, Math. Comput. **44**, 443 (1985)

V. L. Moruzzi, J. F. Janak, and A. R. Williams, *Calculated Electronic Properties of Metals* (Pergamon Press, New York 1978)

Á. Nagy, Phys. Rev. A **55**, 3465 (1997)

V. Natoli and D. M. Ceperley, J. Comput. Phys. **117**, 171 (1995)

W. Nolting, *Grundkurs: Theoretische Physik*, vol 3: *Elektrodynamik* (Springer, Berlin 2004)

H. J. Nowak, O. K. Andersen, T. Fujiwara, O. Jepsen, and P. Vargas, Phys. Rev. B **44**, 3577 (1991)

R. Nozawa, J. Math. Phys. **7**, 1841 (1966)

J. M. Ortega and W. C. Rheinboldt, *Iterative Solution of Nonlinear Equations in Several Variables* (Academic Press, New York 1970)

H. Ou-Yang and M. Levy, Phys. Rev. Lett. **65**, 1036 (1994)

G. S. Painter, Phys. Rev. B **24**, 4264 (1981)

D. A. Papaconstantopoulos, *Handbook of the Band Structure of Elemental Solids* (Plenum Press, New York 1986)

J. P. Perdew, Phys. Rev. B **33**, 8822 (1986)

J. P. Perdew, Physica B **172**, 1 (1991)

J. P. Perdew, K. Burke, and M. Ernzerhof, Phys. Rev. Lett. **80**, 891 (1998)

J. P. Perdew and M. Levy, Phys. Rev. Lett. **51**, 1884 (1983)

J. P. Perdew and Y. Wang, Phys. Rev. B **45**, 13244 (1992)

H. G. Petersen, D. Soelvason, J. W. Perram, and E. R. Smith, J. Chem. Phys. **101**, 8870 (1994)

C. J. Pickard and M. C. Payne, Phys. Rev. B **59**, 4685 (1999)

W. E. Pickett, H. Krakauer, and P. B. Allen, Phys. Rev. B **38**, 2721 (1988)

E. Polak, *Computational Methods in Optimization* (Academic Press, New York 1971)

E. L. Pollock and J. Glosli, Comput. Phys. Commun. **95**, 93 (1996)

M. Posternak, H. Krakauer, A. J. Freeman, and D. D. Koelling, Phys. Rev. B **21**, 5601 (1980)

D. L. Price and B. R. Cooper, Phys. Rev. B **39**, 4945 (1989)

P. Pulay, Mol. Phys. **17**, 197 (1969)

P. Pulay, Mol. Phys. **18**, 473 (1970)

P. Pulay and F. Török, Mol. Phys. **25**, 1153 (1973)

A. K. Rajagopal, J. Phys. C **11**, L943 (1978)

A. K. Rajagopal, Adv. Chem. Phys. **41**, 59 (1980)

A. K. Rajagopal and J. Callaway, Phys. Rev. **B7**, 1912 (1973)

G. Rajagopal and R. J. Needs, J. Comput. Phys. **115**, 399 (1994)

M. V. Ramana and A. K. Rajagopal, J. Phys. C **12**, L845 (1979)

M. V. Ramana and A. K. Rajagopal, Adv. Chem. Phys. **54**, 231 (1983)

M. Rasolt and D. J. W. Geldart, Phys. Rev. B **34**, 1325 (1986)

J. Rath and A. J. Freeman, Phys. Rev. B **11**, 2109 (1975)

M. E. Rose, *Elementary Theory of Angular Momentum* (Wiley, New York 1957)

M. E. Rose, *Relativistic Electron Theory* (Wiley, New York 1961)

A. Savin, A. D. Becke, J. Flad, R. Nesper, H. Preuß, and H. G. von Schnering, Angew. Chem. **103**, 421 (1991); Angew. Chem. Int. Ed. Engl. **30**, 409 (1991)

A. Savin, O. Jepsen, J. Flad, O. K. Andersen, H. Preuß, and H. G. von Schnering, Angew. Chem. **104**, 186 (1992)

S. Y. Savrasov and D. Y. Savrasov, Phys. Rev. B **46**, 12181 (1992)

A. Seidl, A. Görling, P. Vogl, J. A. Majewski, and M. Levy, Phys. Rev. B **53**, 3764 (1996)

W. F. Shadwick, J. D. Talman, and M. R. Norman, Comput. Phys. Commun. **54**, 95 (1989)

L. J. Sham and M. Schlüter, Phys. Rev. Lett. **51**, 1888 (1983)

R. T. Sharp and G. K. Horton, Phys. Rev. **90**, 317 (1953)

C. A. Sholl, Proc. Phys. Soc. **91**, 130 (1967)

R. Sinclair and F. Ninio, J. Phys.: Cond. Matt. **2**, 2143 (1990)

J. Soler and A. R. Williams, Phys. Rev. B **40**, 1560 (1989)

J. Soler and A. R. Williams, Phys. Rev. B **42**, 9728 (1990)

J. Soler and A. R. Williams, Phys. Rev. B **47**, 6784 (1993)

M. Springborg and O. K. Andersen, J. Chem. Phys. **87**, 7125 (1987)

G. P. Srivastava and D. Weaire, Adv. Phys. **36**, 463 (1987)

M. Städele, J. A. Majewski, P. Vogl, and A. Görling, Phys. Rev. Lett. **79**, 2089 (1997)

J. Sticht, Bandstrukturrechnung für Schwere-Fermionen-Systeme. PhD thesis, Technische Hochschule Darmstadt (1989)

J. Stoer, *Numerische Mathematik 1*, (Springer, Berlin 1993)

P. Strange, J. Staunton, B. L. Gyorffy, J. Phys. C **17**, 3355 (1984)

H. Takayama, M. Tsukuda, H. Shiba, F. Yonezawa, M. Imada, and Y. Okabe, *Computational Physics as a New Frontier in Condensed Matter Research* (The Physical Society of Japan, Tokyo 1995)

T. Takeda, Z. Phys. B **32**, 43 (1978)

T. Takeda, J. Phys. F **9**, 815 (1979)

T. Takeda, J. Phys. F **10**, 1135 (1980)

J. D. Talman, Comput. Phys. Commun. **54**, 85 (1989)

J. D. Talman and W. F. Shadwick, Phys. Rev. A **14**, 36 (1976)

M. Taut, Phys. Rev. B **57**, 2217 (1998)

A. Y. Toukmaji and J. A. Board, Jr., Comput. Phys. Commun. **95**, 73 (1996)

G. Vignale and M. Rasolt, Phys. Rev. Lett. **59**, 2360 (1987)

G. Vignale and M. Rasolt, Phys. Rev. B **37**, 10685 (1988)

Y. Wang and J. P. Perdew, Phys. Rev. B **43**, 8911 (1991)

Y. Wang and J. P. Perdew, Phys. Rev. B **44**, 13298 (1991)

J. L. Warren and T. G. Worlton, Comput. Phys. Commun. **8**, 71 (1974)

S.-H. Wei, H. Krakauer, and M. Weinert, Phys. Rev. B **32**, 7792 (1985)

P. Weinberger, *Electron Scattering Theory for Ordered and Disordered Matter* (Clarendon Press, Oxford 1990)

M. Weinert, R. E. Watson, and J. W. Davenport, Phys. Rev. B **32**, 2115 (1985)

M. Weinert, E. Wimmer, and A. J. Freeman, Phys. Rev. B **26**, 4571 (1982)

J. A. White and D. M. Bird, Phys. Rev. B **50**, 4954 (1994)

A. R. Williams and J. van W. Morgan, J. Phys. C **7**, 37 (1974)

A. R. Williams and J. Soler, Bull. Am. Phys. Soc. **32**, 562 (1987)

E. Wimmer, H. Krakauer, M. Weinert, and A. J. Freeman, Phys. Rev. B **24**, 864 (1981)

R. Winkler, J. Phys.: Cond. Matt. **5**, 2321 (1993)

H. Wondratschek and J. Neubüser, Acta. Cryst. **23**, 349 (1967)

T. G. Worlton and J. L. Warren, Comput. Phys. Commun. **3**, 88 (1972)

J. W. Wrench, Jr., Math. Comput. **22**, 617 (1968)

W. Yeung, J. Phys.: Cond. Matt. **4**, L467 (1992)

R. Zeller, Modelling Simul. Mater. Sci. Eng. **1**, 553 (1993)

J. M. Ziman, *Principles of the Theory of Solids* (University Press, Cambridge 1965)

Index